T0141874

Advances in Intelligent Systems and Computing

Volume 912

Series editor

Janusz Kacprzyk, Systems Research Institute, Polish Academy of Sciences, Warsaw, Poland
e-mail: kacprzyk@ibspan.waw.pl

The series "Advances in Intelligent Systems and Computing" contains publications on theory, applications, and design methods of Intelligent Systems and Intelligent Computing. Virtually all disciplines such as engineering, natural sciences, computer and information science, ICT, economics, business, e-commerce, environment, healthcare, life science are covered. The list of topics spans all the areas of modern intelligent systems and computing such as: computational intelligence, soft computing including neural networks, fuzzy systems, evolutionary computing and the fusion of these paradigms, social intelligence, ambient intelligence, computational neuroscience, artificial life, virtual worlds and society, cognitive science and systems, Perception and Vision, DNA and immune based systems, self-organizing and adaptive systems, e-Learning and teaching, human-centered and human-centric computing, recommender systems, intelligent control, robotics and mechatronics including human-machine teaming, knowledge-based paradigms, learning paradigms, machine ethics, intelligent data analysis, knowledge management, intelligent agents, intelligent decision making and support, intelligent network security, trust management, interactive entertainment, Web intelligence and multimedia.

The publications within "Advances in Intelligent Systems and Computing" are primarily proceedings of important conferences, symposia and congresses. They cover significant recent developments in the field, both of a foundational and applicable character. An important characteristic feature of the series is the short publication time and world-wide distribution. This permits a rapid and broad dissemination of research results.

More information about this series at http://www.springer.com/series/11156

Mostafa Ezziyyani
Editor

Advanced Intelligent Systems for Sustainable Development (AI2SD'2018)

Vol 2: Advanced Intelligent Systems Applied to Energy

Editor
Mostafa Ezziyyani
Computer Sciences Department,
Faculty of Sciences and Techniques
of Tangier
Abdelmalek Essaâdi University
Souani Tangier, Morocco

ISSN 2194-5357 ISSN 2194-5365 (electronic)
Advances in Intelligent Systems and Computing
ISBN 978-3-030-12064-1 ISBN 978-3-030-12065-8 (eBook)
https://doi.org/10.1007/978-3-030-12065-8

Library of Congress Control Number: 2019930141

This Springer imprint is published by the registered company Springer Nature Switzerland AG
The registered company address is: Gewerbestrasse 11, 6330 Cham, Switzerland

Preface

Overview

The purpose of this volume is to honour myself and all colleagues around the world that we have been able to collaborate closely for extensive research contributions which have enriched the field of Applied Computer Science. Applied Computer Science presents a appropriate research approach for developing a high-level skill that will encourage various researchers with relevant topics from a variety of disciplines, encourage their natural creativity, and prepare them for independent research projects. We think this volume is a testament to the benefits and future possibilities of this kind of collaboration, the framework for which has been put in place.

About the Editor

Prof. Dr. **Mostafa Ezziyyani,** IEEE and ASTF Member, received the "Licence en Informatique" degree, the "Diplôme de Cycle Supérieur en Informatique" degree and the PhD "Doctorat (1)" degree in Information System Engineering, respectively, in 1994, 1996 and 1999, from Mohammed V University in Rabat, Morocco. Also, he received the second PhD degree "Doctorat (2)" in 2006, from Abdelmalek Essaadi University in Distributed Systems and Web Technologies. In 2008, he received a Researcher Professor **Ability Grade. In 2015, he receives a PES grade —the highest degree at Morocco University.** Now he is a Professor of Computer Engineering and Information System in Faculty of Science and Technologies of Abdelmalek Essaadi University since 1996.

His research activities focus on the modelling databases and integration of heterogeneous and distributed systems (with the various developments to the big data, data sciences, data analytics, system decision support, knowledge management, object DB, active DB, multi-system agents, distributed systems and mediation). This research is at the crossroads of databases, artificial intelligence, software engineering and programming.

Professor at Computer Science Department, Member MA laboratory and responsible of the research direction Information Systems and Technologies, he formed a research team that works around this theme and more particularly in the area of integration of heterogeneous systems and decision support systems using WSN as technology for communication.

He received the first WSIS prize 2018 for the Category C7: ICT applications: E-environment, First prize: MtG—ICC in the regional contest IEEE - London UK Project: "World Talk", The qualification to the final (Teachers-Researchers Category): Business Plan Challenger 2015, EVARECH UAE Morocco. Project: «Lavabo Intégré avec Robinet à Circuit Intelligent pour la préservation de l'eau», First prize: Intel Business, Challenge Middle East and North Africa—IBC-MENA. Project: «Système Intelligent Préventif Pour le Contrôle et le Suivie en temps réel des Plantes Médicinale En cours de Croissance (PCS: Plants Control System)», Best Paper: International Conference on Software Engineering and New Technologies ICSENT'2012, Hammamat-Tunis. Paper: «Disaster Emergency System Application Case Study: Flood Disaster».

He has authored three patents: (1) device and learning process of orchestra conducting (e-Orchestra), (2) built-in washbasin with intelligent circuit tap for water preservation. (LIRCI) (3) Device and method for assisting the driving of vehicles for individuals with hearing loss.

He is the editor and coordinator of several projects with Ministry of Higher Education and Scientific Research and others as international project; he has been involved in several collaborative research projects in the context of ERANETMED3/PRIMA/H2020/FP7 framework programmes including project management activities in the topic modelling of distributed information systems reseed to environment, Health, energy and agriculture. The first project aims to propose an adaptive system for flood evacuation. This system gives the best decisions to be taken in this emergency situation to minimize damages. The second project aims to develop a research dynamic process of the itinerary in an events graph for blind and partially signet users. Moreover, he has been the principal investigator and the project manager for several research projects dealing with several topics concerned with his research interests mentioned above.

He was an invited professor for several countries in the world (France, Spain Belgium, Holland, USA and Tunisia). He is member of USA-J1 programme for TCI Morocco Delegation in 2007. He creates strong collaborations with research centres in databases and telecommunications for students' exchange: LIP6, Valencia, Poitier, Boston, Houston, China.

He is the author of more than 100 papers which appeared in refereed specialized journals and symposia. He was also the editor of the book "New Trends in Biomedical Engineering", AEU Publications, 2004. He was a member of the Organizing and the Scientific Committees of several symposia and conferences dealing with topics related to computer sciences, distributed databases and web technology. He has been actively involved in the research community by serving as reviewer for technical, and as an organizer/co-organizer of numerous international and national conferences and workshops. In addition, he served as a programme committee member for international conferences and workshops.

He was responsible for the formation cycle "Maîtrise de Génie Informatique" in the Faculty of Sciences and Technologies in Tangier since 2006. He is responsible too and coordinator of Tow Master "DCESS - Systèmes Informatique pour Management des Entreprise" and "DCESS - Systèmes Informatique pour Management des Enterprise". He is the coordinator of the computer science modules and responsible for the graduation projects and external relations of the Engineers Cycle "Statistique et Informatique Décisionnelle" in Mathematics Department of the Faculty of Sciences and Technologies in Tangier since 2007. He participates also in the Telecommunications Systems DESA/Masters, "Bio-Informatique" Masters and "Qualité des logiciels" Masters in the Faculty of Science in Tetuan since 2002.

He is also the founder and the current chair of the blinds and partially signet people association. His activity interests focus mainly on the software to help the blinds and partially signet people to use the ICT, specifically in Arabic countries. He is the founder of the private centre of training and education in advanced technologies AC-ETAT, in Tangier since 2000.

Mostafa Ezziyyani

Contents

Predictions and Modeling Energy Consumption for IT Data Center

Merzoug Soltane[1(✉)], Philippe Roose[2], Derdour Makhlouf[3], and Kazar Okba[4]

[1] Department of Computer Sciences, El-Oued University, El Oued, Algeria
merzoug-soltane@univ-eloued.dz
[2] LIUPPA, University of Pau et Pays de l'Adour, Anglet, France
Philippe.roose@univ-pau.fr
[3] Department of Computer Sciences, Tebessa University, Tebessa, Algeria
m.derdour@yahoo.fr
[4] Department of Computer Sciences, Biskra University, Biskra, Algeria
kazarokba@gmail.com

Abstract. Recent statistics of energy consumption by Cloud datacenter show the DCs consumes more and more energy each year that created big challenge in Cloud research. IT industry is keenly aware of the need for Green Cloud solutions that save energy consumption in Cloud DCs. A great deal of attention has been paid to minimize energy consumption in cloud datacenter. However, to understand the relationships between running tasks and energy consumed by hardware we need to propose mathematical models of energy consumption. The models of energy consumption can be help as to saving energy. Both researchers aim to proposed mechanism for energy consumption. In this paper, we analyzed the relationships between Cloud system manager and energy consumption. This paper aims at proposing and designing energy consumption models with mechanism of prediction energy.

Keywords: Energy consumption · Energy modeling · Energy predictions · Datacenter energy consumption

1 Introduction

Today, both researchers and engineers aim to minimize energy consumption based on energy modeling, because is one of the most important researches in Cloud computing [1].

However, the fast evolution of cloud datacenter created really problem of energy consumption. For example a typical datacenter needs 10 megawatts of power to operate [1], in 2013 [2], the Datacenter in the USA consumed 91 billion of electricity (kW/h) costing US businesses 13$ billion per year for electricity bills and generating nearly 100 million tons of carbon pollution (CO_2) per year.

Nowadays, the modeling of energy consumption become very complicated because the scale of datacenter becomes larger and larger, the energy consumption of datacenter becomes bigger, each level of service cloud like (SaaS, PaaS, IaaS) has different energy

M. Ezziyyani (Ed.): AI2SD 2018, AISC 912, pp. 1–11, 2019.
https://doi.org/10.1007/978-3-030-12065-8_1

consumption methods and these methods can't be used or migrated between each level of cloud.

Another problem of energy prediction in cloud datacenter make a lot of researchers and engineers eager to work on the development of predictive energy consumption mechanisms. However, in order to further predict future energy consumption researchers need to build energy model by their own. During the modeling of energy consumption, researchers need to capture the relationships between hardware resources and their energy consumptions.

In this paper, we analyzed the energy consumption of datacenter by each computing resources, analyzed different energy consumption model, and analyzed performance on predicting energy consumption by all datacenter resources. The contribution of the paper are as follows:

i. Proposed datacenter Rack power diagram to show 3 levels of power consumption
ii. We analyzed related work of energy consumption and we proposed energy consumption models
iii. We analyzed prediction energy consumption by all resources of datacenter and we proposed our prediction mechanism
iv. We implemented first part of datacenter infrastructure in cloud simulation

The rest of the paper is organized as follows, Sect. 2 describes related work about energy prediction and energy modeling, In Sect. 3, all detailed of our contribution. Experiments and conclusion are in Sects. 4 and 5, respectively.

2 Related Works

To understand and control the energy consumption of a datacenter, we need first understand energy management mechanisms to identify all sources of energy consumption to help as to building energy model framework. This section of literature review covers and shows the different techniques in many axis of energy consumption model and energy prediction.

2.1 Energy Consumption Module

The difference between power and energy is very important, because reduction of the power consumption does not always reduce the consumed energy, multiple work have been done to model the energy consumption can be classified into (module for hardware components, software components).

For processor energy consumption modeling work, first we find Shao *et al.* developed an instruction-level energy model for Intel Xeon Phi processor, which is the first commercial many core/multi-thread ×86-based processor [3]. They developed an instruction-level energy model for Xeon Phi through the results obtained from an energy per instruction ↼*Epi*↽ψ characterization made on Xeon Phi.

In the same line of research we find, Kliazovich et al. [4] modeled the energy of server by power and frequency of CPU and a constant. In their model, all components expect for CPU where considered constant. Same authors in another paper [5], described an energy model of the whole datacenter, where computing nodes were liners model of CPU utilization, and switch was multivariate function. Another work, which is almost similar. Lee et al. [6] used the maximal power and minimal power, current power of CPU to model the energy consumption.

Up to now, we have focused on energy consumption models based on physical characteristics of the data center. Equally important is to consider the type of software application of a datacenter. After analysis, we can broadly categorized datacenter software into two categories: *the Operating System/Virtualization layer and the applications.*

There has been a number of works on datacenter application energy modeling, like Smith *et al.* [7]. Described "Application Profiles," which is a means for presenting resource utilization of distributed application deployment. The Application Profiles they developed captures usage of the CPU, memory, hard disk, and network accesses. They described "Cloud Monitor," a tool that infers the energy consumption from software alone with computationally generated power models. Krishnan et al. [8] proposed a VM power consumption model using two hardware PMC events such as instruction retired (inst ret/sec) and LLC (Last Level Cache) misses for modeling power consumption of CPU, and memory.

In another hand, it is important to understand the energy consumption of OS level. Did One such characterization was done by Li *et al.* [9], which characterized the behavior of a OS across a large spectrum of applications to identify OS energy profiles and proposed OS energy consumed model. The OS energy consumption profiling gave a breakdown of power dissipation of OS routines, this study show the multiple application execution are found to consume 50% of total power on the examined OS routines. The capacitive load to the network causes significant power consumption about 34%.

2.2 Energy Prediction

In this sub-section, we present relevant approaches proposed in the literature for prediction energy.

Some works of prediction energy addressed to optimize power consumption in cloud datacenter, Hieu et al. [10] proposed Virtual machine consolidation algorithm with usage prediction (VMCUP) for improving the energy efficiency of cloud data centers. The proposed algorithm executed during the virtual machine consolidation process to estimate the short-term future CPU utilization based on the local history of the considered servers. The joint use of current and predicted CPU utilization metrics allows a reliable characterization of overloaded and under-loaded servers, thereby reducing both the load and the power consumption after consolidation.

Farahnakian et al. [11] presented a linear regression based CPU usage prediction (LiRCUP), for VM migration. Specifically, the authors estimated the future CPU usage to predict overloaded and under-loaded hosts; then, some of VMs are migrated to other hosts before a SLA violation occurs. Consequently, such a solution relies on early migration of VMs even when the current resource usage of the considered hosts is still acceptable, thus resulting in unnecessary migrations.

In the same line of research, Dhiman et al. [12] proposed a power modeling methodology based on Gaussian Mixture Models that predicts power consumption by a physical machine running multiple VM instances. The proposed approach requires a training phase to perceive the relationship between the metrics of the workload and the power consumption. The authors have evaluated the proposed model via experimental studies involving different types of the workload. The obtained experimental results have shown that the model predicts the power consumption with high accuracy (<10% prediction error), which is consistent over all the tested workloads.

3 Datacenter Energy Modeling

In this section, we describe how we organized resources of datacenter. We designed the flow of electrical power between all levels of datacenter (DC) resources Fig. 1. A more detailed presented in Fig. 2 the latter figure provides an overview of the organization of DC energy consumption modeling.

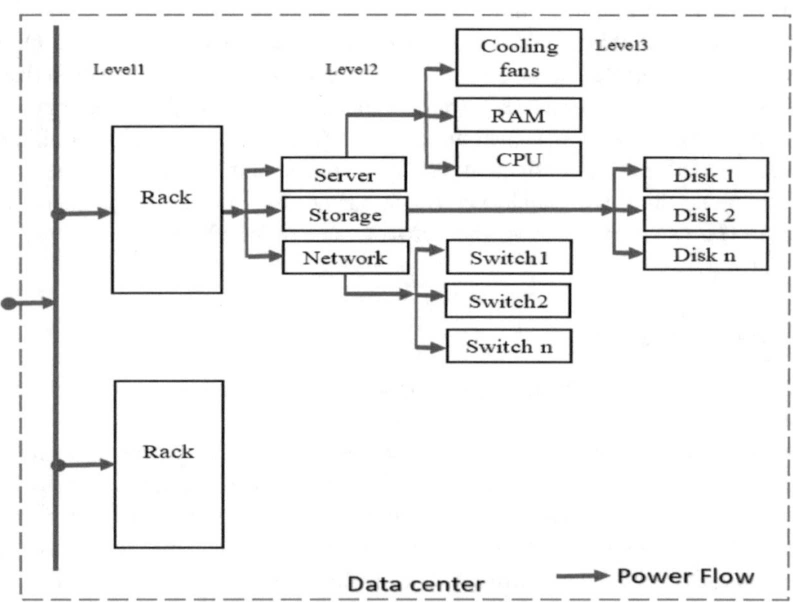

Fig. 1. Rack power flow in datacenter

- Level 1: contained rack installed in a datacenter.
- Level 2: contained servers, storage server, and network equipment.
- Level 3: contained all resources like (CPU, RAM, Disk, Switch).

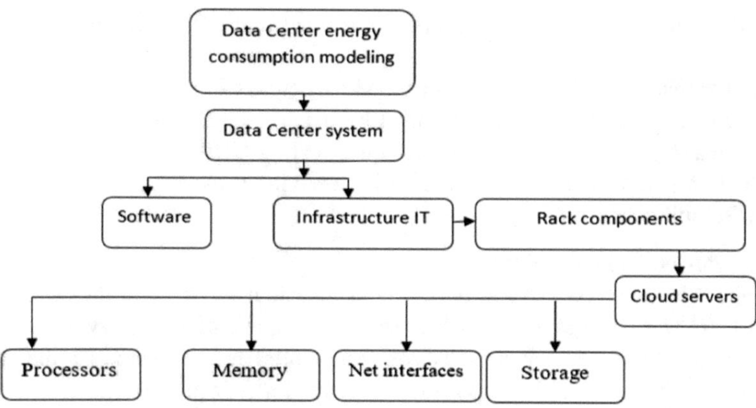

Fig. 2. Rack power flow in datacenter

3.1 Model of Power Consumption

In this sub-section, we describe detailed of our models of energy consumed in each level

Level 1 (Rack power) P. rack = power of all resources installed in the rack = (P.CPU + P.RAM + P.CoolingF) + (P.Disk) + (P.Switch)

Level 2 (power consumption for server) The energy consumed by the server S computes according to the following formula

P.Server = All energy consumed by all components of this server

= (P.CPU) + (P.RAM) + (P.CoolingF)

Model of power consumption for storage components P.Storage = (P. ActiveDisk*M)

m = total number of Active disk storage

Model of power consumption for network equipment

P.Network = (P.Switch*N) n = total number of Active switch

Level 3

Processor Power Consumption

Today, Modern processors such as Intel Xeon consist of multiple billions of transistors, which make them utilize huge amount of energy. Because of these, powerful CPU considered the one of the largest power consumers of a server, the major categories of processors used in current datacenter systems multi-core CPU.

We modeling CPU activity in server (s) associate with allocation new VM its consumption brought to the server. This value represents the increase consumption induced by the hosting of this new VM on this server.

$$P.Cpu(Vm, S) = Cpu(Vm) + \frac{(P.active(S) * \mu) - (P.idel(S) * \mu)}{Cpu\,max(S)}$$

P.Cpu (Vm.s): represents the energy consumed by CPU when server S allocated new VM

P.active (S): represents power consumed by active CPU in server S
P.idel (S): represents power consumed by inactive CPU in server S
CPU max(s): represents max power consumed by CUP in server S
CPU (Vm): represents allocated CPU in new virtual machine
µ: CPU utilization rate

Memory Power Consumption

The memory is the second largest energy consumer in a server consumes about ≈30% of the total power. IT equipment such as servers comprise of a memory hierarchy. The rapid increase of the memory capacity and bandwidth has contributed for memory sub system to consume a significant portion of the total server power.

We modeling RAM energy consumed in server (s) associate with new request of allocation VM its consumption brought to the server. This value represents the increase consumption induced by the hosting of this new VM on this server.

$$P.Ram(Vm, S) = RAM(Vm) + \frac{(P.active(S) * \partial) - (P.idel(S) * \partial)}{RAM\, max(S)}$$

P.Ram (Vm.s): represents the energy consumed by RAM when server S allocated new VM

P.active (S): represents power consumed by active Ram in server S
P.idel (S): represents power consumed by inactive Ram in server S
RAM max(s): represents max power consumed by RAM in server S
RAM (Vm): represents allocated RAM in new virtual machine
ə: RAM utilization rate.

Cooling Fans Power Consumption

We modeling cooling fans energy consumed in server (s) related with application run in the server. This value represents the increase consumption induced by the hosting new run application on this server.

$$P.Cool(S) = \frac{\sum P.ActiveFans}{Pmax(S)}$$

P.cool (S) = represents the energy consumed by cooling system in server S
P.activeFans = represents the energy consumed by each active fans in server S
P max(s): represents max power consumed by server S

Storage Power Consumption

Hard Disk is currently the main type of storage media used in data center servers. The rapid increase of the hard disk capacity impact positively on mass storage of data center servers hi can storage to 20 Terabit.

We modeling energy consumed by storage capacity in server (s) related with the allocation of new VM.

$$P.storag(Vm, S) = Storag(Vm) + \frac{(P.activeDisk(S))}{Pmax(S)}$$

P.storag (Vm.s): represents the energy consumed by storage when server S allocated new VM

P.activeDisk (S): represents power consumed by active disk storage in server S

Storage (Vm): represents storage capacity allocated for a new virtual machine

P max(s): represents max power consumed by server S

Network Power Consumption

Network interface is a significant contributor to the system power consumption, consume between (10% to 7%) can be either in idle mode or in active (either receiving or transmitting packets) mode, the total energy consumption of the network interface can be represented as

We modeling energy consumed by network switch based on stat of each interface, we define tow stat for network interface (active interface, idle interface).

$$\text{P.Switch} = \left(\sum \text{P.Active.interface}\right) + \left(\sum \text{P.idle.interface}\right)$$

3.2 Our Proposed Mechanism of Prediction

The proposed model is a combination of a several sub models (VM resource tracing, workload controller, management system resources and power tracing, base model of prediction, energy calculation model be on our proposed model of energy consumption). To estimate the energy consumption from our predicted mechanism and the remaining measurable parameters of energy consumption. We give detailed overview of the proposed model with its inputs and outputs given in Fig. 3.

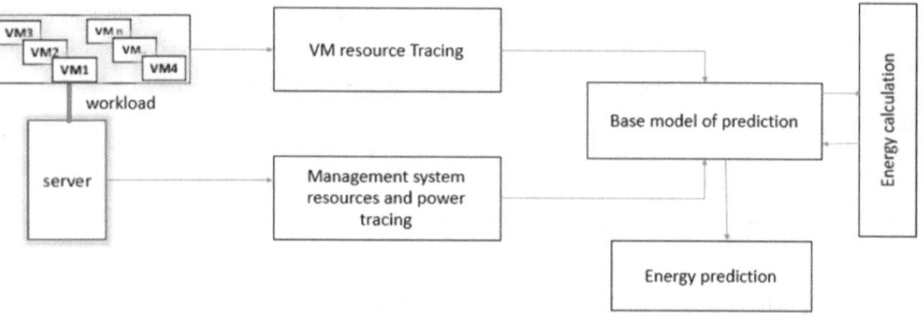

Fig. 3. Detailed of model of prediction for energy consumption

- VM resources tracing: This model tracks the state of all virtual machines allocating in the server. More precisely this model follows the resources allocated for each VM and control their energy consumption.
- Management system resources and power tracing: This model follows the state of all PMs. More precisely, this model tracks the physical resources installed and allocated in each server and controlled the energy consumption by each PM.
- Workload controller: this model controls all allocation requests. Moreover, send information of each allocation request to VM resources tracing model.
- Energy calculation model: this model based on our energy consumption model.
- Base model of prediction: this model base in algorithm of prediction to give estimated energy consumption by the system (Fig. 4).

Algorithm of prediction

Prediction energy

Input: (i) VM resources information (ii) server power information (iii) energy calculation

Output: (o) resources needed, (o) energy needed

1: at the end of each period, checks the latency of each VM

2: calculate the utilization resources rate by each VM, with the sum of the latencies

3: calculate the number of physical CPUs used

4: calculate the number of RAMs used

5: calculate the size of storage space used

6: For a period of time (Ti) do

7: if ((VM latency) is always < predetermined threshold) and (resources allocate < Max servers resources) then

8: calculate energy consumption

9: if (new resources allocate + previous resources allocate < predetermined threshold) then

10: Compare energy consumption results with history of previous energy consumption

11: Dertimneted energy consumption for courant time (T_i) predetermined by our system

12: give prediction energy needed for next T_{i1} based on energy consumption for courant time (T_i)

4 Implementation and Evaluation

To evaluate performance, we simulated our algorithm and model of energy consumption using the ICan Cloud simulator, which is a graphical runtime environment for network simulation. We simulated scenario for allocating resources subsequently, we implemented one cloud computing service provider to offer cloud services to users, and the datacenter implemented for this cloud runs many servers as shown in Table 1. C++ used to evaluate the performances of our prediction strategies. We tested them with a number of user requests to show the performances of our algorithm, such as global energy prediction and energy consumption

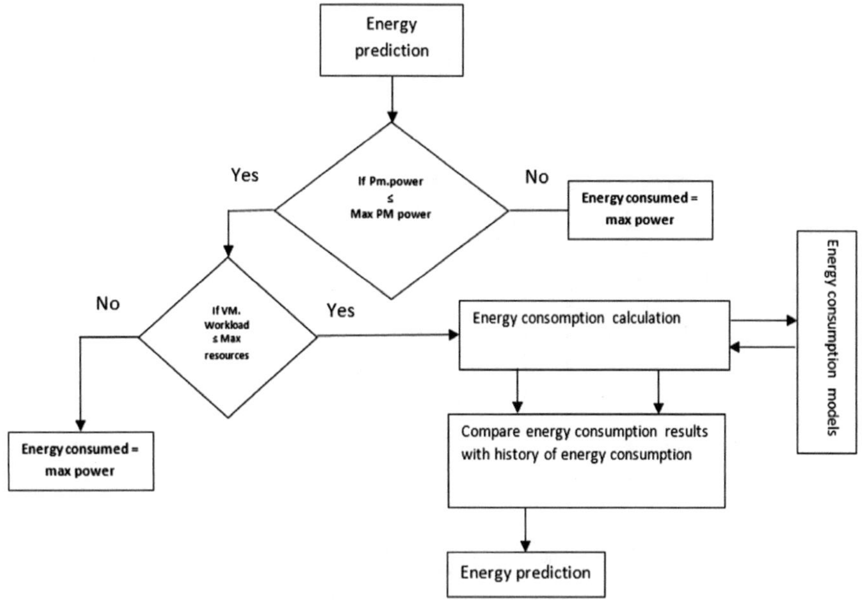

Fig. 4. Chart of prediction algorithm

4.1 Simulation Results

Figure 5 illustrates graphical presentation of simulation our prediction energy strategies and energy consumption by datacenter

Table 1. Datacenter resource

N°DC	Datacenter resource		
	Rack	Server	Storage
1	100	Cpu dual coreRam 16 g	1To

Figure 5 shows the energy consumption rate compared with energy prediction rate. When we apply our proposed algorithm in this work, we observe the remarkable good prediction of energy consumed. This result of prediction shows the performance of our model of power consumption and our prediction strategy and confirms our choice to use modeling approach to measure the energy needed to ruining datacenter resources.

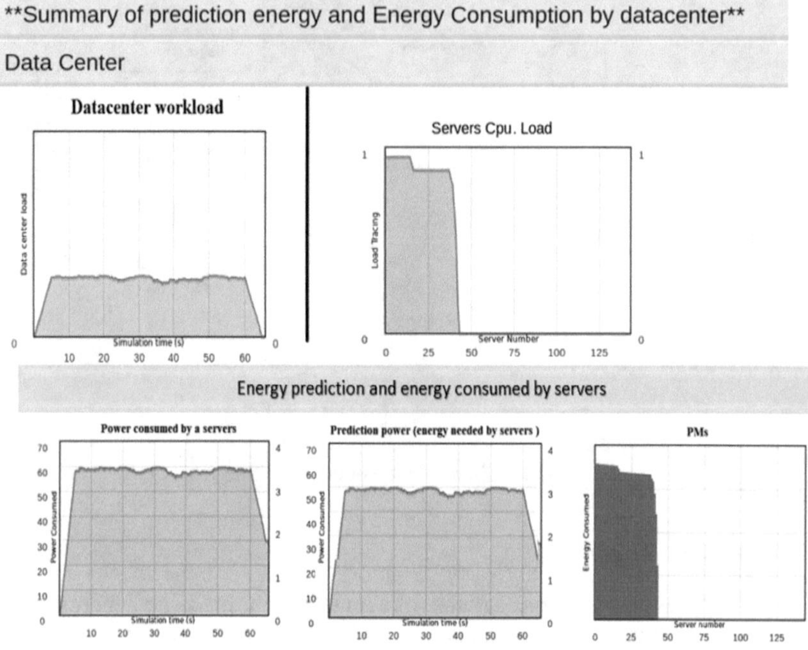

Fig. 5. Simulation result

5 Conclusion

We introduced the problem of power consumption by cloud datacenters. Then, we presented in this paper modeling approach to measured energy consumed by each resources in datacenter. We combined prediction strategies with our model of power to calculate power consumption. The simulation aims to clarify the efficiency of the internal function (tracing energy consumed by each resources), and the external function (the energy consumption and energy predicated). In the future, we plan to develop more models of energy consumption with a version of the prediction algorithm with more intelligence.

References

1. Hooper, A.: Green computing. Commun. ACM **51**(10), 11–13 (2008)
2. https://www.nrdc.org/resources/americas-data-centers-consuming-and-wasting-growing-amounts-energy

3. Shao, Y., Brooks, D.: Energy characterization and instruction-level, energy model of Intel's Xeon Phi processor. In: Proceeding of the IEEE ISLPED, pp. 389–394, September 2013
4. Kliazovich, D., Bouvry, P., Khan, S.U.: GreenCloud: a packet-level simulator of energy-aware cloud computing data centers. J. Supercomput. **62**(3), 1263–1283 (2012)
5. Kliazovich, D., Bouvry, P., Khan, S.U.: DENS: data center energy-efficient network-aware scheduling. Clust. Comput. **16**(1), 65–75 (2013)
6. Lee, Y.C., Zomaya, A.Y.: Energy efficient utilization of resources in cloud computing systems. J. Supercomput. **60**(2), 268–280 (2012)
7. Smith, J., Khajeh-Hosseini, A., Ward, J., Sommerville, I.: Cloud monitor: profiling power usage. In: Proceedings of the IEEE 5th CLOUD Computing, pp. 947–948, June 2012
8. Bhavani, K., Hrishikesh, A., Ada, G., Karsten, S.: VM power metering: feasibility and challenges. ACM SIGMETRICS Perform. Eval. Rev. **38**, 56–60 (2011)
9. Li, T., John, L.K.: Run-time modeling and estimation of operating system power consumption. In: Proceedings of the ACM SIGMETRICS, International Conference on Measuring, Modeling Computing Systems, pp. 160–171 (2003)
10. Hieu, N.T., Di Francesco, M., Ylä-Jääski, A.: Virtual machine consolidation with usage prediction for energy-efficient cloud data centers. In: IEEE 8th International Conference on Cloud Computing (CLOUD), pp. 750–757. IEEE, June 2015
11. Farahnakian, F., Liljeberg, P., Plosila, J.: LiRCUP: linear regression based CPU usage prediction algorithm for live migration of virtual machines in data centers. In: 39th EUROMICRO Conference on Software Engineering and Advanced Applications (SEAA), pp. 357–364. IEEE, September 2013
12. Dhiman, G., Mihic, K., Rosing, T.: A system for online power prediction in virtualized environments using Gaussian mixture models. In: Proceedings of the 47th DAC, 2010, pp. 807–812 (2010)

Experimental Study of the Impact of Drying Parameters on Dandelion Root by a Solar Dryer

Haytem Moussaoui(✉), Hamza Lamsyehe, Ali Idlimam, Abdelkader Lamharrar, and Mounir Kouhila

Team of Solar Energy and Medicinal Plants (EESPAM), Teacher's Training College, Cadi Ayyad University, Marrakesh, Morocco
Haytem.moussaoui@gmail.com

Abstract. The proliferation of microorganisms, as well as the development of various chemical reactions, are the main causes behind the deterioration of the product. Both are caused by the quantity of water held within food. The following study aims at modelling the drying process and optimizing its conditions of use, as well as analyzing the interaction between the factors based on experiments, statistical calculations, and analysis. The drying kinetics of Dandelion root were carried out in an indirect active solar dryer with a separate solar collector and a drying unit. Drying is a process engineering operation that involves the removal of free water and some of the bound water without affecting the chemical structure in such a way as to lighten the weight and minimize the risks of many unwanted chemical transformations. To perform this procedure we use a technique based on solar energy, called solar drying forced convection, which requires sufficient sunlight to give a radiant energy to insulator in order to heat the air sucked by the fan. The modelling of the solar dryer is done in several methods, the plans of experiments, is one of these methods, based on statistical and experimental studies. This work is devoted to optimization of solar dryer setting parameters for plants. The outcome is a model that helps to analyze the impact of the drying parameters on the response, which is the time of drying. A determination of factors that affect the response in order to analyze their impact on the response.

Keywords: Solar energy · Drying parameters · Semi-empirical models · Dandelion root · Full factorial design

1 Introduction

The current growing environment presents a variety of threats to the world, as we know it. Surmounting them necessitates considering semi-permanent acts to attain a sustainable environment. Advanced innovations in renewable energy technologies offer a large number of effective and efficient solutions [1].

Solar drying, as a means of preserving food, was considered the most used system of solar energy. The drying of fruits, vegetables and meats is one of the processes of

M. Ezziyyani (Ed.): AI2SD 2018, AISC 912, pp. 12–20, 2019.
https://doi.org/10.1007/978-3-030-12065-8_2

high-energy consumption in the food processing industry and is a better method of reducing post-harvest expenses and losses.

The dandelion root is mostly used in three major fields. First, it is widely present in traditional uses and food. Since it contains certain crucial vitamins like A, B, C, D; amino acids as well as minerals [2]. It is eaten particularly in salads. What's more, its leaves are used as a diuretic while the root is a powerful natural detoxing. Second, it is used in the field of phytotherapy. Despite being considered a wild plant, the dandelion proves to be an effective cure for a large number of diseases. Many phytotherapist worldwide use it to treat arteriosclerosis, asthenia, cellulite, cholesterol, color, springtime, hepatic, hemorrhoid, liver, gout, malaria, obesity, rheumatism, constipation, urea, varices and warts [3]. Additionally, the dandelion is used to heal the following health problems: fevers, boils, eye problems, fluid retention, liver congestion, heartburn and various skin problems [4]. Third, it is used in pharmacology. The scientific investigations in the pharmacological profile of the dandelion show that the dandelion extracts or the plant itself are used in making drugs for the following activities: diuretic activity, choleretic activity, anti-inflammatory activity, anti-oxidative activity, anti-carcinogenic activity, anti-allergic activity, anti-hyperglycemic activity, anti-coagulatory, anti-thrombotic activity, and prebiotic activity [5]. On the topic, and over recent years, solar drying field has witnessed a growing interest. Indeed, solar drying proves to be the best alternative capable of surmounting the shortcomings of the other type of drying [6]. As such, many have tackled the task designing certain devices that are environment-friendly with a new ferocity. Therefore, forced convection appears as an innovative, time and energy-saving mode in solar crops drying. Furthermore, forced convection is of high predominance in the fields of industry and agriculture.

What's more, the time dedicated to drying is in direct rapport to the availability and the overall equipment effectiveness of dryers. It is of importance to note that achieving optimal drying periods is actually the main factor that affects the drying cost. Therefore, it is vital to model the forced convection solar drying system to the extent that we can analyze the impact of serval factors (dryer parameters) optimize the period. The literature furnishes us with several methods that allow the systems' modeling.

2 Materials and Methods

2.1 Description of Solar Dryer

Conducting this study requires a forced convection solar dryer. This solar dryer makes use of two energy sources (solar energy and electrical energy) to operate the drying process. The device's constituents are the following: a solar collector, aeration duct, a drying chamber, a centrifugal fan, a thermoregulatory besides electrical resistors. A sample product is also needed in addition to a chronometer to measure time [7].

The experimental system is an indirect forced convection solar dryer shown in Fig. 1 composed of several parts that ensure the proper running of the system; the mean components are:

(1) A 2.5 m2 (2.5 m × 1 m) solar captor.
(3) A centrifuge ventilator with debit of 0.083 m3/s.
(4) A ventilation duct with a parallelepiped-shaped tunnel Sect.
(6) An auxiliary source (4 kW Electrical resistances)
(8) A 0.9 depth, and 0.5 width, and 1.4 m height drying cabinet (with 10 floors)

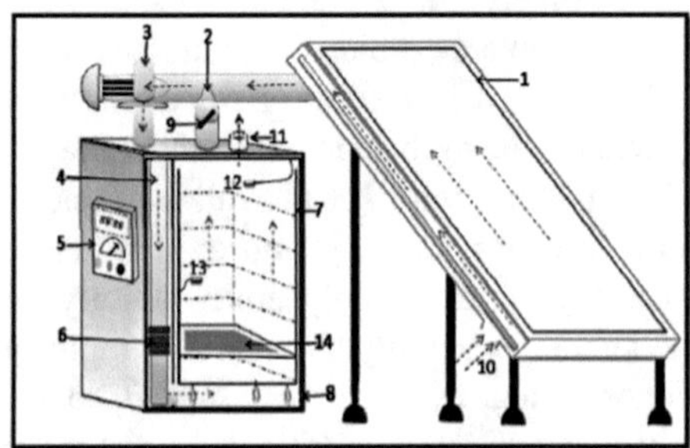

Fig. 1. Convective solar dryer installed in Marrakech (ENS): (1) Solar air collector, (2) ventilation duct, (3) fan, (4) suction duct, (5) control box, (6) electric booster, (7) stages, (8) cabinet dryer, (9) air shutter, (10) air inlet, (11) air outlet, (12) humidity sensor, (13) thermocouple, (14) sample holder.

The solar heated air flow upward the samples; it enters below the trays in the drying cabinet below. To keep a constant air-drying temperature; the use of an auxiliary heater is necessary unavoidable. The calculation of the mass loss of the product on the tray during the experiment is done via a (±0.01 g); the weight was measured by removing the product from the drying cabinet each 10 min until the weight became stable [8].

2.2 Experimental Protocol

In the study, several conditions and measures are respected in order to conduct the experiment. Firstly, it is important to note the collection of the samples used in all experiments must be done in the same zone. These samples are to be measured by using a digital caliper 5 mm with a margin of (+/–) 0.5 mm. After that, the samples are regularly and uniformly put in a porous shelf to guarantee a homogeneous diffusion of the samples during the drying process. The next step is to put the shelf on the drying chamber and measure the masse variation of the samples. When the mass reaches a stable point, it is necessary to calculate the drying time. By so doing, we seek to model the impact of the temperature, the airflow and the air recycling factors on the convective drying of Dandelion root.

2.3 Experimental Design

The technique of DOE is preceded by the simultaneous variation of the levels of the continuous factors, or the modalities of the discrete factors, elaborated from the system studied to each experiment. This procedure will make it possible to minimize the number of experiments to be carried out while increasing the number of factors studied [9].

The domain of a factor is the value given to a particular factor to perform an experiment is called level. This domain is defined by the high and low levels of all factors and possibly by constraints between factors. In this study, the influence of three factors: the temperature (T), the air airflow (F) and plant air recycling (R) on the response (drying time) were investigated.

The factorial design used as well as the parameters studied in their coded and real values are summarized in Table 3. Thus, 27 trial experiments were needed in the investigation undertaken herein. The additive model with interaction proposed contains linear terms as well as those describing the interaction between different experimental variables. This phase is referred to as the determination of the domain of the study. The following equation Eq. (1) is very useful in describing the factors effects and their overall interactions.

$$Y(\%) = a_0 + a_1X_1 + a_2X_2 + a_3X_3 + a_{12}X_1X_2 + a_{13}X_1X_3 + a_{23}X_2X_2 \quad (1)$$

Where Y is the response, a_0 is a constant coefficient, a_1, a_2 and a_3 are linear factors coefficients, a_{12}, a_{23} and a_{13} designate variables interaction effects: X_1, X_2, and X_3 corresponding to T, F and R, respectively. This relationship describes how the experimental variables influence the experimental response.

Finally, the analysis of variance (ANOVA) is a statistical test to compare overall the expectation of several samples. The name of this test is explained by its way of proceeding: one decomposes the total variance of the sample in two partial variances, the variance between the classes and the residual variance, and one compares these two variances [10].

In other words, the ANOVA is a powerful tool that is help us to analyze how the model is suitable to the system behaviour and if the predicted values given by the model are significant or not.

3 Results and Discussion

In order to study the impact of temperature, the airflow and the air recycle factors on the drying time of the Dandelion root, Full Factorial Design was constructed and the ANOVA was calculated using the software JMP from SAS version 10.

3.1 Domain and Level of Factors

The Table 1 summarizes the different levels and domains of each factor. Number 1 is attributed to the highest temperature possible in this system, which is 80 °C. Number –1 is attributed to the lowest temperature, which is 60 °C. The number 1 has also been allocated to the airflow level of 300 while –1 to the level of 150. Finally, the factor air recycling takes the value of 0% when the level corresponds to number 1. This value means that the air coming out of the dryer is not used again in drying process; and it takes the value of 100% mm when the level corresponds to –1, which means that air coming out of the dryer is totally used in the system.

Table 1. Factors values and levels.

	Temperature	Airflow	Air recycling
Level	T (°C)	F ($m^3.h^{-1}$)	R (%)
–1	60	150	0
0	70	225	50
1	80	300	100

3.2 Experimental Design and Results

The Table 2 summarizes the various experiments concerning the full factorial design data. In every experiment, we set the temperature, the airflow and the air recycling factors so that we can calculate the response of the drying system, which is exemplified by the drying time of the Taraxacum Officinale leaves. Additionally, the factorial design data is very useful in the calculation of the predicted model that would represent the impact of the factors investigated. Another necessary point to add is that this empirical model has been obtained by using the JMP software from SAS. For instance, in the first experiment the temperature is set at 60 °C; the airflow is fixed at 225; while, the no air recycling is fixed at 100%. As a result, the required time of the drying of the root is determined; this later is represented by Yexp. On the other hand, the Ymod, which indicates the value of the predicted drying time is also, determined relying on the same software. Finally, the residue is the outcome of the difference between the Yexp and Ymod that were calculated in order to analyze the convenience of the predicted model. In other words, when the residue is minimized, the degree of suitability becomes higher.

Table 2. The full factorial design

Test	Real and coded values						Responses		
	T (°C)	X1	F (m3.h-1)	X2	R (%)	X3	Yexp (%)	Ymod (%)	Residue
1	60	–1	225	0	100	–1	135	138,52	–3,52
2	70	0	225	0	100	–1	100	105,19	–5,19
3	80	1	300	1	100	–1	65	60,88	4,12
4	80	1	225	0	100	–1	70	71,85	–1,85
5	60	–1	300	1	100	–1	125	121,71	3,29
6	60	–1	225	0	0	1	130	133,52	–3,52
7	80	1	150	–1	100	–1	85	82,82	2,18
8	60	–1	150	–1	100	–1	160	155,32	4,68
9	80	1	225	0	50	0	70	71,02	–1,02
10	60	–1	300	1	50	0	120	120,05	–0,05
11	80	1	150	–1	0	1	80	79,49	0,51
12	70	0	150	–1	0	1	110	114,07	–4,07
13	60	–1	150	–1	0	1	150	148,66	1,34
14	70	0	150	–1	100	–1	115	119,07	–4,07
15	60	–1	300	1	0	1	120	118,38	1,62
16	80	1	300	1	50	0	60	60,88	–0,88
17	60	–1	225	0	50	0	135	136,02	–1,02
18	70	0	225	0	50	0	105	103,52	1,48
19	80	1	150	–1	50	0	85	81,16	3,84
20	70	0	300	1	0	1	90	89,63	0,37
21	70	0	300	1	100	–1	90	91,30	–1,30
22	70	0	225	0	0	1	105	101,85	3,15
23	70	0	300	1	50	0	90	90,46	–0,46
24	60	–1	150	–1	50	0	155	151,99	3,01
25	80	1	300	1	0	1	60	60,88	–0,88
26	70	0	150	–1	50	0	115	116,57	–1,57
27	80	1	225	0	0	1	70	70,19	–0,19

3.3 Determination of the Polynomial Model and the Impact of Factors

In order to model the behaviour of the system response via the variation of the drying parameters several model has been elaborated. According to the full factorial design, the polynomial model has been selected. The following equation displays the coefficients value of the empirical model [11].

$$Y(\%) = 368.52 - 3.25X_1 - 0.17X_2 + 0.03X_3 + 3.89 \times 10^{-3}X_1X_2 - 1.67 \times 10^{-3}X_1X_3 - 2.22 \times 10^{-3}X_2X_3 \quad (2)$$

The equation mentioned before shows the value of coefficients, which correlates with the degree of the factor's impact. Where the first coefficient in the equation Eq. (2) is a constant coefficient, the X_1 refers to the temperature (T), the X_2 refers to the airflow (F) and the X_3 refers to the air recycling (R).

3.4 The Analysis of the Factors Impact and their Interaction

The vertical-horizontal bar graph is used in order to have an overview on the influence of the three factors on the system's response understudy in addition to factorial them. The Fig. 2 shows that the bar that corresponds to a1 is the highest one. This means that the factor that correlates with the coefficient a1 has the biggest effect of the response. Relying on the equation Eq. (1) the a1 represents the impact of the temperature (T). Regarding the effect of the interaction between factors on the response of the drying system, the degree of the interaction between the temperature and the debits is found to be higher than that of the rest.

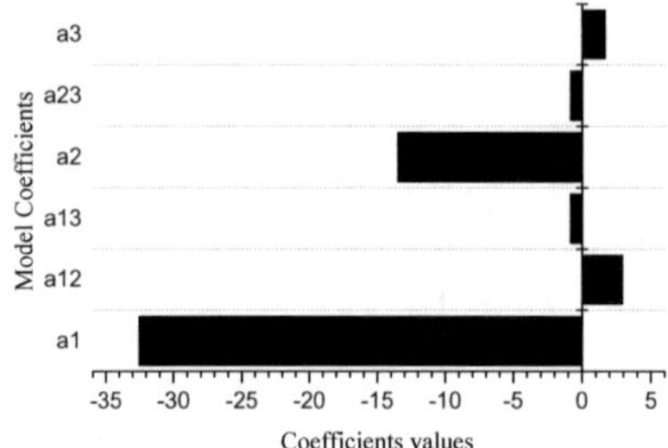

Fig. 2. The coefficients value of the empirical model

3.5 The Analysis of Variance (ANOVA)

The Table 3 summaries the statistical parameters of the analysis of variance (ANOVA). The value of the Ficher-Snedecor (F1) is 387.4123 and its ProbF1 equal to 0.001 < 5%. This later means that the modeling of the factors' impact is found to be a significant. The Fisher test (ProbF1) demonstrate that the polynomial model that describe the system under study is significant because the ProbF1 < 5% [12].

Table 3. The ANOVA of empirical model

Source	Degrees of freedom	Sum squares	Mean square	F_1	$ProbF_1$
Model	6	22249.306	37708.22	387.4123	<0.001*
Error	20	191.435	9.57		
C. Total	26	22440.741			

3.6 The Model Validation Test

Figure 3 displays the relationship between the experimental responses and the predicted responses Yexp = f (Ymod). It also shows that the experimental values closely fit the values predicted by the empirical model represented by the straight line in the graph. Additionally the Fig. 4 shows the residual between the experimental data and the product data. Finally, the R-square is found to be equal to 0.99. That result indicate a good fit of the data to the equation Eq. (1).

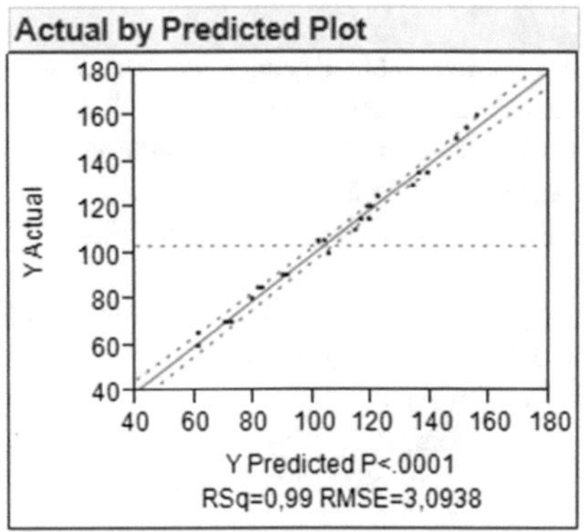

Fig. 3. The representation of the experimental data in function of predicted data

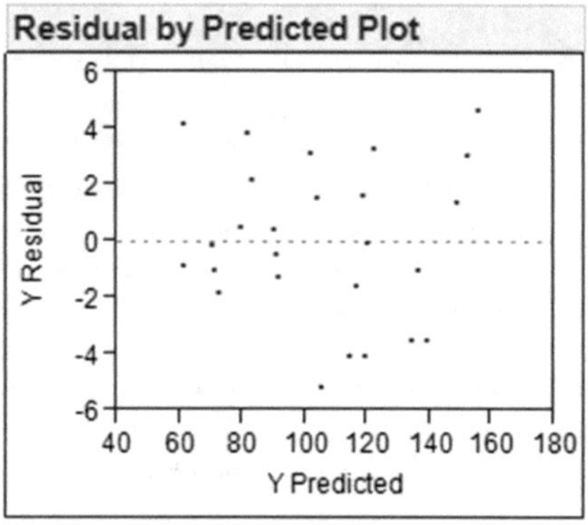

Fig. 4. The residual by predicted curve

4 Conclusion

In a conclusion, in order to model the effect of the factors that influence the drying time of the dandelion root using a solar convective dryer. Therefore, the temperature (T), the airflow (F) and air recycling (R) are selected to be the factors of our study. The outcome is the drying time was found affected by two factors: temperature (X1) and airflow (X2), with lower drying time resulting in at high temperature and high airflow. Additionally we have found that no huge difference between the responses established by the model and the experiment responses. This evidence is represented by a satisfactory coefficient of correlation R2 = 0.99.

References

1. Ibrahim, D.: Renewable energy and sustainable development: a crucial review. Renew. Sustain. Energy Rev. **4**, 157–175 (2000)
2. Michael, T., Murray, N.D.: Taraxacum officinale (Dandelion). In: Pizzorno, J.E., Murray, M. T. (Eds.) Textbook of Natural Medicine. Churchill Livingstone (2013)

3. Chiusoli, A., Goidanich, G., Taddei, I.: Segreti e VirtùdellePianteMedicinali. Selezione dal reader's digest; Ristampa edizione (Chapter Tarassaco) (1980)
4. Escudero, N.L., De Arellano, M.L., Fernández, S., Albarracín, G., Mucciarelli, S.: Taraxacum officinale as a food source. Plant Foods Hum. Nutr. **3**, 1–10 (2003)
5. Martinez, M., Poirrier, P., Chamy, R., Prüfer, D., Schulze-Gronover, C., Jorquera, L., Ruiz, G.: J. Ethnopharmacol. **169**(3), 244–262 (2015)
6. VijayaVenkataRaman, S., Iniyan, S., Goic, R.: A review of solar drying technologies. Renew. Sustain. Energy Rev. **16**, 2652–2670 (2012)
7. Ali, I., Abdelkader, L., El Houssayne, B., Mohamed, K., El Khadir, L.: Solar convective drying in thin layers and modelling of municipal waste. Appl. Therm. Eng. **108**, 41–47 (2016)
8. Safa, M., Mourad, O., Nadia, H., Lamyae, L., Ali, I., Mostafa, M.: Drying characteristics and kinetics solar drying of Moroccan rosemary. Renew. Energy **108**, 303–310 (2017)
9. Emerson, R.W., Cavazzuti, M.: Deterministic optimization, Optimization Methods: From Theory to Design Scientific and Technological Aspects in Mechanics (2013)
10. Atungulu, G.G., Olatunde, G.A.: Assessment of new in-bin drying and storage technology for soybean seed. Drying Technol. **4**, 383–399 (2017)
11. Aït Aghzzaf, A., Rhouta, B., Rocca, E., Khalil, A., Caillet, C., Hakkou, R.: Heptanoic acid adsorption on grafted palygorskite and its application as controlled release corrosion inhibitor of steel. Mater. Chem. Phys. **148**, 335–342 (2014)
12. Lundstedt, T., Seifert, E., Abramo, L., Thelin, B., Nystrom, A., Peterson, J., Bergman, R.: Expiremental design and optimization. Chemom. Intell. Lab. Syst. **42**, 3–40 (1998)

Analytical Modelling and Analysis of Thermal Behavior for Series Resistance of Solar Cell

Mohamed Louzazni[1(✉)], Ahmed Khouya[2], Khalid Amechnoue[1], Marco Mussetta[3], and Rachid Herbazi[1]

[1] Mathematics Informatic and Applications Team, National School of Applied Sciences, Abdelmalek Essaadi University, 1818, Tangier, Morocco
louzazni@msn.com

[2] Laboratory of Innovative Technologies, National School of Applied Sciences, Abdelmalek Essaadi University, 1818, Tangier, Morocco

[3] Department of Energy, Politecnico di Milano, 20156 Milan, Italy

Abstract. In this paper, the analysis of the nonlinear and complex modelling of a solar cell is addressed, to study its thermal behaviour. From literature it is well known that the output power of a solar cell monotonically decreases with the temperature; here, we investigate the specific expression based on derivation of current and power to simplify the equation model. Thus, applying the specific expression of the series resistance, we analyze the relation characteristics between the power and the temperature. Moreover, we aim to present a method to determine the specific theoretical expression of the series resistance with combination of parallel parasitic resistances. The analytical model equation is tested and compared with experimental results to add validity to the model.

Keywords: Power equation · Solar cell · Complex characteristics · Series resistance · Analytical model

1 Introduction

The behaviour of output characteristics of photovoltaic cell plays an important role in solar industry because it reflects the cell or generator performance. Several papers present the complex and nonlinear equations of solar cell and the single and double diode models are the most used models [1]. The single diode is accurate for result and low computational cost [2, 3], the recombination loss is ignored in low irradiance [4]. The diode represents the carriers diffusing across the P-N junction and recombining in the bulk or at surfaces. The second diode is added to improve the accuracy of solar cell model. It is accounts the recombination by traps within the depletion region in more accurate cell behaviour at low irradiance [5], or recombination at passive cell edge [6]. Theoretically, more diodes can be added to better account the correct modelling and localize the effects in solar cells like Auger recombination [7]. The multi diode model is illustrated in Fig. 1.

M. Ezziyyani (Ed.): AI2SD 2018, AISC 912, pp. 21–29, 2019.
https://doi.org/10.1007/978-3-030-12065-8_3

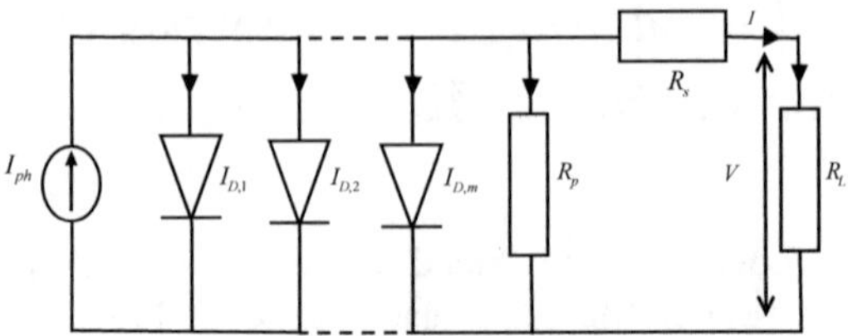

Fig. 1. Equivalent circuit of solar cell model multi-diodes.

The general mathematical description of the diode model in Fig. 1 is given by

$$I = I_{ph} - \sum_{i=1}^{m} I_{S_i} \left[\exp\left(\frac{V + R_s.I}{n_i V_T} \right) - 1 \right] - \frac{V + R_s.I}{R_p} \tag{1}$$

where I_{ph} is the photocurrent proportional to the irradiance; I_{si} the reverse saturation, n_i quality factor for the ith diode, $V_T = KT/q$ is thermal voltage with K is Boltzmann's constant, T is the cell temperature in Kelvin and q is the electronic charge, R_s and R_p are resistances in series and parallel, respectively.

Due to the inherent nonlinearity, it is not straightforward to determine the series resistance and the relationship between the power and the temperature, even for the simplest case of $m = 1$. However, the theory expression of R_s is still unknown and has not been clearly disclosed in previous research. Several methods used the five- and seven-model with $m = 1, 2$ to determine the series resistance and assume that shunt resistance affect is neglected. A review of techniques to determine the series resistance is given in [8, 9] for different solar cell technologies. In [10] the Co-content function was used to extract the series resistance of plastic solar cells. However, the influence of temperature on the behaviour of series resistance decrease and varies exponentially with temperature [11].

The dynamic method presented in this work, does not need any kind of assumptions about the value of the parameters, or use special points. In fact, the required four points can be arbitrarily located along the current-voltage curve. We based on derivation of the nonlinear current-voltage and power characteristics of solar cell to simplify the equation model. Further, it is well known that the output power of a solar cell monotonically decreases with its temperature. The results show that the series resistance decrease with temperature and has an exponential expression such as positive temperature coefficient type compared to experimental data.

2 Dynamic System Formulation

The current-voltage characteristic of solar cell connected to the R_L load resistance presented in Eq. (1) is implicit and nonlinear can be writing as:

$$I = \frac{R_p}{R_s + R_L + R_p} \left[I_{ph} - \sum_{i=1}^{m} I_{S_i} \left(\exp\left(\frac{(R_s + R_L).I}{n_i V_T} \right) - 1 \right) \right] \tag{2}$$

Let use: $x_i = q\frac{R_s+R_L}{n_i.K.T}.I,\;\; x0_i = q\frac{R_s+R_L}{n_i.K.T}.I_{si},\;\; y = \frac{R_p}{R_s+R_L+R_p}$

with the new dynamic parameters, the Eq. (2) of generalized multi-diode model of solar cell connected to load resistance become:

$$I = y.I_{ph} - y\sum_{i=1}^{m} I_{s_i}(e^{x_i}-1) \tag{3}$$

we note that x, y, I and P is the power of solar cell are a continuous function, taking differential once on both sides of equations, respectively, results in;

$$\frac{dx_i}{x_i} = \frac{dR_s+dR_L}{R_s+R_L} + \frac{dI}{I} - \frac{dT}{T} \tag{4}$$

$$\frac{dy}{y} = \frac{dR_p}{R_p} - \frac{dR_s+dR_L+dR_p}{R_s+R_L+R_p} \tag{5}$$

$$dI = y.dI_{ph} + \left[I_{ph} - \sum_{i=1}^{m} I_{s_i}(e^{x_i}-1)\right]dy - y\sum_{i=1}^{m} I_{s_i}e^{x_i}.dx_i \tag{6}$$

$$dP = I^2.dR_L + 2I.R_L.dI \tag{7}$$

Combining the equations, and the parameters I_{ph}, R_p and R_L are constant with respect to the temperature T. we get.

$$\begin{aligned}\left.\frac{dI}{dT}\right|_{I_{ph},R_L,R_p} = &-\frac{1}{R_s+R_L+R_p}.\frac{y.\left[I_{ph}-\sum_{i=1}^{m} I_{s_i}(e^{x_i}-1)\right]}{1+y.\sum_{i=1}^{m} x_{0i}.e^{x_i}}.\frac{dR_s}{dT} \\ &-\frac{I.y.\sum_{i=1}^{m} x_{0i}.e^{x_i}}{1+y.\sum_{i=1}^{m} x_{0i}.e^{x_i}}.\left(\frac{1}{R_s+R_L}.\frac{dR_s}{dT}-\frac{1}{T}\right)\end{aligned} \tag{8}$$

Rewriting Eq. (8) as:

$$\frac{dP}{dT} = 2I.R_L.\frac{dI}{dT} \tag{9}$$

The final equation of power for multi-diodes model solar cell is given as;

$$\left.\frac{dP}{dT}\right|_{I_{ph},R_L,R_p} = \frac{2.R_LI^2.y.\sum_{i=1}^{m} x_{0i}.e^{x_i}}{1+y.\sum_{i=1}^{m} x_{0i}.e^{x_i}}\frac{1}{T} - \left[\frac{2I.R_Ly.\left[I_{ph}-\sum_{i=1}^{m} I_{s_i}(e^{x_i}-1)\right]}{(R_s+R_L+R_p)\left(1+y.\sum_{i=1}^{m} x_{0i}.e^{x_i}\right)} + \frac{2.R_LI^2.y.\sum_{i=1}^{m} x_{0i}.e^{x_i}}{(R_s+R_L)\left(1+y.\sum_{i=1}^{m} x_{0i}.e^{x_i}\right)}\right].\frac{dR_s}{dT} \tag{10}$$

The output power monotonically decreases with the temperature of the solar cell, that is,

$$\frac{dP}{dT} < 0 \tag{11}$$

The series resistances $R_{s.}$ must be relevant to T, that means dR_s/dT must be satisfied. The analysis is significant to determine the specific theory expression of the series resistance.

3 Temperature Coefficient of Series Resistance

The temperature coefficient describes the relative change of a physical property that is associated with a given change in temperature. According to theory, there are three types of thermal sensitive resistances [12]: conductor type, negative temperature coefficient type and positive temperature coefficient type: Consequently, the form of R_s relevant to T, must belong to one of the above three types.

3.1 Conductor Temperature Coefficient Type

The temperature dependence type of conductors is linear and can be described as:

$$R_s = R_0(1 + \alpha T) \tag{12}$$

where α is the conductor temperature coefficient $\alpha > 0$ and R_0 is the condition resistance. Taking differential once on both sides of Eq. (12)

$$\frac{dR_s}{dT} = \alpha R_0 > 0 \tag{13}$$

The equation of power becomes:

$$\left.\frac{dP}{dT}\right|_{I_{ph},R_L,R_p} = \frac{2.R_L I^2.y.\sum_{i=1}^{m} x_{0i}.e^{x_i}\left(R_s + R_L + R_p\right)(R_s + R_0) - 2I.R_L y.T.(R_s + R_L)\left[I_{ph} - \sum_{i=1}^{m} I_{s_i}(e^{x_i} - 1)\right]}{\left(R_s + R_L + R_p\right)\left(1 + y.\sum_{i=1}^{m} x_{0i}.e^{x_i}\right)(R_s + R_L)T}. \tag{14}$$

Using the Taylor's theorem for $e^{x_i} \approx 1$, the Eq. (15) becomes

$$\left.\frac{dP}{dT}\right|_{I_{ph},R_L,R_p} \approx \frac{2.R_L I^2.y.\sum_{i=1}^{m} x_{0i}.e^{x_i}\left(R_s + R_L + R_p\right)(R_s + R_0) - 2I.R_L y.T.(R_s + R_L).I_{ph}}{\left(R_s + R_L + R_p\right)\left(1 + y.\sum_{i=1}^{m} x_{0i}.e^{x_i}\right)(R_s + R_L)T} \tag{15}$$

$$2.R_L I^2.y.\sum_{i=1}^{m} x_{0i}.e^{x_i}\left(R_s+R_L+R_p\right)(R_s+R_0) > 2I.R_L y.T.(R_s+R_L).I_{ph} \tag{16}$$

From Eq. (16) $\frac{dP}{dT}\Big|_{I_{ph},R_L,R_p} > 0$ the contradicts the requirement of Eq. (9), so R_s does not belong to the conductor type.

3.2 Negative Temperature Coefficient Type (NTC)

The NTC effect is governed by an Arrhenius equation can be given as.

$$R_s = R_0 e^{\frac{B}{T}} \tag{17}$$

where B is the semiconductor material coefficient $B > 0$ and R_O is resistance at temperature T_O. The differential of the previous function gives:

$$\frac{dR_s}{dT} = -\frac{BR_s}{T^2} < 0 \tag{18}$$

The equation of power becomes:

$$\frac{dP}{dT}\Big|_{I_{ph},R_L,R_p} = \frac{2.R_L I^2.y.\sum_{i=1}^{m} x_{0i}.e^{x_i}}{1+y.\sum_{i=1}^{m} x_{0i}.e^{x_i}}\frac{1}{T} + \left[\frac{2I.R_L y.\left[I_{ph}-\sum_{i=1}^{m} I_{s_i}(e^{x_i}-1)\right]}{\left(R_s+R_L+R_p\right)\left(1+y.\sum_{i=1}^{m} x_{0i}.e^{x_i}\right)} + \frac{2.R_L I^2.y.\sum_{i=1}^{m} x_{0i}.e^{x_i}}{(R_s+R_L)\left(1+y.\sum_{i=1}^{m} x_{0i}.e^{x_i}\right)}\right].\frac{BR_s}{T^2} > 0 \tag{19}$$

It is obvious that an Eq. (19) contradicts the requirement of Eq. (17). Hence, R_S also does not belong to the negative temperature coefficient type.

3.3 Positive Temperature Coefficient Type (PTC)

A PTC material can be designed to reach a maximum temperature for a given input voltage. The positive temperature coefficient type can be described by the following relation:

$$R_s = R_0 e^{BT} \tag{20}$$

where B is the semiconductor material coefficient $B > 0$ and R_O is resistance at temperature T_O. The differential of the previous function gives:

$$\frac{dR_s}{dT} = BR_s > 0 \tag{21}$$

To satisfy the condition of Eq. (8), the conditions is required that the equation of power presented (15) is.

$$\left.\frac{dP}{dT}\right|_{I_{ph},R_L,R_p} = 2.R_L R_s I^2.y.\sum_{i=1}^{m} x_{0i}.e^{x_i}.\left[I_{ph} - \sum_{i=1}^{m} I_{s_i}(e^{x_i}-1)\right](R_s+R_L).\frac{(R_s+R_L+R_p)(1_s+R_L+T.B) - 2.B.R_s I.R_L.T.y.}{(R_s+R_L+R_p)\left(1+y.\sum_{i=1}^{m} x_{0i}.e^{x_i}\right)(R_s+R_L)T} \tag{22}$$

$$1 < \frac{BT(R_s+R_L)}{(R_s+R_L+R_p)(1_s+R_L+T.B)}\left[\sum_{i=1}^{m}\left(\frac{I_{ph}}{x_i.I_{s_i}.e^{x_i}} - \frac{1}{x_{i.}} - \frac{I_{s_i}}{x_{i.}.e^{x_i}}\right)\right] \tag{23}$$

The Eq. (21) are in accords with the definition of the semiconductor material positive temperature coefficient type with B > 0. Therefore, the series resistance R_s is a positive temperature coefficient type and has a specific form.

4 Applications and Discussion

The numerical value of dynamic series resistance may be derived from the I-V characteristics at a certain temperature. In [13] an analytical relation was proposed to calculate the dynamic series resistance of silicon solar cell for single diode. The relation is generalised for multi-dimension silicon solar cell model for multi-diode with $\sum_{i=1}^{m} n_i = 1.49$. The experimental and positive temperature coefficient (PTC) of R_S values are taking from [13]. The compared experimental and numerical values of R_s using the theory expression are shown in Fig. 2a.

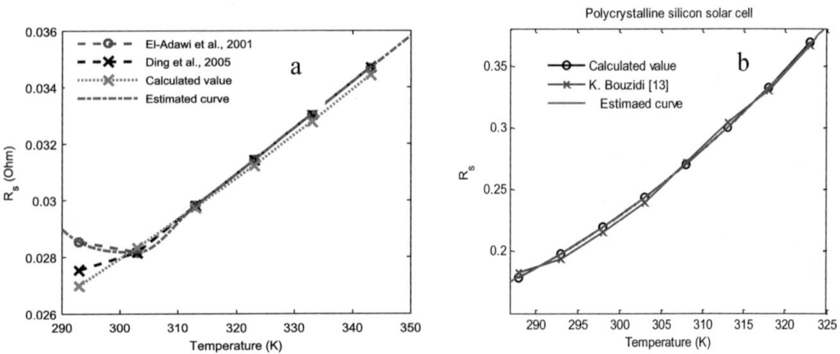

Fig. 2. Curve dynamic series resistance of, (a) the silicon solar cell, (b) polycrystalline silicon solar cell.

The experimental and positive temperature coefficient (PTC) of R_S values with $R_0 = 4.6\ 10^{-4}\Omega cm^2$ and B = 0.0207 K^{-1} of polycrystalline silicon solar cell using in [14] are summarised in Table 1. The numerical values of R_s using the theory expression of are shown in Fig. 2b.

Table 1. Experimental and calculated data of the polycrystalline silicon solar cell.

T	288	293	298	303	313	318	323
R_s	0.1825	0.1938	0.2150	0.2391	0.2715	0.3041	0.3294
R_s (PTC)	0.1786	0.1981	0.2197	0.2436	0.2702	0.2996	0.3323

The Table 2 shows the resulting values of the dynamic series resistance and Fig. 3a the variation of the resistance with estimated value of commercial Siemens single crystalline silicon Solar cell of the surface 10×10 cm^2 used in [12].

Table 2. Experimental data of the commercial siemens single crystalline silicon solar cell.

T (K)	295	310	318	343	350
$R_s(\Omega)$ [15]	0.1115	0.1309	0.1427	0.1559	0.1584

The experimental values of dynamic resistance for photovoltaic panel used in [16] are presented in Table 3 and Fig. 3b shows the variation of the dynamic resistance with temperature.

Table 3. Experimental data of photovoltaic panel.

T (K)	15	25	35	45	55	65
$R_s(\Omega)$	0.6326	0.6347	0.6367	0.6378	0.6388	0.6399

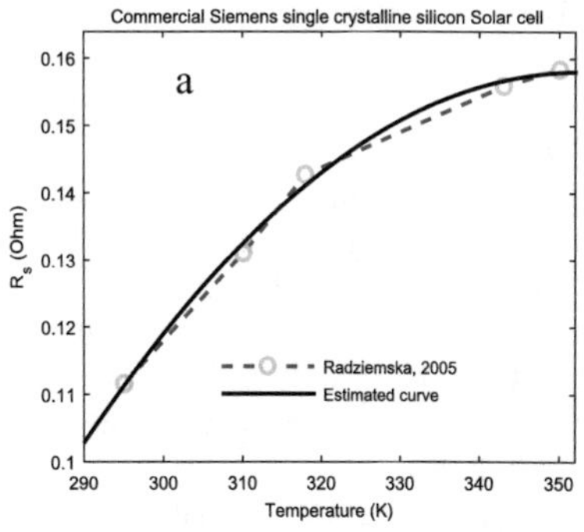

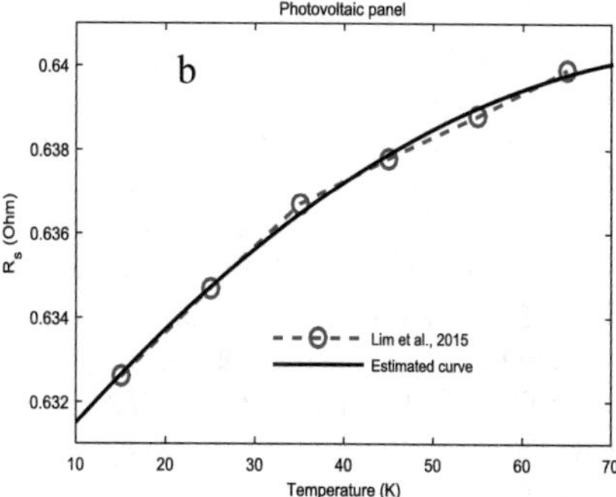

Fig. 3. Curve dynamic series resistance of, (a) Commercial Siemens single crystalline silicon Solar cell of the surface 10×10 cm^2, (b) photovoltaic panel.

5 Conclusion

This paper has presented the dynamic thermal behavior of a multi diode solar cell model with series and shunt resistance. The approach is based on the derivation of the nonlinear power characteristics with respect to temperature of multi diode solar cell model, in order to simplify the equation. This study focused on the polycrystalline silicon solar, commercial Siemens single crystalline silicon Solar cell with surface $10 \times 10\ \text{cm}^2$, OST-80 photovoltaic module and photovoltaic panel. The results showed the series resistance has an exponential expression such as positive temperature coefficient type, and it has been validated on the considered commercial solar cells.

References

1. Soon, J.J., Low, K.S.: Optimizing photovoltaic model for different cell technologies using a generalized multidimension diode model. IEEE Ind. Electron. Soc. **62**(10), 6371–6380 (2015)
2. Celik, A.N., Acikgoz, N.: Modelling and experimental verification of the operating current of mono-crystalline photovoltaic modules using four- and five-parameter models. Appl. Energy **84**(1), 1–15 (2007)
3. Villalva, M.G., Gazoli, J.R., Filho, E.R.: Comprehensive approach to modeling and simulation of photovoltaic arrays. IEEE Power Electron. Soc. **24**(5), 1198–1208 (2009)
4. Louzazni, M., Khouya, A., Amechnoue, K., Gandelli, A., Mussetta, M., Crăciunescu, A.: Metaheuristic algorithm for photovoltaic parameters: comparative study and prediction with a firefly algorithm. Appl. Sci. **8**(3), 339 (2018)
5. Ishaque, K., Salam, Z., Taheri, H.: Simple, fast and accurate two-diode model for photovoltaic modules. Solar Energy Mater. Solar Cells **95**(2), 586–594 (2011)
6. Streetman, B.G., Banerjee, S.: Solid State Electronic Devices, 6th edn. Pearson/Prentice Hall, Upper Saddle River (2006)
7. McIntosh, K.: Lumps, humps and bumps: three detrimental effects in the current-voltage curve of silicon solar cells. Thesis published (2001)
8. Nishioka, K., Sakitani, N., Uraoka, Y., Fuyuki, T.: Analysis of multicrystalline silicon solar cells by modified 3-diode equivalent circuit model taking leakage current through periphery into consideration. Solar Energy Mater. Solar Cells **91**(13), 1222–1227 (2007)
9. Mialhe, P., Khoury, A., Charles, J.P.: A review of techniques to determine the series resistance of solar cells. Solar Energy Mater. Solar Cells **91**(18), 1698–1706 (2007)
10. Bashahu, M., Habyarimana, A.: Review and test of methods for determination of the solar cell series resistance. Renew. Energy **6**(2), 129–138 (1995)
11. Ortiz-Conde, A., García Sánchez, F.J., Muci, J.: New method to extract the model parameters of solar cells from the explicit analytic solutions of their illuminated I-V characteristics. Solar Energy Mater. Solar Cells **90**(3), 352–361 (2006)
12. Radziemska, E.: Dark I-U-T measurements of single crystalline silicon solar cells. Energy Convers. Manage. **46**(9–10), 1485–1494 (2005)
13. El-Adawi, M.K., Al-Nuaim, I.A.: A method to determine the solar cell series resistance from a single I-V Characteristic curve considering its shunt resistance-new approach. Vacuum **64**(1), 33–36 (2001)

14. Bouzidi, K., Chegaar, M., Bouhemadou, A.: Solar cells parameters evaluation considering the series and shunt resistance. Solar Energy Mater. Solar Cells **91**(18), 1647–1651 (2007)
15. Ding, J., Cheng, X., Fu, T.: Analysis of series resistance and P-T characteristics of the solar cell. Vacuum **77**(2), 163–167 (2005)
16. El-Adawi, M.K., Al-Nuaim, I.A.: A method to determine the solar cell series resistance from a single I-V characteristic curve considering its shunt resistance-new approach. Vacuum **64**(1), 33–36 (2001)

Extreme Learning Machine Based Multi-Agent System for Microgrid Energy Management

Dounia El Bourakadi(✉), Ali Yahyaouy, and Jaouad Boumhidi

LIIAN Laboratory, Computer Science Department,
Faculty of Sciences Dhar-Mahraz, Sidi Mohamed Ben Abdellah University,
Fez, Morocco
{dounia.elbourakadi,ali.yahyaouy,
jaouad.boumhidi}@usmba.ac.ma

Abstract. In this paper, an intelligent energy management system is presented for distributed structure like a smart microgrid. To model the microgrid, a Multi-Agent System is proposed based on Extreme Learning Machine algorithm to estimate the wind and photovoltaic power output from weather data. In this study a microgrid, with different generation units and storage units is considered. Provision of utility grid insertion is also given if the total energy produced by microgrid falls short of supplying the total load or if there is an excess of energy produced instead of to be wasted. Thus the goal of our Multi-Agent System is to control the amount of power delivered or taken from the main grid in order to reduce the electricity bill and make profit by selling the surplus in the energy market. After supplying the load requirements, Extreme Learning Machine algorithm for classification is used to make decision about selling/purchasing electricity from the main grid, and charging/discharging batteries. Finally for simulation, the Java Agent Development Framework platform is used to implement the approach and analyze the results.

Keywords: Renewable energy · Microgrid · Prediction · Extreme Learning Machine · Multi-Agent System

1 Introduction

At present, the major source of global energy of the world comes from fossil fuels; which are flammable, dangerous and exhaustible. Renewable energies constitute the alternative to fossil energies because they disturb less the environment and do not emit a gas with greenhouse effect. For it, many countries tend to the new concept of microgrid (MG) that mainly comprise renewable and/or conventional sources, including photovoltaic (PV) power, wind power, internal combustion engine, gas turbine, and micro-turbine together with a cluster of loads [1]. MG can be connected directly to the distribution network (DN) or operate in isolated mode. However, this system type is touched by the site climatic data change, that's why the different sources must be controlled by an intelligent energy management system (EMS). One of its main objectives is to achieve a high level of flexibility, not only during operation, but also during outages and during all its life cycle. Classical control methods are not

M. Ezziyyani (Ed.): AI2SD 2018, AISC 912, pp. 30–39, 2019.
https://doi.org/10.1007/978-3-030-12065-8_4

efficient to manage and control the different operations for such system. For this reason, in the past few years the paradigm of Multi-Agent System (MAS) has been invented and used in power systems in order to ensure the more powerful results, thanks to the advantages which characterize this approach. Although, we can found many works, in the literature, based on MAS to manage and control the MG [2–7].

In MGs, the main energy sources are the renewable sources like wind and solar energies. The integration of the forecast of whether data and output power is necessary and there are many proposed works to predict the wind and solar power [8–12]. However, most of them used artificial neural network (ANN) algorithm for prediction that has some drawbacks such as the convergence time which is slow and the weakness to find the global optimistic result.

In this paper we propose an optimal dispatch strategy of a MG that consists of wind and solar power and a battery energy storage system. For maximum utilization of renewable energy sources, an intelligent EMS is presented, which optimizes the MG system operations on an hour-by hour time scale using a MAS scheduling energy system based on the Extreme Learning (ELM) algorithm for the prediction of power produced by renewable sources for the next hour, and for the decision making while respecting MG constraints. The goal of our MAS is to control the amount of power delivered or taken from the main grid in order to reduce the electricity bill and make profit by selling the surplus in the energy market. In our study, the system attempts to supply its costumers locally. The primary contributions of this paper include the following: supply the power requirements; increase battery life by maximizing battery charging or discharging for a continuous number of states and finally maximize the utilization of renewable energy resources, and minimize the import/export from the DN which means more environmental friendly and sustainable operation.

The remainder of this paper is organized as follows. In the Sect. 2, we present the adopted MG and its components. Then a detailed multi-agent architecture which involves ELM for prediction and for decision making is proposed in the Sect. 3. In the Sect. 4, we present the simulation and we discuss the results. We conclude the paper in the Sect. 5.

2 Adopted Microgrid

MG consists in general of distributed generators, renewable generation, energy storage and local demands. Figure 1 presents the adopted MG which is on connected-mode. Taking into account the unpredictable weather conditions and to assume a continuous availability of energy, PV panels are supplied with wind turbine. The system also includes a number of batteries to either store the excess of energy produced by renewable energy sources or provide the energy demanded by the load when there is low renewable energy produced.

An optimal dispatch strategy of such a system requires that the load is primarily met by the renewable resources ($P_T = P_{PV} + P_W$) and the battery (P_B) is dispatched to meet the load only when P_T is less than the load demanded (P_L). Battery charging may take place only when P_T is greater than the consumer load such that the battery acts as the storage of surplus renewable energy. In case if the total power generated is not enough to cover the supplying load, we discharge the batteries, otherwise the energy is bought from the DN.

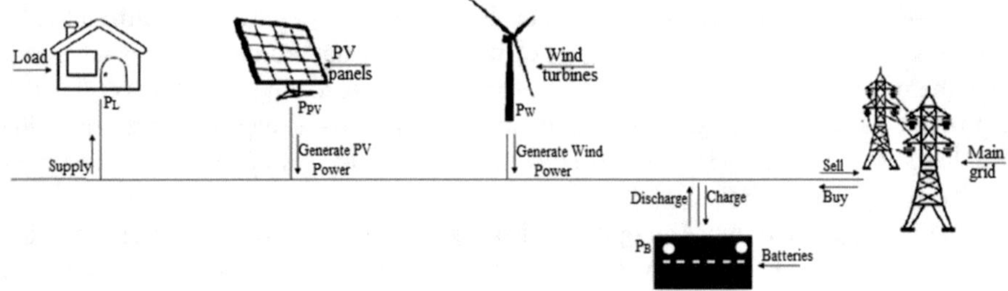

Fig. 1. Diagram overview of the adopted MG with renewable energy sources.

2.1 Wind Turbine

The wind turbine produces electricity from wind. It is considered as a primary source in our MG. The output power generated by the wind turbine can be considered as a function of the wind velocity [13] and can be calculated as follow:

$$\begin{cases} P_w = 0, & V_{ac} < V_{ci} \\ P_w = aV_{ac}^2 + bV_{ac} + c, & V_{ci} \leq V_{ac} < V_r \\ P_w = P_{wr}, & V_r \leq V_{ac} > V_{co} \end{cases} \quad (1)$$

Where P_{wr}, V_{ci}, and V_{co} are the rated power, cut-in and cut-out wind speed respectively. Furthermore V_r and V_{ac} are the rated and actual wind speed. Constants *a*, *b*, and *c* depend on the type of the wind turbine.

We assume *AIR403* wind turbine model in this paper. According to the data from the manufacturer, their characteristics are as follow [14]:

a = 3.4; b = −12; c = 9.2; P_{wr}= 130 Watt; V_{ci}= 3.5 m/s; V_{co}= 18 m/s; V_r= 17.5 m/s.

2.2 Photovoltaic Module

The PV module ensures the production of electrical energy during daylight hours. It is considered as the main source of energy. The output power of the module can be calculated as follow [14]:

$$P_{PV} = P_{STC} \frac{G_{ING}}{G_{STC}} [1 + k(T_c - T_r)] \quad (2)$$

where:

P_{PV} The output power of the module at Irradiance G_{ING};

P_{STC} The Module maximum power at Standard Test Condition (STC);

G_{ING} Incident Irradiance;
G_{STC} Irradiance at STC (1000 W/m^2);
k Temperature coefficient of power;
T_c The cell temperature;
T_r The reference temperature.

We assume that *SOLAREX MSX-83* modules are used in this paper. Their output characteristics are: peak power = 83 W, voltage at peak power = 17.1 V, current at peak power = 4.84A, short circuit current = 5.27A, and open circuit voltage = 21.2 V at STC.

2.3 Energy Storage

A battery bank is a group of batteries connected together using series or parallel wiring. It allows storing electricity generated by renewable system for later use at any time. The power from the battery is needed whenever the renewable power produced is insufficient to supply the load demand. In addition, the energy is stored whenever the supply from the renewable sources exceeds the load demand. The State Of Charge (SOC) of a battery is defined as its available capacity expressed as a percentage. It can be calculated using the following equation where C is the capacity of battery and C_{Ref} is the reference capacity of battery:

$$SOC = \frac{C \times 100}{C_{Ref}} \tag{3}$$

It is important that the SOC of the battery prevents the battery from overcharging or undercharging. The associated constraints can be formulated by comparing the battery SOC with the battery $\mathrm{SOC_{min}}$ and the battery $\mathrm{SOC_{max}}$, as shown in Eq. (4). This paper assumes that $\mathrm{SOC_{min}}$ and $\mathrm{SOC_{max}}$ equal 20% and 80%. It is also assumed that the initial SOC of the battery is 80% at the beginning of the simulation.

$$SOC_{\min} \leq SOC \leq SOC_{max} \tag{4}$$

3 Proposed Energy Management System

The energy management in flexible structure such as MG is a complex problem. This complexity comes from the uncertainty of the environment and its variations. EMS should consider these constraints and attempt several objectives at the same time. In this paper, we present an intelligent EMS for MG. Its goal is to control the amount of power delivered or taken from the DN in order to maximize the benefit on the one hand, by selling electricity to the DN, after filling the local demand of MG and charging batteries. On the other hand, it must reduce the cost by purchasing electricity from the DN only if the load demand is greater than the produced energy and batteries are discharged. To achieve these goals, we divide the energy management strategy into two main phases (see Fig. 2):

- **Prediction phase:** We opt to use ELM algorithm to predict the amount of PV and wind power produced for the next hour.
- **Decision making phase:** The decision making with ELM algorithm for classification is adopted to determine which auxiliary energy source (batteries or DN) will fill the need energetic and to take a reasonable decision about storing or selling electricity.

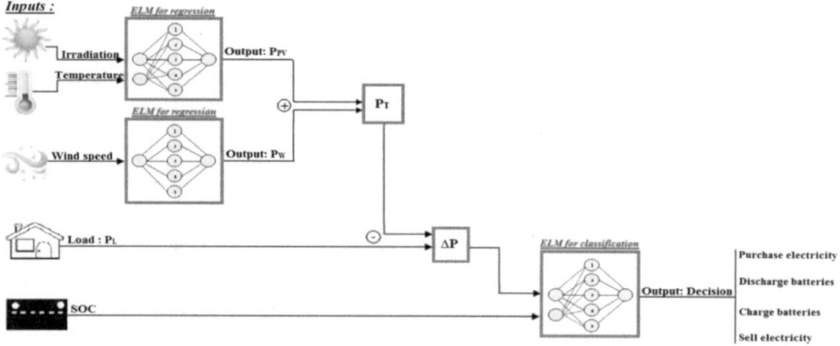

Fig. 2. Proposed energy management diagram.

3.1 MAS Architecture of the Adopted MG

MAS is a distributed system consisting of a set of agents. Unlike artificial intelligence systems that simulate human reasoning capabilities, MAS are ideally designed and implemented as a set of interacting agents, according to modes of cooperation, competition or coexistence [15].

In this study we use MAS to model the adopted MG. Six agents are established in the entire system as mentioned in Fig. 3:

- **AgentWindTurbine**: It predicts the wind power produced at time t by the wind turbine. It uses ELM algorithm to predict the wind power for the next hour.
- **AgentPV:** This agent estimates the PV power produced by solar panels at time t. It uses ELM algorithm to predict the PV power for the next hour.
- **AgentBatteryBank**: It is the agent who is responsible for the storage of energy. Among its roles: gives information on the SOC of batteries and provides power to the MG when trigger request comes from the AgentController.
- **AgentLoad**: It gives information of load demanded in MG at time t.
- **AgentMainGrid**: This agent offers the different selling or purchase price of electricity over the time and provides power to the MG when trigger request comes from the AgentController.
- **AgentController**: This agent is the responsible for the coordination between the other agents and the decision making.

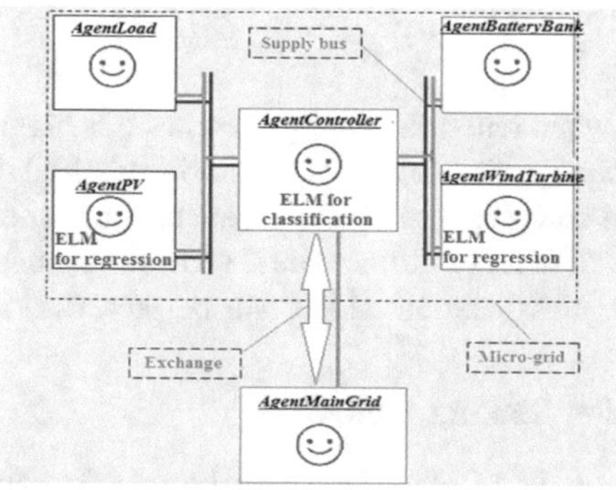

Fig. 3. MAS architecture of adopted MG.

3.2 Proposed Prediction Method Adopted

In MG, the main energy sources are renewable energy like wind and PV energy. The integration of the forecast of whether data and output power is necessary and there are many proposed works to predict the wind and solar power. However, most related works that used ANN algorithm for prediction have some problems such as the convergence time which is slow, the weakness to find the global optimistic result and the difficulty to calculate the quadratic convex programming modelling the real problem. Besides, these techniques are unable to cope with the strong fluctuation of observations. Taking into account these problems mentioned above, we opt to use ELM algorithm for the prediction of power produced by renewable sources, because it overcomes the shortcomings of several other machine learning techniques.

In our previous work [16], ELM algorithm was chosen to predict PV and wind power produced using a preparing learning dataset. In this work, to learn the ELM algorithm we use real observations of weather data. ELM was introduced by Huang in [17]. It overcomes the shortcomings of several other machine learning techniques. Besides its flexibility, the algorithm has also shown a good generalization ability on benchmark tasks, both in classification and in function regression [18]. It requires less human interference and can run thousands times faster than those conventional methods. Hidden nodes are chosen randomly and the output weights are determined analytically by determining the optimal combination of the output signals of the hidden layer [18].

In this paper, we use two predictors ELM. The first predict the PV power produced (P_{PV}) based on irradiation and temperature as inputs. The second predict the wind power produced (P_W) based on the wind speed as input. Thus we can obtain the total energy produced (P_T) by renewable energy sources in the MG at each period of time t.

$$P_T(t) = P_{PV}(t) + P_W(t) \tag{5}$$

After getting the total power produced by renewable sources P_T, we calculate ΔP, the difference between P_T and P_L (load demand).

$$\Delta P(t) = P_T(t) - P_L(t) \tag{6}$$

ΔP can be positive when renewable energy produced is bigger than load or negative when load is bigger than renewable energy produced. If ΔP is negative we must complete the need from batteries when batteries are charged or from DN when batteries are discharged. To decide which auxiliary source will fill the need (batteries or DN) we use ELM algorithm for classification which will be presented in the next subsection.

3.3 Proposed Decision Making Method

In this paper, we use an ELM classifier to make decision about selling/purchasing electricity from the main grid, or charging/discharging batteries. Thus, to establish an efficient energy management strategy, decision is taken based on the status of loads, renewable energy sources and batteries (see Fig. 4). The inputs of the classifier are ΔP and SOC. The output is Decision which represents a class among for classes numerated from 1 to 4. In fact, according to the values of the inputs we make decision among four choices as follow:

- **Class 1:** Purchase necessary power when ΔP is negative and SOC $\leq$ 20%.
- **Class 2:** Discharge batteries when ΔP is negative and SOC > 20%.
- **Class 3:** Charge batteries when ΔP is positive and SOC $\leq$ 80%.
- **Class 4:** Sell excess power to DN when ΔP is positive and SOC > 80%.

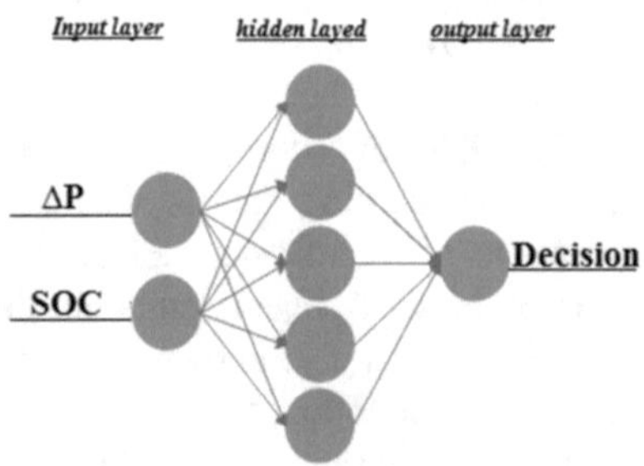

Fig. 4. Illustration of ELM's classifier.

4 Simulation and Results

MAS can be implemented based on a number of open-source agent platforms that can aid developers to build a complex agent system in a simplified fashion [19]. Our MAS was developed in the Java Agent Development Framework platform (JADE), an open source IEEE FIPA compliant platform [20]. It is employed to perform some simulations.

To learn the two ELM predictors we used observations of irradiation, temperature and wind speed. Figures 5 and 6 present results of PV power and wind power predicted for 24 h (one day). The curves compare the results predicted using ELM algorithm and the results obtained using mathematical formulas (Eqs. (1) and (2) cited in the Sect. 2).

The algorithm was trained and simulated using java. After the analysis, the obtained results are quite acceptable, and then ELM algorithm can predict the PV and wind power for the next hour taking into account the irradiation, temperature and wind speed.

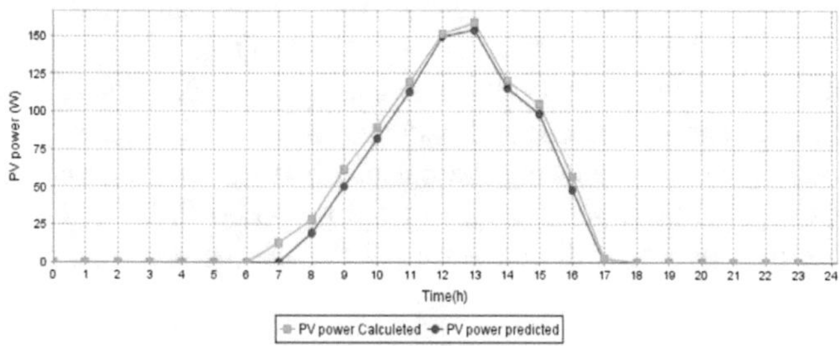

Fig. 5. PV power predicted by ELM (red) and calculated by mathematical formula (green).

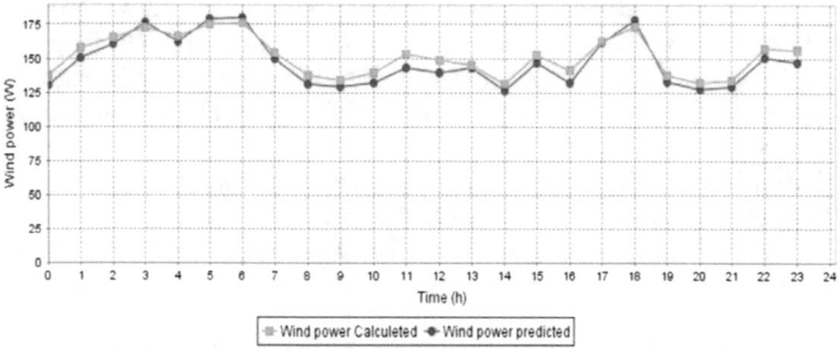

Fig. 6. Wind power predicted by ELM (red) and calculated by mathematical formula (green).

In order to ensure a long lifetime for the battery, its SOC should be as much as possible more than 20%. ELM classifier adopted allowed to obtain this level of guard. Figure 7 presents the SOC of the battery during the period test of 2000 h which is almost equivalent to three months. The variation of SOC remains over 20% with only two deep discharges in all the test period.

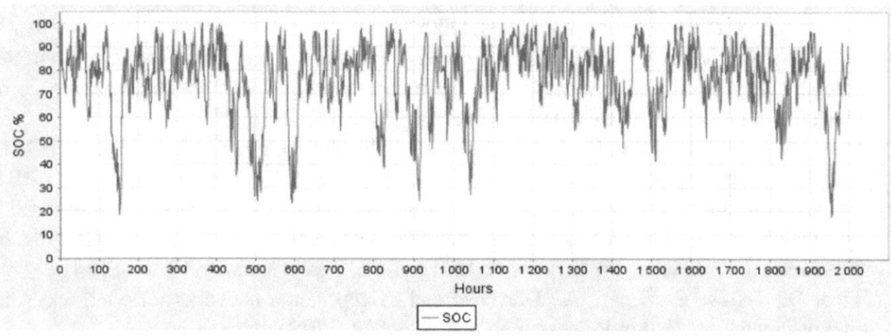

Fig. 7. SOC of batteries during the period test.

5 Conclusion

In this paper a MAS was proposed to model the adopted MG. The power generated from the renewable sources was predicted using ELM algorithm to overcome problems related to complex mathematical modeling. The results confirm that ELM can be applied in a very flexible way and shows several advantages.

To establish an efficient energy management strategy, decision is taken based on the status of loads, energy sources and batteries using an ELM classifier, in order to satisfy economic goals and reducing the balance between cost and benefit. The simulation results highlight the performance of the proposed EMS as the SOC remains over 20% with only two deep discharges in all the test period.

References

1. Lasseter, R., Paigi, P.: Microgrid: a conceptual solution. In: Proceedings of the IEEE 35th Annual Power Electronics Specialists Conference, Germany, pp. 4285–4290 (2004)
2. Hommelberg, M., Warmer, C., Kamphuis, I., Kok, J., Schaeffer, G.: Distributed control concepts using multi-agent technology and automatic markets: an indispensable feature of smart power grids. In: Proceedings of the IEEE Power Engineering Society General Meeting, USA, pp. 1–7 (2007)
3. Fu-Dong, L., Min, W., Yong, H., Xin, C.: Optimal control in microgrid using multi-agent reinforcement learning. ISA Trans. **51**, 743–751 (2012)
4. Didi, E., Boumhidi, J.: Multi agent system based on law of gravity and fuzzy logic for coalition formation in multi micro-grids environment. J. Ambient Intell. Humanized Comput. **9**, 337–349 (2016)
5. Bui, V., Hussain, A., Kim, H.: A multiagent-based hierarchical energy management strategy for multi-microgrids considering adjustable power and demand response. IEEE Trans. Smart Grid **9**, 1323–1333 (2016)
6. Didi, E., Serraji, M., Nfaoui, E., Boumhidi, J.: Multi-agent architecture for optimal energy management of a smart micro-grid using a weighted hybrid BP-PSO algorithm for wind power prediction. Int. J. Technol. Intell. Plann. **11**, 20–35 (2016)
7. Khatibzadeh, A., Besmi, M., Mahabadi, A., Haghifam, M.: Multi-agent-based controller for voltage enhancement in AC/DC hybrid microgrid using energy storages. Energies **10**, 169 (2017)
8. Pan, D., Liu, H., Li, Y.: Optimization algorithm of short-term multi-step wind speed forecast. Proc. Chin. Soc. Electr. Eng. **26**, 87–91 (2008)
9. Chao, W., Wenjun, Y.: Short-term wind speed prediction of wind farms based on improved particle swarn optimization algorithm and neural network. In: International Conference on Mechanic Automation and Control Engineering, China (2010)
10. Juan, M., Jos, L., Rodolfo, D., Jos, A.: Forecast of hourly average wind speed using ARMA model with discrete probability transformation. Electr. Eng. Control **98**, 1003–1010 (2011)
11. Di Piazza, A., Di Piazza, M.C., Vitale, G.: Estimation and forecast of wind power generation by FTDNN and NARX-net based models for energy management purpose in smart grids. In: International Conference on Renewable Energies and Power Quality, Spain (2014)
12. Vico, D., Terres, A., Omari, A., Dorronsoro, J.: Deep neural networks for wind and solar energy prediction. Neural Process. Lett. **46**, 829–844 (2017)

13. Chedid, R., Akiki, H., Rahman, S.: A decision support technique for the design of hybrid solar- wind power systems. IEEE Trans. Energy Convers. **13**, 76–83 (1998)
14. Faisal, M., Koivo, H.: Modelling and environmental/economic power dispatch of microgrid using multiobjective genetic algorithm optimization. In: Fundamental and Advanced Topics in Wind Power, pp. 361–378 (2011)
15. Chaib-draa, B., Ken, A., Williams, J., Hall, C., Kent, R.: Distributed artificial intelligence: an overview. Encycl. Comput. Sci. Technol. **31**, 215–243 (1994)
16. El Bourakadi, D., Yahyaouy, A., Boumhidi, J.: Multi-agent system based on the fuzzy control and extreme learning machine for intelligent management in hybrid energy system. In: Proceeding of the Intelligent Systems and Computer Vision conference, Morocco (2017)
17. Huang, G., Zhu, Q., Siew, C.: Extreme learning machine: a new learning scheme of feedforward neural networks. In: Proceeding of the IEEE International Joint Conference on Neural Networks, Hungary (2004)
18. Huang, G., Zhu, Q., Siew, C.: Extreme learning machine: theory and applications. Neurocomputing **70**, 489–501 (2006)
19. Pipattanasomporn, M., Feroze, H., Rahman, S.: Multi-agent systems in a distributed smart grid: design and implementation. In: Proceeding of the Power Systems Conference and Exposition, USA (2009)
20. Bellifemine, F., Caire, G., Greenwood, D.: Developing Multi-Agent Systems with JADE, p. 303. Wiley (2007)

The Experimental Study and Modeling the Drying Kinetics of Mediterranean Mussel (Mytilus Galloprovincilis) Type by Convective Solar Energy

M. Kouhila(✉), A. Idlimam, A. Lamharrar, H. Lamsyehe, and H. Moussaoui

Team of Solar Energy and Medicinal Plants, Cadi Ayyad University, High School of Trainee Teachers, Marrakech, Morocco
kouhilamounir@gmail.com

Abstract. Drying is a process of hydration and Elimination of water which allows the proliferation of microorganisms and development of chemical reactions without influencing morphological structure of Food Material, this research focused on the influence of temperature on drying kinetics of the Mediterranean mussels (mytilus galloprovincilis) as per the requirement for storage seafood. Convective drying kinetics and hygroscopic behavior of Mytilus Galloprovincilis was carried out in a solar dryer operating in forced convection. Experimental drying kinetics were measured at three air temperatures (50, 60, and 70 °C), and two air flow rates fixed at (300 and 150 $m^3.h^{-1}$) with ambient air temperature in the range of 22 to 38 °C, 18 to 27% for ambient humidity, 422 to 788 w/m^2 and 530 to 898 w/m^2 for flow solar irradiation inclined and horizontal respectively. Experimental data of Drying are collected to plot the characteristic curve drying. Nine mathematical models available in the literature using simulation for describing drying curve. The logarithmic model showed the best fitting of experimental data with a highest value of correlation coefficient (r), and lowest value of reduced chi-square (χ^2).

Keywords: Drying kinetics · Mathematical model · Mediterranean mussel · Solar dryer

1 Introduction

Drying is considered the oldest food preservation technique, and a common unit operation in many chemical and process industries. The removal of moisture prevents the growth and reproduction of microorganisms causing decay and minimizes many of the moisture mediated deteriorative reactions. Drying is also particularly considered an important technique to conserve the perishable foods [1].

Mytilus Galloprovincilis contain up to 60% of water and are thus highly perishable were cold preservation techniques are often missing, Preservation as cooling is necessary because it is a widely used to maintain quality of fresh product and prevent high enzymatic reaction and bacterial activity in fresh fish [2], most of mussels (Mytilus

M. Ezziyyani (Ed.): AI2SD 2018, AISC 912, pp. 40–53, 2019.
https://doi.org/10.1007/978-3-030-12065-8_5

Galloprovincilis) from Essaouira region (Morocco) is not consumed by merchandiser and transformed into uncultivated product, then it remains superfluous and goes waste. the same at india, around 20% of fresh seafood was attenuated due to deficient technique of cold storage and irregular postharvest practices [3]. Small fish spices in the same of crap (chelwa) and prawn are dried in northern India [4]. Solar dryer is very frequent practice of aquaculture product in many developing countries such as morocco.

Drying of Mussels is primarily carried out by traditionally under open sun. Sun drying produce as clean energy and represents a lowest process on energy consumption to preserve seafood. Natural or direct sun drying has been employed afterward time immemorial, however this technic has disadvantage in way to determine the drying parameter, Weather trouble, insect infection, contamination owing to air pollution.

In spite of to preserve aquaculture and agriculture food, open sun drying is mainly expert in subtropical countries where solar radiation is available [5]. On the other hand; the drying technic can develop product quality, enhance shelf life and support their processing [6, 7].

The knowledge of all experiment data of drying in different air condition as temperature, humidity and flow rate, gives the homogeneity and correlation points gathered into global curve especially named characteristic drying curve (CCD). the main interest of CCD curve is reduced all the experiment data and it could be examined in the other convective air condition in a usable form different to the experiment himself.

A deep knowledge of heat transfer parameters, mass, diffusion and the drying behaviour of the product are considered indispensable for the conceptualization, simulation and optimization of the drying process. It is then necessary to have an accurate model that can predict rates of elimination of water and describe the drying of the product, nine theoretical models were studied to predict and validate experimental data, those statistical model are: Logarithmic, Midilli-Kucuk, Newton, page, Handerson and Pabis, Tow-term, Diffusion Approach, Wang and singh, Modified Handerson and Pabis.

To our knowledge, there are no published works concerning drying of Mytilus Galloprovincilis, therefore, in this study is focused on:

- Study the influence impact of drying air temperature on drying kinetics of Mediterranean mussels;
- Establish a characteristic drying curve of Mediterranean mussel corresponding mainly to an overall data of the experiment;
- Validate the experimental data using nine statistical models extract from the literature for predicting and modelling drying curve.

2 Material and Method

2.1 Description of the Drying System

The experimental apparatus used for the drying of Mediterranean mussels is an indirect forced convection solar dryer (Fig. 1) at the laboratory of the "High School of Trainee Teachers, Marrakech, Morocco". The dryer is the type "cupboard" to activate total or partial recycling drying air with ten polyvalent shelves. This type of dryer provides a hot air flow rate characterized with aero-thermal condition (temperature, humidity and flow rate) [8].

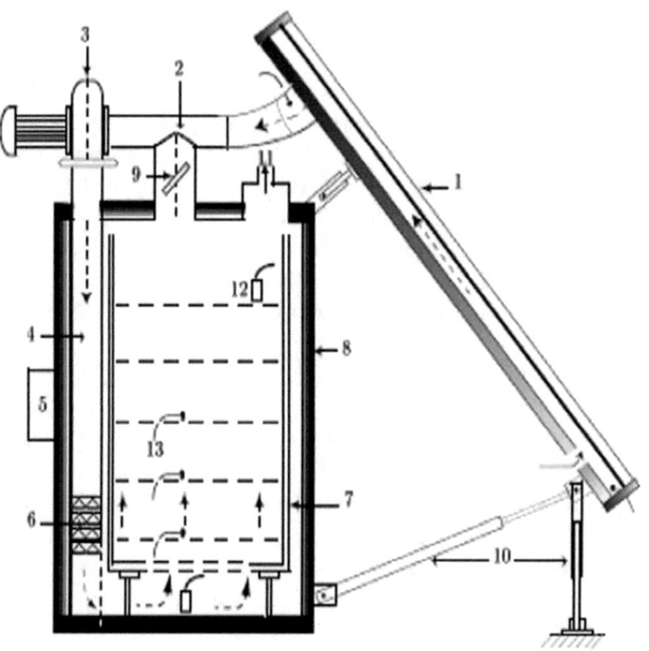

(1) solar collector, (2) ventilation duct, (3) fan, (4) suction line, (5) control box, (6) power supply, (7) floors, (8) drying cabinet, (9) air valve, (10) air inlet, (11) air outlet, (12) humidity probe, (13) thermocouple

Fig. 1. Dispositive of solar dryer

The mass of the product used in drying experiments was (30.0 ± 0.01) g, simples were uniformly spread evenly on a drying tray, that was then placed on the first shelf of the drying cabinet [9], Moreover, ambient air is preheated in a single glazing connected to a solar collector and allow The heated air enters the drying cabinet below the trays and flows upwards through the simples. Auxiliary heater was used for controlling the drying air temperature constant. Solarmeter was used to measure the amounts of daily solar radiation. Temperature measurements at different points and recordings in the solar drying were made by Cr-Alumel thermocouples (0.2 mm diameter) connected to a data logger enabling ± 0.1 °C accuracy and the outlet temperatures were measured with thermometers. The relative humidities were detected bay capacitance sensors and determined by Probes Humicolor ± 2% [10]. A digital weighing apparatus (± 0.001 g) measures the mass loss of the product during the drying process.

The experiments were carried out at three air drying temperatures (50, 60 and 70 °C) and tow drying air flow rates $\left(150 \text{ and } 300 \text{ m}^3.\text{h}^{-1}\right)$ as shown in Table 1. The ambient temperature during the drying period varied from 36 to 42 ± 1 °C, ambient air humidity from 10.52 to 18.86% ± 2%.

Table 1. Drying conditions during experiments in the solar dryer

Exp N°	Air flow rate $Dv(m^3/s)$	Drying temperature T ± 0.1 (°C)	Relative humidity $Rh\pm$ (%)	Ambient temperature (°C)	t (min)
1	300	50	10.52	41	320
2	300	60	18.86	36	230
3	300	70	9.82	42	210
4	150	50	13.69	39	525
5	150	60	13.92	39	330
6	150	70	14.71	38	140

During drying of Mytilus Galloprovincilis for each test, the weight of the sample on the tray was measured undertaken each 5 min until the variation of mass get constant. For each product, it is possible to define an optimal value of Mass wet (Me) for which the product is not deteriorated and maintains its nutritional and organoleptic properties [11]. Indeed, In the first step of this work, desorption isotherms were evaluated for determining the optimum relative humidity (Hr) for predicting the wet mass of sample were drying experiment should be completed. For our research in Mytilus Galloprovincilis, desorption isotherm technique envisages an optimum relative humidity (Hr) of 40%. The final dry matter mass of each sample was determined by a drying oven whose temperature was fixed at 105 °C for 24 h.

This measurement allows as to calculate equilibrium moisture content per dry basis for each temperature (50, 60 and 70 °C) and tow air flow rate, then presenting the drying curves for studying the influence of various air thermal condition (temperature and flow rate) on the drying kinetics of Mediterranean mussels. According to the all experimental data, drying rate versus time was consequently observed from the characteristic drying curve which form of a polynomial at degree 3. In the Subsequent, different equilibrium moisture content obtained and projected on drying curve for various thermal condition were approached and modelled by nine statistical mathematical models.

2.2 Determination of Drying Characteristic Curve

Drying Curve

The drying curves are the curves representing the variation of the equilibrium water content Me as a function of drying time t, or those giving the drying rate as a function of time versus the water content. These curves are obtained experimentally by following the evolution of the wet mass of the product Mh during the drying process by successive weighing until reaching the final moisture content.

To obtain the dry mass Ms, the product is placed at the end of each test in an oven heated to a temperature of 105 °C for 24 h. This curve contains all the experimental information [8, 11]. Installation and drying conditions must make it possible to achieve this optimum value.

Characteristic Drying Curve
The Van Meel [12] transformation is applied for determining the characteristic drying curve allows to study the influence of the drying air temperature for drying kinetics of Mediterranean musel (Mytilus Galloprovincilis). Drying rate curve obtained for different air condition by a single normalized drying rate curve [13]. The moisture content was converted to the moisture ratio MR and calculate coefficient f dimensionless drying rate expressed with form below:

$$MR(t) = \frac{M_h - M_s}{M_s} \tag{1}$$

$$f = \frac{\left(-\frac{dMR}{dt}\right)_t}{\left(-\frac{dMR}{dt}\right)_0} \tag{2}$$

Where f = the dimensionless drying rate
MR is defined from Eq. (1)
$\left(-\frac{dM}{dt}\right)_t$ = the drying rate at any time of drying (Kg water/Kg d.m min)

In the case were the constants as (humidity, temperature, velocity, size of the product to dry) consisting a tolerable range, the characteristic drying curve has the validate the proprieties following:

$$\begin{cases} f = 0 & \text{for} \quad MR = 0 \\ 0 \leq f \leq 1 & \text{for} \quad 0 \leq MR \leq 1 \\ f = 1 & \text{for} \quad MR \geq 1 \end{cases}$$

The general form of characteristic drying curve is used for gathered and generalized experimental point data for drying kinetics of Mediterranean Mussels in order to define a polynomial equation at degree 3.

2.3 Fitting of the Solar Drying Curve

To predict the most suitable drying model for describing the solar drying curve obtained for Mediterranean mussels, nine empirical and semi- empirical models was given by different authors [14–22] and used to fitted solar drying curve (Table 2).

The correlation coefficient r, the reduced chi-square and the Mean Bias Error (MBE) were the statistical parameters used for selecting the best equation describe the thin draying curves of Mediterranean mussels [23], thus, According the highest coefficient

Table 2. Mathematical models applied to the drying curve

Model name	Model equation	Coefficients
Newton [14]	$MR^* = \exp(-kt)$	K
Page [15]	$MR^* = \exp(-kt^n)$	k,n
Henderson and Pabis [16]	$MR^* = a\exp(-kt)$	a,k
Logarithmic [17]	$MR^* = a\exp(-kt) + c$	a,k,c
Two-term [18]	$MR^* = a\exp(-k_0 t) + b\exp(-k_1 t)$	a,k_0,b,k_1
Wang and Singh [19]	$MR^* = 1 + at + bt^2$	a,b
Diffusion Approximation [20]	$MR^* = a\exp(-kt) + (1-a)\exp(-kbt)$	a,b,k
Modified Henderson and Pabis [21]	$MR^* = a\exp(-kt) + b\exp(-k_0 t) + c\exp(-k_1 t)$	a,b,k,k_0,k_1
Midilli-Kucuk [22]	$MR^* = a\exp(-kt^n) + bt$	a,b,k,n

(r) value, the lowest (MBE) and the lowest (χ2), the suitable model could be selected. These parameters are given as following:

$$r = \frac{\sum_{i=1}^{N}\left(MR_{eq_{\exp,i}} - \overline{M}R_{eq_{\exp,i}}\right) \times \left(MR_{eq_{pred,i}} - \overline{M}R_{eq_{pred,i}}\right)}{\sqrt{\sum_{i=1}^{N}\left(MR_{eq_{\exp,i}} - \overline{M}_{eq_{\exp,i}}\right)^2} \times \sqrt{\sum_{i=1}^{N}\left(MR_{eq_{pred,i}} - \overline{MR}_{eq_{pred,i}}\right)^2}} \tag{3}$$

$$\overline{MR}_{eq_{\exp,i}} = \frac{1}{N}\sum_{i=1}^{N} MR_{eq_{\exp,i}} \tag{4}$$

$$\overline{MR}_{eq_{pred,i}} = \frac{1}{N}\sum_{i=1}^{N} MR_{eq_{pred,i}} \tag{5}$$

$$\chi^2 = \frac{\sum_{i=1}^{N}\left(MR_{eq_{\exp,i}} - MR_{eq_{pred,i}}\right)^2}{df} \tag{6}$$

$$MBE = \sqrt{\frac{1}{N}\sum_{i=1}^{N}\left(MR_{eq_{\exp,i}} - MR_{eq_{pred,i}}\right)^2} \tag{7}$$

$$\overline{M}R_{eq_{\exp,i}} = \frac{1}{N}\sum_{i=1}^{N} MR_{eq_{\exp,i}} \tag{8}$$

Where:

$\overline{MR}_{eq_{exp,i}}$	: Equilibrium experimental water content
$\overline{MR}_{eq_{pred,i}}$	: Equilibrium predict water content
N	: number of experimental points
df	: Degree of regression model

Computer programs (Curve Expert 4.1 and Origin 8.0) were applied for calculate the coefficients for each model describe the drying curve, this program applied by Marquardt-Levenberg nonlinear optimization method.

3 Results and Discussion

3.1 Determination of Drying Kinetics

Drying curve represent to evolution of moisture content as function of time, it signified the amount of moisture containing within sample. The drying kinetic of sample indicate the loss of mass and could be presented by drying rate. Drying experiment are performed at three drying air temperatures (50, 60 and 70°C) and tow air flow rate (300 and 150 $m^3.h^{-1}$) as shown in (Table 1).

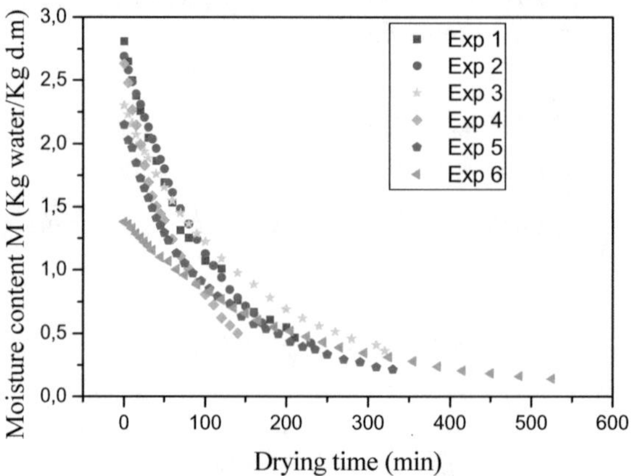

Fig. 2. Evolution of the moisture content of Mediterranean mussel versus the drying time for different conditions of drying air

Six experiment (Table 1) was tested to representing the drying curve, the Fig. 2 show the evolution of the moisture content as function of time from different air condition, it is observed on the drying curve that the moisture content decrease considerably with time and it decrease when the temperature of the drying air increase. Indeed, we observed an absence of initial period (phase 0), the increasing drying rate period, and (phase 1) the constant drying rate period. There is Only the presence of the falling drying rate period (phase 2). this phenomenon was interpreted that amount of free moisture availability is higher in the initial stage of drying, hence it removed considerably [24].

During phase 2, the enquired energy applied to drying has used to evaporate moisture from the product in both vapour and liquid state, this effect involves increasing temperature of the product and is represented by the decrease in drying rate [25, 26]. This mechanism is governed by the effective diffusion in materials, which is often depends on pressure, temperature and moisture content of the product. These results in agreement for other products [27–29].

According analysis for drying rate as shown Fig. 3, it represents the ratio $\left(\frac{dM}{dt}\right)$ versus time (min).

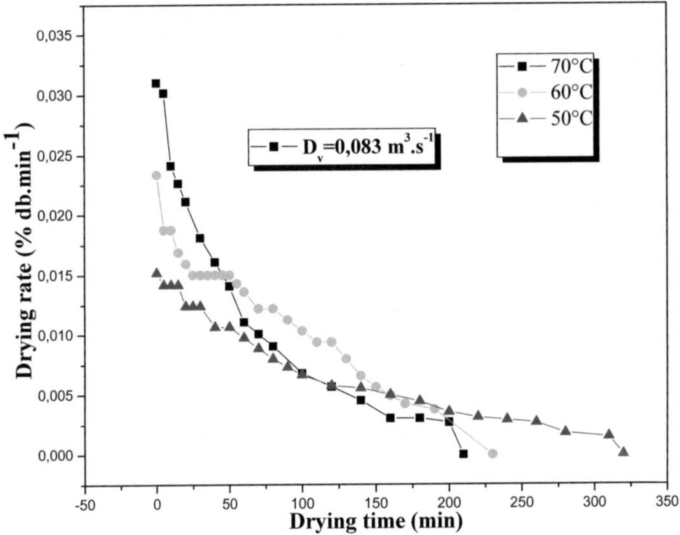

Fig. 3. Evolution of drying rate versus time of the same flow rate at 300 m^3.h^{-1}

This figure mentioned clearly the decreasing rate period (phase 2) as indicated previously at 70, 60 and 50 °C from the same flow rate rate 300 m^3.h^{-1}. It was also noted the presence of irregularities as peaks in the form of plateau in the kinetics drying curve of Mediterranean mussels (Fig. 3), this is clearly appeared from temperature 60 °C exactly at 0.015 (% db/min) flow rate, and 50 °C in the range of flow rate from 0.01 to 0.015 (% db/min). this irregulates envisage alternation of tow drying period, firstly in decrease and followed by a phase constant, indeed, there are produced usually from the complex mechanism to transform free moisture from inside to the surface of the product to be evaporated, generally Mediterranean mussel dispose a particularity body different to thin layer, it contains pores that moisture reside on and it irregular to be evaporated totally, this processes continue until the partial moisture residual be eliminated. Hence the appearance of constant phases in drying rate curve. These observations are considerable from the same type of layers and in agreement for other seafood product such as (Pink shrimps, prawn) that dispose the same comportment [2, 4].

3.2 Influence of the Drying Air Temperature on the Drying Kinetics

According to the results illustrate in Figs. 4 and 5 for three temperatures (50, 60 and 70 °C) and for the same air flow rate 300 and 150 $m^3.h^{-1}$ respectively, it is observed that, For the same drying air flow rate, the drying rate increases when the temperature of the drying air increases with moisture content. Consequently, the drying kinetics of

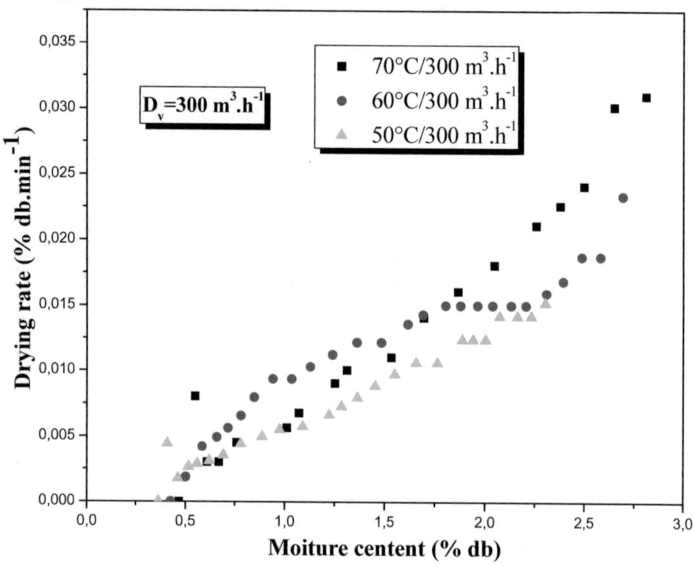

Fig. 4. Influence of the temperature on Mediterranean Mussels at D_v = 300 $m^3.h^{-1}$

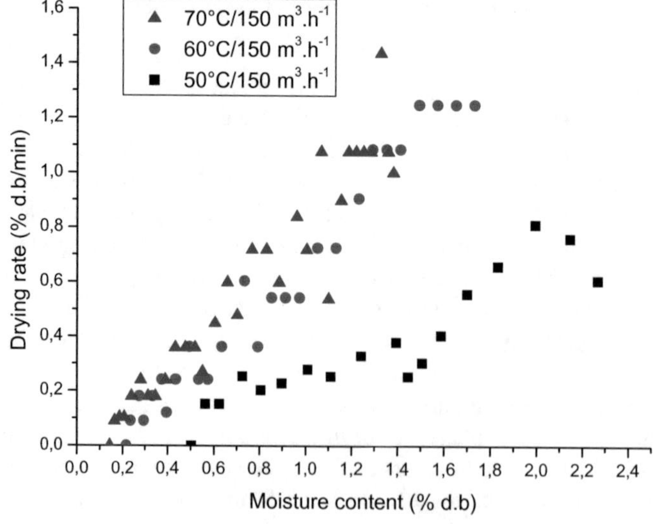

Fig. 5. Influence of the temperature on Mediterranean Mussels at D_v = 150 $m^3.h^{-1}$

Mediterranean mussels is influenced by drying air temperature. Generally, this influence depends for the majority of porous hygroscopic in product. Similar result is noted in other work on solar drying [30, 31].

3.3 Characteristic Drying Curve of Mytilus Galloprovincilis

The results obtained for the six performed experiment are synthetized in order to establish a law of drying that we call characteristic drying curve, it consisting in normalizing the kinetics of drying from experimentation in a theoretical model. The method adopted allow to establish a correlation based to the nonlinear optimization method of Levenberg-Marquad in the form of a third-degree polynomial equation, we used Origin 8.0 Software to obtain the characteristics drying curve for Mediterranean Mussels.

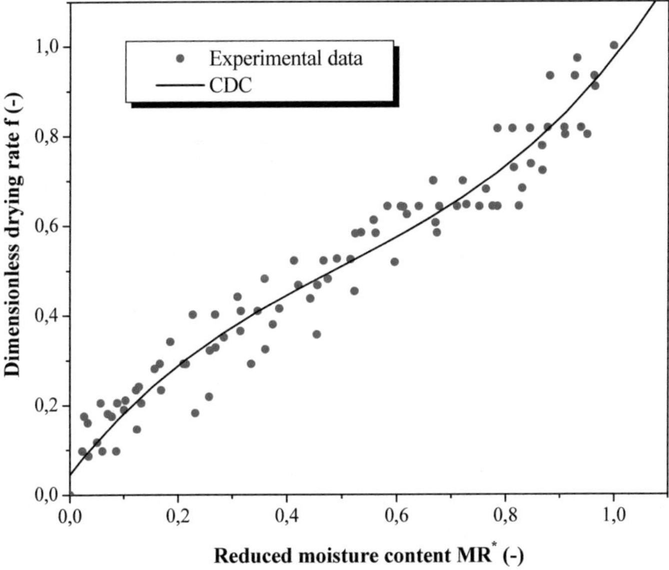

Fig. 6. Characteristic drying curve of Mytilus Galloprovincilis

Experimental drying data are plotted in Fig. 6 to represent Drying rate (f) versus Reduced moisture content f = f(MR*). This figure shows that all experimental points in different conditions of air temperature and flow rate, its gathered correctly on the CDC curve [30].

$$f = 0,0467 + 1.509MR^{*} - 1.751MR*^{2} + 1.168MR*^{3} \tag{9}$$

The best smoothing used to choose two criteria for evaluate goodness of fit, it concerning the standard error (ESM = 0.017) and the correlation coefficient (r = 0.956).

3.4 Modeling of Drying Curves

In drying treatments, modelling of drying curve is necessary to predict moisture content in a product [32]. Nine mathematical models or empirical equation was used to describe the shape of drying Mediterranean Mussels and tested for applied to the drying curve. All experimental data and constants drying (r,MSE and χ^2) was adjusted using non-linear optimization based on the Levenberg-Marquad algorithms with Origin 8 and Curve Expert 4.1 software.

The appropriate model for describing the shape of drying kinetics is selected according to the following criteria: high correlation coefficient (r), minimal average system error (MSE) and minimal ki-square (χ^2).

The average values of constants of these models were determined and summarized in Table 3.

Table 3. Coefficients of models describing the drying kinetics

Model name	Coefficients	r	χ^2	MSE
Newton	k = 0.0137	0.9949	1.162 10^{-3}	3.2014 10^{-2}
Handerson and Pabis	a = 1.053 k = 0.01238	0.9947	6.802 10^{-4}	2.4066 10^{-2}
Logarithmic	a = 1.053 k = 0.01238	0.9947	6.802 10^{-4}	2.4066 10^{-2}
Tow-term	a = 0.5031 k0 = 0.0138 b = 0.5031 k1 = 0.0138	0.9958	7.548 10^{-4}	2.4066 10^{-2}
Wang and Singh	a = −0.0120 b = 2.6945*10^{-5}	0.9969	2.164 10^{-2}	1.0842 10^{-2}
Midilli-kucuk	a = 0.9980 k = 0.0135 n = 0.9779 b = −0.0002	0.9996	5.004 10^{-4}	1.7688 10^{-2}
Page	k = 0.011 n = 1.0518	0.9981	3.487 10^{-4}	1.829 10^{-2}
Diffusion Approach	a = 1.0002 k = 0.0072 b = 1	0.9523	1.962 10^{-2}	1.0542 10^{-1}

From Table 4, The Logarithmic model seems to best model describing the drying kinetics of Mediterranean mussels because it has the highest value of r and the lowest valuWes of ki-square χ^2 and MBE. Indeed, Fig. 7 shows an excellent agreement between the reduced experimental water contents and those predicted by the Logarithmic model.

Table 4. Results of statistical analysis of the mathematical models

Models	R	χ^2	MSE
Newton	0.9949	$1.1484\ 10^{-3}$	$3.2069\ 10^{-2}$
Page	0.9981	$3.4875\ 10^{-4}$	$1.8290\ 10^{-2}$
Handerson and Pabis	0.9947	$6.8020\ 10^{-4}$	$2.4066\ 10^{-2}$
Logarithmic	0.9995	$1.4352\ 10^{-4}$	$1.0449\ 10^{-2}$
Tow-term	0.9958	$7.5481\ 10^{-4}$	$2.4066\ 10^{-2}$
Wang and Singh	0.9969	$2.1644\ 10^{-2}$	$1.0842\ 10^{-2}$
Diffusion Approximation	0.9523	$1.9627\ 10^{-2}$	$1.0542\ 10^{-1}$
Modified Handerson and Pabis	0.9985	$8.5195\ 10^{-4}$	$2.4544\ 10^{-2}$
Midilli and Kucuk	0.9996	$3.6918\ 10^{-4}$	$1.4793\ 10^{-2}$

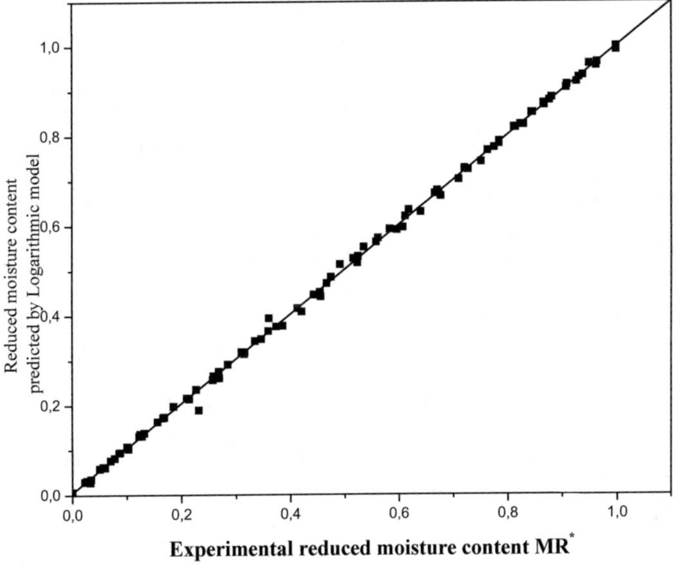

Fig. 7. Reduced water content predicted by the Logarithmic model as a function of the reduced experimental water Content of Mediterranean mussels

4 Conclusion

The convective solar drying experiments were conducted in the temperature range of 50–70 °C and tow drying air flow rates (300 and 150 $m^3.h^{-1}$) in a convective partial solar dryer. The kinetics drying for Mytilus Galloprovincilis was studied and must be known for understand the characteristics of drying process.

The evolution of moisture content for different air conditions was checked, the shape of drying curve was indicated only the presence of phase 2 which corresponding of the falling drying rate period.

According to the various aero-thermal condition of solar dryer (temperature and velocities), it was concluded that the drying air temperature was found to be the main factor influencing the drying kinetics of Mytilus Galloprovinvilis. The drying rate increases when the temperature of the drying air increases with raise of moisture content.

The concept of characteristic drying curve obtained by Van Meel transformation [] and provided valuable information for predicting the drying rate for other experimental conditions other than those in which our performed experiments.

In this research, the goodness of fit of the experimental data of drying with nine drying nine models was evaluated. The statistical parameters were analyzed, The Logarithmic equation was the best model to fit the experimental data and was recommended as the drying model for Mytilus Galloprovincilis.

References

1. Dincer, I.: Sun drying of sultana grapes. Dry. Technol. **14**, 1827–1838 (1996)
2. Boudinar, B., Mouhani, H., Bendou, A., Idlimam, A., Lamharrar, A.: Experimental study of the drying kinetics of the pink shrimps: Parapenaeus Longirostris type. J. Appl. Phys. **8**(5), 23–28 (2016)
3. Prakash, V., Lonsane, B.K., Kumar P.R.: Fish processing of value-added products: an efficient tool for nutrition security and techno-economic development of the nation. In: Proceeding of the National Seminar on Fish for All National Launch, Kolkata, India, pp. 99–107 (2003)
4. Punekar S., Mandape, M.K.: Socio economic status of prawn in relation to post harvest technology and its impact in local fish market, Fresh Water Prawns. In: International Symposium, Kochi, India, pp. 146–150 (2003)
5. Szulmayer, W.: From Sun drying to solar dehydration. Food Technol. Aust. **23**(9,10) (1971)
6. Petrova, I., Banthle, M., Eikevik, T.M.: Manufacturing of dry-cured ham: a review, part 2 drying kinetics, modeling and equipment. Eur. Food Res. Technol. **241**(4), 447–458 (2015). https://doi.org/10.1007/s00217-015-2485-z
7. Galvao, M., Déborah, M., Pena, R.: Drying kinetics and hycroscopic behavior of pirarucu (Arapaima Gigas) fillet with different salt contents. Food Sci. Technol. **62**(1), 144–151 (2015)
8. Idlimam, A., Lamharrar, A., El Houssayne, B., Kouhila, M., El Khadir, L.: Solar convective drying in thin layers and modeling of municipal waste at three temperatures. Appl. Therm. Eng. **108**, 41–47 (2016)
9. Kouhila, M.: Étude expérimentale et théorique de cinétiques de séchage convective partiellement solaire des plantes médicinales et aromatiques (menthe. Verveine. Sauge et eucalyptus), Thèse de Doctorat d'Etat. Université Cadi Ayyad Marrakech, Morocco (2001)
10. Kouhila, M., Belghit, A., Daguenet, M.: Détermination expérimentale et théorique des courbes de sorption et de la cinétique de séchage de la menthe verte. Entropie, 20–31 (2001)
11. AitMohamed, L., Kouhila, M., Jamali, A., Lahsani, S., Kechaou, N., Mahrouz, M.: Single layer solar drying behaviour of citrus aurantium leaves under forced Convection. Energy Convers. Manag. **46**, 1473–1483 (2005)
12. Van Meel, D.A.: Adiabatic convection batch drying with recirculation of air. Chem. Eng. Sci. **9**, 36–44 (1958)

13. Timoumi, S., Zagrouba, F.: Water sorption and dehydration kinetics of Tunisian rosemary leaves. Desalination **185**(1–3), 517–521 (2005)
14. Lewis, W.K.: The rate of drying of solid materials, factors influencing the maximum rates of air drying of shelled corn in thin layer. J. Ind. Eng. **5,** 427–433, 1949–1959 (1921)
15. Page, C.: Factors influencing the maximum rates of air drying of shalled corn in thin layer. Unpublished M.S. Thesis, Purdue University, Lafayette, IN (1949)
16. laninié, M.P., Velié, D., Tomas, S., Bilié, M., Bucié, A.: Modeling of drying and rehydration of carrots using peleg's model. Eur. Food Res. Technol. **221**, 446–451 (2005)
17. Yagcioglu, A.: Drying characteristics of laurel leaves under different conditions, Faculty of Agriculture; Cukurova University editor, Adana, Turkey (1998)
18. Henderson, S.: Progress in developing the thin layer drying equation. Trans. Am. Soc. Agric. Eng. **17**, 1167–1168 (1947)
19. Wang, G.Y., Sing, R.P.: Single layer drying equation for rough rice. Am. Soc. Agric. Eng. **78**, 3001 (1978)
20. Kassem, A.S.: Comparative studies on thin layer drying models for wheat. In: 13th International Congress on Agricultural Engineering, Morocco, vol. 6 (1998)
21. Karathanos, V.T.: Determination of water content of dried fruits by drying kinetics. J. Food Eng. **39**, 337–344 (1999)
22. Midilli, A., Kucu, H., Yaper, Z.: A new model for single layer drying. Dry. Technol. **20**, 1503–1513 (2002)
23. Togrul, L., Pehlivan, D.: Mathematical modelling of solar drying of apricot in thin layers. J. Food Eng. **55**, 209–216 (2003)
24. Garware, T., Sutar, N., Thorat B.: Drying of tomato using different drying methods: comparison of drying kinetics and rehydration ratio, In: Proceedings of the 16th International drying symposium, Hyderabad, India, pp. 1427–1432 (2008)
25. Idlimam, A., Ethmane Kane, C.S., Kouhila, M.: Single layer drying behaviour of grenade peel in a forced convective solar dryer. Rev. Ener. Renouv. **10**(2), 191–203 (2007)
26. Lahsasni, S., Kouhila, M., Mahrouz, M., Jaouhari, J.T.: Drying kinetics of prickly pear fruit (Opuntia ficus indica). J. Food Eng. **61**(2), 173–179 (2004)
27. Yaldiz, O., Ertekin, C.: Thin layer solar drying of some vegetables. Dry. Technol. **19**(3–4), 83–97 (2001)
28. Kaymak Ertekin, F.: Drying and rehydrating kinetics of green and red peppers. J. Food Sci. **67**(1), 168–175 (2002)
29. Passamai, V., Sravia, L.: Relationship between a solar drying model of red pepper and the kinetics of pure water evaporation. Dry. Technol. **15**(5), 1419–1432 (1997)
30. Kouhila, M., Kechaou, N., Otmani, M., Fliyou, M., Lahsasni, S.: Experimental study of sorption isotherms and drying kinetics of Moroccan Eucalyptus globulus. Dry. Technol. **20**, 2027–2039 (2002)
31. Belghit, A., Kouhila, M., Boutaleb, B.C.: Experimental Study of Drying Kinetics by Forced Convection of Aromatic Plants. Energy Convers. Manag. **41**, 1303–1321 (2000)
32. Xiao, H.W., Pang, C.L., Wang, L.H., Bai, J.W., Yang, W.X., Gao, Z.J.: Drying kinetics and quality of Monukka seedless grapes dried in an air-impingement jet dryer. Biosyst. Eng. **105**, 233–240 (2010)

Wind and Photovoltaic Energy Availability and Its Cost Estimation for Tangier Region

Lamyae Mellouk[1,2(✉)], Khalid Zine-Dine[2], Mohamed Boulmalf[1], Abdessadek Aaroud[2], and Driss Benhaddou[3]

[1] FIL, TIC Lab, Université Internationale de Rabat, Rabat, Morocco
mellouk.lamyae@gmail.com
[2] Faculty of Sciences, LAROSERI Lab, El Jadida, Morocco
[3] University of Houston, Houston, USA

Abstract. This work presents methodology and results used to estimate both wind and solar energies availability and their cost. The objective is to analyze the potential of these two major sources of alternative energy in Tangier region and to show their ability to replace fossil fuel energy sources especially for remote areas and agriculture applications. Thus, energy availability and its cost for these technologies are determined based on meteorological data, and their engineering and technical characteristics. The obtained results provide the hourly average energy production (in kWh) and its cost (in USD/kWh) for a maximum designed power output of 5 kW for each technology and for each month in the year. Thus, these results show that energy production cost is ranging between 0.01 and 0.3 USD/kWh for solar energy using photovoltaic panels and it is ranging between 0.05 and 0.35 USD/kWh for wind energy using wind turbines with rated power equals to 1 kW. Results of this work could be used as data estimation reference for engineering and research works related to renewable energy applications in the studied region.

Keywords: Photovoltaic energy · Wind energy · Energy availability · Cost estimation

1 Introduction

In Morocco country, many remote agriculture areas do not have access to the existing electrical grids where almost 1.3 million Moroccans live without electricity [1]. One of very promising ways to electrify these regions is constructing new decentralized power systems, which are generally based on using hybrid renewable-fossil fuel energy sources [2]. This is because extending the existing grids could result on huge investment cost especially for regions located in difficult and remote areas [3]. Using decentralized energy systems represents an innovative approach which is needed to satisfy people lacking electricity with reasonable cost, while, in the same time, promoting the replacement of fossil fuel utilization and reducing pollutant emissions [4]. In this regards, Morocco government is accelerating several projects with the aim of strengthening and promoting the integration of renewable energy sources trough the construction, usage and development of decentralized energy sources. Solar and wind

M. Ezziyyani (Ed.): AI2SD 2018, AISC 912, pp. 54–61, 2019.
https://doi.org/10.1007/978-3-030-12065-8_6

are the two major renewable energy sources in Morocco. Wind energy potential is excellent in the northern and southern regions, while the sunshine hours could achieve 3000 h/year in all the country. For these reasons, Moroccan government has set up an ambitious target of meeting 42% of its energy requirements by 2020 using essentially these two renewable energy sources [5]. As part of this initiative, in this paper, we propose new study dealing with wind and solar energy availability estimation and their cost evaluation. Thus, two renewable energy technologies are considered in this work, which are photovoltaic technology and wind turbine technology. This work is organized according to three complementary sections. First section presents meteorological data for Tangier region and technical specification of the considered technologies. Second section deals with modeling and sizing of the considered renewable energy technologies for a maximum power output of 5 kWh and economic evaluation. Finally, the third section presents the obtained results, discussion and their analysis.

2 Related Work

The evaluation of specific cost of electricity (US$/kWh) for both Photovoltaic and wind technologies in order to be compared with fossil fuel energy sources is largely studied in literature. Unfortunately, there are wide dispersions on reporting assumptions and the methodology used to evaluate the specific cost of these two renewable energies leading to a wide dispersion and spread out of the exact value of cost of renewable energy. Indeed, reported energy cost using photovoltaic technologies in literature is ranging between 0.062 US$/kWh [6] and 0.49 US$/kWh [7] and it is ranging between 0.29 US$/kWh and 30 US$/kWh [8] for Wind energy. For this reason the common asked questions about the published papers in renewable energy sources including wind and photovoltaic technologies are: how these reported calculations on cost evaluation of energy were made?, what is the current specific cost?, and how it is changing?. The estimation of the specific energy cost is based on the following major assumptions:

- Lifetime is ranging between 20 and 30 years [9] for photovoltaic technologies and between 20 and 40 years for Wind turbine [10].
- The interest rate is taken equal to 5% for both technologies.
- Maintenance and operation cost is considered to be between 15 and 20% for wind technology [11] and 0% for photovoltaic technology [12, 13] of the annual investment cost.

Other costs including manpower cost, spare parts, are estimated to be 2% for Photovoltaic and 3% for wind technology.

It is worth quoting that the spread out and confusing dispersion on photovoltaic and wind energy costs is also caused by other important factors listed as follow [14]:

- Geographical location and weather conditions.
- Market situation.
- Production capacity and the scale of the generation unit.
- Political issues.
- And engineering performance and specifications of technology.

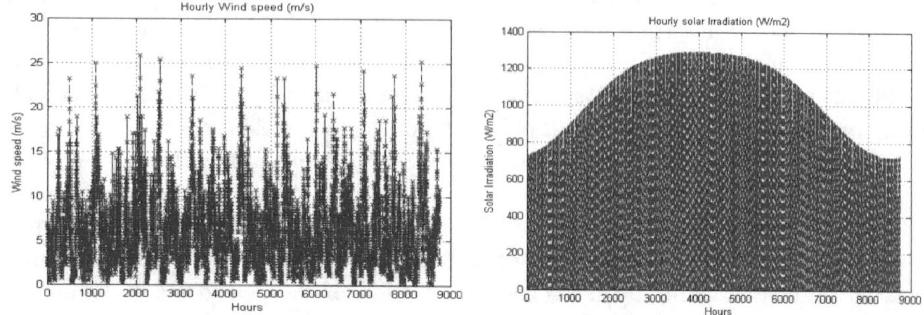

Fig. 1. Hourly wind speed (m/s) and solar irradiation (W/m^2) during one year

3 Meteorological Data and Specifications of the Considered Technologies

3.1 Meteorological Data Estimation

The average hourly wind speed and solar irradiation during one year, as presented in Fig. 1, are obtained from Typical Meteorological Year data (TMY) [15]. These data present the average wind speed (in m/s) and solar irradiation (DNI in W/m^2) measurements taking during each hour and during several years. These data are arranged for each season where the average hourly data are calculated and presented in Fig. 2 for each day (24 h). These data will be used as data based in next section of this paper dealing with design, sizing and economic evaluation of energy production using Photovoltaic panels and Wind turbines.

3.2 Specifications of PV Panels and Wind Turbines

Making an appropriate choice of each single component is of great importance to ensure proper operation and to avoid as much as possible problems that could limit their performances. Thus, in this study, the selection of solar panel and wind turbines type depends essentially on their availability in Morocco market, cost, estimated life time, efficiency and estimated maintenance rate. Their technical and engineering characteristics are provided in Tables 1 and 2.

4 Modeling, Sizing and Economic Evaluation

Photovoltaic panels (PV) generate electrical power by converting solar radiation into electricity. The conversion efficiency, the absorber area and the power produced by each single PV panel are related using the following equation:

$$P_{pv} = \eta_{pv} A_{panel} E_g \tag{1}$$

To account for losses caused by dust deposition, cables, inverters and losses related to temperature, the performance ratio (PR_{PV}) is introduced in the expression (1), thus, the power output of each (PV) panel is estimated using Eq. (2):

$$P_{pv,out} = PR_{pv}\eta_{pv}A_{panel}E_g \tag{2}$$

Where E_g is solar irradiation (DNI). The relationship between the electrical power output (P_w) produced by a single wind turbine and wind velocity (v) could be expressed by the following correlation [12]:

$$P_w = \begin{cases} 0 & \text{if } v > v_f \text{ or } v < v_c \\ P_r\left(av^3 - b\right) & \text{if } v_c < v < v_r \\ P_r & \text{if } v_r < v < v_f \end{cases} \tag{3}$$

Where: $b = \frac{(V_c/V_r)^3}{1-(V_c/V_r)^3}$ And $a = \frac{1+b}{V_r^3}$

And v_c is cut-in wind speed (m/s), vr is the rated wind speed (m/s), v_f is the cut-out wind speed (m/s) and Pr is the rated power (Watt).

Cost estimation of electrical energy produced by both technologies takes into account the investment cost of components, cleaning and maintenance cost, manpower cost, spare parts and auxiliary equipment cost, and installation cost. The investment cost of each panel is provided in Table 1, in addition to 40 USD/panel is included as an annual maintenance and cleaning cost and 90 USD/panel to account for auxiliary equipment such as inverters. The investment cost of wind turbines is given in Table 2. The annual maintenance cost for wind turbines is estimated equal to 100 USD/turbine. Life time for both wind turbines and solar photovoltaic panels is estimated equal to 20 years. Land cost is not considered in this work; however, 6% of the total cost is estimated as an additional cost for the manpower, spare parts and installation costs. The annual amortization factor F is calculated using Eq. (4):

$$F = \frac{i(1+i)^n}{(1+i)^n - 1} \tag{4}$$

Where 'n' is the number of years (life time), and 'i' is the interest rate.

The hourly energy cost estimation is done using the following steps:

- 1st step: Sizing of Wind turbines and Photovoltaic panels for a maximum power production of 5 kW, in this step, the sizing considers the case when the natural energy sources are at their maximum value (wind speed and solar irradiation).
- 2nd step: Calculation of the annual amortized total cost of photovoltaic panels' field and wind farm previously designed in the first step, and calculation of the daily and hourly total cost.
- 3rd step: Calculation of the hourly energy cost for both technologies using the following equation:

$$C_{energy} = C_{hour}/E_{hour} \tag{5}$$

In addition to meteorological data, all these three steps are programmed using Matlab software. The obtained results are provided in the following section.

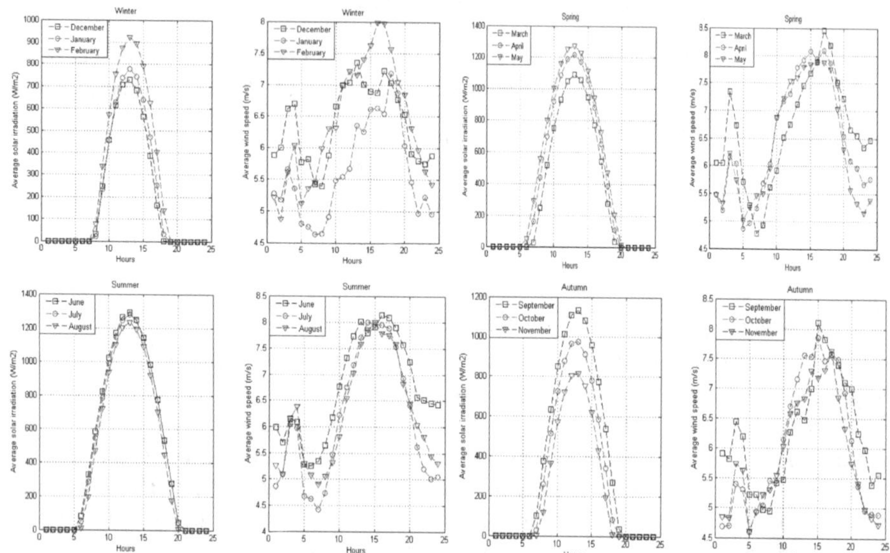

Fig. 2. Average hourly solar Irradiation (W/m^2) and wind speed (m/s) during each season in the year

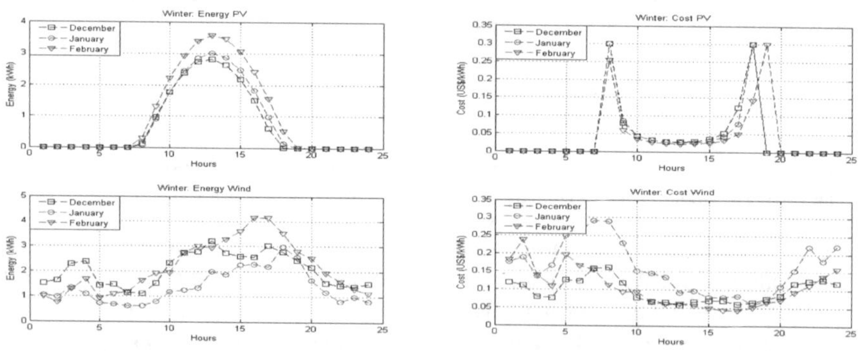

Fig. 3. Results obtained for winter periods - Energy output (kWh) and its Cost (USD/kWh)

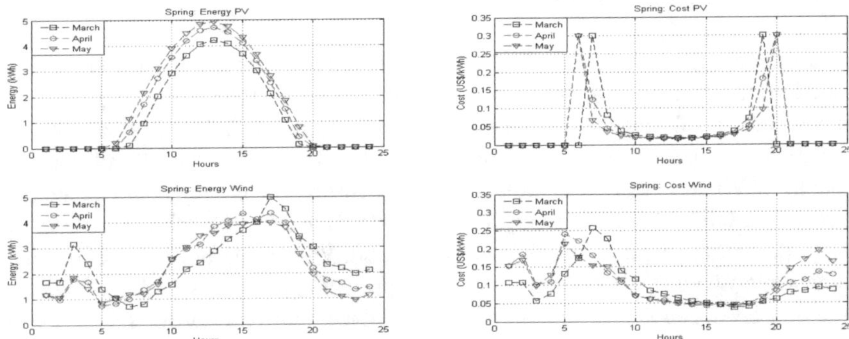

Fig. 4. Results obtained for spring periods - Energy output (kWh) and its Cost (USD/kWh)

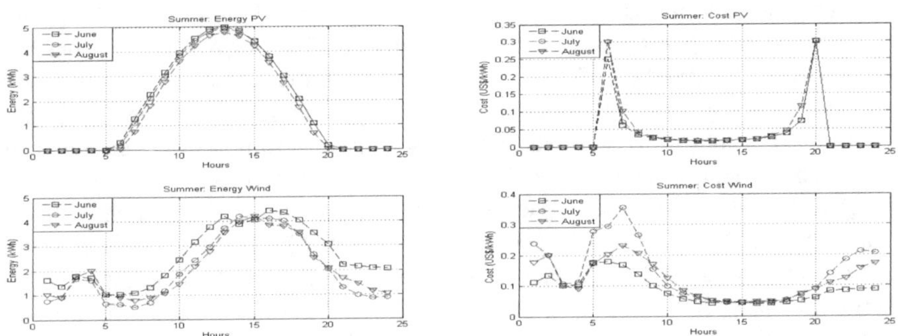

Fig. 5. Results obtained for summer periods - Energy output (kWh) and its Cost (USD/kWh)

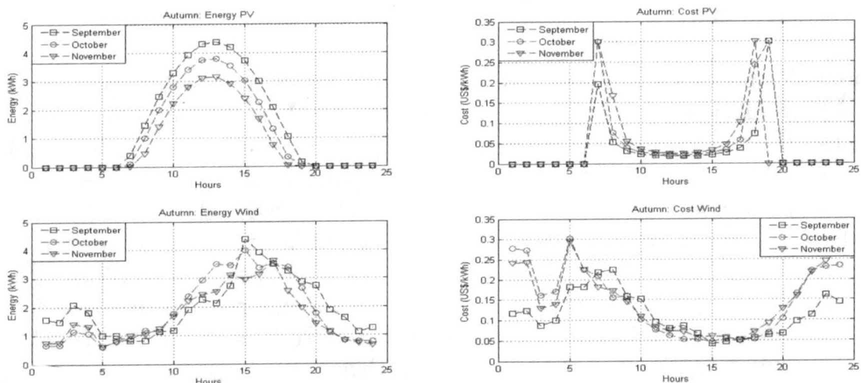

Fig. 6. Results obtained for autumn periods - Energy output (kWh) and its Cost (USD/kWh)

5 Results and Analysis

In this section, the average hourly energy availability output and its cost are obtained based on a design power output of 5 kW for wind turbines' farm and solar photovoltaic panels. These results are presented for each month. As shown in Fig. 3, during January, the maximum energy output that could be achieved by both PV and Wind systems is equal to 3 kWh/h, during this period, the energy cost production by photovoltaic could be less than 0.05 US$/kWh during seven hours of the middle of the day while wind energy cost is ranging between 0.06 and 0.3 US$/kWh. During February, as illustrated in the same figure, the maximum energy output could achieve 4 and 3.5 kWh/h for wind and PV technologies respectively, while the cost of energy for PV is approximately similar to January month, while it is ranging between 0.04 and 0.23 US$/kWh for wind technology. Figure 4 shows that for March, the cost of PV energy could be less than 0.03 for the seven hours of the middle of the day. Figures 4 and 5 show that during the period between April and August, solar energy potential became more important, thus, the maximum designed power output could be achieved during this period and the PV energy cost could be less than 0.05 US$/kWh during more than seven hours of the middle of the day, during this the same period the average cost of wind energy is approximately equal to 0.1 USD/kWh. Figure 6 and again Fig. 3 show that the potential of PV energy starts to be reduce, where the number of hours where the cost of photovoltaic energy is less than 0.05 US$/kWh does not exceed 7 h. Wind energy technology achieves its maximum energy output only during March where the minimum cost is approximately equal to 0.04 US$/kWh. These results, show that Photovoltaic panels are more advantageous than Wind technology in terms of cost of energy produced, however, Photovoltaic panels could only perform during sunshine hours with an average duration of seven hours, while during night, solar photovoltaic energy needs to be combined with large capacity energy storage systems leading to increasing the cost of energy production and maintenance rate.

6 Conclusion

In this, the potential of the two major renewable energy sources namely, wind energy and photovoltaic solar energy for Tangier region is evaluated based on the meteorological data of tangier, engineering specifications and economic evaluation of the considered technologies. The obtained results are provided for each month of the year and they are presented for each hour in the day. These results shows that both technologies could be more advantageous in terms of cost in comparison with fossil fuel energy, thus, the average cost of wind energy is approximately 0.1 US$/kWh which is approximately equal to the average cost of electrical energy produced based on fossil fuel energy source, while the cost of solar energy could be less than 0.05 US$/kWh during more than 7 hours per day during the whole year. In addition to low cost of energy production, these two technologies also present the advantage of replacing fossil fuel energy and then contributing also in the reduction of pollutant emissions.

Results of this work could be also used as data base for topics related to renewable energy applications such as Micro-grids studies and fossil fuel energy replacement in the agriculture applications. The extension of this work will be the study, analysis, comparison and optimization of different energy storage systems to be combined with these two major renewable energy sources in order to construct hybrid wind-photovoltaic-fossil fuel Microgrid and energy systems.

Acknowledgment. Financial and Technical support of the "Université Internationale de Rabat (UIR)- TIC LAb" is gratefully acknowledged.

References

1. Access to electricity (% of population)-Data. https://data.worldbank.org/indicator/EG.ELC.ACCS.ZS
2. Chakrabarti, S., Chakrabarti, S.: Rural electrification programme with solar energy in remote region–a case study in an island. Energy Policy **30**(1), 33–42 (2002)
3. Dagnachew, A.G., Lucas, P.L., Hof, A.F., Gernaat, D.E.H.J., de Boer, H.-S., van Vuuren, D. P.: The role of decentralized systems in providing universal electricity access in Sub-Saharan Africa – a model-based approach. Energy **139**, 184–195 (2017)
4. Javidsharifi, M., Niknam, T., Aghaei, J., Mokryani, G.: Multi-objective short-term scheduling of a renewable-based microgrid in the presence of tidal resources and storage devices. Appl. Energy **216**(15), 367–381 (2018)
5. Choukri, K., Naddami, A., Hayani, S.: Renewable energy in emergent countries: lessons from energy transition in Morocco. Energy Sustain Soc. **7**, 25 (2017). https://doi.org/10.1186/s13705-017-0131-2
6. Moore, L.M., Post, H.N.: Five years of operating experience at a large, utility-scale photovoltaic generating plant. Prog. Photovolt. Res. Appl. **16**, 249–259 (2008)
7. Yang, C.: Reconsidering solar grid parity. Energy Policy **38**, 3270–3273 (2010)
8. Gökçek, M., Genç, M.S.: Evaluation of electricity generation and energy cost of wind energy conversion systems (WECSs) in Central Turkey. Appl. Energy **86**, 2731–2739 (2009)
9. Branker, K., Pathak, M.J.M., Pearce, J.M.: A review of solar photovoltaic levelized cost of electricity. Renew. Sustain. Energy Rev. **15**, 4470–4482 (2011)
10. Ortegon, K., Nies, L.F., Sutherland, J.W.: Preparing for end of service life of wind turbines. J. Clean. Prod. **39**, 191–199 (2013)
11. Blanco, M.I.: The economics of wind energy. Renew. Sustain. Energy Rev. **13**, 1372–1382 (2009)
12. Nelson, D.B., Nehrir, M.H., Wang, C.: Technical note, unit sizing and cost analysis of stand-alone hybrid wind/PV/fuel cell power generation systems. Renew. Energy **31**, 1641–1656 (2006)
13. Shaahid, S.M., El-Amin, I.: Techno-economic evaluation of off-grid hybrid photovoltaic–diesel–battery power systems for rural electrification in Saudi Arabia—a way forward for sustainable development. Renew. Sustain. Energy Rev. **13**, 625–633 (2009)
14. Hammons, T.J.: Integrating renewable energy sources into European grids. Int. J. Electr. Power Energy Syst. **30**(8), 462–475 (2008)
15. http://rredc.nrel.gov/solar/old_data/nsrdb/1961-1990/tmy2/

Structural and Vibrational Study of Hydroxyapatite Bio-ceramic Pigments with Chromophore Ions (Co^{2+}, Ni^{2+}, Cu^{2+}, Mn^{2+})

Eddya Mohammed, Tbib Bouazza, and El-Hami Khalil(✉)

Laboratory of Nanosciences and Modeling, Faculty of Khouribga, University of Hassan 1st, Khouribga, Morocco
khalil.elhami@uhp.ac.ma

Abstract. Incorporating metal ions into a calcium hydroxyapatite structure is a successful pathway to increase their physical, chemical and biological properties. The calcium hydroxyapatite was obtained by solid state method at a high temperature, using $CaCO_3$ and $(NH_4)_2HPO_4$ as sources of calcium and phosphorus. Metal ion (Mn^{2+}, Co^{2+}, Ni^{2+}, Cu^{2+}) incorporation was carried out by dint of grinding and high temperature effect to remove all the impurity. The Hydroxyapatite powders that doped with metal ions were characterized by X-ray diffraction (XRD), and Fourier transforms infrared spectroscopy (FTIR) analysis to evaluate the structural and compositional changes. The only phase that is presented in pure hydroxyapatite sample was the hexagonal system. A Rietveld refinement has shown that doping with these ions affects the volume unit cell of HAP-M and it will be changed. We found that the samples doped HAP-M (M = Mn^{2+}, Co^{2+}, Ni^{2+}, Cu^{2+}) stabilizes only in the monoclinic phase.

Keywords: Hydroxyapatite · X-ray diffraction · Rietveld refinement

1 Introduction

The crystalline structure arranges the atoms in crystal. These atoms are repeated periodically in space under the action of symmetry operations of the space group and thusly form the crystalline structure. This structure is a fundamental concept for many areas of science and technology. It is completely described by the crystal lattice parameters, its Bravais lattice, its space group and the position of the atoms in the asymmetric unit.

The hydroxyapatite (HAP) with general formula $Ca_{10}(PO_4)_6(OH)_2$ is a phosphorus apatite which crystallizes natural state in the hexagonal system [1, 2]. It is the main constituent of bones and teeth, it has excellent affinity with bone tissue. So its advantage is to create strong chemical bond with the bone [3]. The hydroxyapatite can be used also for the purification of aqueous media [4, 5] and the immobilization of uranium in nuclear waste [6].

Around the world, many studies and substitution experiments have been carried out on hydroxyapatites to increase its physical, chemical and biological properties [7–10].

M. Ezziyyani (Ed.): AI2SD 2018, AISC 912, pp. 62–70, 2019.
https://doi.org/10.1007/978-3-030-12065-8_7

The synthetic hydroxyapatite used for orthopedic and dental application may be satisfactory in terms of mechanical properties [11, 12], but the problem of quick dislocation of crystalline structure is also present and causes damage of materials. By this study, we want to see the effect of weak Ca substitution of hydroxyapatite by a metal ion (Mn^{2+}, Co^{2+}, Ni^{2+}, Cu^{2+}) on grain size and structural properties for increasing the mechanical properties of hydroxyapatite.

When the atoms come together to form a solid, their valence electrons interact because of Coulomb forces, and they also sense the electric field produced by their own nucleus and that of the other atoms. In addition, two specific quantum mechanical effects occur. First, by Heisenberg's principle of uncertainty, constraining electrons to a small volume increases their energy.

2 Materials and Methods

Crystalline powders of Hydroxyapatite (HAP) with general chemical formula $Ca_{10}(PO_4)_6(OH)_2$ were prepared by a solid state method, all reagents are in the solid state. We mix and grind directly and manually with mortar and pestle a 1.5773 g of ammonium phosphate $(NH_4)_2HPO_4$ and 1.9927 g of calcium carbonate $CaCO_3$, then we put it in the oven at 200 °C and we raise the temperature by 200 °C and achieve the inter grinding after each 24 h duration. Until we reach 1000 °C.

Hydroxyapatite- metal-doped HAP-M, with chemical formula $Ca_{9.5}M_{0.5}(PO_4)_6(OH)_2$ where (M = Mn^{2+}, Co^{2+}, Ni^{2+}, Cu^{2+}) were also prepared by a solid state method, and all reagents precursors are in the solid state with well determined masses for each sample. See Table 1. We mix and grind directly and manually with mortar and pestle an ammonium phosphate $(NH_4)_2HPO_4$ and calcium carbonate $CaCO_3$ and precursor that contains metal for each sample as the following:

- Nickelapyrroline C_4H_6Ni for Hydroxyapatite- Nickel (HAP-Ni): $Ca_{9.5}Ni_{0.5}(PO_4)_6(OH)_2$.
- Copper (II) Chloride Dihydrate $CuCl_2.2H_2O$ for Hydroxyapatite-Copper(HAP-Cu): $Ca_{9.5}Cu_{0.5}(PO_4)_6(OH)_2$.
- Cobalt (II) nitrate hexahydrate $CoN_2O_6.6H_2O$ for Hydroxyapatite- Cobalt (HAP-Co): $Ca_{9.5}Co_{0.5}(PO_4)_6(OH)_2$.
- Manganese (II) carbonate CMnO3 for Hydroxyapatite- Manganese (HAP-Mn): $Ca_{9.5}Mn_{0.5}(PO_4)_6(OH)_2$.

Each sample was prepared independently of the others. And we follow the same thermic treatment than the HAP in the oven from 200 °C to 1000 °C and we raise the temperature by 200 °C realizing an inter grinding every 24 h duration.

The synthesis process as resumed in graphical representation as shown in (Fig. 1).

Table 1. Mass of reagents for synthesis of Hydroxyapatite HAP, HAP-M (M = Ni, Cu, Co, Mn).

Precursor sample	$(NH_4)_2HPO_4$ (g)	$CaCO_3$ (g)	Precursor of metal (g)
HAP	1.5773 g	1.9927 g	0 g
HAP-Ni	1.5627 g	1.8755 g	m(C_4H_6Ni) = 0.2454 g
HAP-Cu	1.5627 g	1.8710 g	m($CuCl_2.2H_2O$) = 0.1677 g
HAP-Co	1.5627 g	1.8753 g	m($CoN_2O_6.6H_2O$) = 0.1853 g
HAP-Mn	1.5627 g	1.8790 g	m($CMnO_3$) = 0.1136 g

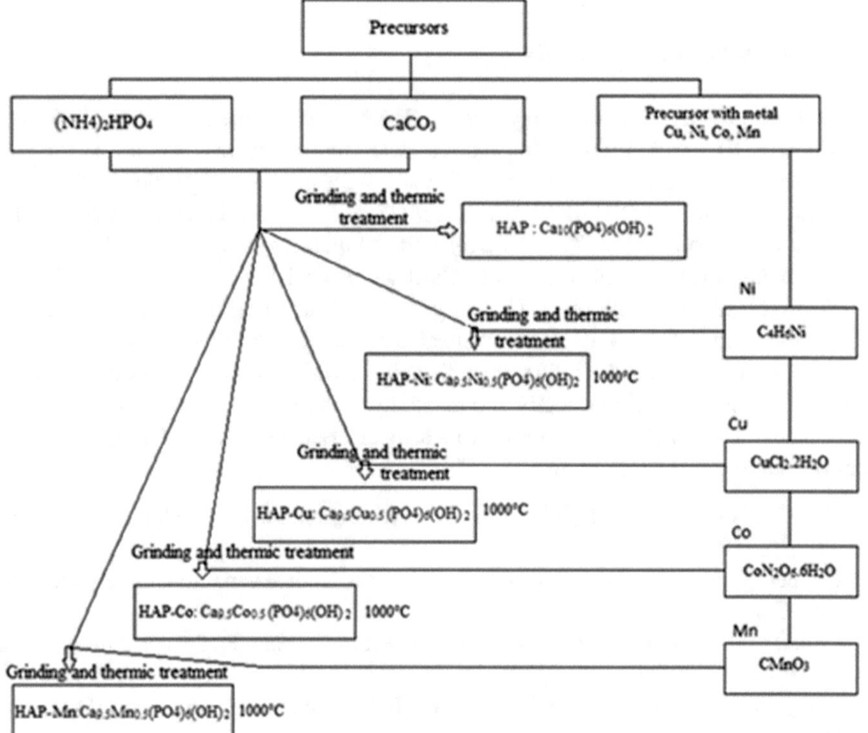

Fig. 1. The synthesis process for HAP, HAP-M (M = Ni, Cu, Co, Mn) by solid state method.

3 Results and Discussion

3.1 X-Ray Diffraction (XRD)

The structure of HAP has examined using XRD Bruker D8 ADVANCE. X-rays are electromagnetic radiation with photon energies in the range of 100 eV–100 keV. For diffraction applications, only short wavelength x-rays (hard x-rays) in the range of a few angstroms (Å) to 0.1 angstrom (1 keV–120 keV) are used. X-rays are ideally suited for probing the atomic structure of solids because their wavelengths (0.1–2 Å) are of comparable length to the radii of atoms and they are sufficiently energetic to penetrate most solid materials to provide information about the bulk structure [13].

XRD pattern of the sample HAP showed the structure of the prepared sample was similar to the hydroxyapatite as shown in (Fig. 2). There is a high consistency between the data of HAP and that from of the standard database, with lattice dimensions of a = b = 0.120632 nm, c = 0.168488 nm. None of the other impurity was observed in the XRD pattern, indicating that the chief inorganic phase of the sample is hydroxyapatite crystal. The result obtain was similar to [14] as reported. When the initial phase is replaced by the transition elements, it has been found that the cell changes the hexagonal structure to the monoclinic. Also the size of the crystallites in nanometers of the samples which are synthesized by the solid state method is found. The sizes of the crystallites has an effect on the cell volume. Table 2 shows the parameters of the cell of each structure.

Table 2. The structural data of the HAP AND HAP-M (M = Ni, Co, Cu and Mn).

Samples	HAP	HAP-Ni	HAP-Cu	HAP-Co	HAP-Mn
Cristal system	Hexagonal	Monoclinic	Monoclinic	Monoclinic	Monoclinic
S.Group	P 6/m mm	P 2/m	P 2/m	P 2/m	P 2/m
a(Å)	12.0632	14.1032	10.9300	19.3960	12.2833
b(Å)	12.0632	16.1747	12.0486	3.6470	13.8610
c(Å)	16.8488	8.7496	5.4934	12.4599	6.474
v(Å)	2123.3755	1984.6949	718.6627	881.1540	1100.522
Alpha(°)	90	90.00	90.00	90.00	90.00'
Beta(°)	90	96.078	96.583	91.263	93.297
Gama(°)	120	90.00	90.00	90.00	90.00
D(nm)	3,6060	4,5321	2,7058	4,7782	5,3022

The grain size of the different Samples was determined using the spectra of the DRX and the debye-Scherrer formula [15]:

$$D = k * \lambda / B.cos(\theta) \quad (1)$$

D (nm): the grain size = 1 the constant λ: the beam wavelength of the X-rays used, B the width at mid-height of the most peak, θ the diffraction angle expressed in radian. The width at half height and the size of the grains are inversely proportional and their variation as a function of the doping rate. The steric influence of the M^{2+} (Mn^{2+}, Co^{2+}, Ni^{2+}, Cu^{2+}) cation can be studied comparing the cell parameters.

The compounds HAP crystallize in the hexagonal cristal system with the P6/mmm space group like the work in reference [16], and all other HAP-M (M = Ni, Cu, Co, Mn) crystallize in the monoclinic cristal system with the P2/m space group like the work in reference [17].

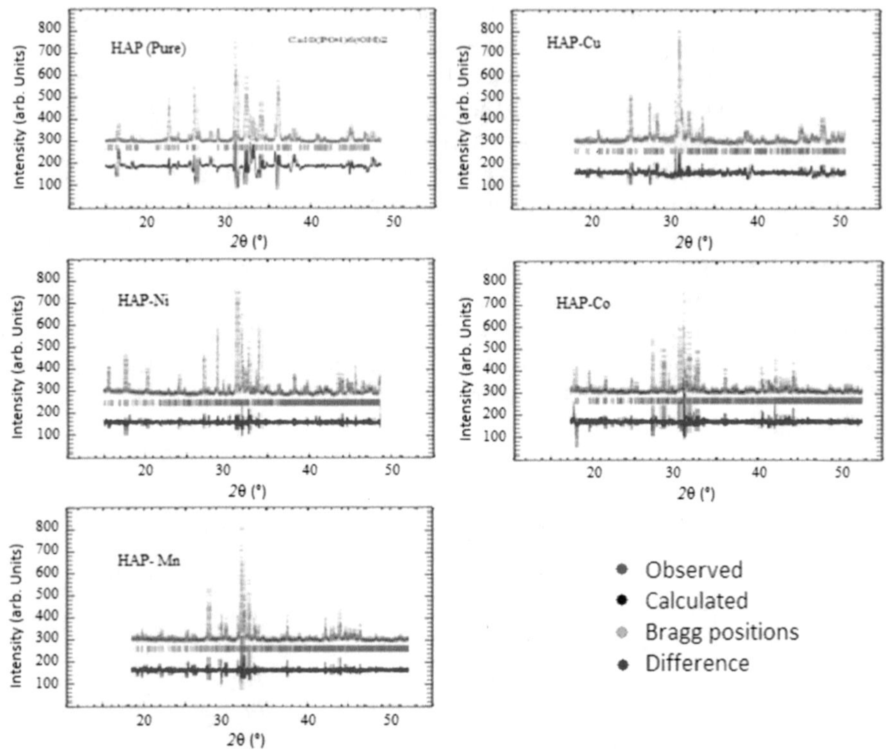

Fig. 2. The XRD spectra HAP and HAP-M (M = Ni, Cu, Co, Mn).

This change of crystal system has an influence on the structural parameters of each hydroxyapatite [18], which also have an influence on the optical and electronic properties of each hydroxyapatite [19, 20] according to the incorporated metal Ni, Cu, Co and Mn. We see that each cell of hydroxyapatite has a unique structural parameter values different to others according to the inserted metal. To have more structural stability, the atoms reorganizes themselves for minimizing interatomic vibration [21] and electrical interactions [22]. In this case, the pure hydroxyapatite stabilizes in the hexagonal system, but after the incorporation of the metal ions in its structure, the hydroxyapatite is stabilized in the monoclinic system with unique cell volume for each HAP-M product. These cell volume variation results for each product can explain the variation in urbach energy [23]. Less the cell volume of Hydroxyapatite decreases after the metal inserter, the structural disorder is lower, and the urbach energy remains small and near to that of the initial hydroxyapatite phase. The greatest urbach energies in this study are those of HAP-Co and HAP-Cu which have the smallest cell volume which causes a great variation of the structural parameters a, b and c of the cell of hydroxyapatite, this variation is shown in (Fig. 3). This parameter variation has an influence on the diameter of the hydroxyapatite's grains as shown in (Fig. 4) and shows that the grain size of all HAP-M has increased compared to the initial size of pure HAP except in the case of HAP-Cu which has the smallest grain size with smaller cell volume.

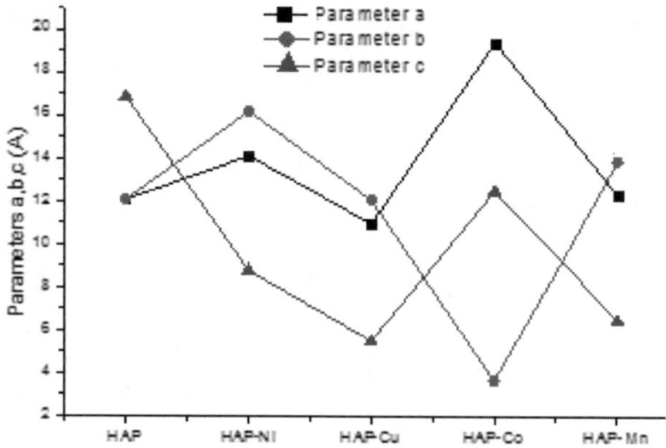

Fig. 3. A Variation of the lattice parameters of HAP and HAP-M (M = Ni, Cu, Co, Mn).

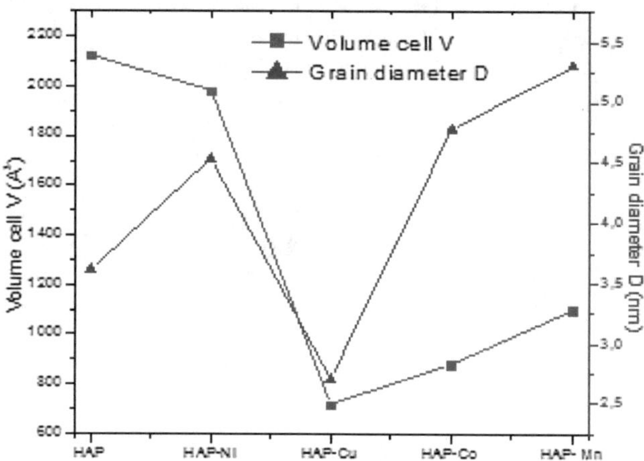

Fig. 4. Volume cell and grain diameter of HAP and HAP-M (M = Ni, Cu, Co, Mn).

3.2 Fourier-Transformed Infrared Spectroscopy (FTIR)

FTIR spectroscopy is an Optical technique that detects molecular bond vibrations and rotations upon absorption of infrared light. Because different chemical functional groups absorb IR light at different frequencies [24], FTIR spectroscopy can be used for chemical structure analysis, chemical fingerprinting and chemical imaging [25]. This absorption has shown the existence of the bonds of samples, the infrared spectrum represents the transmittance (T) on the ordinate, it is expressed as a percentage (%) according to the waves number, and is expressed in (cm^{-1}). We can see that the absorption strips are pointing downwards a low transmission value corresponds to a high absorption and each band is characterized by its position, width and intensity [24]. FTIR spectroscopy thus provides both qualitative and quantitative information and is shown to be an effective companion technique to X-ray diffraction, The FTIR spectra show the bands arising from inter atomic vibrations.

The FTIR spectrum was scanned from 4000–400 cm^{-1} by FTIR Perkin Elmer apparatus using the KBr-disc method for identified the functional groups of the hydroxyapatite. The broadband at 3250 and 1500 cm^{-1} were attributed at absorbed water, while sharp peak at 3500 cm^{-1} was attributable to the stretching vibration of the lattice HO^- ion and medium sharp peak at 700 cm^{-1} was assigned to the OH deformation mode. The characterization band for PO_4^{3-} appear at 470, 610, 1030 and

Table 3. The FT-IR band assignments for HAP and HAP-M (M = Ni, Cu, Co, Mn).

Number of peak	Frequency ν (cm^{-1})	Assigned vibration mode	Band shape HAP (Pure)	Band shape HAP-Ni	Band shape HAP-Cu	Band shape HAP-Co	Band shape HAP-Mn
1	470	PO_4^{-3}	Low Intense	Low intense	Low intense	Low intense	Low intense
2	610	PO_4^{-3}	Intense	Wide	Wide	Wide	Wide
3	700	OH^- out – of – plane bending	Low intense	Low intense	Low intense	Low intense	Low intense
4	1030	PO_4^{-3}	Intense	Wide	Wide	Wide	Wide
5	1070	PO_4^{-3}	Low intense	Low intense	Low intense	Low intense	Low intense
6	1350	PO_4^{-3}	Low intense	Low intense	Low intense	Low intense	Low intense
7	1500	H_2O	Low intense	Low intense	Low intense	Low intense	Low intense
8	1625	H_2O	Intense	Intense	Intense	Intense	Intense
9	2010	H_2O	Low intense	Low intense	Low intense	Low intense	Low intense
10	2375	H_2O	Low intense	Low intense	Low intense	Low intense	Low intense
11	2880	H_2O	Low intense	Low intense	Low intense	Low intense	Low intense
12	3250	H_2O	Low intense	Low intense	Low intense	Low intense	Low intense
13	3500	H_2O and HO^{-1}	Wide	Wide	Wide	Wide	Wide
14	3800	Structural stretching	Low intense	Low intense	Low intense	Low intense	Low intense

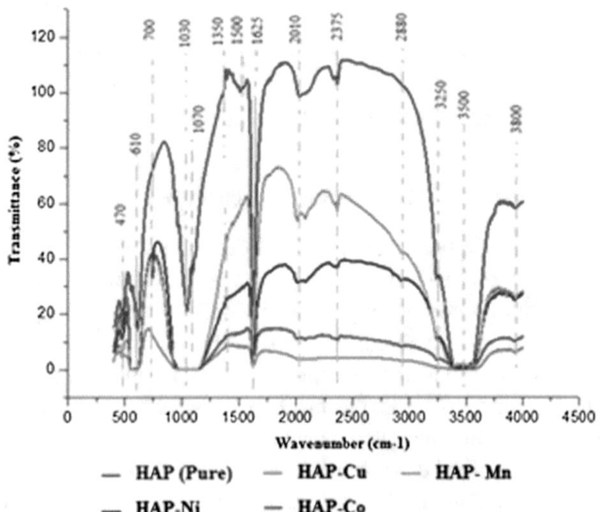

Fig. 5. The FTIR spectra of pure HAP and HAP substituted by transition elements (Ni, Cu, Co and Mn).

1070 cm^{-1} and 1350 cm^{-1} for samples substituted by transition elements. The observation of the asymmetric P-O stretching vibration at the PO_4^{3-} band at 1070 cm^{-1} for HAP no substituted and 1350 cm^{-1} as a distinguishable peak, together while the sharp peaks at 700,610 and 470 cm^{-1} correspond to the triple degenerate bending vibration of PO_4^{3-} in hydroxyapatite. Our FTIR results was similar to those reported [14, 26]. All FTIR spectrums of all our samples are shown in (Fig. 5) and the band shape and assigned vibration mode function frequency as reported in Table 3.

4 Conclusion

The results obtained from the characterization of X-ray diffraction made it possible to observe the presence of the single phases of the hydroxyapatite powder. The substitution of Ca^{2+} by M^{2+} (M^{2+} = Mn^{2+}, Co^{2+}, Ni^{2+}, Cu^{2+}) exhibit a decrease in the cell volume of all powders HAP-M (M = Mn^{2+}, Co^{2+}, Ni^{2+}, Cu^{2+}) and a decrease in transmittance of infrared light. The most striking aspect of the set of samples is their high crystallinity, preferential orientation of the crystallographic planes powders HAP-M (M = Mn^{2+}, Co^{2+}, Ni^{2+}, Cu^{2+}) influenced by the preparation technique itself and the operating conditions (chemicals, annealing temperature, etc.). It should also be noted, the widening of the feet of diffraction lines for all the powders explored.

The incorporation of Co, Ni, Cu and Mn metal ions in hydroxyapatites gave us four biomaterials for the use in dental amalgams, with more mechanical and electronic performance, and less risk of toxicity compared to metal amalgams that cause problems of toxicity and inflammation. These results can also be used in the orthopedic application to solve the problem of increasing concentration of cobalt after installation cobalt chromium alloy of Hip metal implant that causes inflammation problems.

References

1. Ma, G., Liu, X.Y.: Hydroxyapatite: hexagonal or monoclinic? Cryst. Growth Des. **9**(7), 2991–2994 (2009)
2. Bahrololoom, M.E., Javidi, M., Javadpoura, S., Ma, J.: Characterisation of natural hydroxyapatite extracted from bovine cortical bone ash. J. Ceram. Process. Res. **10**(2), 129–138 (2009)
3. Sopyan, I., Mel, M., Ramesh, S., Khalid, K.A.: Porous hydroxyapatite for artificial bone applications. Sci. Technol. Adv. Mater. **8**(1), 116–123 (2007)
4. Zhou, H., Wu, T., Dong, X., Wang, Q., Shen, J.: Adsorption mechanism of BMP-7 on hydroxyapatite (001) surfaces. Biochem. Biophys. Res. Commun. **361**(1), 91–96 (2007)
5. Rusu, V.M., Ng, C.H., Wilke, M., Tiersch, B., Fratzl, P., Peter, M.G.: Size-controlled hydroxyapatite nanoparticles as self-organized organic–inorganic composite materials. Biomaterials **26**(26), 5414–5426 (2005)
6. Arey, J.S., Seaman, J.C., Bertsch, P.M.: Immobilization of uranium in contaminated sediments by hydroxyapatite addition. Environ. Sci. Technol. **33**(2), 337–342 (1998)
7. Gibson, I.R., Bonfield, W.: Novel synthesis and characterization of an AB-type carbonate-substituted hydroxyapatite. J. Biomed. Mater. Res. Part A **59**(4), 697–708 (2002)

8. Miao, X., Tan, D.M., Li, J., Xiao, Y., Crawford, R.: Mechanical and biological properties of hydroxyapatite/tricalcium phosphate scaffolds coated with poly (lactic-co-glycolic acid). Acta Biomater. **4**(3), 638–645 (2008)
9. Suchanek, W., Yoshimura, M.: Processing and properties of hydroxyapatite-based biomaterials for use as hard tissue replacement implants. J. Mater. Res. **13**(1), 94–117 (1998)
10. Ito, M., Hidaka, Y., Nakajima, M., Yagasaki, H., Kafrawy, A.H.: Effect of hydroxyapatite content on physical properties and connective tissue reactions to a chitosan–hydroxyapatite composite membrane. J. Biomed. Mater. Res. Part A **45**(3), 204–208 (1999)
11. Akao, M., Aoki, H., Kato, K.: Mechanical properties of sintered hydroxyapatite for prosthetic applications. J. Mater. Sci. **16**(3), 809–812 (1981)
12. Hedia, H.S., Mahmoud, N.A.: Design optimization of functionally graded dental implant. Bio-Med. Mater. Eng. **14**(2), 133–143 (2004)
13. Pizzini, S., Roberts, K.J., Dring, I.S., Oldman, R.J., Cupertino, D.C.: Application of X-ray absorption spectroscopy to the structural characterisation of monodispersed benzotriazole coatings on partly oxidised copper thin films. J. Mater. Chem. **3**(8), 811–819 (1993)
14. Nejati, E., Firouzdor, V., Eslaminejad, M.B., Bagheri, F.: Needle-like nano hydroxyapatite/poly (l-lactide acid) composite scaffold for bone tissue engineering application. Mater. Sci. Eng. C **29**(3), 942–949 (2009)
15. Holzwarth, U., Gibson, N.: The Scherrer equation versus the 'Debye-Scherrer equation'. Nat. Nanotechnol. **6**(9), 534 (2011)
16. Zhou, J., Zhang, X., Chen, J., Zeng, S., De Groot, K.: High temperature characteristics of synthetic hydroxyapatite. J. Mater. Sci. Mater. Med. **4**(1), 83–85 (1993)
17. Elliott, J.C.: Monoclinic space group of hydroxyapatite. Nature **230**(11), 72 (1971)
18. Kraus, W., Nolze, G.: POWDER CELL–a program for the representation and manipulation of crystal structures and calculation of the resulting X-ray powder patterns. J. Appl. Crystallogr. **29**(3), 301–303 (1996)
19. Zhu, M., Aikens, C.M., Hollander, F.J., Schatz, G.C., Jin, R.: Correlating the crystal structure of a thiol-protected Au25 cluster and optical properties. J. Am. Chem. Soc. **130** (18), 5883–5885 (2008)
20. French, R.H., Glass, S.J., Ohuchi, F.S., Xu, Y.N., Ching, W.Y.: Experimental and theoretical determination of the electronic structure and optical properties of three phases of ZrO 2. Phys. Rev. B **49**(8), 5133 (1994)
21. Young, R.A., Elliott, J.C.: Atomic-scale bases for several properties of apatites. Arch. Oral Biol. **11**(7), 699–707 (1966)
22. Pedone, A., Corno, M., Civalleri, B., Malavasi, G., Menziani, M., Segrea, U., Ugliengo, P.: An ab initio parameterized interatomic force field for hydroxyapatite. J. Mater. Chem. **17** (20), 2061–2068 (2007)
23. Benramache, S., Benhaoua, B.: Influence of annealing temperature on structural and optical properties of ZnO: in thin films prepared by ultrasonic spray technique. Superlattices Microstruct. **52**(6), 1062–1070 (2012)
24. Fu, B., Sun, X., Qian, W., Shen, Y., Chen, R., Hannig, M.: Evidence of chemical bonding to hydroxyapatite by phosphoric acid esters. Biomaterials **26**(25), 5104–5110 (2005)
25. Ling, Y., Rios, H.F., Myers, E.R., Lu, Y., Fezng, J.Q., Boskey, A.L.: DMP1 depletion decreases bone mineralization in vivo: an FTIR imaging analysis. J. Bone Miner. Res. **20** (12), 2169–2177 (2005)
26. Wang, A., Liu, D., Yin, H., Wu, H., Wada, Y., Ren, M., Jiang, T., Cheng, X., Xu, Y.: Size-controlled synthesis of hydroxyapatite nanorods by chemical precipitation in the presence of organic modifiers. Mater. Sci. Eng. C **27**(4), 865–869 (2007)

The Behavior of a Photovoltaic Module Under Shading, in the Presence of a Faulty Bypass Diode

Mohamed Zebiri(✉), Mohamed Mediouni, and Hicham Idadoub

National School of Applied Sciences (ENSA), BP 1136, Agadir, Morocco
{m.zebiri,m.mediouni}@uiz.ac.ma, hicham8ida@gmail.com

Abstract. Keeping low-cost industrial systems in operational condition has become a critical factor in business performance. At the moment, the forecast maintenance proves to be an essential activity in order not to incur untimely maintenance costs.

In photovoltaic and wind renewable energy production systems where production is dependent on meteorological conditions, the study of the failures of these systems is essential in order to identify them and to be able to develop a working methodology to predict degradation and thus be able to maximize energy production.

In this paper we will study the behavior of a photovoltaic (PV) generator composed of two modules which are M1 and M2. Since M1 is unshaded, we focus on M2 which is shaded and work at different irradiations levels with a bypass diode failure using the Power-Voltage (P-V) characteristics. Bypass diodes are critical components in PV modules as they provide protection against the shading effect. Failure of bypass diode in short circuit reduces the PV module power, while diode failure in open circuit leaves the module susceptible for extreme hotspot heating and potentially fire hazard.

This study will enable us to be able to prematurely detect and locate these failures and thus guarantee a good efficiency in the maintenance interventions, a reduction in costs and, consequently, a better productivity by increasing the rate of availability of the installations. For that, we will simulate the electric model of a module under Psim software which is a complete modeling tool oriented towards electrical engineering and compare the results obtained with the model of the panel given by Psim library.

Keywords: Photovoltaic · Bypass diode failure · Psim based simulation

1 Introduction

The electrical power generated by a renewable energy system can be greatly reduced in relation to the optimum production conditions (maximum power point) by the influence of meteorological conditions given by many factors, such as mismatch effects, shading, wind speed, ambient temperature rise, etc..... [1].

M. Ezziyyani (Ed.): AI2SD 2018, AISC 912, pp. 71–80, 2019.
https://doi.org/10.1007/978-3-030-12065-8_8

In the case of a photovoltaic generator, influencing the production, we encounter the effect of shading which result in non-uniform insolation conditions, the failure of the bypass diode, the blocking diode and the connectors… [2].

We will present the behavior of a module during bypass diode failures, which in its good state is conductive when the sum of the voltage of the cells that protect it is negative, and therefore the diode is blocked in the opposite case. In its failing state, this protective role is no longer assured. The electrical faults associated with this diode are: the short circuit diode, the diode failure in open circuit [3] and the inverted diode. In addition to these electrical faults, this diode could possibly be slammed during the operation and act as an impedance of any value [2]. We will present the behavior of this system in the presence of a slammed bypass diode and see the characteristics P-V according to values of the impedance of the slammed bypass diode. To carry out the simulations we use the PSIM software and the electrical characteristics of the Solarex MSX-60 solar panel.

2 PV Cell Model Used

To model the characteristics current-voltage (I-V) and P-V of photovoltaic cells, several models are used in the literature and vary in complexity and precision, operating under various conditions. The model with one diode is the most widely used to represent a PV cell, thanks to its simplicity and precision Fig. 1a [4, 5].

In the normal operating mode of the photovoltaic cell, the open circuit voltage is in the order of 0.6 V for the crystalline cells, whereas in the reverse bias operation the voltages can reach more than −20 V [6]. We will use the model with one diode protected by the bypass diode. Figure 1b presents a circuit model of a PV panel protected by bypass diode (D_b) [7, 8].

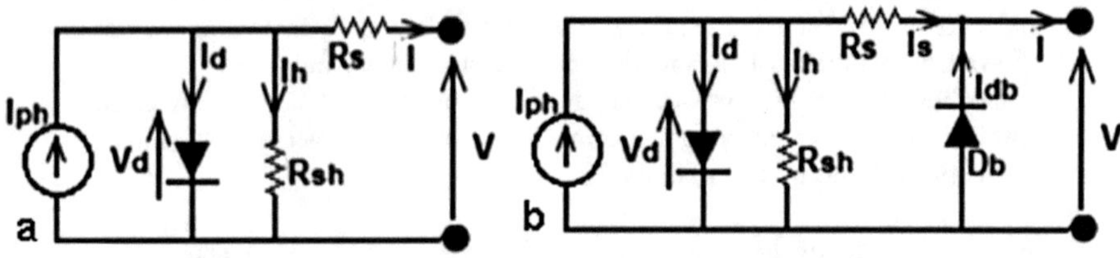

Fig. 1. Solar cell equivalent circuit (a) and equivalent circuit with the bypass diode Db (b)

In parallel to a group of solar cells (around 18 cells each group) bypass diodes are integrated into the PV module. These bypass diodes reduce the power loss caused by partial shading on the PV module. Besides the power loss, the bypass diode avoids the reverse bias of the individual solar cells higher than the allowed inverse polarization voltage of the solar cells. If a cell is inverted with a voltage greater than that predicted for the cell, hot spots may appear, which may cause browning, burning marks or, in the worst case, a fire.

The current generated by the PV cell is given by (1) applying Kirchhoff's law.

$$I = I_{ph} - I_d - I_h + I_{db} \tag{1}$$

Where:

$$I_{ph} = (I_{sc} + K_i(T - 298.15))G/1000 \tag{2}$$

$$I_d = \left(I_{sat,d}\left(e^{V_d/V_{t,d}} - 1\right)\right) \tag{3}$$

$$I_{sat,d} = \frac{I_{sc} + K_i(T - 298.15)}{\exp\left(\frac{q(V_{oc} + K_v(T - 298.15))}{aKTN_s}\right) - 1} \tag{4}$$

$$I_h = (V + I_s R_s)/R_{sh} = (V + R_s(I - I_{db}))/R_{sh} \tag{5}$$

With $V_{t,d} = N_s aKT/q$ the thermal voltage and I_{ph} is the photo-current of a cell, G is the sun irradiation, *Ki* is the temperature coefficient of I_{sc}, I_d is the current of the diode of a cell, $I_{sat,d}$ is its inverse saturation current, K_v is the temperature coefficient of V_{oc}, Ns is the number of cells connected in series, I_{db} is the current of the bypass diode, *q, a, k* and *T* respectively denote the charge of the cell electron, diode ideality factor, Boltzmann constant and p-n junction temperature, *V* is the voltage across the cell, *I* is the cell current, R_s and R_{sh} are the series and shunt resistors of the cell.

Figure 2 shows the electrical model used for the simulation which consists of two modules with bypass diode.

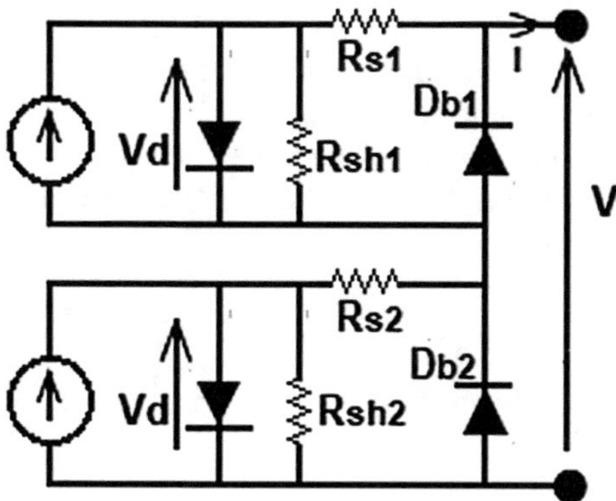

Fig. 2. Electrical model used for simulation

3 Simulations

The photovoltaic system can be simulated with an equivalent circuit model based on the photovoltaic model Fig. 1a and the solar module used in the Psim library. The Psim software offers its parameters and its characteristics I-V and P-V, so we can validate our modeling based on the calculations made under Psim. In this study, we use Solarex MSX-60 solar panel.

Table 1 presents the specifications of one PV module under standard test conditions (STC, 25 °C and 1000 W/m^2). Figure 3 shows the specifications of two modules and parameters of one single module.

Table 1. Solarex MSX-60 PV module specifications

Parameter	Value
Maximum power, Pmax	60 W
Voltage at Pmax, Vmp	17.1 V
Current at Pmax, Imp	3.5 A
Short-circuit current, Isc	3.8 A
Open-circuit voltage, Vco	21.1 V
Temperature coefficient of Voc, Kv	−80 mV/°C
Temperature coefficient of Isc, Ki	2.4 mA/°C
Number of series connected cells	36

3.1 Simulation of the Equivalent Model

The simulation is performed under Psim environment [9, 10]. To study the shading effect, we used a string with two PV panels connected in series. The solar panels were considered identical in terms of manufacturing parameters and working temperature (T = 298.15 K). The first panel M1 is subjected to irradiation of 1000 W/m^2 while the second M2 will undergo a variation of 1000, 800, 400 and 200 W/m^2 with a normal bypass diode. For bypass diode faults we use the same modules with the same irradiations and the bypass diode will be used in the following cases: (short-circuited, in open circuit and slammed, in this case it is replaced by an impedance of 10, 20, 40 and 80 Ω. Figure 4 shows the mathematical model realized under Psim using (2), (3) and (4). This model is used to realize the photovoltaic system of the two modules M1 and M2 for the simulation Figs. 5 and 6. The module M1 is unshaded, while M2 is shaded.

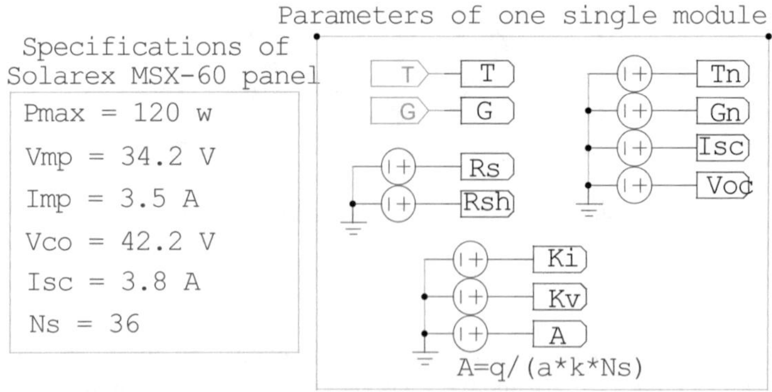

Fig. 3. Parameters and specifications used.

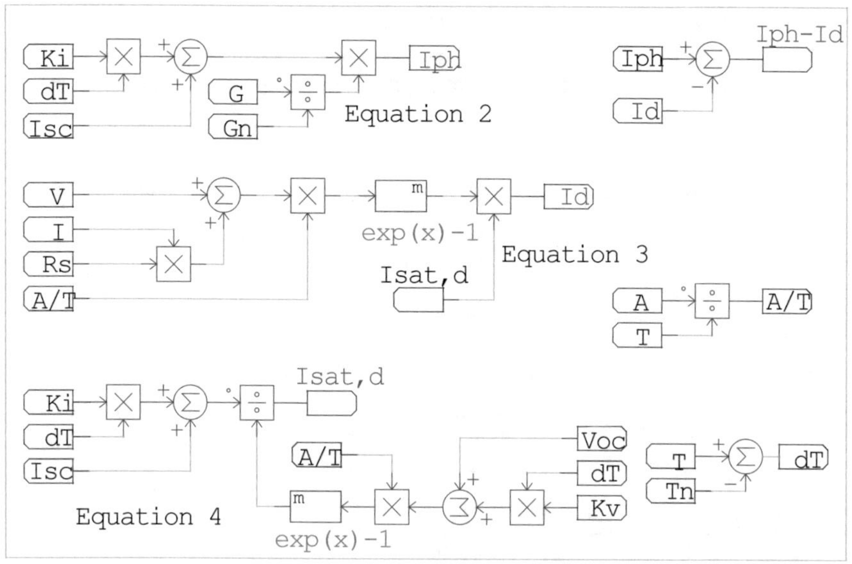

Fig. 4. Mathematical model built under Psim.

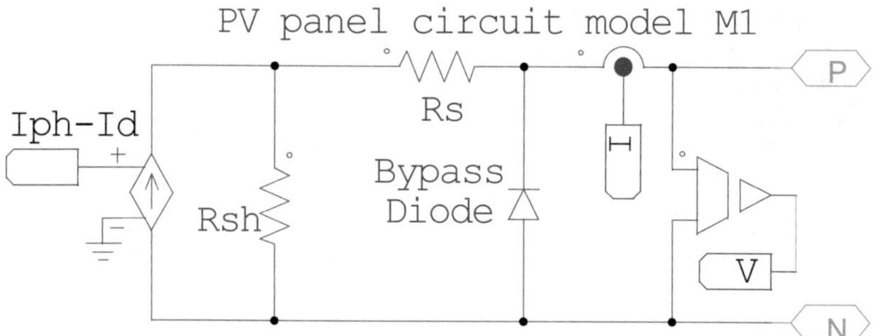

Fig. 5. PV panel circuit model M1 with bypass diode.

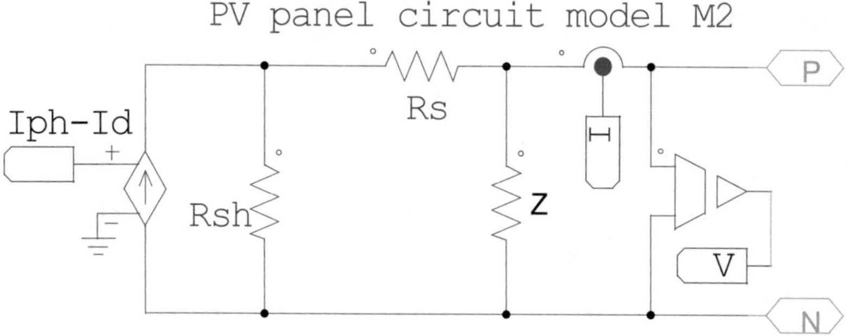

Fig. 6. PV panel circuit M2 with an impedance Z of a slammed bypass diode.

Using the utility provided by Psim, we were able to extract the parameters of the R_s, R_{sh} and a presented in Table 2 as well as the I-V and P-V characteristics. We used these results to compare them with the characteristics obtained by our mathematical model in order to adjust it (Fig. 7).

Table 2. Parameters extracted from a Solarex MSX-60 module

R_s	0,008 Ω
R_{sh}	1000 Ω
a	1,2

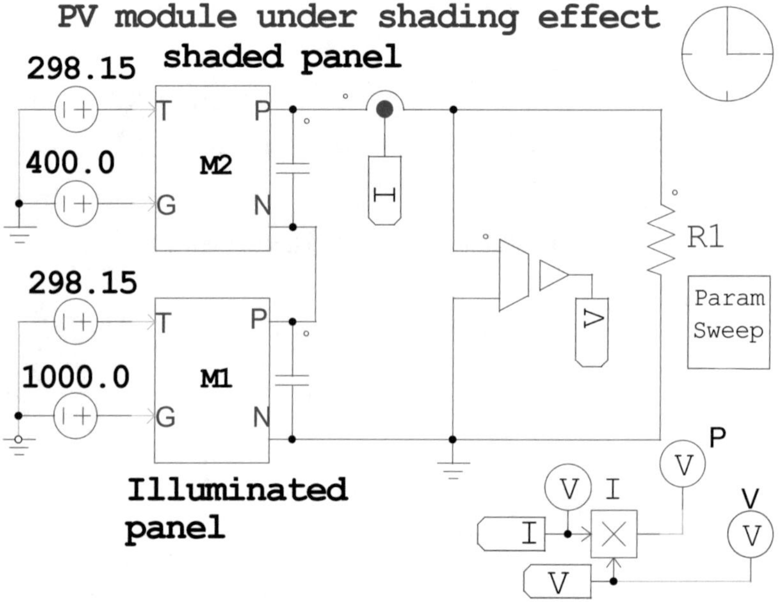

Fig. 7. Two modules based on the mathematical model in Fig. 4

The M1 module is subjected to radiation of 1000 W/m^2 while the M2 module is subjected to different radiation (1000, 800, 400 and 200 W/m^2) for each radiation we change the impedance Z of the bypass diode according to Table 3.

Table 3. The scenarios used. (X: the test is performed)

Module	Radiance W/m^2	Failure				
		Short circuit	Z = 10 Ω	Z = 20 Ω	Z = 40 Ω	Z = 80 Ω
M2	1000	X	X	X	X	X
M2	800	X	X	X	X	X
M2	400	X	X	X	X	X
M2	200	X	X	X	X	X

First, we present the P-V characteristics of the shaded system with normal bypass diode and without bypass diode. Then we present P-V characteristics with a faulty bypass diode and having an impedance that we increased (Z = 10, 20, 40, 80 Ω). For each value of Z we vary the irradiation from 200 to 1000 W/m^2 by incrementing in steps of 200 W/m^2 each time.

From Figs. 8, 9 and 10 we have an illustration of the characteristics P-V resulting from these manipulations. Figure 8a represents the P-V characteristics of the shaded system with an operational bypass diode, while Fig. 8b the bypass diode is not used.

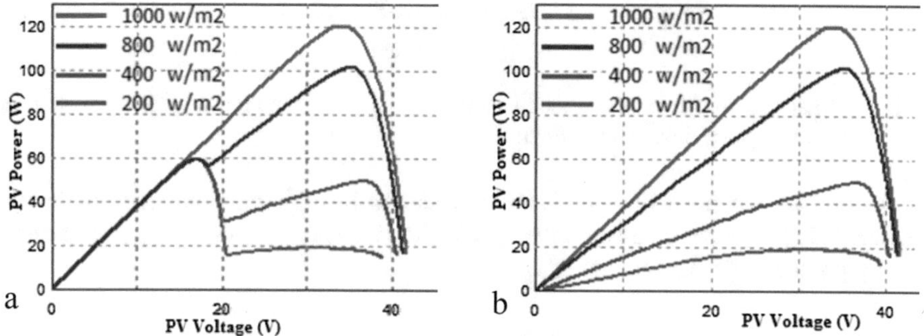

Fig. 8. Characteristics P-V of two modules M1 to 1000 W/m^2 and M2 to 1000, 800, 400 and 200 W/m^2 with operational bypass diode (a) and without bypass diode (b).

Figure 9a until d show the P-V characteristics under shading system with a slammed bypass diode that have impedance which we vary from 10 to 80 Ω.

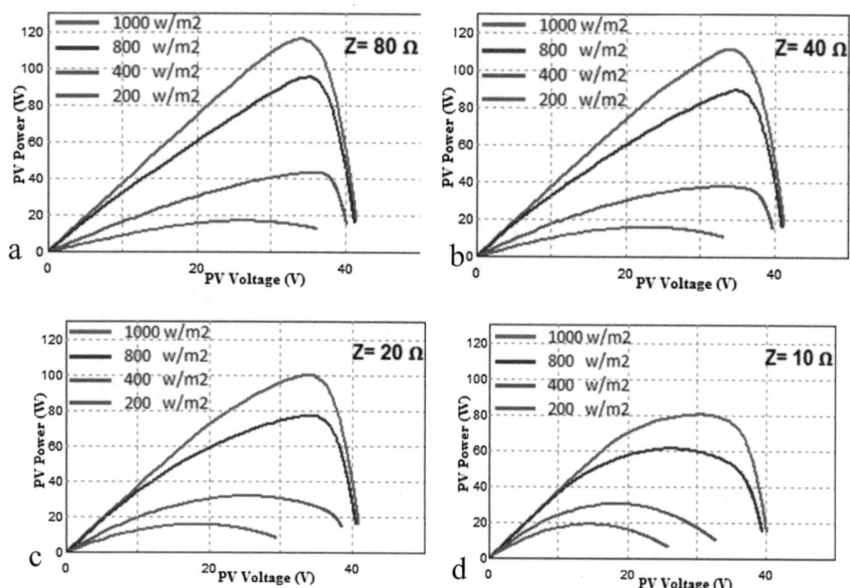

Fig. 9. Characteristic P-V of two modules M1 to 1000 W/m^2 and M2 to 1000, 800, 400 and 200 W/m^2 with an impedance Z = 80 Ω (a), 40 Ω (b), 20 Ω (c) and 10 Ω (d).

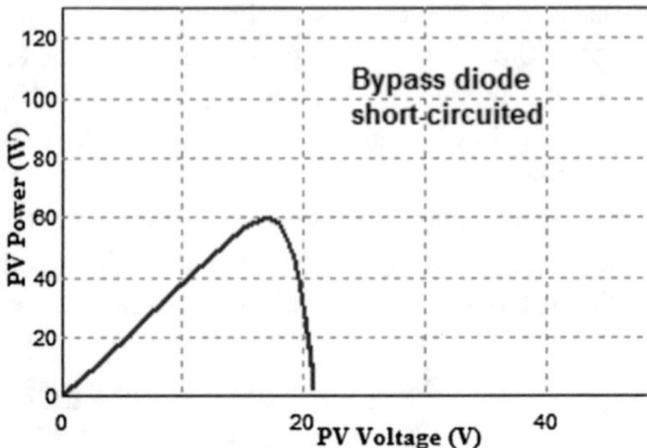

Fig. 10. Characteristics P-V of two modules M1 to 1000 W/m^2 and M2 to 1000, 800, 400 and 200 W/m^2 with bypass diode short-circuited

In Fig. 10 we have the same curve whatever the illumination received by the module M2, this is because when the bypass diode is short circuited we only have the energy produced by the module M1.

3.2 Simulation with the Module Proposed by Psim

We used the modules offered in the Psim library (Psim 11.03 Demo version examples) Fig. 11 to compare the simulations made with the mathematical model we use. We have used for this simulation only two values of the impedance Z of the slammed bypass diode (80 and 20 Ω). Figures 12a and b present the P-V characteristics obtained that we can compare with those on Fig. 9.

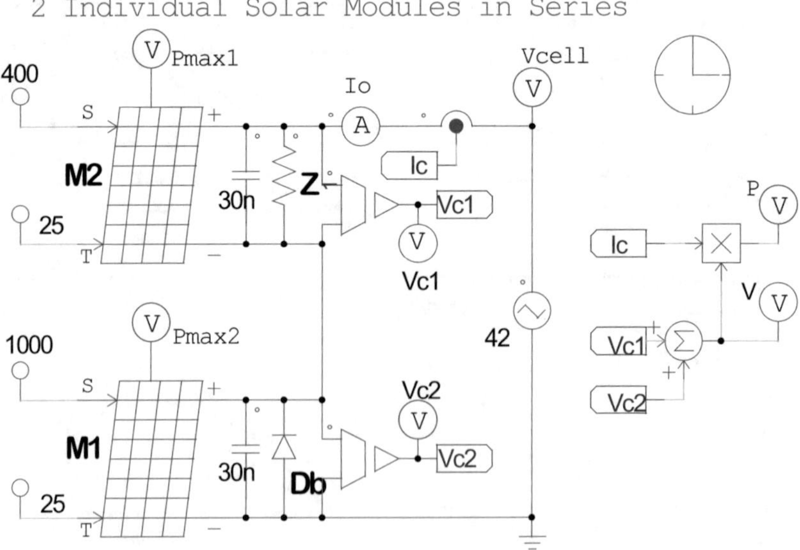

Fig. 11. Solar modules in series (Source: Psim 11.03 Demo version examples)

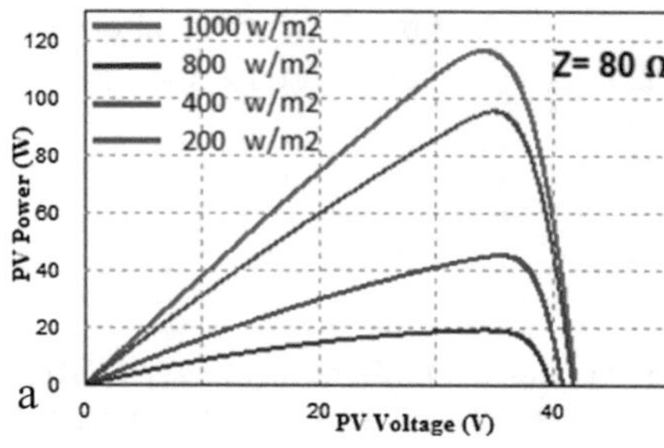

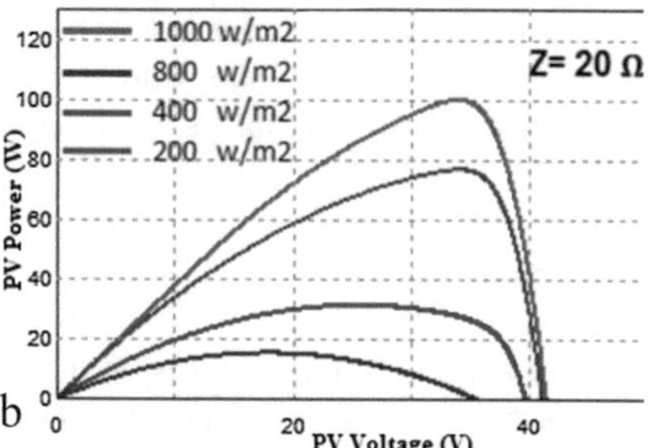

Fig. 12. Characteristics P-V of two modules M1 to 1000 W/m^2 and M2 to 1000, 800, 400 and 200 W/m^2 with an impedance Z = 80 Ω (a) and Z = 80 Ω (b) with the solar model in Psim library.

4 Results and Interpretations

From this Figs. 8, 9 and 10 we were able to compare the appearance of the characteristic P-V of a shaded PV system for different values of Z and different levels of radiation. In the absence of by-pass diode Fig. 8b, the power decreases considerably with the intensity of the radiation. With bypass diodes Fig. 8a the power output gets higher comparing with the system without these diodes as in Fig. 8b. In the case where the diode is slammed and has impedance the shape of the curve depends on the value of this impedance. With a high impedance 80 Ω Fig. 9a it tends to look like a characteristic without a bypass diode, while with a small impedance of 10 Ω Fig. 9d the maximum power point decreases considerably as well as the voltage at the maximum power point and also we can see a strong reduction of V_{oc}.

On the one hand when the bypass diode is slammed and behaves like an impedance the current received by the load depends on this impedance. If it tends to zero all the power produced which should normally be supplied to the load is absorbed by this impedance. On the other hand, if the value of the impedance increases the power produced by the system is almost supplied to the load. A part of this power is passed into the module under shading and the load receives only a part of the current lower than the current it should receive.

5 Conclusion

In this paper, we have presented the behavior of a photovoltaic system of energy production composed of two modules, M1 and M2 which is under shadow.

We have compared the results of simulations obtained using the mathematical equivalent model to form the two modules with the model presented in the Psim library.

We have shown that it is possible to use the characteristic P-V of a photovoltaic system to detect failures of the bypass diode, in the case where it is slammed and causes impedance. To do this we have to compare the PV characteristic of the system under study with the PV characteristic of the safe one. This detection allows the maintenance manager to intervene and thus avoid a decrease in energy production and the destruction of the module by the hotspot effect.

References

1. Dhimish, M., Holmes, V., Mehrdadi, B., Dales, M., Mather, P.: Detecting defective bypass diodes in photovoltaic modules using Mamdani fuzzy logic system. Glob. J. Res. Eng. (F) Electr. Electron. Eng. **17**(5), 33–44 (2017). Version 1.0
2. Bun, L.: Détection et localisation de défauts pour un système PV. Thèse de doctorat, Laboratoire de génie électrique de Grenoble (G2ELAB), Université de Grenoble (2011)
3. Shiradkar, N.: Predictive modeling for assessing the reliability of bypass diodes in photovoltaic modules. Electronic Theses and dissertations. 1401. University of Central Florida (2015). http://stars.library.ucf.edu/etd/1401
4. Matter, K., El-Khozondar, H.J., El-Khozondar, R.J., Suntio, T.: Matlab/Simulink modeling to study the effect of partially shaded condition on photovoltaic array's maximum power point. Int. Res. J. Eng. Technol. **2**(2), 697–703 (2015)
5. Silvestre, S., Boronat, A., Chouder, A.: Study of bypass diodes configuration on PV modules. Appl. Energy **86**, 1632–1640 (2009)
6. Notton, G., Caluianu, I., Colda, I., Caluianu, S.: Influence d'un ombrage partiel sur la production électrique d'un module photovoltaïque en silicium monocristallin. Rev. Energ. Renouvelables **13**(1), 49–62 (2010)
7. Orozco-Gutierrez, M.L., Ramirez-Scarpetta, J.M., Spagnuolo, G., Ramos-Paja, C.A.: A technique for mismatched PV array simulation. Renew. Energy **55**, 417–427 (2013)
8. Petrone, G., Spagnuolo, G., Vitelli, M.: Analytical model of mismatched photovoltaic fields by means of Lambert W-function. Sol. Energy Mater. Sol. Cells **91**, 1652–1657 (2007)
9. Motahhir, S., El Ghzizal, A., Sebti, S., Derouich, A.: Shading effect to energy withdrawn from the photovoltaic panel and implementation of DMPPT using C language. Int. Rev. Autom. Control (I.RE.A.CO.) **9**(2), 88–94 (2016)
10. Villalva, M.G., Gazoli, J.R., Ruppert, E.F.: Modeling and circuit-based simulation of photovoltaic array. Braz. J. Power Electron. **14**(1), 35–45 (2009)

Towards a Realistic Design of Hybrid Micro-grid Systems Based on Double Auction Markets

Imane Worighi[1,2,3](✉), Abdelilah Maach[1], Joeri Van Mierlo[2,3], and Omar Hegazy[2,3]

[1] Mohammadia School of Engineers, Mohammed V University in Rabat, Rabat, Morocco
imaneworighi@gmail.com
[2] ETEC Department and MOBI Research Group, Vrije Universiteit Brussel (VUB), Pleinlaan 2, 1050 Brussels, Belgium
[3] Flanders Make, 3001 Heverlee, Belgium

Abstract. Renewable Energy Sources (RESs) are playing an important role in reducing carbon emissions of conventional fossil fuel, greenhouse gases and other pollutants. Moreover, they have a potential benefit of reducing energy cost and providing power to meet the energy demand. Nevertheless, grid stability is criticized via intermittent and non-controllable behavior of RESs. In this regards, coupling power generation units such as solar and wind with Energy Storage Systems (ESSs) is the most likely solution to pave the way for the outstanding transition from conventional power system to smart grid infrastructure. Nonetheless, this infrastructure requires models and controls for the implementation of next-generation grid architecture. Therefore, a virtualization of the above infrastructure could be overriding and cost-effective for implementation phase. In that regard, the present research proposes a hybrid micro-grid model including wind and solar energy, loads, a double auction market and ESSs. The proposed model is tested and validated using virtualization of the above proposed grid structure. The implementation of the virtualized system enhances the role of ESSs leading the power system to be stable. ESSs allow accumulating the surplus energy for later use in those periods in which wind and solar contribute to overproduction. Furthermore, it enables delivering back the extra energy in peak periods. Simulation results show the effectiveness of the proposed model and highlight the role of ESSs in stabilization. As a result, the proposed model can be used to test other scenarios such as implementation of controls and smart appliances of the new and improved power system augmented with RESs.

Keywords: Renewable Energy Sources · Smart grid · Energy demand · Virtualization · Energy Storage Systems · Peak periods

Nomenclature

Photovoltaic system

I_{PV} photovoltaic panel current
V_{PV} photovoltaic panel voltage
I_{sc} photovoltaic cell short-circuit current

M. Ezziyyani (Ed.): AI2SD 2018, AISC 912, pp. 81–96, 2019.
https://doi.org/10.1007/978-3-030-12065-8_9

I_o saturation current
R_S series cell resistance
R_p parallel cell resistance
V_T thermal voltage
N_{cell} number of cells for a single panel

Wind turbine
v_{WT} wind speed at wind turbine hub height
v_{AN} wind speed recorded from anemometer
z_{WT} height of the turbine hub
z_{AN} height of the anemometer

Batteries
P_B total output power of batteries
P_{WT} total output power of wind
P_{PV} total output power of PV system
P_L total consumption loads

1 Introduction

Industrialization and modernization have led to the rapid decline fossil-fuel reserves and growth of hydrocarbon-based energy consumption, which is one of the most formidable challenges for the human life and the environment [1]. Moreover, the nearby depletion of fossil fuels reserves, increasing the cost of fuel for conventional power generation imposes the existing grid to hasten the energy transition to alternative energies such as wind and Photovoltaic (PV) [2]. In this regard, more sustainable and RESs are being integrated into the power systems to address the increasing concerns of energy demand, energy cost, and greenhouse gas emissions [3]. Furthermore, the use of renewable energy is projected to increase substantially in the European Union to reach a share of 20% in final energy consumption and 10% renewable energy in transport by 2020 [4]. The renewable energy contribution is further expected to increase to 55%–75% of gross final energy consumption in 2050 [4]. Indeed, the Paris Agreement went into force on 4 November 2016 will be another accelerating factor for the use of electricity from renewable energy sources [5]. Nonetheless, instability and inherent intermittent of renewable power generation units due to weather conditions pose tremendous operational challenges in power systems. The power output of such intermittent sources, is influenced significantly by season and weather, and can vary abruptly and frequently [6]. Besides that, this intermittency creates fluctuation of power and later increase system instability and voltage fluctuation [7]. For instance, integration of RE in Pakistan reveals problems such as thermal overloading, voltage violations and generators' oscillations [8]. Indeed, authors in [9] highlighted the impact of this integration on system operations including overloading of substation transformers, current and voltage harmonics, and PQ in general. Subsequently, the overall performance of the power systems is influenced. To overcome this problem, effective coordination between RESs and ESSs is highly recommended and considered as the most likely solution to the aforementioned challenges [10]. Further, this combination will improve system efficiency and reliability [2]. Consequently, a wind-solar-ESS hybrid power system is the key technology to tackle the power fluctuation problem [11]. This hybrid power system would be double-beneficial to the next-generation grid, especially when wind and solar power generation are employed to feed the grid, and storage systems to smooth the wind and solar power fluctuations [12]. In that regard, storage can fill in the supply gaps as well as absorb excess production, and since storage responds quickly, it can adjust the power output to respond to fluctuation of wind or solar output [13]. In this context, the role of battery energy storage system (BESS) has been highlighted in a context of hybrid wind-solar structure along with an optimal capacity of the BESS [14]. Moreover, in Ref. [14], an optimization strategy has been defined to combine the generation of a solar PV power plant and a wind power plant primarily to match the power demand at a particular location and additionally to suppress the effect of inherent intermittency in renewable energy (RE) resources with the support from a BESS. As a result, this strategy leads to improvement of system reliability and minimization of unit cost of power. Furthermore, Mamen et al. [15] proposed Hybrid Energy Storage Systems (HESS); electrical storage devices and electrochemical energy storage are combined to minimize, smooth out the mismatch of demand and supply. Besides that, Zade et al. [16] emphasized the role of Hybrid solar and wind energy systems used for rural electrification and modernization of remote area to improve power quality. The authors proposed a hybrid power system and used controllers to regulate source voltage, source current and load voltage [16]. Furthermore, a flexible market that govern how plants are scheduled and dispatched, how reliability is assured, and how customers are billed, can allow access to significant existing flexibility, often at lower economic costs than options requiring new sources of physical flexibility [17]. Therefore, the existence of a market in next-generation power grid is mandatory to help customers in controlling their bill and using the electricity wisely. In that regard, attention should be given to market rules, design, and operations when integrating new technologies.

In that regard, analysis and design of power electronics and power systems are required to support integration of new technologies to the new power grid. This can only be performed by the presence of smart grid pattern to design and control different technologies and power source structures. However, to enable such a huge infrastructure, it is needed to go through existing smart grid models in a context of a hybrid micro-grid model. In that regard, a hybrid micro-grid model can reduce system complexity and give a clear picture into smart grid applications studies. Furthermore, a micro-grid is locally controlled energy system that uses different types of RESs (solar, wind, biomass, hydro, ocean), energy generators (Diesel, gasoline, biogas, biodiesel), ESSs (batteries, flywheel, hydrogen, thermal), loads (residential, commercial and industrial) and control equipment (inverters and converters) [18]. Moreover, a micro-grid can be connected to the utility grid or disconnected from the grid so that distributed generators continue autonomously to power the users and operate separately off grid without obtaining power from outside [19].

In this context, properly modelling and simulating the hybrid micro-grid system has become one of the primary objectives. In this regard, Dawoud et al. [20] reviewed new ways of energy practice of hybrid sources. Using optimal placing of components, the authors presented a general hybrid RE resources architecture and the physical modelling of the RE resources with numerous methodologies and principles of the optimization for the hybrid micro-grid [20]. Similarly, Khisa et al. [21] presented a model of a hybrid wind-solar and battery system installed in a school in Naivasha-Kenya to analyze the Load-Power profile. The proposed model has been used to understand the nature of dynamic characteristics of a hybrid wind-solar generation unit and to study the impact of such dynamic phenomena on the operation, stability and power quality of the overall system [21]. Moreover, Wang et al. [22] proposed a micro-grid model including a wind turbine, a PV cell, loads and Composite Energy Storage System (CESS) using the simulation tool DIgSILENT/PowerFactory. They highlighted the role of CESS which can effectively contribute to power flow control. Additionally, Louie in Ref. [23] proposed a high-level schematic of hybrid micro-grid architecture and examined the operation of hybrid solar/wind micro-grids using measured data from a 5-kW system in Muhuru Bay, Kenya. In this regard, Louie highlighted the role of such architecture combined with controllers in influencing the operational behavior of the micro-grid and prioritization of energy sources [23].

Besides modelling the hybrid micro-grid system, virtualization is needed to test such system before implementation phase. Furthermore, the real power grid cannot be used for tests and validation, and the virtual system is accurate as the real system. Therefore, this virtualization could provide better insight into cost-effectiveness of hybrid micro-grid models. Moreover, it would make reconfiguring and orchestrating a network of field devices relatively straightforward [24]. In the present article, the main contributions can be categorized as follows:

1. Presenting a hybrid micro-grid model;
2. Introducing components used in the proposed model;
3. Defining controllers used in the proposed model;
4. Emphasizing market design and its role in reducing load demand;
5. Coupling of the proposed grid structure with solar and wind power generation units, and also the effect of their penetration on power system is discussed;
6. The proposed model is optimally designed and accurately modeled in the power-systems simulation tool GridLab-D;
7. Integrating ESSs into the proposed grid structure to maintain system stability.

To sum up, in the following paragraphs, a hybrid micro-grid model is presented. The main objective of introducing such a model is to reduce system complexity and to study effectiveness of the proposed model. The introduced smart micro-grid is composed of renewable energy generations, ESSs, a double auction market and loads, which can operate in grid-connected and stand-alone modes. Then, the proposed micro-grid model is implemented to emphasize market role, test integration of RESs with a variable output that depends on weather conditions, study effect of this integration on system stability and highlight the role of ESSs in maintaining system stability and improving system quality and sustainability.

In addition, the rest of the article is organized as follows: a proposed micro-grid model is presented in Sect. 2. Results and discussion are provided in Sect. 3. Finally, the outcome of the article is summarized and concluded in Sect. 4.

2 Proposed System Model

Hybrid micro-grid systems (HMGS) comprise of several parallel connected distributed resources with electronically controlled strategies, which are capable to operate in both islanded and grid connected mode [25]. Furthermore, as these systems are based on RES, they provide a cost-effective solution for providing energy in remote areas where transmission and distribution facilities are not available or expensive to do so [26].

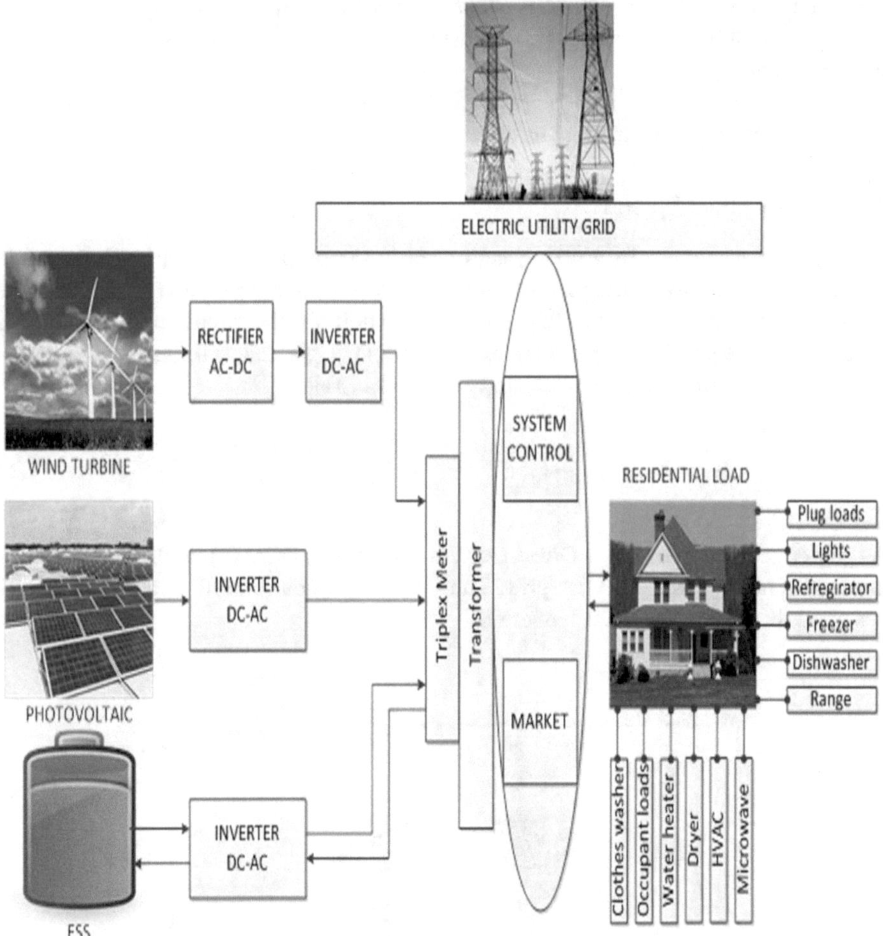

Fig. 1. The proposed wind and PV hybrid micro-grid model

Moreover, combining PV and wind power systems provide numerous benefits such as achieving better quality and reliability for end users [27], improving system performance, decreasing fluctuation of generation, investment costs, and the size of storage systems [28]. Otherwise, if PV systems or wind turbines are used independently, size of system and investment cost will increase substantially. Besides that, the performance of such hybrid systems depends highly on management and control of storage systems which can be used to smooth the intermittency of power-based RESs generation units and enhance system efficiency [29]. Further, the fast deployment of distributed energy resources in the electric power system has highlighted the need for an efficient energy trading transactive model [30].

In this context, a hybrid micro-grid model is required to deal with challenges of RES integration into the next-generation grid, reduce energy cost, perform load management and maintain grid stability. The proposed hybrid micro-grid model is presented in Fig. 1. The model consists of PV generation system, doubly-fed induction generator (DFIG) wind turbine, ESSs, a Market, a system control, residential loads as well as the utility grid.

To have better insight into the proposed hybrid micro-grid model, in the following paragraphs, each component that has interaction with others is described and discussed.

2.1 Photovoltaic System

The equivalent circuit of the PV system used in the proposed model is illustrated in Fig. 2 [31]. Where I_{SC} is the current source, indicating solar irradiation in parallel with the diode characterized as the PN junction. In addition, the circuit contains a series resistance R_S representing material resistance and a parallel leakage resistance R_P. Thus, the PV current can be calculated as follows [31]:

$$I_{PV} = N_{cell}\left[I_{sc} - I_o\left(e^{\frac{V_{PV}-R_S I_{PV}}{V_T}} - 1\right) - \left(\frac{V_{PV} + R_s I_{PV}}{R_p}\right)\right] \tag{1}$$

As power generation of PV is Direct Current (DC) form, DC/AC inverter is required to inject generated power into the grid. Thus, the PV system interface with the grid is designed through a DC to AC inverter.

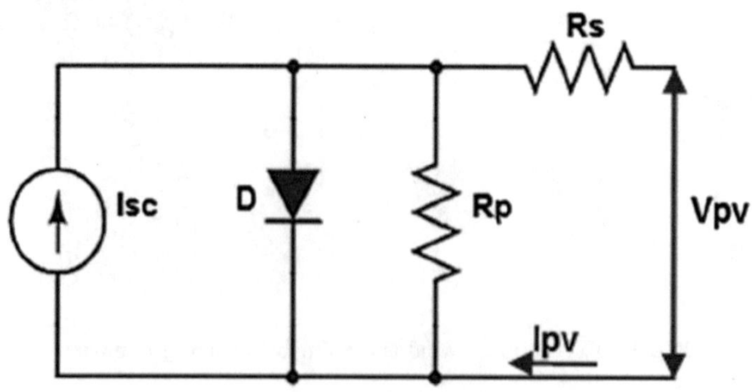

Fig. 2. Equivalent circuit of the PV cell [32].

2.2 Wind Turbine

The output power of the wind turbine is a function of the wind speed at turbine hub height [28]. The wind speed at turbine hub altitude can be obtained as a function of the wind speed recorded from anemometer. The function used to calculate v_{WT} as a function of v_{AN} is indicated in Eq. 2 [31]:

$$v_{WT} = v_{AN} * \frac{\ln\left(\frac{Z_{WT}}{Z_o}\right)}{\ln\left(\frac{Z_{AN}}{Z_o}\right)} \tag{2}$$

The wind turbine used is a DFIG due to its various advantages such as its low cost, flexible active and reactive power control capabilities, reduced converter costs and size, and lower power losses [32, 33]. Moreover, the wind turbine outputs three-phase AC, which is rectified to DC before connecting to the inverter that converts DC to AC.

2.3 Storage System

The storage system used in the proposed model is based on batteries. BESSs are connected to bi-directional inverter to store the surplus of energy produced in low-demand or high-generation periods and to supply energy when the demand exceeds generation. Therefore, the input power of the batteries can be either positive or negative depending on whether the battery is being charged or discharged. The input power of the batteries can be written as follows [28]:

$$P_B = P_{WT} + P_{PV} - P_L/\eta_{inv} \tag{3}$$

If P_B is null, then the battery is neither in charge nor in discharge modes. If it is positive, then the battery is being charged due to an excess of generated power. Finally, if P_B is negative, then the battery is being discharged due to a deficiency of generated power in the micro-grid.

Besides that, the battery is characterized by an important parameter which is the state of charge (SOC). It is defined as the ratio between the stored energy and its nominal storage capacity [31].

2.4 Market System

The proposed model highlights the importance of energy trading and demand response by implementing a market. In this regard, a market design with demand response strategy are highly recommended, as consumers can reduce the amount of electricity at the peak when the electricity price is higher, leading to the distributed generation (DG) priority, and sell excess electricity to the grid to gain economic benefits [34]. Especially a double auction market has been used in the proposed model to allow consumer to change its energy consumption based on the market price. The consumer can determine the appropriate price to pay and bid its desired demand for electricity [35].

In this context, the model contains a double-auction market that accepts demands, supply bids and clears on five minutes intervals as Real Time Price (RTP) has been used for billing. The market design used is illustrated in Fig. 3 as follows:

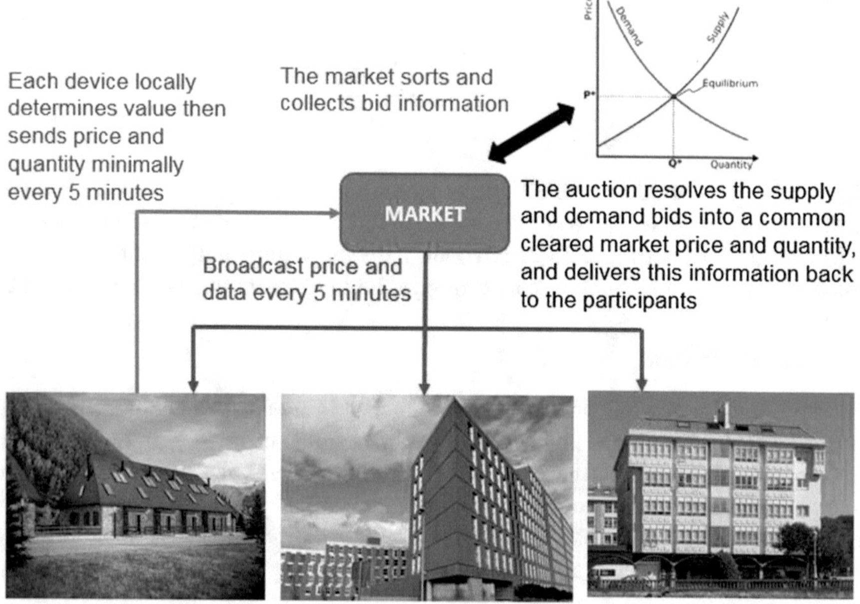

Fig. 3. Demand response and RTP under a double-auction market

2.5 Load System

The proposed model consists of residential loads. Each single-family residence is equipped with a physical model of the Heating, Ventilations, and Air Conditioning (HVAC) system, a water heater, Plug loads, lights, refrigerator, dishwasher and ZIP models of other appliances. Moreover, in each residential model, a transactive controller is implemented as an interactive controller, reacting to price changes and returns information back to the central controller. Indeed, these controllers can bid into the market and vary the setpoints as a function of price [35].

3 Simulation and Performance Evaluation

In order to evaluate the implementation of the proposed model, a small residential power systems with renewable sources (solar and wind), storage, market and domestic load are modeled via the power system modeling tool GridLab-D within IEEE 13 node [36]. The standard system is augmented with power generation units, BESSs, a market, smart appliances and controllers. In this regard, smart appliances such as HVAC systems and water heaters, which can interact with the market, are implemented to each house. Moreover, other loads are added, namely plug loads, lights, refrigerators, dishwashers, microwaves, dryers, occupant loads and ZIP models of other appliances. Furthermore, houses are equipped with smart meters and connected to transformer through a triplex meter. A triplex line connects transformer triplex meter to house triplex meter. Indeed, each step-down transformer is connected to one of the IEEE 13 nodes. In this regard, 1230 houses are distributed among the nodes of the IEEE 13 node feeder. Different house types are distinguished: houses with lower power consumption, houses with medium power consumption and bigger houses with higher power consumption. Figure 4 depicts the structure of the residential micro-grid model.

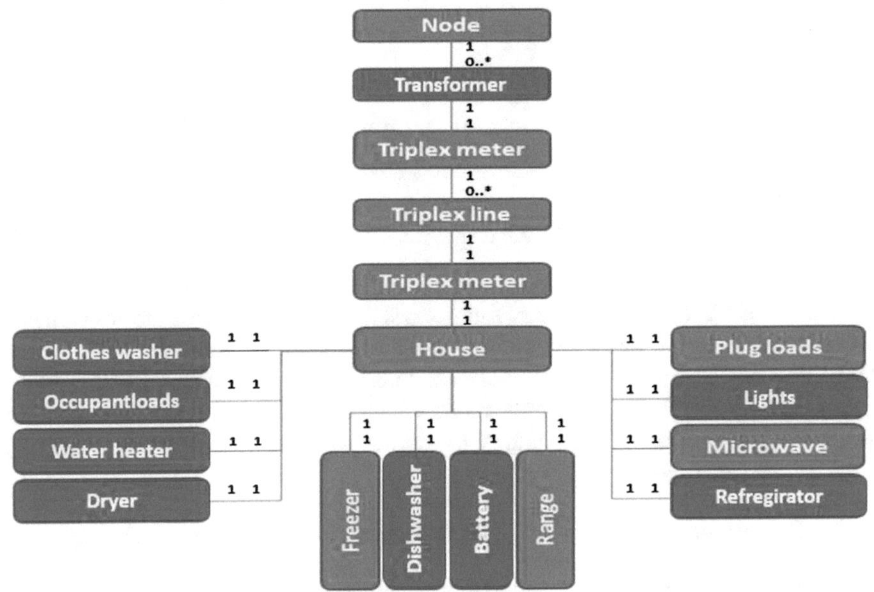

Fig. 4. Residential micro-grid structure

First, a double-auction market is implemented in the proposed system, and transactive energy controllers are added to each house. Then Simulation results highlights the effect of controllers on energy demand, depicts how the controllers can affect the residential load and how a customer can participate in a double-auction market in order to change its energy consumption based on the cleared market price. In this regard, the customer's bill was evaluated through the integration of RTP where the price can change every 5 min.

Second, power generation sources are integrated into the proposed system, which are wind and solar. Their impact on the system is emphasized considering both demand and cost of energy. The wind power generation system consists of a DFIG which is rated at 10 kW. Figure 5 presents the wind speed of the DFIG system.

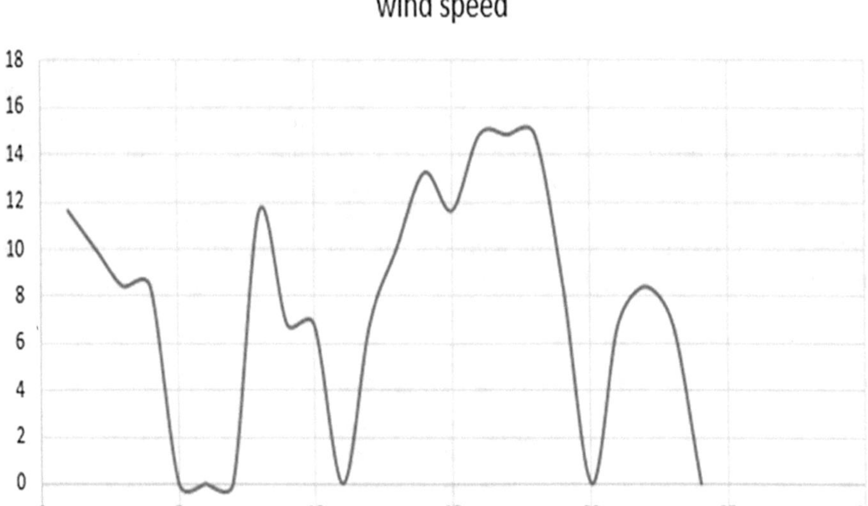

Fig. 5. Wind speed of DFIG system

Regarding PV system, Nineteen single-crystal silicon PV panels are implemented in this study. Each PV has 1 kW/m^2 for irradiance and 65 m^2 area. The inverter used for PV system to convert DC to AC is a Pulse width modulation (PWM) with a constant power factor (PF) generator mode which is equal to 1, and the defined efficiency of PV equals to 17%. Therefore, the output power of one PV per unit is 11 kW based on the given Eq. 4:

$$\text{Power Output} = \text{Area} \times \text{Solar Irradiance} \times \text{Conversion Efficiency} \tag{4}$$

Eventually, ESSs consisting of batteries with inverters are implemented to maintain system stability and mitigate intermittency of renewable energies. The batteries are Lithium-ion with rated capacity of 100 kWh. Furthermore, they are connected to the grid through inverters rated at 10 kVA.

Finally, by following the proposed hybrid micro-grid model and implementing the above technologies, a virtual micro-grid system is achieved. This virtual system can be used to test smart grid applications and integration of other technologies and controls.

3.1 Simulation Results and Analysis

The simulation was run for one month by using the weather information from Typical Meteorological Yearly (TMY) data for Atlanta, GA corresponding to August [37]. The primary focus has been given to measurements on node 645. The system was first implemented by employing a double auction market and transactive controllers. Figure 6 shows the effect of the market with RTP billing mode and house controllers on energy demand. The peak load is distinctly reduced by 13%.

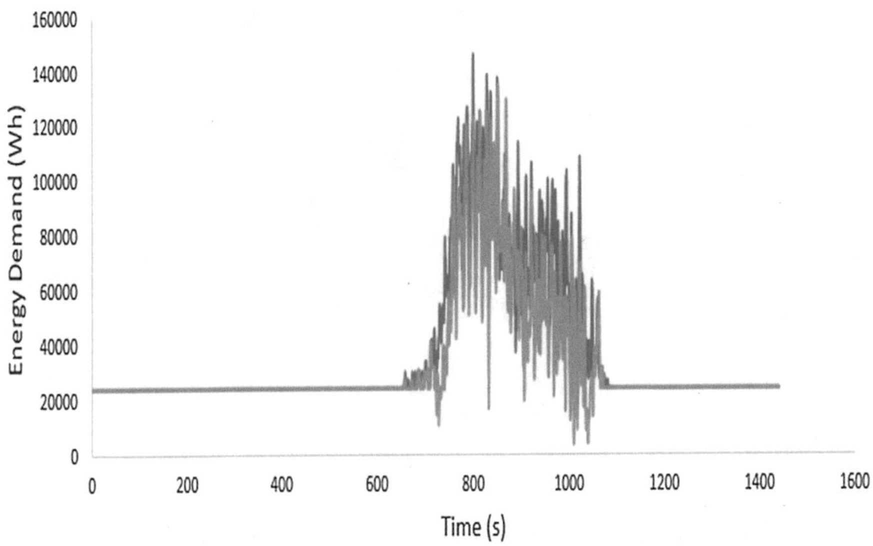

Fig. 6. System with and without market and home controllers

Afterwards, wind and PV generation systems have been implemented. The effect of introducing intermittent energy sources is depicted in Fig. 7. The introduction of wind and solar panels impacts the energy demand curve. However, generation of wind and solar energies exceeds the load which can induces overvoltage. Furthermore, such an excess causes electrical power to flow in the opposite direction from its usual flow [38]. This phenomenon occurs when PV power generation is excessive during the middle of the day where the sun irradiance is at its highest level or when wind speed is prominent due to weather conditions. This issue can affect power quality and reliability resulting in voltage deviations [38]. In Fig. 7 the effect of introducing renewable energies is apparent. The Figure shows the excess of generation over load, thus a significant demand reduction occurs in this period.

In addition, the introduction of wind and solar energies induces significant cost reduction. Figure 8 shows energy cost reduction after integration of renewable energies.

To handle the generation excess, it is needed to reduce the effect of high wind and PV generation, taking into consideration atmospheric data, solar irradiation and wind speed. Therefore, flowing the proposed hybrid model, BESSs have been integrated which are scheduled to charge when generation is prominent, and discharge when load is at its highest level and generation is deficient. Figure 9 shows energy demand after implementation of BESSs. The excess of generation has been removed. In that regard, BESS is known as a very efficient technology and now being considered in many applications related to the distributed generations such as over voltage issues [39]. Moreover, the introduction of BESSs' results in cost reduction shown in Fig. 10.

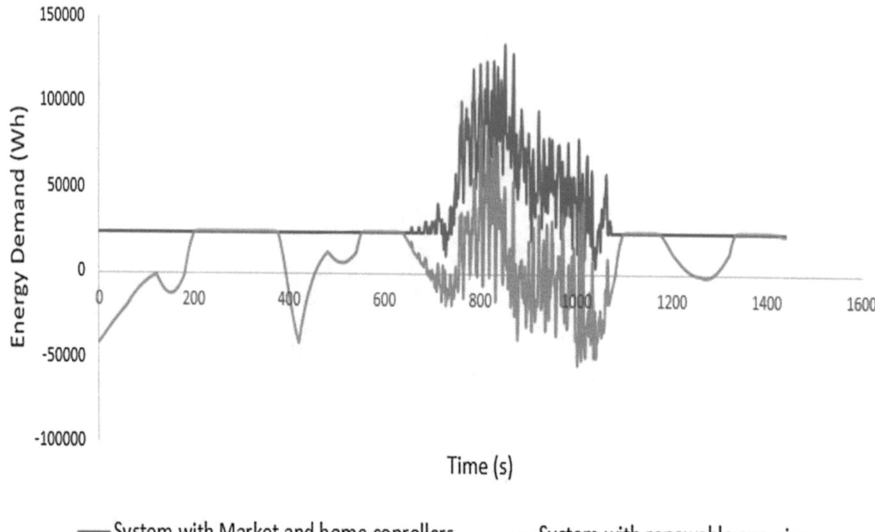

Fig. 7. System with and without renewable energies

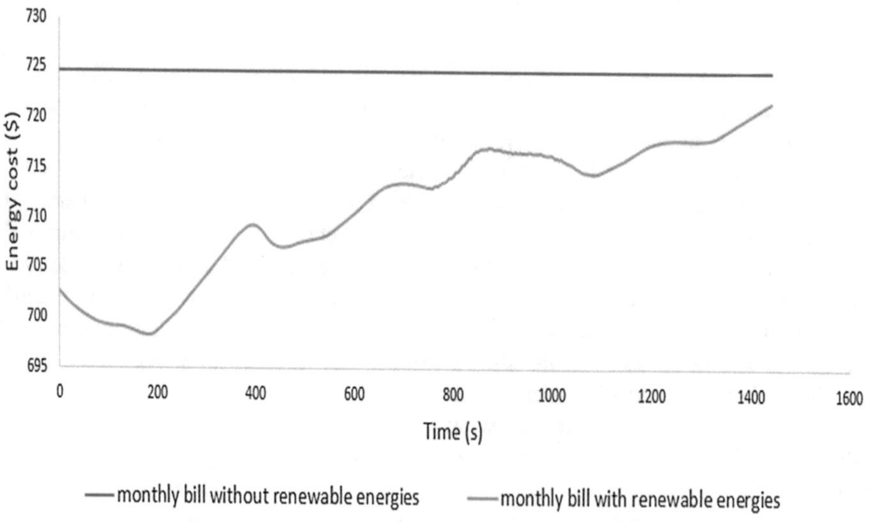

Fig. 8. Cost of energy with and without renewable energies

Ultimately, the simulation results show the benefit of implementing hybrid micro-grid system. The virtualization of the proposed model provides adequate results and achieves system stability. The energy demand is managed using ESSs. Furthermore, the energy cost is reduced with the proposed hybrid micro-grid model. Consequently, the proposed hybrid model can be used to test other smart grid applications and implementation of other new technologies and controls.

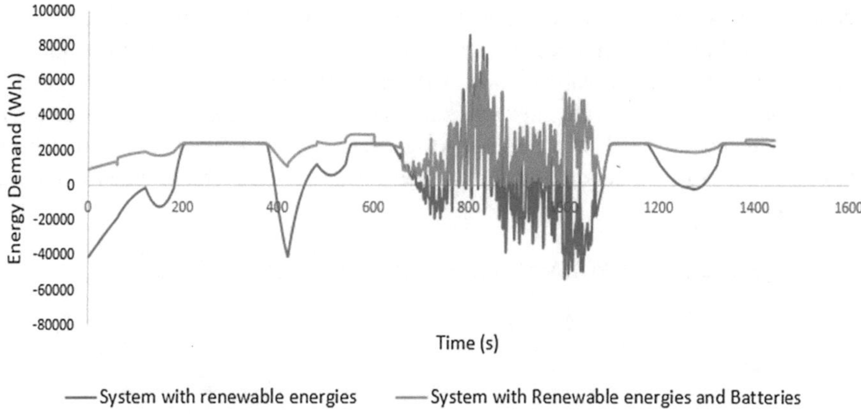

Fig. 9. Effect of BESSs integration on energy demand

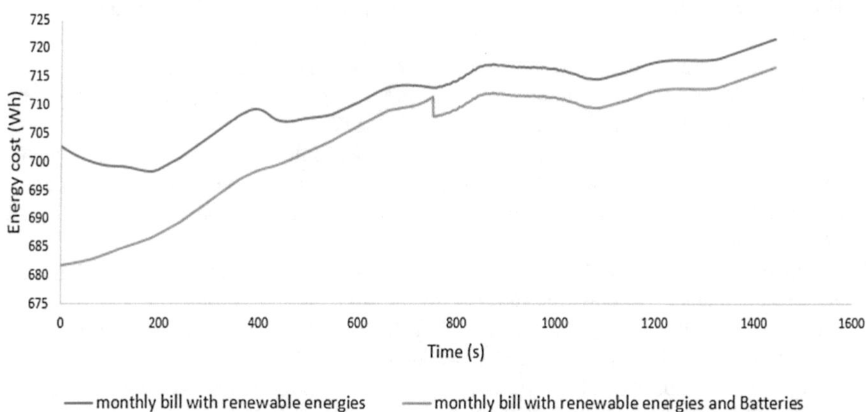

Fig. 10. Effect of BESSs integration on energy cost

4 Conclusion

In this article, a hybrid micro-grid model including wind and solar energies, ESSs, a market and residential loads was proposed. The proposed model aims to virtualize smart grid system to test smart grid applications and analyze grid behavior after integration of new technologies and controllers to the power system. To validate the proposed model, grid stability was also analyzed. First, the effect of demand response was highlighted by utilizing transactive controllers in homes, RTP billing for residential end-users and a double auction market to enable interaction of controllers with market prices. Both energy demand and cost were reduced. Furthermore, the effect of renewable energies integration was pointed out. High PV and wind generation resulted in excess of generation over load and contributed to system destabilization. To overcome this drawback, BESSs were implemented to the grid. Subsequently, these storage systems performed grid stability and energy cost depletion. The results showed that the combination of the above technologies as mentioned in the proposed hybrid micro-grid model, contribute to grid stability and perform a reduction in demand and cost of energy. The results are expected in a real smart grid system and perform a validation of the proposed model which can be used to test other smart grid applications such as reliability, blackout detection, advanced controls, electric vehicle integration and management.

Acknowledgements. We acknowledge Flanders Make for the support to our team.

References

1. Cosgrove, C.: Energy consumption in GCC countries. Power on, power up. In: 4th IEEE International Conference on Engineering Technologies and Applied Sciences (ICETAS) (2017)
2. Kharrich, M., Akherraz, A., Sayouti, Y.: Optimal sizing and cost of a Microgrid based in PV, WIND and BESS for a School of Engineering. In: International Conference on Wireless Technologies, Embedded and Intelligent Systems (WITS), pp. 1–5 (2017)
3. Li, Y.R., Hurley, W.G.: Editorial special issue on sustainable energy systems integration. IEEE J. Emerg. Sel. Top. Power Electron. **3**, 854–857 (2015)
4. Scarlat, N., Dallemand, J.F., Ferrario, F.M., Banja, M., Motola, V.: Renewable energy policy framework and bioenergy contribution in the European Union – an overview from National Renewable Energy Action Plans and Progress Reports. Renew. Sustain. Energy Rev. **51**, 969–985 (2015)
5. Arantegui, R.L., Waldau, A.J.: Photovoltaics and wind status in the European Union after the Paris agreement. Renew. Sustain. Energy Rev. **81**(Part 2), 2460–2471 (2018)
6. Kiang, T.H., Koon, C.H., Peng, T.S., Rong, L.J., Jin, T.P.: Application of hybrid generator system in smart grid. Energy Procedia **143**, 686–692 (2017)
7. Dougal, R.A., Liu, S., White, R.E.: Power and life extension of battery–ultracapacitor hybrids. IEEE Trans. Compon. Packag. Technol. **25**(1), 120–131 (2002)
8. Ahmed, S., Zafran, M., Khan, F.M., Waqar, M.A., Hasan, Q.U.: Impact of integrating wind and solar energy on vulnerable power systems. In: International Multi-topic Conference (INMIC) (2017)
9. Shafiullah, G.: Impacts of renewable energy integration into the high voltage (HV) networks. In: 4th International Conference on the Development in the Renewable Energy Technology (ICDRET) (2016)
10. Worighi, I., Maach, A., Abdelhakim, H.: Smart grid architecture and impact analysis of a residential microgrid, pp. 1–6, October 2016
11. Sinha, S., Chandel, S.S.: Review of recent trends in optimization techniques for solar photovoltaic–wind based hybrid energy systems. Renew. Sustain. Energy Rev. **50**, 755–769 (2015)
12. Peng, C., Zou, J., Zhang, Z., Han, L., Liu, M.: An ultra-short-term pre-plan power curve based smoothing control approach for grid-connected wind-solar-battery hybrid power system. IFAC-PapersOnLine **50**(1), 7711–7716 (2017)
13. Budischak, C., Sewell, D., Thomson, H., Mach, L., Veron, D.E., Kempton, W.: Cost-minimized combinations of wind power, solar power and electrochemical storage, powering the grid up to 99.9% of the time. J. Power Sour. **225**, 60–74 (2013)

14. Khalid, M., Savkin, A.V., Agelidis, V.G.: Optimization of a power system consisting of wind and solar power plants and battery energy storage for optimal matching of supply and demand. In: IEEE Conference on Control Applications (CCA) (2015)
15. Mamen, A., Supatti, U.: A survey of hybrid energy storage systems applied for intermittent renewable energy systems. In: 14th International Conference on Electrical Engineering/Electronics (2017)
16. Zade, A.B., Gaikwad, A., Jeevane, K.P.M.: Hybrid solar and wind power generation with grid interconnection system for improving power quality. In: IEEE 1st International Conference on Power Electronics, Intelligent Control and Energy Systems (ICPEICES) (2016)
17. Cochran, J., Denholm, P., Speer, B., Miller, M.: Grid integration and the carrying capacity of the U.S. grid to incorporate variable renewable energy. National Renewable Energy Lab. (NREL), Golden, CO, United States (2015)
18. Ghenai, C., Janajreh, I.: Design of solar-biomass hybrid microgrid system in Sharjah. Energy Procedia **103**, 357–362 (2016)
19. Worighi, I., Maach, A., Abdelhakim, H.: Modeling a smart grid using objects interaction, pp. 1–6, October 2015
20. Dawoud, S.M., Lin, X., Okba, M.I.: Hybrid renewable microgrid optimization techniques: a review. Renew. Sustain. Energy Rev. **82**(Part 3), 2039–2052 (2018)
21. Khisa, S., Ebihara, R., Dei, T.: Dynamics of a grid connected hybrid wind-solar and battery system: case study in Naivasha-Kenya. Energy Procedia **138**, 680–685 (2017)
22. Wang, B., Zhang, B., Hao, Z.: Control of composite energy storage system in wind and PV hybrid microgrid. In: IEEE International Conference of IEEE Region 10 (TENCON 2013) (2013)
23. Louie, H.: Operational analysis of hybrid solar/wind microgrids using measured data. Energy Sustain. Dev. **31**, 108–117 (2016)
24. Cahn, A., Hoyos, J., Hulse, M., Keller, E.: Software-defined energy communication networks: from substation automation to future smart grids. In: 2013 IEEE International Conference on Smart Grid Communications (SmartGridComm), Vancouver, BC, pp. 558–563 (2013)
25. Jayachandran, M., Ravi, G.: Design and optimization of hybrid micro-grid system. Energy Procedia **117**, 95–103 (2017)
26. Azaza, M., Wallin, F.: Multi objective particle swarm optimization of hybrid micro-grid system: a case study in Sweden. Energy **123**, 108–118 (2017)
27. Ambia, M.N., Durra, A.A., Caruana, C., Muyeen, S.M.: Power management of hybrid micro-grid system by a generic centralized supervisory control scheme. Sustain. Energy Technol. Assess. **8**, 57–65 (2014)
28. Amrollahi, M.H., Bathaee, S.M.T.: Techno-economic optimization of hybrid photovoltaic/wind generation together with energy storage system in a stand-alone micro-grid subjected to demand response. Appl. Energy **202**, 66–77 (2017)
29. Cen, Z., Kubiak, P., López, C.M., Belharouak, I.: Demonstration study of hybrid solar power generation/storage micro-grid system under Qatar climate conditions. Solar Energy Mater. Solar Cells **180**, 280–288 (2018)
30. Liu, Y., Zuo, K., Liu, X.A., Liu, J., Kennedy, J.M.: Dynamic pricing for decentralized energy trading in micro-grids. Appl. Energy **228**, 689–699 (2018)
31. Cau, G., Cocco, D., Petrollese, M.: Modeling and simulation of an isolated hybrid micro-grid with hydrogen production and storage. Energy Procedia **45**, 12–21 (2014)

32. Chen, Y., Wen, M., Yin, X., Cai, Y., Zheng, J.: Distance protection for transmission lines of DFIG-based wind power integration system. Int. J. Electr. Power Energy Syst. **100**, 438–448 (2018)
33. Ezzahi, M., Khafallah, M., Majid, F.: Structural integrity of a doubly fed induction generator (DFIG) of a wind power system (WPS). Procedia Struct. Integr. **9**, 221–228 (2018)
34. Wang, Y., Huang, Y., Wang, Y., Zeng, M., Li, F., Wang, Y., Zhang, Y.: Energy management of smart micro-grid with response loads and distributed generation considering demand response. J. Clean. Prod. **197**(Part 1), 1069–1083 (2018)
35. Worighi, I., Maach, A.: Virtualization of the smart grid using entity/relation model. In: International Conference on Advanced Communication Technologies and Networking (CommNet), Marrakech, pp. 1–9 (2018)
36. IEEE Power and Energy Society, IEEE PES Test Feeders. http://ewh.ieee.org/soc/pes/dsacom/testfeeders.html
37. Marion, W., Urban, K.: Users Manual for TMY2s – Typical Meteorological Years Derived from the 1961-1990 National SolarRadiation Data Base, National Renewable Energy Lab, Golden, CO, NREL/TP-463-7668 (1995)
38. Aryanezhad, M.: Management and coordination of LTC, SVR, shunt capacitor and energy storage with high PV penetration in power distribution system for voltage regulation and power loss minimization. Int. J. Electr. Power Energy Syst. **100**, 178–192 (2018)
39. Chaudhary, P., Rizwan, M.: Voltage regulation mitigation techniques in distribution system with high PV penetration: a review. Renew. Sustain. Energy Rev. **82**(Part 3), 3279–3287 (2018)

Effect of Erbium Addition on Optical and Electrical Properties of Polytetrafluoroethylene

Abdelghafour El Moutarajji, Bouazza Tbib, and Khalil El-Hami(✉)

Laboratory of Nanosciences and Modeling, University of Hassan 1st FPK, Khouribga, Morocco
khalil.elhami@uhp.ac.ma

Abstract. Our aim focused on semiconductor materials based on polytetrafluoroethylene (PTFE) doped by various concentrations of Erbium element (Er). The materials have been characterized by X-ray diffraction, optical properties of materials have been studied by UV-visible spectroscopy and spectroscopy of Fourier transform infrared (FT-IR). Our results showed a change of direct cell parameters of samples and information on micro strain of samples. It has been found that the increase in the percentage absorbance of Erbium grafted on PTFE that resulted in the increasing the absorption coefficient α, electrical conductivity and gap energy.

Keywords: Semiconductor · X-ray diffraction · FTIR · Optical properties · Microstrain · Electrical conductivity and gap energy

1 Introduction

Conductive materials, including lanthanides are extremely dynamic because of their interesting properties; for example: Short band gap, high absorption coefficients, bonding energy, chemical stability and environmentally friendly applications, the other hand insulators such as Teflon represent a very high electrical resistivity (transverse volume resistivity > 1018 Ω.cm) and thermal (maximum operating temperature peak 300 °C and thermal conductivity 23° equal 0.23 W/(Km). Teflon also represents a variety of applications namely Bearings, pads, soles Electrical Insulating Joints, anti-corrosion, the protection of cells solar panels [1]. Our aim is to make the material a semiconductor for use both for the protection of solar panel cells and the manufacture of photovoltaic cells (solar cells based on organic semiconductors) which may be flexible and easily manufactured by printing and coating techniques [2]. Organic semiconductors have the advantage of chemical adaptation; they also have flexibility of their properties. An example of this type of solar cells is based on mixtures of poly {[2-methoxy-5 - [(3,7-dimethyloctyl) oxy] phenylene] vinylene} (MDMO-PPV) and 1- [3-(methoxycarbonyl) propyl] -1-phenyl- (6,6) -C61 (PCBM) that has been recently manufactured and extensively studied with power conversion efficiencies up to 3% [3]. In this work, polytetrafluoroethylene (PTFE) is doped by a transition element such as Erbium (Er) which is part of the lanthanides. In the following paragraph we will detail

M. Ezziyyani (Ed.): AI2SD 2018, AISC 912, pp. 97–110, 2019.
https://doi.org/10.1007/978-3-030-12065-8_10

the method of development and the characterization technique of our product to determine the optical, electronic and structural parameters.

2 Experimental Method of Elaboration

The Teflon based semiconductor synthesis was prepared by solid state with various percent of Erbium is shown in Table 1.

Table 1. Represent teflon and teflon doped on function of various percent of Erbium (Er).

	Mass (g) of	
	PTFE on (g)	Erbium on (g)
PTFE	1	0
PTFE+Er (1%)	0,99	0,01
PTFE+Er (2%)	0,98	0,02
PTFE+Er (3%)	0,97	0,03
PTFE+Er (4%)	0,96	0,04
PTFE+Er (5%)	0,95	0,05

Samples are placed in alumina burrows after 1 mm grain size burring, after which they are placed in at 100 °C for 45 min and we have increase the temperature oven 500 °C, it's the transition of temperature of Teflon and Teflon doped by Erbium.

3 Results and Discussions

3.1 Characterization by UV-Visible

Our work that we have chosen concerning the electrical and optical characterizations of Teflon and Teflon doped by Erbium (Er), this study consist of effect Erbium (Er) on the electrical and optical properties of Teflon. The samples were characterized by UV-Visible.

3.1.1 Transmittance

The Fig. 1 shows the transmission as a function of the wavelength (λ), we see five spectra with different transmission, the latter has started from the ultra violet to the infrared domain so the minimal transmission of the spectra or the common point between they took the value 5% and 250 cm^{-1}, at this point the polymer transmitted all intensities even though it doped with Erbium besides the 1% doped teflon transmission value is 65% in the visible then 60% in the ultra violet, the 2% doped Teflon transmission value is 40% in the visible and 35% in the ultra violet as well as the transmission value the Teflon doped 4% is 15% in the visible and 10% in the ultra violet, as well as the Teflon transmission value doped with 5% of Erbium is 20% in the visible range and 12% in the ultra violet field. Indeed, when the percentage of Erbium is important, the transmission decreases in the visible range and in the ultra-violet field, we also see an accelerated increase in transmission in the infrared field of around 825 cm^{-1}. This increase will continue up to 96% since teflon is much more transparent in the infrared range than glass. These differences in Erbium composition give rise to systematic changes in optical properties such as refractive index and absorption coefficient not to mention the interesting fact to note is the rapid growth of absorption towards the violet. This absorption must be even greater in the ultraviolet; Erbium also has a high transmittance in the infrared. Moreover, we see bands of the same wave-length in the transmission spectra, the width and the height of the bands are varied during doping with Erbium. We take the case of teflon doped with 5% Erbium we find three bands, one in the field of ultra violet to 375 cm^{-1} and the others in the visible range of 525 cm^{-1} and 650 cm^{-1} corresponding transitions in the doped polymer, they are induced the values of the optical gap energy (E_{gap}^{Opt}).

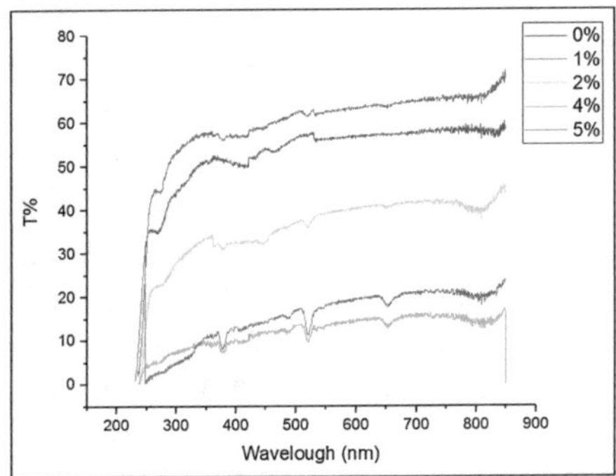

Fig. 1. The spectra of transmission for Teflon $(C_2F_4)_n$ doped by Erbium (Er)

3.1.2 Absorbance

The Fig. 2, represent the absorption spectrum as a function of the wavelength (λ), this spectrum only shows the electronic absorption, and it is measured between 200 and 850 nm in using Teflon doped with Erbium (Er) in the solid state. The ambient temperature we notice that the five compounds have a strong absorption in the region of ultra violet and against the compounds absorbs only weakly between 400 and 850 nm with different values of the absorbance of more when increasing the concentration of Erbium (Er) in teflon the absorbance decreases, we have noted that teflon is a highly coordinating polymer and may be capable of complexing on Erbium (Er^{+5}) ions and when Erbium (Er) is depleted in electrons, to promote complex formation teflon-Erbium (Er) although the uptake hardly increases anymore which indicates that the modified Teflon surface, It should be noted that no band is observed in the spectrum of Teflon.

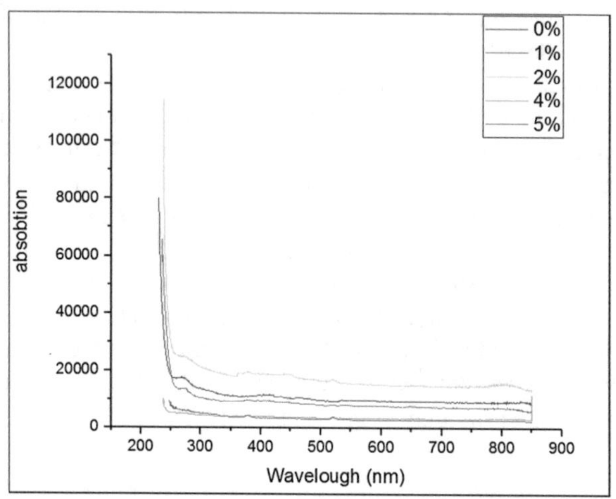

Fig. 2. The absorbance of Teflon $(C_2F_4)_n$ doped by Erbium (Er)

3.1.3 Determination of Gap Energy

By experiment, to determine the gap energy, Teflon doped by deferential percentage of Erbium (Er), one can plot $(\alpha h\nu)^2$ as a function of the photonic energy ($h\nu$) based on the values of the optical transmission as shown in Fig. 3. We notice a straight line in a certain region. In order to obtain the value of the optical gap energy (Eg^{opt}) of Teflon doped with erbium (Er) by increasing the percentage up to 5% as shown in Table 1, the tangent with each curve can be traced to the intersection with the abscissa axis. The estimated values of the gap energy were written in Table 1. It was observed that the optical gap energy (Eg^{opt}) decreased with the increase in the percentage of Erbium (Er) in the polymer. This result showed that erbium (Er) has a significant effect on the crystal lattice, a very high electrostatic interaction. This observed trend can also be correlated with the atom distribution defects due to the average atomization energy which is increased with the addition of erbium moreover it absorbed by Teflon, so the gap band is due to the coherent diffusion of the electrons by an effective periodic potential which depends on the state in the ground state of the system, Table 1 illustrates the variation of the optical band gap energy (Eg) with respect to the Erbium (Er) compositions in Teflon. The relation obtained is not linear; the non-linearity of the band gap variation with the compositions has already been reported for many published works. This result confirmed that this polymer can be used as semiconductors and photovoltaic and optoelectronic devices. When certain additives are added in high concentration, a new alloy or compound is formed and a new band gap is observed. In conclusion, the gap energy is an energy gap for the electrons in the crystal lattice that requires some kind of energy for the conduction to be done by an electric field in the semiconductors.

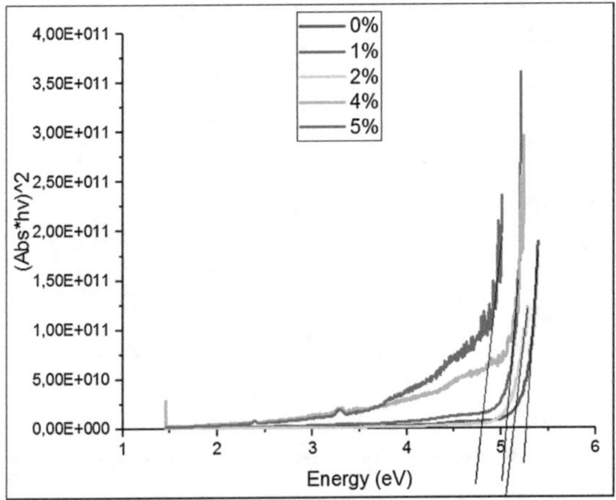

Fig. 3. The variation of the dependence $(\alpha h\nu)^2$ as a function of photon energy (E_{ph})

Table 2. Represent the gap energy (E_g) of Teflon and Teflon doped by Erbium (Er)

Percentage	Gap energy on (eV)
Téflon+(0%) d'Erbium	5,25
Téflon+(1%) d'Erbium	5,15
Téflon+(4%) d'Erbium	5,05
Téflon+(5%) d'Erbium	4,75

3.1.4 Determination of the Energy of Urbach

The Fig. 4 shows the evolution of gap energy (E_g) and that of Urbach (E_u) measured on our samples according to the percentage used during the doping. The values of energy (E_g) were determined from the formula of Tauc [4] and those of Urbach (E_u) using the Urbach's law [5]. We note that the two quantities E_g and E_u evolve in strictly opposite directions. The Fig. 4 shows the next description: When one increases the other decreases, and when the first one reaches the second reaches its minimum. This clearly confirms the very close correlation which exists between these two quantities and which is logically due to the presence of a more or less significant density of dangling bonds in the material. For low percentage values, there is a large, steady and rapid increase in the Urbach energy (E_u) as shown in Table 2.

The Fig. 5 shows the Evolution of the optical gap (E_g) and the Urbach energy (E_u) measured on our samples according to the percentage used during the doping. The values of E_g were determined from the formula of Tauc [4] and those of E_u using Urbach's law [5]. We note that the two quantities E_g and E_u evolve in strictly opposite directions (Table 3).

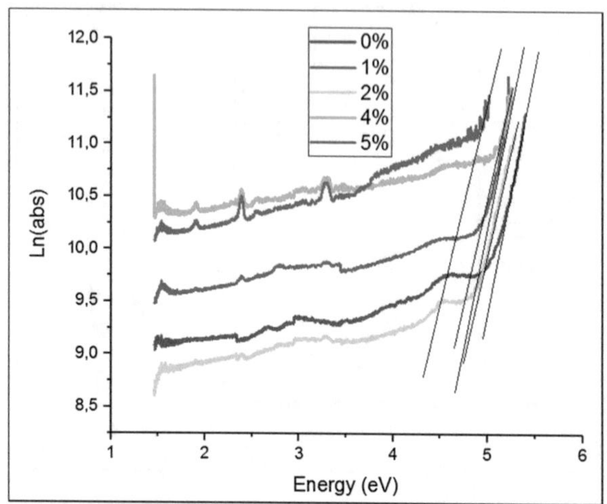

Fig. 4. The variation of dependence of absorption as a function of photon energy (E_{ph})

Table 3. Represent the urbach energies and teflon doped with erbium (Er)

Percentage	Urbach energy (eV)
Téflon+(0%) d'Erbium	0,233
Téflon+(1%) d'Erbium	0,24
Téflon+(3%) d'Erbium	0,25
Téflon+(4%) d'Erbium	0,355
Téflon+(5%) d'Erbium	0,58

4 Characterization by X-Ray Diffraction Analysis

Teflon-based compounds have been characterized by ADVANCE D8 X-Ray diffraction, we have noticed a more intense peak in the crystallized phase, this peak represents the probability of finding the phase, moreover the purity of the material as well. Other peaks appeared. The effect of the Erbium element (Er) on the studied phase gave information on the displacement of the peaks as well as the intensity, it was deduced that the erbium reacted with the Teflon. Figure 6 shows the X-ray diffraction spectra of Teflon and doped Teflon. All the reflections peaks were indexed in orthorhombic system Space group (P m m m). The parameters generated by Fullprof program as shown in Table 4. The effect of Erbium (Er) on the Tetrafluoroethylene $(C_2F_4)_n$, such as the grains size, displacement of position and intensity, have been studied in the percentage range 1% to 5% is shown in the Fig. 6.

The Fig. 8 shows the variation of cell parameters of tetrafluoroethylene doped by Erbium, according to the percentage. We observe an increase of the parameters (a, b and c) between 0 and 1%, then a decrease of the parameters (b, c) between 1% and 4% for returns in increase but the parameter (a) continue to increase until 2% then we

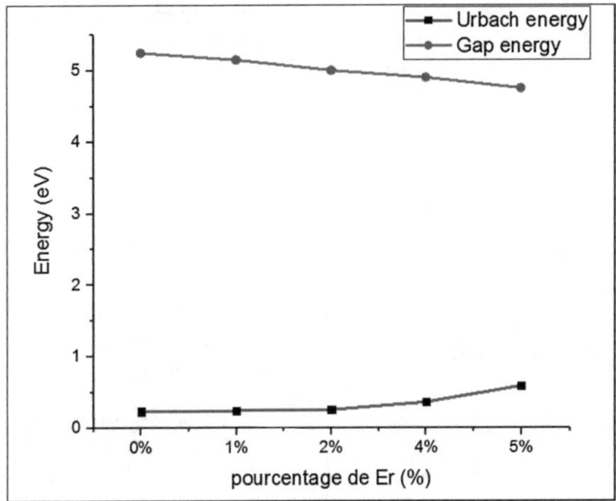

Fig. 5. The evolution of the optical gap (E_g) and the Urbach energy (E_u) according to the percentage.

observe a high an increase between 3% and 5%. These variations can give some explanation of the crystalline structure of the samples formed. The decrease of parameterizes b and c can be connected the grain size of Teflon doped, the increase or the decrease of the parameters of the cell can be bound to the elasticity of the network trained by groupings bound only by summits (Fig. 7).

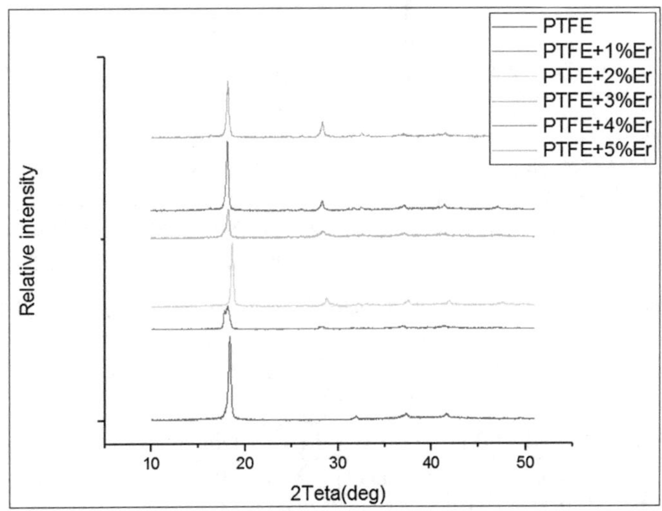

Fig. 6. X-ray diffractograms of PTFE and different percentage of Erbium (Er) doped in the Teflon material

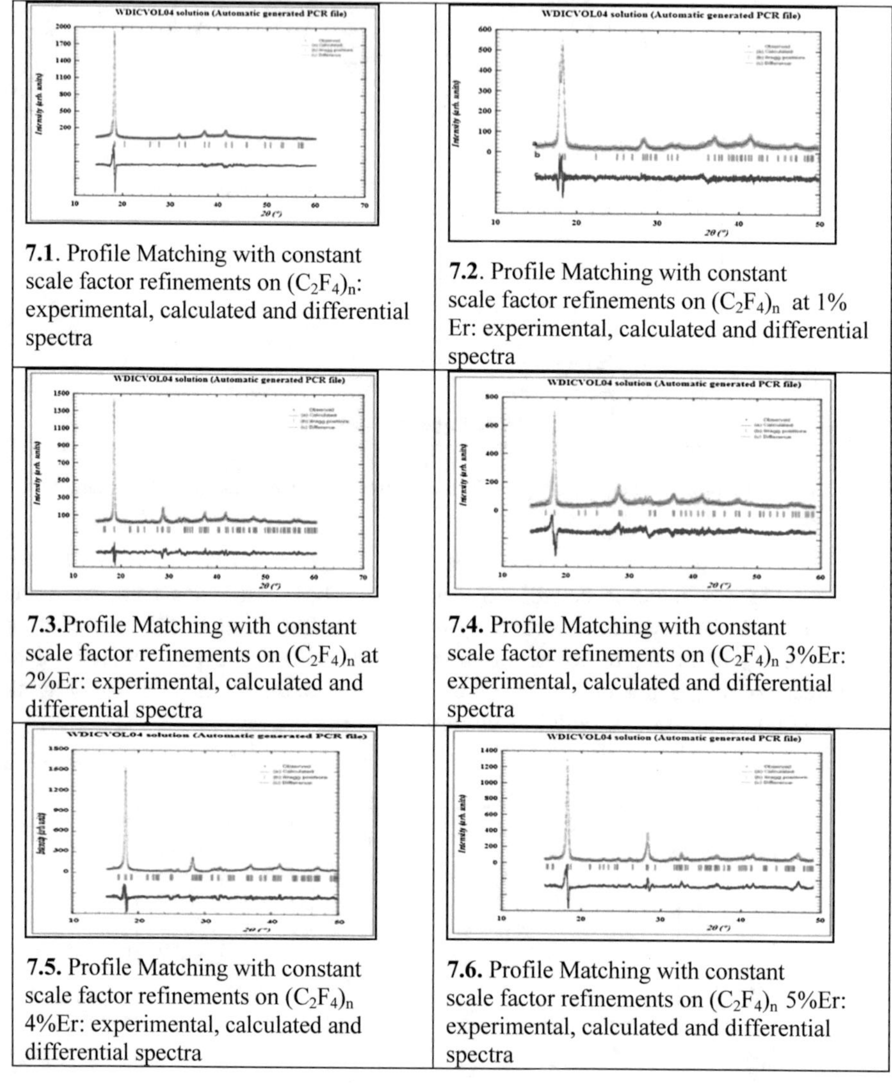

7.1. Profile Matching with constant scale factor refinements on $(C_2F_4)_n$: experimental, calculated and differential spectra

7.2. Profile Matching with constant scale factor refinements on $(C_2F_4)_n$ at 1% Er: experimental, calculated and differential spectra

7.3. Profile Matching with constant scale factor refinements on $(C_2F_4)_n$ at 2%Er: experimental, calculated and differential spectra

7.4. Profile Matching with constant scale factor refinements on $(C_2F_4)_n$ 3%Er: experimental, calculated and differential spectra

7.5. Profile Matching with constant scale factor refinements on $(C_2F_4)_n$ 4%Er: experimental, calculated and differential spectra

7.6. Profile Matching with constant scale factor refinements on $(C_2F_4)_n$ 5%Er: experimental, calculated and differential spectra

Fig. 7. The figures of X-ray diffractograms of various temperatures of Teflon and Teflon doped

Table 4. Cell parameters extracted from full pattern refinements of various percentage of erbium on Teflon and the Teflon doped

Temperature	A [Å]	B [Å]	C [Å]	V [Å^3]	Space group	Crystal system
PTFE	4.8326	4.3457	3.4599	72.6615	P m m m	Orthorhombic
PTFE+1%Er	7.1251	6.9111	6.6416	327.0475	P m m m	Orthorhombic
PTFE+2%Er	10.7686	6.2683	5.3154	358.7940	P m m m	Orthorhombic
PTFE+3%Er	6.2885	5.2723	4.8830	161.8952	P m m m	Orthorhombic
PTFE+4%Er	15.6326	4.9031	3.9878	305.6577	P m m m	Orthorhombic
PTFE+5%Er	18.9923	5.6523	5.6064	601.8480	P m m m	Orthorhombic

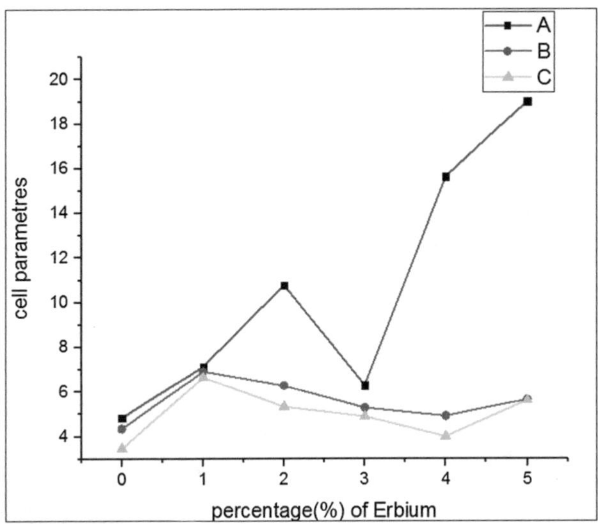

Fig. 8. The variation of cell parameters of tetrafluoroethylene and tetrafluoroethylene doped by Erbium, according to the percentage.

4.1 Determination of Micro Structural Parameters

The mechanism of X-rays Diffraction in the crystalline materials is that the X-rays scatter from crystals because their electric fields interact with the clouds of electrons of atoms in crystals. The X-rays scattered from the periodic neighboring atoms interfere and give rise to a plan of diffraction. The plan of diffraction is modulated by the function of transfer of the detector which in his turn modifies the shape of the profile of diffraction of the X-rays. So, a profile of line of diffraction is the result of the convolution of a number of independent variables contributing to the form profile to know the instrumental variables and the micro structural effects. The instrumental variable includes the width of crack of reception, the transparency of the sample, the nature of the source of X-rays, the axial difference of the incidental beam and the flat sample geometry [6]. The micro structural effects responsible for the profile of shape of the peaks of diffraction are the following ones. Size finished crystals or domains and micro strain the crystal contains defects of network. These profiles are equipped with functions of shape of adapted profile in a way that the functions asymmetric peaks and it should be so simple mathematically as possible to do the calculation of all the by products have variables.

4.2 Crystallite Size and Microstrain with William-Hall

From the X-rays in the polycrystalline material is due to the presence of crystallites (effect of size) as well as the micro strain effect. The slopes of the plan W-H represent the internal average microstructure. While the opposite of the orderly at the origin of the axis $(\beta^*\cos(\theta)/\lambda)^2$ give the size of crystallites according to the relation. While the opposite of the orderly at the origin of the axis gives the size of crystallites according to the relation.

$$\left(\frac{\beta\cos(\theta)}{\lambda}\right)^2 = \frac{1}{D^2} + \left(\frac{4\varepsilon\sin(\theta)}{\lambda}\right)^2 \tag{1}$$

Where ß = instrumental corrected extension (expressed in radians) θ = angle of diffraction of Bragg, D = crystallite seize (Å), ε = micro strain and λ = wavelength (Å) [7] (Fig. 9).

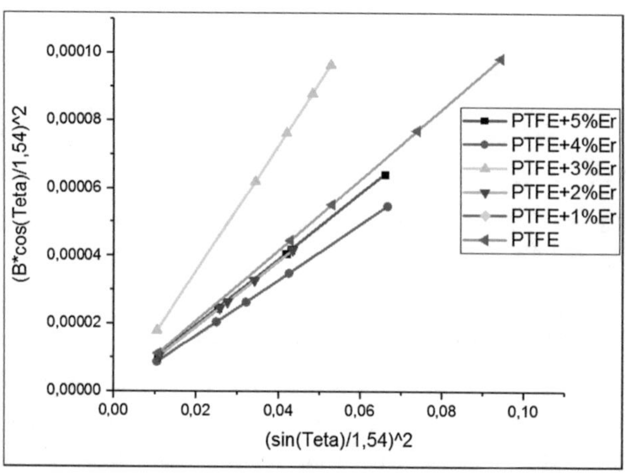

Fig. 9. Williamson–Hall plots of $(\beta*\cos\theta/\lambda)^2$ and $(\sin\theta/\lambda)^2$ of PTFE alloys for different deformation percentage of Er.

Table 5. Represent the grain size, micro-strain and volume of Teflon and Teflon doped

Percentage	Grain seize	Microstrain	Volume
PTFE+(0%) d'Erbium	2,1138	0,50369	72,6615
PTFE+(1%) d'Erbium	4,1348	0,47546	327,0475
PTFE+(4%) d'Erbium	4,86325	0,57477	358,794
PTFE+(3%) d'Erbium	3,69735	0,54984	161,8952
PTFE+(4%) d'Erbium	5,32972	0,485	305,6577
PTFE+(5%) d'Erbium	2,45907	0,50636	305,6577

The Fig. 10 shows the variation of the grain size, micro-strain and volume of tetrafluoroethylene and tetrafluoroethylene doped by Erbium, according to the percentage. We note for each graph four parts, two parties increases and the others decreases. "parte 1 [0:2%]; parte 2 [2%:3%]; parte 3 [3%:4%]; parte 4 [4%:5%]". We observe that when the volume and the size of the grains increase the micro strain decrease also a high

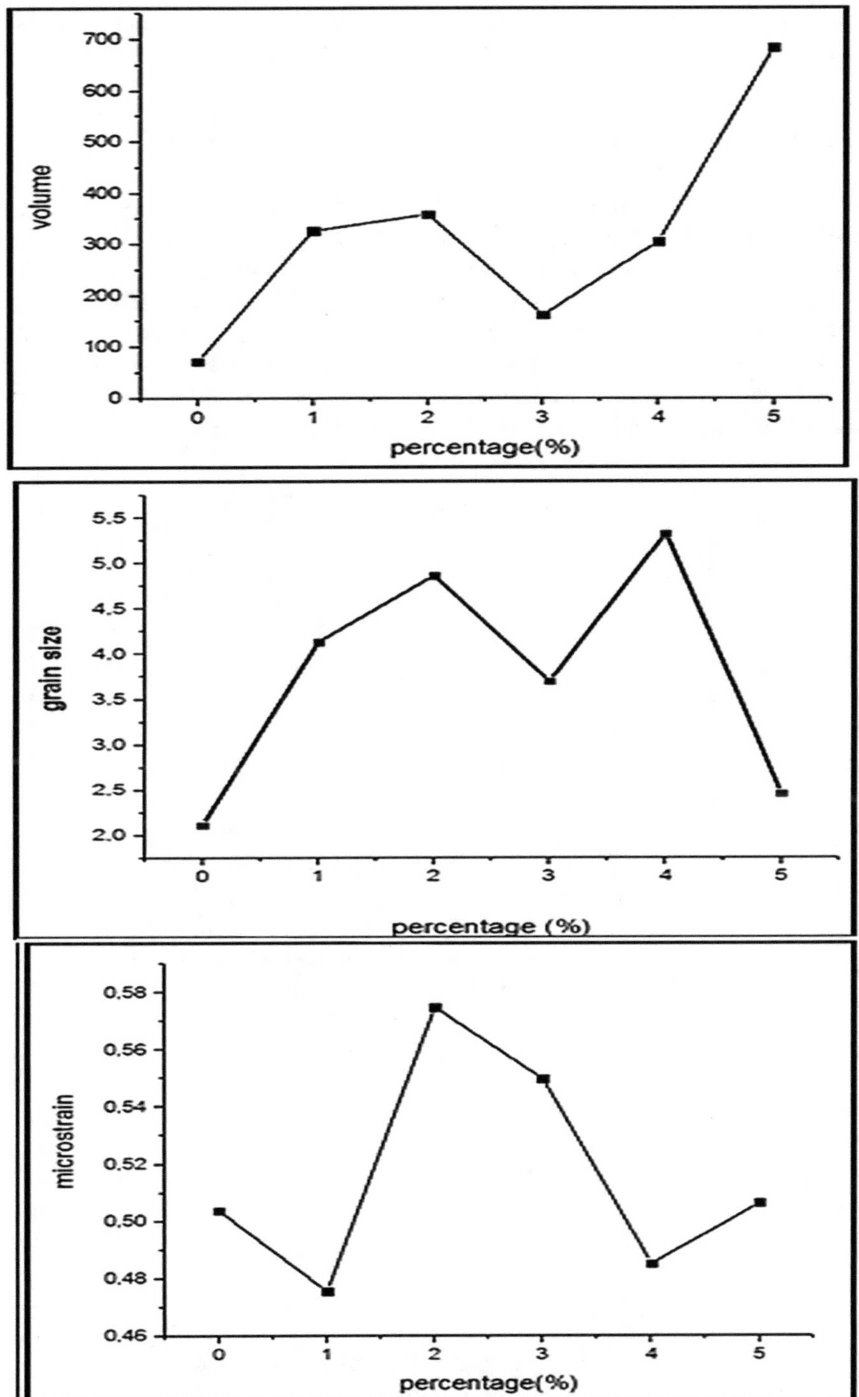

Fig. 10. Variation of the grain size, micro-strain and volume of tetrafluoroethylene and tetrafluoroethylene doped by Erbium, according to the percentage

increase of the volume in the parte 4. So we can conclude that the materials are elastic because the properties of the grain form are sensitive to micro strain.

5 Fourier Transforms Infrared Spectroscopy (FTIR)

A Fig. 11 shows the FTIR spectra of the pure Teflon and the Teflon doped by erbium. As Shown in Figures relatively strong twin peaks at 1200 cm^{-1} and 1125 cm^{-1} corresponding to CF2 asymmetrical stretching and symmetrical stretching modes, respectively, appeared in the absorption spectra of the Teflon doped. Additionally, the weak peak of CF2 wagging mode appeared at 1000 cm^{-1}. However, the weak peak of CF3 stretching mode appeared at 981 cm^{-1}, so the very weak bands present must be due to impurities or surface contamination, it is the effect the Erbium element, in all spectra of the Teflon and Teflon doped. As summarized in Tables 5, 6 and 7).

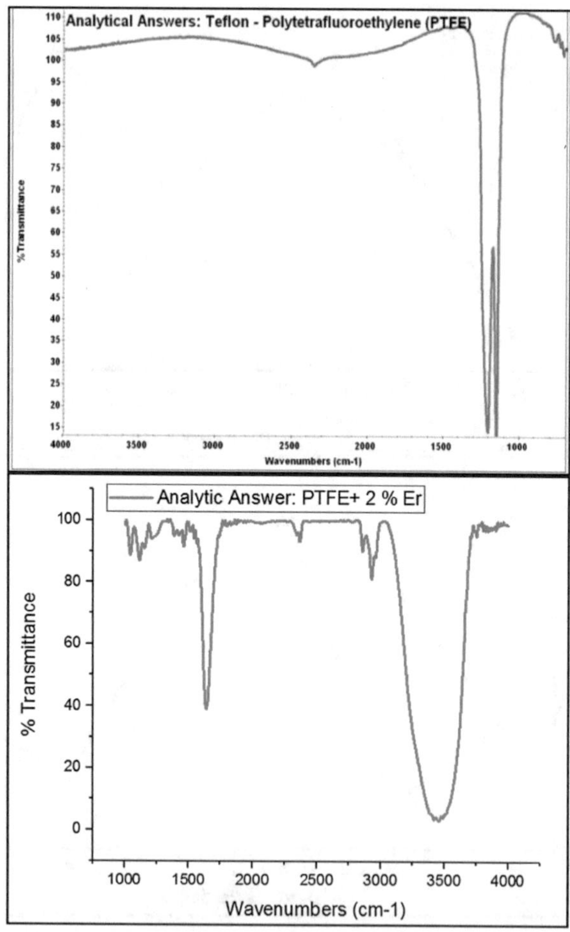

Fig. 11. FT-IR spectra of the Polytetrafluoroethylene and Teflon doped by Erbium (2%)

Table 6. Represent the positions of the spectra with the form and transmittance of the polytetrafluoroethylene

Polytetrafluoroethylene		
Wavenumber (cm^{-1})	Transmittance (%)	Spectral form
1125	13	Delicate
1200	14	Delicate

Table 7. Represent the positions of the spectra with the form and transmittance of the polytetrafluoroethylene doped by erbium (2%).

Polytetrafluoroethylene+2% Erbium		
Wavenumber (cm^{-1})	Transmittance (%)	Spectral form
1043,6616	84,70	Delicate
1111,4911	83,85	Delicate
1457,0223	88,47	Delicate
1639,76	35,43	Delicate
2364,34	89,89	Delicate
2926,12	77,50	Delicate
3460,78	0,25	Wide

6 Conclusion

Teflon and Teflon doped have been studied to measure optical and electrical magnitudes. Various concentrations of Erbium (Er) have given a change in these properties such as gap energy, urbach energy and micro strain. Comparison of the absorbance and the adjusted Teflon Urbach absorption edge and doped Teflon suggests that the extrinsic absorbers may represent a slight increase in absorbance over what might be the intrinsic absorbance of the Teflon polymer. The X-ray diffraction line broadening analysis has been used for measuring the parameters of microstructure of Teflon and Teflon doped. FT-IR spectroscopy is a powerful tool for polymer analysis, and a range of sampling methods are available with varying degrees of sample compatibility and time requirements.

References

1. Brabec, C.: Sol. Energy Mater. Sol. Cells **83**, 273 (2004)
2. Adamopoulos, G.: Electronic transport properties aspects and structure of polymer-fullerene based organic semiconductors for photovoltaic devices. Thin Solid Films **511–512**, 371–376 (2006)
3. Ouali, L., Krasnikov, V.V., Stalmach, U., Hadziioannou, G.: Adv. Mater. **11**(18), 1515 (1999)

4. Tauc, J. (ed.): Amorphous and Liquid Semiconductors. Plenum Press, London and New York (1974)
5. Demichelis, F., Kanidakis, G., Tagliferro, A., Tresso, E.: Appl. Opt. **9**(26), 1737 (1987)
6. Akl, A.A., Hassanien, A.S.: Microstructure characterization of Al-Mg alloys by X-ray diffraction line profile analysis. Int. J. Adv. Res. **2**(11), 1–9 (2014)
7. Rama Rao, P., Anantharaman, T.R.: Z. Metallk **54**, 658 (1963)

Ant Colony Optimization for Cryptanalysis of Simplified-DES

Hicham Grari(✉), Ahmed Azouaoui, and Khalid Zine-Dine

LAROSERI Laboratory, FS, Chouaib Doukkali University, El Jadida, Morocco
grari.hicham@gmail.com,
{azouaoui.a, zinedine}@ucd.ac.ma

Abstract. Ant Colony Optimization is a search meta-heuristic inspired by the foraging behavior of real ant, having a very wide applicability. Especially, it can be applied to different combinatorial optimization problem. In this paper, we present a novel Ant Colony Optimization (ACO) based attack for cryptanalysis of Simplified Data Standard Encryption (S-DES). A known Plaintext attack is used to recover the secret key requiring only two Plaintext-Ciphertext pairs. Moreover, our approach allows us to break S-DES encryption system in a minimum search space when compared with other techniques. Experimental results prove that ACO can be considered as a convincing tool to attack the key used in S-DES.

Keywords: Cryptanalysis · ACO · S-DES · Pheromone

1 Introduction

With the increase of information traffic on communications networks, information security has become a prime concerns. One of the reliable used approaches for information security is cryptology, which has become a scientific discipline dealing with Confidentiality, Integrity, and Authentication.

Cryptology is divided into two complementary fields. They are cryptography and cryptanalysis. Cryptography is the science of building new powerful and efficient cryptosystem. Cryptanalysis is the science and study of method to break cryptographic techniques i.e. ciphers. It is used to find loopholes in the design of ciphers.

Actually, the use of optimisation algorithm in cryptology has become an active research area. It may appear an efficiency way to break complex ciphers. Ant Colony Optimization (ACO) [1] is a well-known meta-heuristics that were successfully used to produce approximate solutions for a large variety of optimization problems. Ant colony Optimization are used to attack DES (Data Encryption System) by Khan, Armughan and Durrani [5]. Also, Grari, Azouaoui, Zine-Dine [6] proposed Cryptanalysis of Simple Substitution Ciphers Using ACO. In this paper we investigate the use of ACO in automated cryptanalysis of Simplified Data Encryption Standard S-DES. We will show that our approach is significantly faster and requires a smaller number of plaintext-ciphertext pairs, when compared to others attacks.

In the previous work, the simplified Data Encryption standard was attacked by Nalini, Raghavendra Rao [7] using different Optimisation heuristics (Genetic

M. Ezziyyani (Ed.): AI2SD 2018, AISC 912, pp. 111–121, 2019.
https://doi.org/10.1007/978-3-030-12065-8_11

algorithm, simulated annealing and tabu search), they demonstrated that optimisation heuristic are ideally suited to break SDES. A cryptanalysis of S-DES is done by Sharma, Pathak, and Sharma [8] using Genetic Algorithm with Ring Crossover and others Operators. Also Garg, Varshney and Bhardwaj [9] are using GA to break S-DES. A ciphertext only attack is used by Al Adwan, Al Shraideh and Rasol [10], to prove the success of GA over brute in cryptanalysis of S-DES. In [11] a novel approach called Genetic Swarm Optimization (GSO) has proved by combining the effectiveness of GA and Particle Swarm Optimization (PSO) to attack S-DES. The results showed that GSO reduces the key search space by the factor of 5.6. Garg [12] attacked S-DES via Evolutionary Computation techniques with comparison between Memetic and genetic algorithms.

The remainder of this paper is organized as follows. In the next section, we introduce Simplified Data Encryption Standard. In Sect. 3, we present the basic and background of Ant colony optimization meta-heuristic. The fully automated attack is given in Sect. 4, with experimental results in Sect. 5. Finally, conclusions are given in Sect. 6.

2 Simplified Data Encryption Standard (S-DES)

The SDES encryption algorithm takes an *8-bit* block of plaintext and a *10-bit* key as input and produces an *8-bit* block of ciphertext as output. The decryption algorithm takes an *8-bit* block of ciphertext and the same *10-bit* key used as input to produce the original 8-bit block of plaintext. The encryption algorithm involves five functions; an initial permutation (***IP***), a complex function called $\boldsymbol{f_K}$ which involves both permutation and substitution operations and depends on a key input; a simple permutation function that switches (***SW***) the two halves of the data; the function f_K again, and a permutation function that is the inverse of the initial permutation ($\boldsymbol{IP^{-1}}$) as shown in Fig. 1.

2.1 S-DES Description

Encryption involves the application of five functions:

(1) *S-DES Key Generation*

Two *8-bit* sub keys are generated from the main Key to be used in the f_K function. Moreover, the key is first subjected to a permutation $P_{10} = [3\ 5\ 2\ 7\ 4\ 10\ 1\ 9\ 8\ 6]$, then a shift operation is performed. The output of the shift operation then passes through a permutation function that produces an *8-bit* output $P_8 = [6\ 3\ 7\ 4\ 8\ 5\ 10\ 9]$ for the first sub key (K_1). The output of the shift operation also feeds into another shift and another instance of P_8 to produce the second sub key (K_2).

(2) *Initial and Final Permutations*

The input to the algorithm is an *8-bit* block of plaintext, which we first permute using the *IP* function $IP = [2\ 6\ 3\ 1\ 4\ 8\ 5\ 7]$. This retains all 8-bits of the plaintext but mixes them up. At the end of the algorithm, the inverse permutation is applied. The inverse permutation is done by applying $IP^{-1} = [4\ 1\ 3\ 5\ 7\ 2\ 8\ 6]$ where we have $IP^{-1}(IP(X)) = X$.

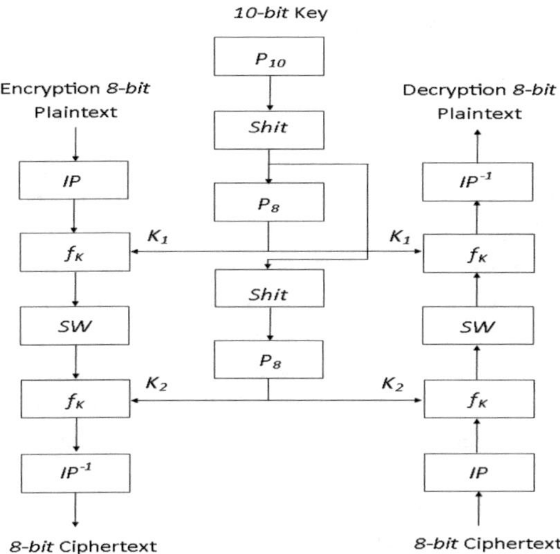

Fig. 1. S-DES encryption and decryption

(3) *The Function fK*

The function f_K, which is the complex component of S-DES, consists of a combination of permutation and substitution functions. The functions are given as follows:

Let *L*, *R* be the left *4-bits* and right *4-bits* of the *8-bits* input to f_K.

Then, $f_K\ (L,\ R) = (L \oplus f(R,\ key),\ R)$

Where ⊕ is the exclusive-OR operation and key is a sub-key. Computation of *f(R, key)* is done as follows:

(a) *Apply expansion/permutation E/P = [4 1 2 3 2 3 4 1] to input 4-bits.*
(b) *Add the 8-bit key (XOR).*
(c) *Pass the left 4-bits through S-Box S0 and the right 4-bits through S-Box S1.*
(d) *Apply permutation* $P_4 = [2\ 4\ 3\ 1]$.

The first and fourth input bits are treated as *2-bit* numbers that specify a row of the *S-box* and the second and third input bits specify a column of the *S-box*. The entry in that row and column in base 2 is the *2-bit* output.

(4) *The Switch Function*

The Switch Function (*SW*) interchanges the left and right *4-bits* of the input, so that the second instance of f_K operates on a different *4-bit*. In this second instance, the *E/P, S0, S1*, and *P4* functions are the same. The key input is *K2*.

3 Ant Colony Optimization ACO

Ant colony Optimization is class of population-based metaheuristic inspired by the foraging behavior of real ants, in particular, their ability to find shortest paths between food sources and their nest using pheromone communication. This simple idea is implemented by the ACO methods to resolve and address hard combinatorial problems such as traveling salesman problems, quadratic assignment problems, vehicle routing problems, or constraint satisfaction problems. The ACO metaheuristic is characterized as being a distributed and stochastic search method based on the indirect communication of a colony of (artificial) ants, mediated by (artificial) pheromone trails. The pheromone trails in ACO serve as distributed numerical information used by the ants to probabilistically construct solutions to the problem under consideration. The ants modify the pheromone trails during the algorithm's execution to reflect their search experience.

The first ACO algorithm, called Ant System (AS) introduced by Dorigo, Maniezzo, Colorni [2] was applied to Travelling Salesmen Problem. Other ACO variants mostly differ in the rule used for the solution construction and the pheromone update, including Ant Colony System (ACS) presented by Dorigo, Gambardella [3], and Min Max Ant System (MMAS) given by Stutzle and Hoos [4].

4 Proposed Approach

Generally, the main procedure to apply ACO to a combinatorial optimization problem works as follows: at each cycle, every ant build a tour (feasible solution) and then pheromone trails are modified based on global updating rule. The algorithm stops iterating when a termination condition is met. To respect this scheme in our proposed algorithm, several components depending on the characteristics of our problem must be defined. In the following, we will describe the details about constructing a solution, and then explain how to define the heuristic information and the fitness function. Finally, we will present the updating pheromone rule.

4.1 Solution Construction

In our approach the search space is modeled as two layers of 10 nodes (length of secret key used by S-DES). The nodes on the top layer are equal to '1' and the bottom layer nodes equal to '0' with a start node N0, and a final node N11.

The search space is a grid of two rows and 10 columns. Every node in a column is connected to the two nodes in the next column (Fig. 2).

An ant starts it tour from the node N_0, moving from left to right; its tour is finished at the last node N_{11}. In each column, an ant can only select a single vertex. At the end, when the tour is completed, it will consist of *10-bit* binary string. This binary string is a candidate or guessed key that will be applied to the original ciphertext and a candidate plaintext is calculated.

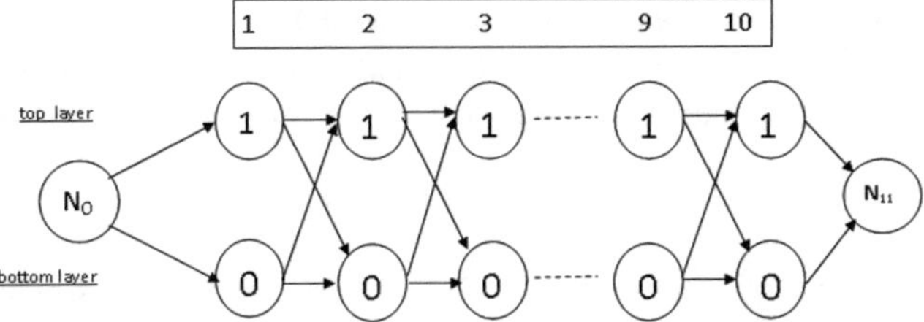

Fig. 2. Search space for cryptanalysis of SDES

Each ant constructs a key using the repeatedly applying a stochastic greedy rule based on a probabilistic move of ants across the nodes. An ant moves from node i to node j with a probability $P(i,j)$ given by Eq. 1.

$$P(i,j) = \begin{cases} 0 & \textit{if already visited} \\ \frac{\tau(i,j)^{\alpha}\rho(i,j)^{\beta}}{\sum_{i \in S} \tau(i,j)^{\alpha}\rho(i,j)^{\beta}} & \textit{Otherwise} \end{cases} \quad (1)$$

This probability is based on two factors; first, the amount of pheromone $\tau(i,j)$ between edge. And second the heuristic value $\rho(i,j)$ representing some priori knowledge of desirability of the choice. The parameters α and β are influencing factors of pheromone and heuristic value, respectively.

4.2 Heuristic Value

Heuristic value helps find acceptable solutions in the early stages of the search process, it is fundamental in making the algorithm find good solutions. Static and dynamic heuristic information are the main types of heuristic information used by ACO algorithms. In the static case, the value of ρ remains unchanged throughout the whole algorithm's run. In the dynamic case, the heuristic information depends on the partial solution constructed so far and therefore has to be computed at each step of an ant's walk.

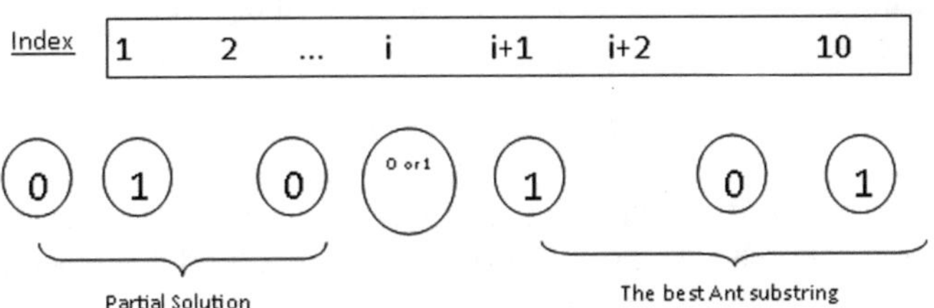

Fig. 3. Heuristic value calculation

In our approach, the candidate key with the best fitness value using Eq. (2) is saved as a global best ant. Now, in the subsequent iterations, at every decision point, the heuristic value is calculated as follows:

Let ‘i’ the vertex in the *ith* column (in the search space). As shown in Fig. 3, at which ant has to decide which node to move next. The *‘j’* is the vertex in next column where an ant can move, only two values are possible i.e. either *‘0’* or *‘1’*.

$$E_j = \{\mathrm{P}_{(1\,\mathrm{to}\,\mathrm{i})}\,|\mathrm{j}|\,\mathrm{BestA}_{(\mathrm{i}+2\,\mathrm{to}\,10)}\}$$

So, the concatenated binary string E_j becomes a guessed key which is evaluated using the fitness Function be used as a heuristic value in Eq. (1).

4.3 Fitness Function

An important step in the formulation of the optimization heuristics is the choice of a suitable Fitness function (Or cost Function). The cost function is an indicator of how close the possible solution obtained during search is to the optimal solution. A good fitness function helps the search algorithm to converge efficiently to the optimal solution.

Let c the number of known plaintext-ciphertext ($\mathrm{P_{Si}}$, $\mathrm{C_{Si}}$) pairs. And P_{Ti} is the candidates plaintext generated using trial key ‘K’. The fitness function F of the Key *‘K’* is calculated as follows:

$$F(K) = \frac{\sum_{i=1}^{c} \#(\mathrm{PSi} \oplus \mathrm{PTi})}{c * 8} \tag{2}$$

$\oplus$ represents the Xor operation, and # denotes the number of zeros in $P_{Si} \oplus P_{Ti.}$

The range of F is (0, 1). Particularly, F is 1, if all bits in P_{Si} are identical to P_{Ti} for any i in [1−c]. The actual key is found when each P_{Si} are identical to P_{Ti}, so our goal is to maximize the fitness function.

4.4 Pheromone Update

The amount of pheromone on edges constituting the tour is modified by applying a global updating rule, only the best ant solution in a particular Run (R) is allowed to update the pheromone. The ants also update the best ant information based on their tours fitness values. The pheromones over the edges constituting the tour of the best ant is updated using Eq. (3), so larger the fitness value, the greater is the amount pheromone concentrated.

$$\tau_{i,j} = \tau_{i,j} + \mathrm{Q} * F(K) \tag{3}$$

With the passage of time, the concentration of pheromone decreases due to diffusion affects; a natural phenomenon known as evaporation. This also ensures that old pheromone should not have too strong influence on the future. So, with evaporation, chances to get stuck at local minima are minimized in ACO. This evaporation can be performed as:

$$\tau_{i,j} = \tau_{i,j} \times \sigma\{which\,\sigma\,will\,be\,between\,0\,et\,1\} \tag{4}$$

4.5 Proposed Algorithm

```
1. Perform initialization of pheromone
2. Complete the tours of (N) ants by making the
decisions using probability equation (1)
3. Calculate fitness value for the tours of (N)
ants according to equation (2)
4. Update best ant information.
5. Update pheromone values on edges
constituting the tours using equation (3)
6. Perform evaporation using equation (4)
7. Repeat the steps from 2 - 5 until a maximum
number of run (R) have been attained or
threshold of Fitness Function is reached.
```

5 Experimental Results and Discussions

In this section, some experiments have been carried out to evaluate the performance of the proposed algorithm. One of the main challenges is to find the optimal setting parameter values. Therefore, the values of parameters assumed in this paper such as α, β (weight of pheromone and heuristic value), *N* (Number of Ants), '*Q*' and σ were fine-tuned by a combination of several experiments in order to optimize the cryptanalysis process. The default value of the parameters was $\alpha = 1$, $\beta = 1$, $Q = 5$, $\sigma = 0.97$ and $\tau_0 = 10$ (initial pheromone value). We have implemented our algorithm with *C++* language.

5.1 Key Space Analysis

We ran a first set of experiments in which we studied the influence that the number of ants used have on the behavior of our algorithm. The first aim is to determine the number of ants to be used in order to find the real key by browsing a minimum key space.

The results obtained after carrying experiments are illustrated in Table 1, showing the number of keys searched before locating the real key, and the average number of run (The average is calculated on 100 launch) needed for different values of N.

As shown in the Table 1, with the values $N = 15$ we need 14 Run de locate the real key (R = 14), the real key is found in a minimum search space. The maximum fitness value is reached after checking 210 keys, thus being considerably lower (by a factor of 5) than the brute-force search space size (which is equal to 2^{10} possible keys).

Table 1. Experimental results for different values of N

Number of ants (N) used	Number of runs (N) needed	Number of keys browsed
5	102	510
10	38	380
15	14	210
20	13	260
25	13	325
30	12	360
35	11	385
40	11	440
45	10	450
50	10	500

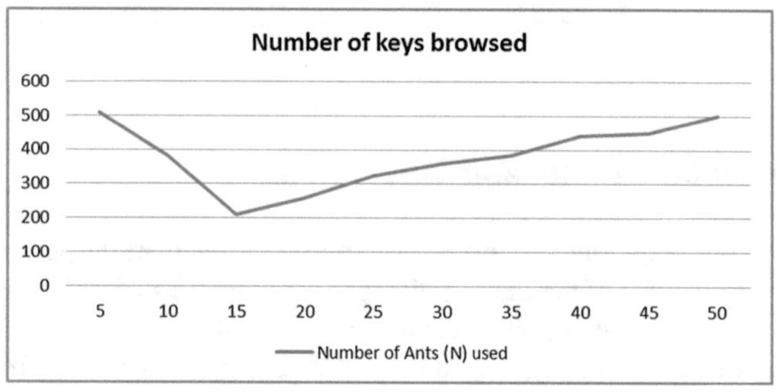

Fig. 4. Number of Key searched evolution

Figure 4 shows the evolution of the number of key browsed before locating the real key under the number of ants used (N), when the number of Ants Exceeds 15 the search space increase rapidly and the number of run needed become stagnant.

Table 2. Comparison results with GA and GSO

Number of keys browsed		
GSO [11]	GA [7]	Our algorithm
182	240	210

Table 2 depicts the comparison of our algorithm with GA used by Nalini [7] and GSO used Vimalathithan [11]. As we can see, our algorithm can recover the secret key after browsing 210 keys, which is better than GA (with 240).

5.2 Number of Plaintext-Ciphertext Pairs Analysis

In Order to assess the relevance of our Fitness Function, the correlation coefficient 'ρ' between the cost function (2) and the number of corrected key elements is calculated, for different values of c (number of pairs known plaintext-ciphertext used).

This coefficient is used to evaluate the relationship between the cost function and the number of recovered key elements. In the case of a perfect direct (increasing) linear relationship we have $\rho = 1$, its means that we have a good mechanism for evaluation of generated key, therefore the convergence of our algorithm to the best key is most guaranteed. Inversely, in a perfect decreasing linear relationship we have $\rho = -1$, and some value in the open interval $(-1, 1)$, indicating the degree of linear dependence between the variables. As it approaches zero there is less of a relationship. To calculate this coefficient, we have used 100 keys as sample, by calculating the fitness and the correct number of bits for each one using formula defined in (5).

$$\rho(x, y) = \frac{n\sum x_i y_i - \sum x_i \sum y_i}{\sqrt{n\sum x_i^2 - (\sum x_i)^2}\sqrt{n\sum y_i^2 - (\sum y_i)^2}} \tag{5}$$

Where x and y are two arrays of n elements.

Figure 5 show the value of the correlation coefficient between our Fitness function and the number of corrected key elements, for different number of pairs plaintext-ciphertext. This graph shows that the correlation coefficient increase with the number of pair used, but from 3 pairs required, this coefficient begins to stagnate and any more gain is observed in the value of the correlation coefficient.

The use of 3 pairs in the fitness function gives almost the same results as $c = 2$, while the use of a single pair can drive the algorithm to a solution with a maximum Fitness Function value without being the right key.

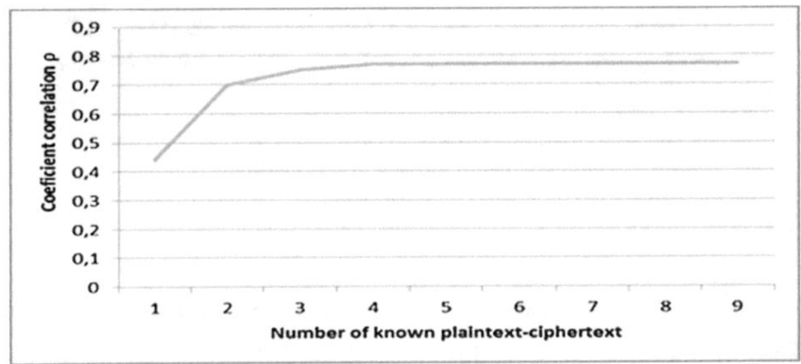

Fig. 5. Correlation coefficient between fitness function value and the number of known plaintext-ciphertext

If we use a fitness function based on only one pair of known plaintext-ciphertext ($c = 1$ in the Eq. 2), we can find a solution that maximizes the Fitness Function but without being the real key. It is for this reason that the need for a second pair proves necessary to confirm that this is the right key.

5.3 Parametric Sensitivity Analysis

In order to analyze the parametric sensitivity, the number of Ant (N) is set to *15* and β is fixed to *1*. With good parameter settings, the long-term effect of the pheromone trails is to progressively reduce the size of the explored search space so that the search concentrates on a small number of promising areas. Yet, this behavior may become undesirable, if the concentration is so strong, as we can see in the Fig. 6 with $2 < \alpha$, that it results in an early stagnation of the search. In such an undesirable situation the system has ceased to explore new possibilities and no better solution is likely to be found anymore.

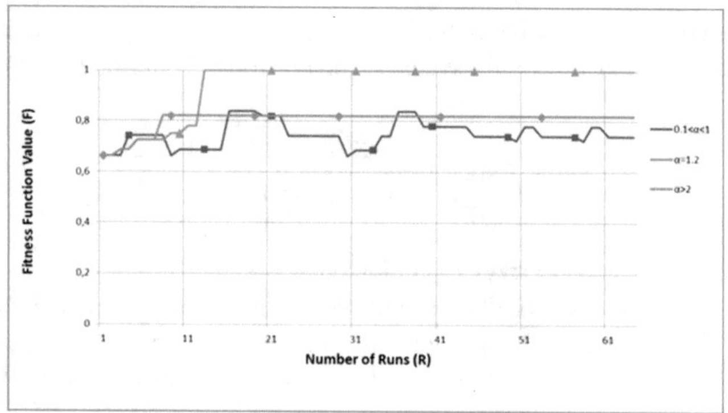

Fig. 6. Fitness Function evolution for different values of α

The overall result suggests that for ACO good parameter settings are those that find a reasonable balance between a too narrow focus of the search process, as we can see in the Fig. 6 with $\alpha = 1.2$, which in the worst case may lead to stagnation behavior, and a too weak guidance of the search, which can cause excessive exploration, and ants never converge to a common solution, this behavior is illustrated in Fig. 6 with α value between 0.1 and 1.

6 Conclusion

This article proposed a new approach of cryptanalysis algorithm for Simplified-DES using ant colony optimization. The experimental results show that our approach is significantly faster when compared to others attacks, and requires a small number of known plaintext-ciphertext pairs.

ACO provides a very powerful tool for the cryptanalysis of Simplified-DES, it is interesting to be applied to cryptanalysis of some others strong encryption algorithms like DES (Data Encryption Standard) or AES (Advanced Encryption Standard).

References

1. Dorigo, M.: Optimization, Learning and Natural Algorithms. Ph.D. thesis (1992)
2. Dorigo, M., Maniezzo, V., Colorni, A.: The ant system: optimization by a colony of cooperating agents. IEEE Trans. Syst. Man Cybern. Part B **26**(1), 29–41 (1996)
3. Dorigo, M., Gambardella, L.: Ant colony system: a cooperative learning approach to the traveling salesman problem. IEEE Trans. Evol. Comput. **1**(1), 53–66 (1997)
4. Stutzle, T., Hoos, H.: Improvements on the ant system, introducing the MAX-MIN ant system. In: Proceedings ICANNGA97—Third International Conference Artificial Neural Networks and Genetic Algorithms. Springer, Wien (1997)
5. Khan, S., Ali, A., Durrani, M.Y.: Ant-crypto, a cryptographer for data encryption standard. IJCSI **10**(1) (2013)
6. Grari, H., Azouaoui, A., Zine-Dine, K.: A novel ant colony optimization based cryptanalysis of substitution cipher. In: International Afro-European Conference for Industrial Advancement AECIA (2016)
7. Nalini, N., Raghavendra Rao, G.: Cryptanalysis of simplified data encryption standard via optimisation heuristics. IJCSNS Int. J. Comput. Sci. Netw. Secur. **6**(1B) (2006)
8. Sharma, L., Pathak, B.K., Sharma, R.G.: Breaking of simplified data encryption standard using genetic algorithm. Glob. J. Comput. Sci. Technol. **12**(5) (2012)
9. Garg, P., Varshney, S., Bhardwaj, M.: Cryptanalysis of simplified data encryption standard using genetic algorithm. Am. J. Netw. Commun. **4**(3), 32–36 (2015)
10. Al Adwan, F., Al Shraideh, M., Al Saidat, M.R.S.: A genetic algorithm approach for breaking of simplified data encryption standard. Int. J. Secur. Appl. **9**(9), 295–304 (2015)
11. Vimalathithan, R., Valarmathi, M.L.: Cryptanalysis of simplified-DES using computational intelligence. WSEAS Trans. Comput. **10**(7), 210–219 (2011)
12. Garg, P.: Cryptanalysis of SDES via evolutionary computation techniques. IJCSIS Int. J. Comput. Sci. Inf. Secur. **1**(1) (2009)

Elephants Herding Optimization for Solving the Travelling Salesman Problem

Anass Hossam(✉), Abdelhamid Bouzidi, and Mohammed Essaid Riffi

Laboratory LAROSERI, Department of Computer Science,
Faculty of Science, Chouaib Doukkali University, EI Jadida, Morocco
anass.hossam@gmail.com,
mr.abdelhamid.bouzidi@gmail.com, said@riffi.fr

Abstract. This paper proposes a novel metaheuristic called Elephant Herding Optimization (EHO) to solve the Travelling Salesman Problem (TSP), which is a combinatorial optimization problem classified as NP-Hard. The EHO algorithm is bio-inspired from the natural herding behavior of elephants groups, which proved its efficiency to solve continued optimization problems. To extend the application of this algorithm, we had proposed a novel adaptation of the EHO by respecting the natural herding behavior of elephants. To test the efficiency of our proposal adaptation, we applied the adapted EHO algorithm on some benchmark instances of TSPLIB. The obtained results shows the excellent performance of the proposed method.

Keywords: Travelling Salesman Problem · Elephants Herding Optimization · Combinatorial optimization · Metaheuristic · Nature-inspired

1 Introduction

The Travelling Salesman Problem (TSP) [1] is a classic problem in combinatorial optimization that can be described as follow, given a list of cities and the distances between each pair of cities, the travelling salesman should find the shortest possible route that visits each city by starting and returns to the first city. This problem classified as NP-hard [2] problem who's the computational complexity increases exponentially with the number of cities.

The importance of the TSP appears in many applications in real world such as electronics, telecommunications, transportation, industry, logistics, and astronomy. No exact algorithm exists for solving the TSP. The need to find an efficient solution to this problem, in a reasonable execution time, has led researchers to propose a various approximation algorithms like heuristics such as Local Search [3], Simulated Annealing [4], Tabu Search [5] etc., and metaheuristics such as Genetic Algorithm (GA) [6], Ant Colony Optimization (ACO) [7], Particle Swarm Optimization (PSO) [8], Bee Colony Optimization (BCO) [9], Discrete novel hybrid PSO [10], Discrete Cat Swarm Optimization [11], Discrete Penguins Search Optimization [12], Discrete Swallow Swarm Optimization [13] etc.

The objective of this work is to adapt the EHO algorithm [14], introduced in 2016 by Gai-Ge Wang et al., to solve the TSP.

M. Ezziyyani (Ed.): AI2SD 2018, AISC 912, pp. 122–130, 2019.
https://doi.org/10.1007/978-3-030-12065-8_12

The rest of this paper is organized as follows: Sect. 2 a presentation of the EHO algorithm introduced to solve continued optimization problems. Section 3 description of the TSP. Section 4 describes the proposed adaptation of EHO to solve the TSP. Section 5 presents in detail the results of numerical experiments on a set of benchmarks from the TSPLIB library [15]. Finally, comes the conclusion in the last section.

2 EHO Algorithm

Gai-Ge Wang et al. introduced the EHO in 2016 [14] to solve continued optimization problems. This method, as indicated by its name, was inspired from the natural herding behavior of elephants, and it was modeled as EHO algorithm, the authors of this algorithm preferred to simplify it into the following idealized rules:

- The elephant population is composed of some clans, and each clan has exactly the same number of elephant.
- At the start of each generation, a number of male elephants will leave their family group.
- In each clan, elephants live together under the leadership of a matriarch. For optimization problem, a matriarch is the fittest elephant in this clan.

2.1 Clan Updating Operator

As mentioned before, the elephants in each clan live under the leadership of a matriarch. Therefore, the next position of each elephant in clan ci is influenced by the position of the matriarch in clan ci. Then, elephant positions are updating as follows:

$$x_{new,ci,j} = x_{ci,j} + \alpha \times \left(x_{best,ci} - x_{ci,j}\right) \times r \tag{1}$$

$x_{new,ci,j}$ is the new position of elephant j in clan ci, $x_{ci,j}$ is the old position of elephant j in clan ci. $x_{best,ci}$ is the matriarch position in clan ci. $\alpha \in [0, 1]$ is a factor that determines the influence of $x_{best,ci}$ on $x_{ci,j}$. $r \in [0, 1]$ is a random number from uniform distribution.

The matriarch position is updated as:

$$x_{new,ci,j} = \beta \times x_{center,ci} \tag{2}$$

$x_{new,ci,j}$ is the new position of the matriarch in clan ci, β is a factor in the range [0,1] that controls the influence of the $x_{center,ci}$ on $x_{new,ci,j}$. $x_{center,ci}$ is the center position in clan ci, and it can be defined as:

$$x_{center,ci} = \frac{1}{n_{ci}} \times \sum_{j=1}^{n_{ci}} x_{ci,j} \tag{3}$$

Where n_{ci} is the number of elephants in clan ci.

2.2 Separating Operator

When mature, male elephants will leave their clans. At each generation, the elephant with the worst fitness in clan ci will be replaced via the separating operator as follow:

$$x_{worst,ci} = x_{min} + (x_{max} - x_{min}) \times \text{rand} \tag{4}$$

x_{max} and x_{min} represent respectively upper position and lower position of the elephant population, $x_{worst,ci}$ is the elephant with the worst fitness in clan ci, rand $\in$ [0,1] is a random number from uniform distribution in the range [0, 1].

2.3 Elephant Herding Optimization Algorithm

```
Begin
  Step 1: Initialization. Set the generation counter t=1;
  initialize the population, set the maximum generation
  MaxGen, the scale factor α and β, the number of clan
  nClan and the number of elephants for each clan n_ci.
  Step 2: Fitness evaluation. Evaluate each elephant in-
  dividual according to its position.
  Step 3: while t < MaxGen do
  Sort all the elephant according to their fitness.
  Implement clan updating operator:
    for ci=1 to nClan do
      for j=1 to n_ci do
        Update x_ci,j and generate x_new,ci,j by equation(1).
        if x_ci,j=x_best,ci then
          Update x_ci,j and generate x_new,ci,j by equation(2).
        end if
      end for
    end for
  Implement separating operator:
  for ci=1 to nClan do
    Replace the elephant with the worst fitness in clan
  ci by equation(4).
  end for
  Evaluate the population according to the newly updated
  positions.
  Update the generation counter, t=t+1.
  Step 4: end while
  Step 5: Return the best solution found.
End.
```

3 The Travelling Salesman Problem

The TSP is defined by the number of cities (N) and the distances between all pairs of cities (dij for i, j ∈ [1, N]). In TSP, the challenge is to find the shortest possible tour that the salesman can follow to visits each city exactly once and return to the origin city.

The TSP was modelled as a Hamiltonian graph where vertices represent cities, and edges correspond to the distances between cities. A tour is now a Hamiltonian cycle (tour that passes through all the vertices), and optimal tour is the Hamiltonian cycle with the minimum cost (the shortest).

4 Use EHO to Solve the TSP

This section describes the proposal method to solve the TSP into two versions, the first one propose an adaptation, that was able to solve the TSP but in important execution time, after that this paper proposes an improved version of the adapted EHO, which had to prove its efficiency to solve the randomly chosen benchmarks instances of the TSPLIB.

4.1 Adapted Discrete EHO to Solve the TSP

The EHO method proposed by Wang et al. [14] was defined to solve continued optimization problems, so it cannot be applied to solve combinatorial optimization problems, given that in continued optimization, a number represent the solution, but in the case of combinatorial optimization, the solution is represented by an order that can be modeled as a vector. By considering that, we should adapt the operations and operators by respecting the real behavior of elephants:

- The position of an elephant represents a solution (Hamiltonian cycle).
- Subtraction between each two position (x1 – x2) present, the set permutation apply to x2 to obtain x1.
- Addition between the set of permutations and a position is the inversion of subtraction. The addition (x + sp) applies the set of permutations to the x position.

The initialization of solutions has a great impact on the functioning of the algorithm, in our case, we preferred to generate solutions (elephant's positions) in a random way.

After the initialization is completed, we perform the evaluation of the generated solutions according to their fitness, and at the end of this step, we determine the matriarch position (solution with the best fitness) for each clan.

Then, the clan-updating operator is implemented with a small modification in the update position of the matriarch. In the case of the TSP, we cannot apply Eq. (3) to calculate the position of $x_{center,ci}$, to solve this problem, we calculate the average fitness in the clan ci ($f_{moy,ci}$) as shown in Eq. (5), and we take the solution who has the closest fitness to this value as $x_{center,ci}$.

$$f_{moy,ci} = \frac{1}{n_{ci}} \times \sum\nolimits_{j=1}^{n_{ci}} f(x_{ci,j}) \quad (5)$$

Finally, we implement the separation operator to replace the solution with the worst fitness in each clan using Eq. (4), and we increment the generation counter.

4.2 Improved Discrete EHO to Solve TSP

Since the results obtained after the execution of the adapted EHO algorithm, are not good enough as shown in Table 1, we decided to hybridize this algorithm with another heuristic such 2-opt [16] to obtain better results.

At the first generation, we determine the matriarch for each clan by taking the elephant with the best position in the clan, and as mentioned before, the elephant's positions are generated in a random way, so the probability to have good matriarch's positions in this generation is too low. For this, we implement 2-opt in a first place to improve the matriarch position of each clan.

Then, in each generation, we apply 2-opt after the implementation of the clan updating operator. The following flowchart describes the improved adapted EHO algorithm (Fig. 1):

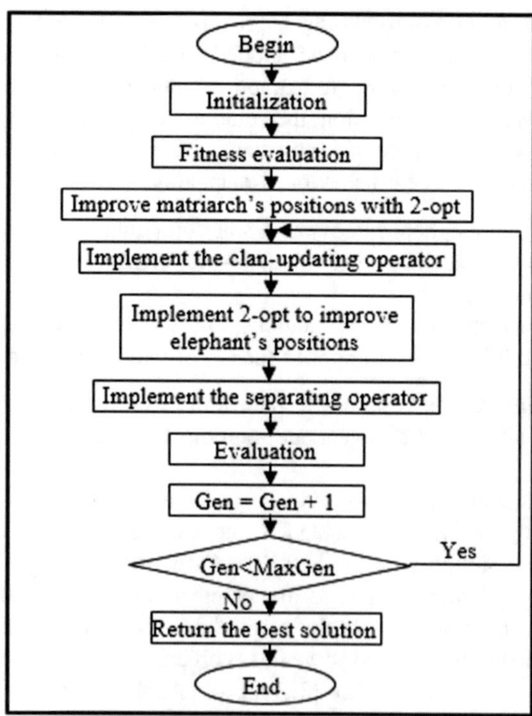

Fig. 1. Flowchart of the improved adapted EHO algorithm.

5 Results and Comparison

The implementation of the adapted Elephants Herding Optimization algorithm and the improved version to solve the TSP were realized on C programming language, and simulations were performed on a personal computer equipped with CORE i5-3210M CPU at 2.50 GHz, 4 Gb RAM, and Windows 7 (64 bits) Professional operating system. Tables 1 and 2 shows the results of executions of these two algorithms on fifteen different benchmark instance of TSPLIB. For both algorithms, the parameters have been fixed as follows: the number of clan nClan was set to 5, the number of elephants in each clan was set to 1000, α was set to 0.9 and β was set to 0.3.

Each table display the following information:

- Inst: name of benchmark instance in TSPLIB library.
- Nb.node: number of nodes.
- Opt: the best solution known for the instance.
- BestR: the best solution found by the algorithm after ten different runs.
- WorstR: the worst solution found by the algorithm after ten different runs.
- Av: the average of ten different runs of the algorithm.
- Time: shows the average time in seconds of ten different runs of the algorithm.
- Err: the percentage of error is calculated by $\{\frac{Av-Opt}{Opt} \times 100\}$.

Table 1. Results obtained by applying Discrete EHO on TSP.

Inst	Nb_node	Opt	BestR	WorstR	Av	Time	Err
Eil51	51	426	630	760	704.8	6.430	65.44
Berlin52	52	7542	11150	12707	12160.7	5.148	61.23
St70	70	675	1286	1646	1476	14.820	118.66
Pr76	76	108159	222020	241240	230424.6	19.094	113.04
Eil76	76	538	1032	1142	1091.3	22.074	102.84
Kroa100	100	21282	60823	69348	66744.5	101.010	213.61
Krob100	100	22141	54181	71827	66078.5	85.199	198.44
Kroc100	100	20749	60773	75063	67603.2	85.815	225.81
Krod100	100	21294	62331	71031	65650.3	75.878	208.30
Kroe100	100	22068	59785	83388	69416.3	80.699	214.55
Eil101	101	629	1400	1588	1514.1	74.209	140.69
Lin105	105	14379	46095	50404	48498.8	111.494	237.28
Pr107	107	44303	177953	216406	201403.4	139.495	354.60
Pr124	124	59030	228204	280386	253235	193.206	328.99
Bier127	127	118282	242103	273829	260442.2	157.139	120.18
Ch130	130	6110	15558	21007	19522.9	235.779	219.52
Pr136	136	96772	342107	396634	362987.2	311.517	275.09
Ch150	150	6528	22027	25044	24233.6	375.680	271.22
Kroa150	150	26524	97372	112590	108611.5	466.316	309.48
Krob150	150	26130	107750	122903	109367	412.023	318.54

Table 1 presents the results obtained after ten different runs of the adapted EHO algorithm on fifteen instances of TSPLIB. This algorithm gives acceptable results but not good enough, and the execution time is quite long (Fig. 2).

Table 2. Results obtained by applying the Improved Discrete EHO on TSP.

Inst	Nb_node	Opt	BestR	WorstR	Av	Time	Err
Eil51	51	426	426	426	426	6.519	0.00
Berlin52	52	7542	7542	7542	7542	0.255	0.00
St70	70	675	675	675	675	4.792	0.00
Pr76	76	108159	108159	108159	108159	9.203	0.00
Eil76	76	538	538	539	538.5	189.903	0.09
Kroa100	100	21282	21282	21282	21282	8.634	0.00
Krob100	100	22141	22141	22141	22141	33.069	0.00
Kroc100	100	20749	20749	20749	20749	6.912	0.00
Krod100	100	21294	21294	21294	21294	18.074	0.00
Kroe100	100	22068	22073	22121	22100	58.125	0.14
Eil101	101	629	630	634	632.6	517.912	0.57
Lin105	105	14379	14379	14379	14379	9.714	0.00
Pr107	107	44303	44303	44303	44303	8.876	0.00
Pr124	124	59030	59030	59030	59030	4.165	0.00
Bier127	127	118282	118282	118392	118321.6	495.519	0.03
Ch130	130	6110	6110	6113	6112.25	658.553	0.03
Pr136	136	96772	96772	96781	96775	175.447	0.00
Ch150	150	6528	6550	6575	6564.2	1867.273	0.55
Kroa150	150	26524	26524	26524	26524	721.539	0.00
Krob150	150	26130	26132	26132	26132	163.478	0.00

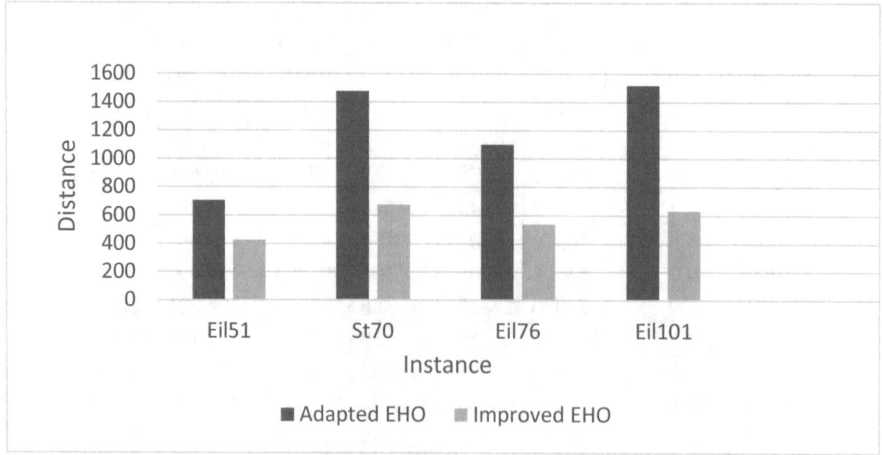

Fig. 2. Comparison between the results obtained after ten different runs of the adapted EHO algorithm and the improved adapted EHO algorithm on these instances: Eil51, St70, Eil76 and Eil101.

Table 2 presents the results obtained after ten different runs of the improved adapted EHO algorithm on fifteen instances of TSPLIB. This algorithm gives best results in a short execution time.

6 Conclusion

In this paper, we have first presented an adaptation of EHO algorithm via Gai-Ge Wang, without modification or hybridization, to solve the symmetric TSP, this adaptation did not give good results compared to the results obtained for the continuous problems. Secondly, we have improved the adapted EHO algorithm by hybridization with 2-opt to obtain best solutions. The simulation results were compared to the best solution, the worst solution and the average after ten different runs on each instance. The results of the comparison have proved the performance of the improved adapted EHO algorithm. In the future, we will aim to extend the proposed algorithm to be applied on various real applications area based on TSP.

References

1. Flood, M.M.: The travelling-salesman problem. Oper. Res. **4**(1), 61–75 (1956)
2. Hochbaum, D.S.: Approximation Algorithms for NP-Hard Problems. PWS Publishing Co., Boston (1996)
3. Korupolu, M.R., Plaxton, C.G., Rajaraman, R.: Analysis of a local search heuristic for facility location problems. J. Algorithms **37**(1), 146–188 (2000)
4. Aarts, E., Korst, J., Michiels, W.: Simulated annealing. In: Search Methodologies, p. 187–210. Springer, Boston (2005)
5. Gendreau, M., Hertz, A., Laporte, G.: A tabu search heuristic for the vehicle routing problem. Manag. Sci. **40**(10), 1276–1290 (1994)
6. Fonseca, C.M., Fleming, Peter, J., et al.: Genetic Algorithms for Multiobjective Optimization: Formulation, Discussion and Generalization. In: ICGA, pp. 416–423 (1993)
7. Dorigo, M., Birattari, M.: Ant colony optimization. In: Encyclopedia of Machine Learning, pp. 36–39. Springer, Boston (2011)
8. Poli, R., Kennedy, J., Blackwell, T.: Particle swarm optimization. Swarm Intell. **1**(1), 33–57 (2007)
9. Wong, L.-P., Low, M.Y.H., Chong, C.S.: A bee colony optimization algorithm for travelling salesman problem. In: Second Asia International Conference on Modeling & Simulation, AICMS 2008, pp. 818–823. IEEE (2008)
10. Bouzidi, M., Riffi, M.E.: Discrete novel hybrid particle swarm optimization to solve travelling salesman problem. In: 5th Workshop on Codes, Cryptography and Communication Systems (WCCCS), pp. 17–20. IEEE (2014)
11. Bouzidi, A., Riffi, M.E.: Discrete cat swarm optimization to resolve the travelling salesman problem. Int. J. Adv. Res. Comput. Sci. Softw. Eng. **3**(9), 13–18 (2013)
12. Mzili, I., Riffi, M.E.: Discrete penguins search optimization algorithm to solve the travelling salesman problem. J. Theor. Appl. Inf. Technol. **72**(3), 331–336 (2015)
13. Bouzidi, S., Riffi, M.E.: Discrete swallow swarm optimization algorithm for travelling salesman problem. In: Proceedings of the 2017 International Conference on Smart Digital Environment, pp. 80–84. ACM (2017)

14. Wang, G.-G., Deb, S., Gao, X.-Z., et al.: A new metaheuristic optimisation algorithm motivated by elephant herding behaviour. Int. J. Bio-Inspired Comput. **8**(6), 394–409 (2016)
15. Reinelt, G.: TSPLIB—a travelling salesman problem library. ORSA J. Comput. **3**(4), 376–384 (1991)
16. Mcgovern, S.M., Gupta, S.M.: 2-opt heuristic for the disassembly line balancing problem. In: Environmentally Conscious Manufacturing III, pp. 71–85. International Society for Optics and Photonics (2004)

Ontology-Based Context Agent for Building Energy Management Systems

Youssef Hamdaoui(✉) and Abdelilah Maach

Department of Computer Science, Mohammadia School of Engineers (EMI), Mohammed V University, Rabat, Morocco
youssefhamdaoui@research.emi.ac.ma, maach@emi.ac.ma

Abstract. Replacing existing electrical grids by future smart grids opens a several opportunities for energy-efficient operation of buildings and cities as well as improved coordination of energy load and supply. Future energy systems will be managed by smart controllers. These smart devices will have to interoperate, they need to exchange information with each other in order to cooperate over complex control tasks. Current data and communication technology provides a suitable basis for the bidirectional flow of information between smart buildings and other smart grid stakeholders. Interoperability will only be achieved when Smart Grid appliances share common semantics on the data they exchange. However, a common concept of shared real energy state and knowledge is essential in order to unify heterogeneous grid contexts, incorporate information of smart grid participants, and process this information in building energy management systems.

In this work, a context agent based on an OWL ontology is presented that enables semantic representation of knowledge for interaction between building energy management systems and smart grids and end prosumers. A well-proven methodology is used to develop this ontology. Furthermore, the ontology application into building energy management systems and smart grid environments is figured, and the functional capabilities of this approach are shown.

Keywords: Ontology · Smart grid · MicroGrid · Buildings · Energy management systems · Efficiency · Multi agent

1 Introduction

The term Smart Grid (SG) is used for future energy systems. The former centralized infrastructure with its unidirectional power flow from large power plants via the consumers, turns into a full-meshed topology including bidirectional power flows. Consequently, interoperability issues appear as a real challenge. Thus, numerous SG devices need to transmit and exchange information with each other in order to cooperate over complex control tasks. The need for new energy strategies that are efficient, stable and sustainable has stimulated the SG development where energy generation is gradually shifting from few large centralized generators to many decentralized distributed power generators. Future SGs will be composed of a mesh of networked clusters and Microgrids (MG) collaborating to deliver electricity to consumers.

M. Ezziyyani (Ed.): AI2SD 2018, AISC 912, pp. 131–140, 2019.
https://doi.org/10.1007/978-3-030-12065-8_13

Microgrids, will be composed of collaborating distributed energy resources (DER). SGs are also composed of smart IT infrastructure and communication technologies subsystem that enable it to disseminate necessary data in a timely manner to be able to make decision and take proper control actions. These new generations of power energy systems transform electrical systems into a large "information system" computing in a pervasive system, where the characteristics of distributed control and knowledge of the state of the network would enable interesting functionalities such as [1–3]:

- Auto-recovery system to ensure a stable power quality.
- High quality power delivery.
- Resistance to cyber-attacks.
- Allow Integration of DER.
- Accommodation of energy storage.
- Monitoring and manage user's consumption.
- Estimation of user's load distribution.
- Real-time islanding in outage case to re distribute energy using online and quasi-real time measurement.
- Operation and maintenance cost optimization.

The major of powr needs is requested by residential and commercial buildings with nearly 40% of final energy consumption [4]. Thus, energy efficiency and optimization of energy consumption become increasingly important and must be rethinking in order to reduce costs and reduce detrimental effects on the environment. For this purpose, coordination of energy supply and load and increased use of decentralized energy resources are essential measures. As traditional power grids do not provide the required infrastructure, they are going to be replaced by SG, which are characterized by (a) a bidirectional flow of information and energy (b) self-healing capabilities, (c) adaptiveness and support islanding and emergency cases [5]. Self-healing is ascribed to a grid capable of automatically anticipating and responding to power system disturbances, while continually optimizing its own performance [6] to guarantee adequacy and security of supply. The vision of a self-healing grid is shared by:

(1) Dynamic optimization of grid performance and efficiency;
(2) Resolve islanding issue based on smart fast reaction to disturbances and minimization of their impacts;
(3) Fast restoration to a stable operation point with little or no human intervention.

An efficient interplay of grid stakeholders is based on the integration of communication infrastructure, a Building energy management system (BEMS) has to combine multiple areas, such as usage profiles, environmental influences, smart grid interactions, or building automation systems, in order to optimize local power consumption. This set of information and knowledge has to be managed in a structured way to aid BEMS functionality.

Using databases and ontologies can present an appropriate solution for such an information and knowledge representation supporting BEMS interaction with smart grid entities. Ontologies are new in the field of computer science. Databases are focused on meeting requirements of a specific application or organization. Stimulated by the Semantic Web, ontologies are suitable for semantic modeling and inference of new knowledge by means of reasoners. Furthermore, ontologies are favored over databases for describing interaction between BEMSs and smart grids and they are more beneficial for sharing knowledge among different domains and linking distributed knowledge [7]. Despite of databases provide efficient information management, advantages of ontologies, like a higher level of abstraction, implementation independence are prevailing regarding application in an open, stable, flexible, and evolving environment as the smart grid. Additionally, ontologies are identified as promising information modeling techniques for the SG enabling formal semantics in combination with a shared understanding [8]. In the domain of smart buildings and smart houses, literature already describes different ontologies. For example, ontology covers building functionality, hardware, users profiles, and context parameters (e.g. location, rooms, position, temperature) [9]. Besides processes, resources, user comfort, or external influences, energy-related information is part of ThinkHome ontology [10–13].

2 Smart Grids

Characteristics of Smart Grids are [14]:

- Promote the participation of costumers in the grid. Prosumers can consumer and produce energy through different DER.
- Accommodates all types of generations and storage options in a plug and play mode. Generation can be either centralized, such as nuclear that we typically have in a plant, or distributed using different natural resources.
- Enable introduction of new product services in the market. New real time market place of buying and selling electricity and services are possible.
- Optimize the operation of electricity in the grid system.
- Mainline supplying power in outage mode.
- Self-healing properties and are resilient to attacks such as cyber and physical attacks. They are expected to recover quickly from natural disasters like tornadoes and hurricanes.

These new generations of power systems transform electrical systems into a large information system computing, where the aforementioned characteristics of distributed control and knowledge of the state of the network would enable interesting functionalities [14]:

SG can be through of as a layered system where communication layer is overlaid over the power layer to implement different needed functionalities. Figure 1 shows SG functionalities divided into layers, namely the consumer, communication, and the distribution layers. The information flows in bi-directions between layers. Each layer has its own devices and technologies.

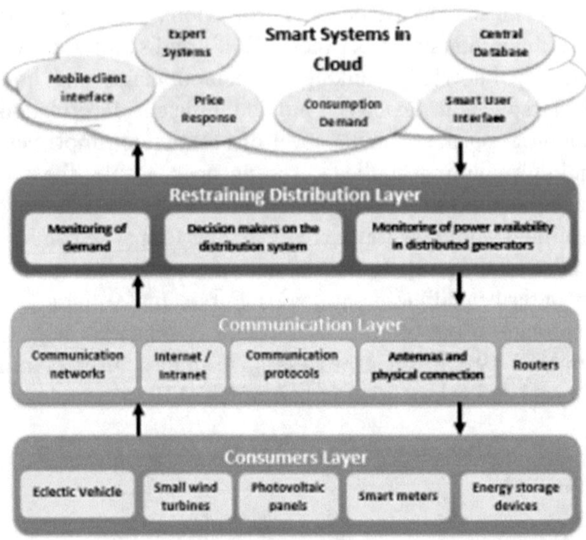

Fig. 1. Smart grid layers

3 Multi-agent Systems

3.1 Advantages of Multi-agent Systems (MAS)

The goal of SG is to develop grid modernization technologies, techniques for Demand/Response with the ability to dynamically optimize grid operations and resources costs and incorporate Demand/Response and end user participation [15]. To achieve this goal, it is important to understand demand behaviors and predict how they might change [16]. Consumers energy demand is complex, given the wide range of interlinking behavioral and IT infrastructure factors combined in many different configurations. The relationship between costumer factors and power consumption, and the wider context of public power supply and society is complex on a range of different spatial and temporal scales. The SG is made of millions of controllable devices; the control system should work efficiently on a large scale and be fault tolerant [17]. We present here comparison between SG characteristics/requirements and their analogies:

- The components of SG are distributed and the power management system is tightly associated with the communications between stakeholders and agents to exchange data. MAS are by nature distributed and concurrent, they are independent entities engaged in the system, they have their own perception of the environment, goal and they try to achieve the best for themselves while behaving strategically. Therefore, in that case when using MAS the amount of information will be greatly reduced in comparison to other in-depth communication methods.
- SG is a holistic system and the failure of some part of it (the breakdown of a transmission line or cut down of a substation, transformer. …) mustn't affect the whole activities. Flexibility denotes the ability of the system to be adaptable to

different situations. The flexibility of MAS is given by the fact that entities are independent and able to perceive and adapt their context.

- SG should demonstrate the plug-and-play concept for integrating energy storage, loads, and sources at the building level with the external utility grid. Plug and play adaptability is widely proven by nature MAS that are able to be scaled by dispersing or by adding other entities.
- As SG will be composed for an aggregate of Micro grid, the control can be delegated to MG. The MAS can perform tasks locally if they have sufficient knowledge and resources, and they can interact with other surrounding entities to help in the completion of tasks or decisions. The problem of control can be approached from a variety of perspectives including cognitive science, heuristic search, and machine learning.
- SG leverage the widespread of the information and communication technologies, these smart technologies are platform-independent and language free. In this perspective, an agent can be developed by different programmed languages and communicate with the other agents in the system.

3.2 Applications of Multi-agent Systems

In existing electrical grid, losses are very high. There is loss in power plants, in transmission lines, in distribution, in consumption units such as residences and companies and also due to human habits that affect the loss significantly. The new grid is in need of an power management system which is a key component of a SG that reduces the wasted power by adjusting heating and cooling usage through collecting information from sensors and report best slot of time for optimizing power. Power efficiency helps the consumers save money by consuming less and also by reducing the global load in the grid that results in the decrease of price of electricity in the grid [23–26]. Therefore, power efficiency helps the users and the grid in achieving global energy and also increasing the grid's reliability. A multi-agent Energy Management System (EMS) is among several EMS architectures that have been proposed to cope with heterogeneity of SGs; several research projects have been released to validate the efficiency of a multi-agent EMS. The latter proved that it can lead to important power savings and efficiently improve energy usage [18].

4 Ontology Concept

4.1 Ontology of Context

It is important to understand what context is and what role context and context modeling play in current systems. A widely accepted definition of context is [19]: Context is any data that can be used to characterize the status of an entity, where an entity may be a person, a place, or an object that is considered relevant to the interaction between a user and an application. Commonly used contexts consist of location, identity, time, temperature, activity…

A system is then seen to be context-aware if it uses context to provide relevant data and services to the user, where relevancy depends on the users' task. A widely accepted definition of context- awareness [19]: "Context awareness is the ability to sense and use different contexts, any application that takes advantage of context is a context-aware application". Context-aware computing [20] is the ability of computing devices to detect, interpret, and respond to changes of system and environment. The ontology uses OWL [21] language for its implementation.

Case representation allows for better assessment of the similarities of current problems compared to past modes. The knowledge base saves data about conditions (problems) and actions (solutions) for previous control situations. Case Ontology is used to represent knowledge of different cases and hold new learning experiences. As a matter of fact, representing cases with ontological model leads to their easy selection owing to the fact that syntactic matching provided by triples form allows for high accuracy and also the search of data using SPARQL language is straight forward and optimized by nature. Case ontology represents successful experienced controls by description taken decision according to their context. Figure 2 shows ontology with case ontology part. Case is primary defined by an auto incremental ID, the other data in the header are Environment ID, Control Entity, Timestamp and Survival Value. The context space' features describe the context's attributes in the form (subjects, predicates, objects) as in the triples store. Predicate is a property for the subject of the statement, the subject is the concept involved and object is the range of the predicate. The predicates are biased by weights to express their influence on the control since attributes may not have the same.

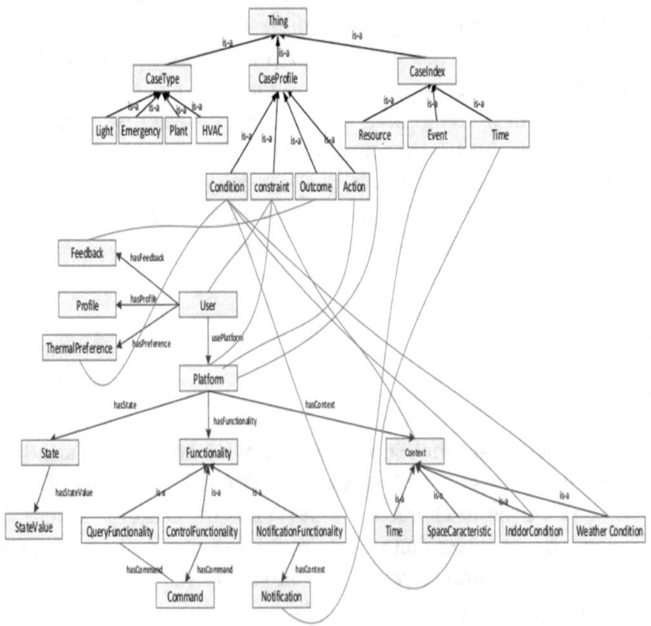

Fig. 2. Model representation

Matching Cases:
Matching is the step that serves to measure the degree of similarity between case and context by comparing their triples and features in order to retrieve best cases for a given context. The similarity measure is performed by comparing between case's features and context's data both saved.

Case Adaptation:
Adaptation is the most important step, it adds smartness and intelligence to what would otherwise be simple pattern matchers. The adaptation means developing a solution to find a best match set based on existing cases, the best match sometimes is not a single case, but a combination of cases. For high precision of control, the best match can be the first nearest neighbor of the current context if it exists; elsewhere the best match is a combination of number of solutions which show a similarity tradeoff between the current context and previous cases.

Figure 3 shows different types of agents ranging from central energy generators to customers and their buildings. The location of agents is visualized by the underlying map. Interaction scenarios form an important part of this ontology. Besides definition of processes, like trading of flexibilities or exchange of pricing data, involved agents and their roles are taken into consideration. Semantically enriched modeling facilitates automatic processing of smart grid interaction at the BEMS level.

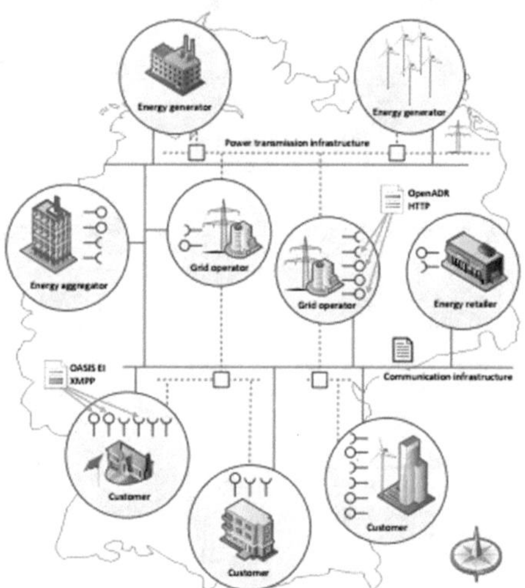

Fig. 3. Ontology domains

5 Ontology Application

5.1 Implementation

In this work, The implantation is done with Web Ontology Language (OWL the proposed ontology as there are several expressiveness issues and syntax problems in OWL [22]. Class hierarchy, object and information properties, and constraints on classes and properties are created by means of the open source ontology editor. Figure 4 visualizes an excerpt of the developed ontology. Besides classes, the figure contains instances, information values, and property relations in order to show ontology utilization. Classes are marked with circles, defined classes have three additional lines, and individuals are tagged with a diamond. Data values show the used data type. The illustrated example contains power retailer that offers the service Energy Price. Besides the parameter configuration Config 1, the service is related to a technology adapter Adapter1 specifying the communication technology for accessing the service. The energy retailer is connected to one of two available, adjacent substations. In order to distinguish between standardized properties and own properties, different namespaces are used. Solid lines represent asserted relations, and dashed lines are used for relations that are inferred by the reasoner. The figure is intended to give an overview on the modeling of instances based on the defined ontology. Thus, it is not complete, and many classes and properties are omitted. For example, location of one substation is left out although this information can be used to infer further knowledge, such as the distance between grid nodes.

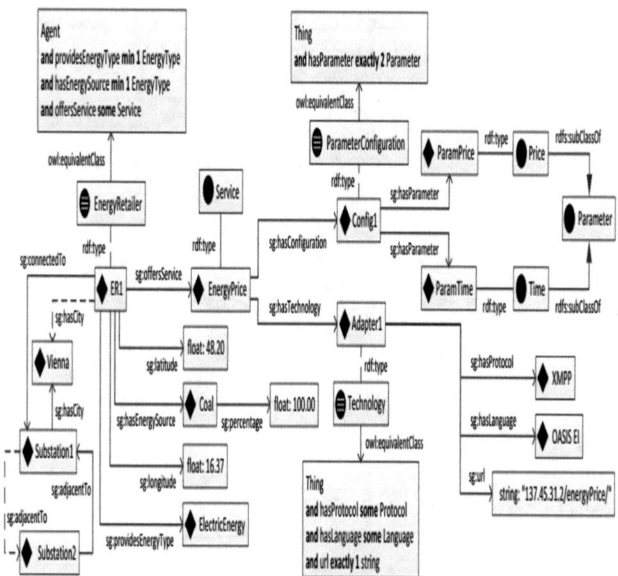

Fig. 4. Classes and instances

6 Concusion

Interaction between BEMSs and smart grids becomes increasingly important to balance energy demand and supply or to incorporate distributed energy resources. Thus, this work presents ontology as part of an abstraction layer that enables semantically enriched description of smart grids and supports operation of BEMSs. The development focuses on four main parts covering grid agents, communication technologies, service interfaces, and grid structure. The proposed approach is implemented in the form of OWL ontology, and its application and integration into BEMSs are described. Furthermore, modeling capabilities, ontology reasoning, and functionality are evaluated using a smart grid test bed.

The development of SG and related technologies combine advances in distributed systems, Artificial Intelligence, control, and information and communications technologies. SGs are expected to exhibit high level of autonomy, self-healing, and reliability and to provide features such as reconfiguration, protection, restoration, and interaction with users through demand response.

References

1. Hamdaoui, Y., Maach, A.: Smart islanding in smart grids. In: Smart Energy Grid Engineering (SEGE), 2016 IEEE, pp. 175–180 (2016). https://doi.org/10.1109/SEGE.2016.7589521
2. Hamdaoui, Y., Maach, A.: An intelligent islanding selection algorithm for optimizing the distribution network based on emergency classification. In: 2017 International Conference on Wireless Technologies, Embedded and Intelligent Systems (WITS), pp. 1–7 (2017). https://doi.org/10.1109/WITS.2017.7934627
3. Hamdaoui, Y., Maach, A.: A smart approach for intentional islanding based on dynamic selection algorithm in Microgrid with distributed generation. In: 2017 International Conference on Big Data, Cloud and Application (BDCA), pp. 1–7. ACM (2017). https://doi.org/10.1145/3090354.3090410
4. Hamdaoui, Y., Maach, A.: A cyber-physical power distribution management system for smart buildings. In: Ben Ahmed, M., Boudhir, A.A. (eds.) SCAMS 2017. LNNS, vol. 37, pp. 538–550. Springer, Cham (2018). https://doi.org/10.1007/978-3-319-74500-8_50
5. Farhangi, H.: The path of the smart grid. IEEE Power Energ. Mag. **8**(1), 18–28 (2010)
6. Hamdaoui, Y., Maach, A.: A novel smart distribution system for an islanded region. In: Ezziyyani, M., Bahaj, M., Khoukhi, F. (eds.) AIT2S 2017. LNNS, vol. 25, pp. 269–279. Springer, Cham (2018). https://doi.org/10.1007/978-3-319-69137-4_24
7. Martinez-Cruz, C., Blanco, I.J., Vila, M.A.: Ontologies versus relational databases: are they so different? A comparison. Artif. Intell. Rev. **38**(4), 271–290 (2012)
8. Fang, X., Misra, S., Xue, G., Yang, D.: Smart grid - the new and improved power grid: a survey. IEEE Commun. Surv. Tutorials **14**(4), 944–980 (2012)
9. Stavropoulos, T.G., Vrakas, D., Vlachava, D., Bassiliades, N.: BOnSAI: a smart building ontology for ambient intelligence. In: Proceedings of the ACM International Conference on Web Intelligence, Mining and Semantics, no. 30, pp. 1–12 (2012)
10. Kofler, M.J., Reinisch, C., Kastner, W.: A semantic representation of energy-related information in future smart homes. Energy Build. **47**, 169–179 (2012)
11. Shah, N., Chao, K.-M., Zlamaniec, T., Matei, A.: Ontology for home energy management domain. In: Digital Information and Communication Technology and Its Applications, pp. 337–347. Springer (2011)
12. Grassi, M., Nucci, M., Piazza, F.: Ontologies for smart homes and energy management: an implementation-driven survey. In: Proceedings of the IEEE Workshop on Modeling and Simulation of Cyber-Physical Energy Systems, pp. 1–3 (2013)

13. Zhou, Q., Natarajan, S., Simmhan, Y., Prasanna, V.: Semantic information modeling for emerging applications in smart grid. In: Proceedings of the International Conference on Information Technology: New Generations, pp. 775–782 (2012)
14. Hamdaoui, Y., Maach, A.: Dynamic balancing of powers in islanded microgrid using distributed energy resources and prosumers for efficient energy management. In: 2017 IEEE Smart Energy Grid Engineering (SEGE), IEEE (2017). https://doi.org/10.1109/SEGE.2017.8052792
15. US DoE: SMART GRID and Demand Response. US Departement of Energy (2010). http://energy.gov/oe/technology-development/smart-grid/demand-response
16. Spataru, C., Barrett, M.: Smart consumers, smart controls, smart grid. In: Sustainability in Energy and Buildings, vol. 22, pp. 381–389. Springer, Heidelberg (2013)
17. Roche, R., Blunier, B., Miraoui, A., Hilaire, V., Koukam, A.: Multi-agent systems for grid energy management: a short review. In: 36th Annual Conference IEEE Industrial Electronics Society, IECON 2010, pp. 3341–3346, November 2010
18. Junwei, L., Bo, Z.: Research on energy management system based on multi-agent, pp. 253–255, July 2010
19. Dey, A.K.: Understanding and using context. Pers. Ubiquit. Comput. PUC **5**(1), 4–7 (2001)
20. Salber, D., Dey, A.K., Abowd, G.D.: Designing and building context aware applications. http://www.cc.gatech.edu/fce/contexttoolkit
21. Web Ontology Language. http://www.w3.org/TR/owl-features/
22. Grau, B.C., Horrocks, I., Motik, B., Parsia, B., Patel-Schneider, P., Sattler, U.: OWL 2: the next step for OWL. Web Semantics Sci. Serv. Agents World Wide Web **6**(4), 309–322 (2008)
23. Hamdaoui Y., Maach A.: Optimized energy management of electric vehicles connected to microgrid. In: M. Ezziyyani, M. (ed.) AI2SD 2018, vol. 912, pp. 371–387. Springer, Cham, (2019). https://doi.org/10.1007/978-3-030-12065-8_34
24. Hamdaoui, Y., Maach, A.: Energy efficiency approach for smart building in islanding mode based on distributed energy resources. In: Ezziyyani, M., Bahaj, M., Khoukhi, F. (eds.) AIT2S 2017. LNNS, vol. 25, pp. 36–49. Springer, Cham (2018). https://doi.org/10.1007/978-3-319-69137-4_4
25. Hamdaoui, Y., Maach, A., El Hadri, A.: Autonomous power distribution system through smart dynamic selection model using islanded micro grid context parameters and based on renewable resources. Int. J. Mech. Eng. Technol. (IJMET) **9**(11), 1755–1780 (2018)
26. Hamdaoui, Y., Maach, A.: Prosumers integration and the hybrid communication in smart grid context. In: Networked Systems. LNCS, vol. 9466. Springer, Cham (2015)

Performance of DFIG Wind Turbine Using Dynamic Voltage Restorer Based Fuzzy Logic Controller

Kaoutar Rabyi(✉) and Hassane Mahmoudi

Electrical Engineering Department, Mohammadia School of Engineers, Mohammed V University, Rabat, Morocco
rabyi.kaoutar@gmail.com

Abstract. In this paper a Doubly Fed Induction Generator (DFIG) based Dynamic Voltage Restorer (DVR) is proposed to handle voltage sag and swell when the DFIG is connected to the electrical grid, the DVR which is a constitution of three H-bridge inverters supplied through a common DC capacitor, can recover sag to 50%, alternatively it operates as an uninterruptable power supply when a failure occurs in the grid supply. The Fuzzy Logic Controller (FLC) is proposed to be used instead of proportional controllers PI of the DVR because of its efficiency in nonlinear systems; there are two controllers in the grid side converter of the DVR which are replaced by two FLC, the simulations were carried out in simulink/Matlab software, the results show the reliability of the proposed DVR based FLC in terms of voltage sag and synchronization time.

Keywords: Dynamic Voltage Restorer · Voltage sag · Doubly fed induction generator · Fuzzy logic controller · Power quality

1 Introduction

In recent years wind energy production has been growing fast. However, many challenges appeared concerning the integration of the wind energy into the electrical grid. The wind variation doesn't work in harmony with the grid. Among the wind energy new grid code, the control of reactive power and the Fault Ride Through (FRT) presents noticeable challenges for wind turbines with variable speed [1].

For a good production, wind farms must operate as a conventional power plants, they are required to overcome variable disconnection during faults and voltage sag conditions. Variable speed wind turbines are widely used and more attractive than fixed-speed generators because of their quality and performance during grid faults; that's why the most frequent technology used is the Doubly Fed Induction Generator (DFIG) based wind turbine (WT) [2, 3].

During existence of the faults, the rotor current is increased to compensate the active power through the rotor side converter (RSC) and the voltage drops to zero. It follows an occurrence of overvoltage in the DC-Link due to the rotor voltage increase. Crowbars are a conventional solution to protect the electronic converter from the over rotor

M. Ezziyyani (Ed.): AI2SD 2018, AISC 912, pp. 141–148, 2019.
https://doi.org/10.1007/978-3-030-12065-8_14

currents flow, but the DFIG acts like an induction machine; it absorbs the reactive power instead of offering it, because the rotor winding is circuited by shunt resistors [4].

The use of a Dynamic Voltage Restorer (DVR) is a great solution because it doesn't require other protective circuit when power generation [5], the schematic of the DVR connected to the DFIG is shown in Fig. 1.

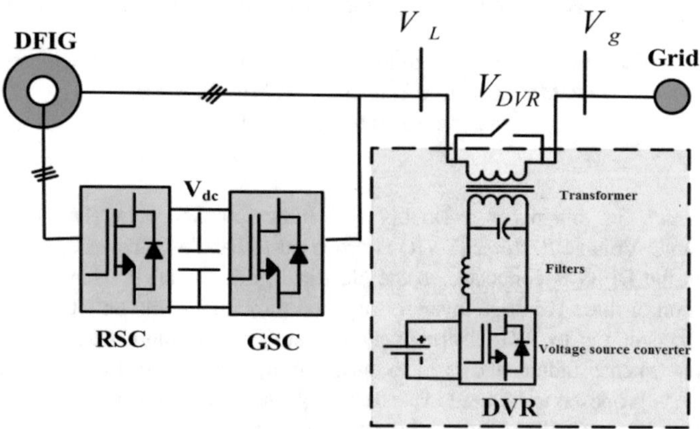

Fig. 1. DVR connected to the DFIG schematic

The efficiency of the DVR for the DFIG FRT capability depends on its control algorithm, the process of the DVR control is carried out in many steps; from input data, the operation mode selection and voltage disturbance detection, reference generation, voltage and current control, modulation and pulse generation for the converter [5].

This paper concentrates on the control system of the DVR, the proposed control strategy combines three control strategy; phase lock loop (PLL), PI controller and Fuzzy Logic Controller (FLC), the control is discussed in this paper under sag conditions, This paper is structured as follows, Sect. 2 discusses the modeling of the DFIG, Sect. 3 explains the DVR control strategy, the voltage reference calculation, the FLC rule bases, Sect. 4 includes the simulation of the DFIG connected to the grid and its improvement when the DVR based FLC is connected, the last section end with a conclusion.

2 DFIG Modeling

Figure 2 shows the equivalent circuit of the DFIG, the rotor is connected to the grid via slip rings via the RSC and GSC, the stator is connected directly (Fig. 1).

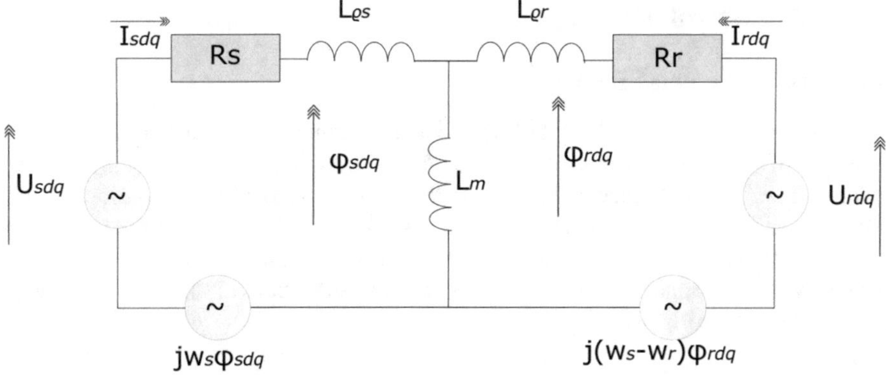

Fig. 2. T-representation of the DFIG equivalent circuit in dq reference frame

The doubly fed induction generator [7] is modeled in reference Park, leading to the following equations:

$$V_{ds} = R_s i_{ds} + \frac{d\phi_{ds}}{dt} - w_s \phi_{qs} \tag{1}$$

$$V_{qs} = R_s i_{qs} + \frac{d\phi_{qs}}{dt} + w_s \phi_{ds} \tag{2}$$

$$V_{dr} = R_r i_{dr} + \frac{d\phi_{dr}}{d_t} - (w_s - w_m)\phi_{qr} \tag{3}$$

$$V_{qr} = R_r i_{qr} + \frac{d\phi_{qr}}{d_t} + (w_s - w_m)\phi_{dr} \tag{4}$$

- $[\phi_{ds}, \phi_{qs}]$, $[V_{ds}, V_{qs}]$, $[i_{ds}, i_{qs}]$ the components of the flux, voltage and current of the stator.
- $[\phi_{dr}, \phi_{qr}]$, $[V_{dr}, V_{qr}]$, $[i_{dr}, i_{qr}]$ the components of the flux, voltage and current of the rotor.

The active and reactive power equations at the stator and rotor windings are written as follows:

$$P_s = V_{ds} I_{ds} + V_{qs} I_{qs} \tag{5}$$

$$Q_s = V_{qs} I_{ds} - V_{ds} I_{qs} \tag{6}$$

$$P_r = V_{qr} I_{dr} + V_{qr} I_{dr} \tag{7}$$

$$Q_r = V_{qr} I_{dr} - V_{qr} I_{dr} \tag{8}$$

3 DVR Control Strategy

3.1 Dynamic Voltage Restorer

DVR is a constitution of three H-bridge inverters supplied through a common DC capacitor [8], the voltage capacitor Vdc is considered as the supply voltage input for the inverter. The DVR is connected in series to the DFIG and the grid, it is a voltage source converter, at the point of common coupling (PCC), it injects the appropriate voltage to overcome the voltage sag, swell and harmonics, and to obtain a nominal stator voltage, for the switching signals of the voltage source converter, the technique of pulse width modulation (PWM) is used [9] (Fig. 3).

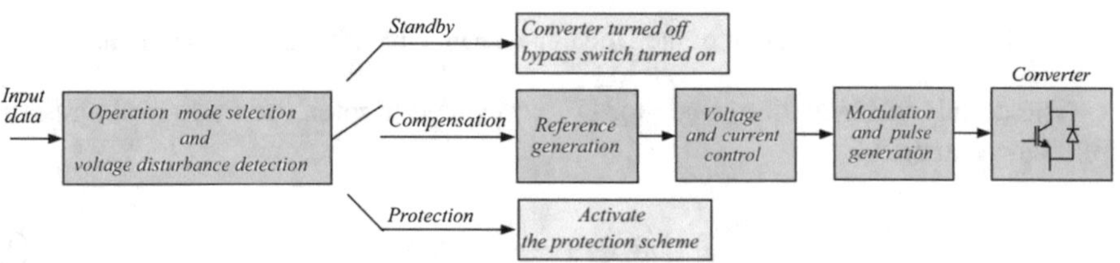

Fig. 3. Schematic of the control system of DVR

Figure 4 shows the control scheme to generate the reference voltage for the control purpose, under voltage sag and swell conditions, the DVR must inject rapidly an AC voltage into the grid, there are many techniques to generate the reference voltage as it is cited in [6], one of these techniques is the synchronous reference frame (SRF); it is based on instantaneous supply voltage value, two reference voltage value are produced Vd and Vq (Fig. 6).

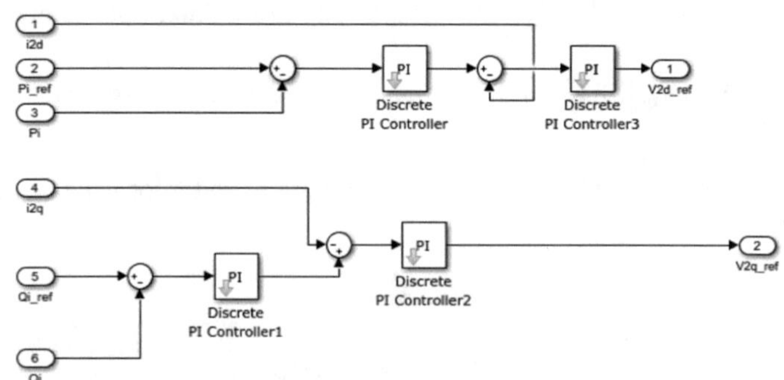

Fig. 4. Schematic of voltage reference calculation

3-phase locked loop (PLL) detects the voltage grid to synchronize the grid phase angle, the grid side inverter is controlled by conventional PI regulators as it shown in Fig. 4, the control generates d-q reference, which is transformed to a-b-c reference (three phases stationary frame), these a-b-c values generate the PWM signals as it shown in Fig. 5. The transformers inject the compensation voltages to the grid at PCC.

The supplied voltage of DV can be written as:

$$V_{SOURCE} = V_{DVR} + Z_{LINE}I_{LOAD} + V_{LOAD} \tag{9}$$

$$V_{DVR} = V_{SOURCE} - V_{LOAD} - Z_{LINE}I_{LOAD} \tag{10}$$

Where V_{SOURCE} is the system voltage, V_{DVR} is the injected voltage by DVR, Z_{LINE} is the DVR line impedance, I_{LOAD}, V_{LOAD} are respectively the DFIG current and voltage.

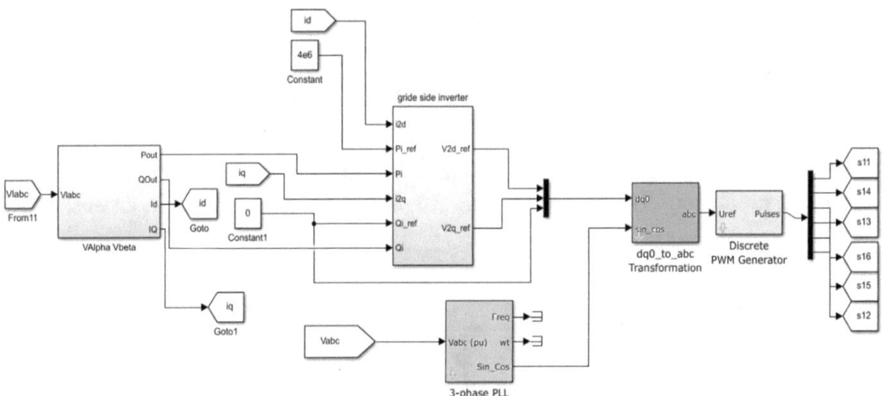

Fig. 5. Schematic of DVR control

3.2 DVR Based Fuzzy Logic Controller

To avoid mathematical calculation the use of FLC is the appropriate choice. The FL is a combination of probability theory, valued logic and artificial intelligence. The FLC has a great time response time, it is sensitive against small variation in the system. There are two types of FLC; the Mamdani and Sugeno fuzzy models, the Mamdani model is more intuitive and is like more human intelligence, it consists of significant computational concern. The Sugeno model is more advanced and very useful for nonlinear systems and is applied in different control subjects [10–12].

The components of the FLC system are:

- The Knowledge Base which contains the knowledge of the input and output
- The membership functions corresponding to the input and outputs, they define the fuzzy "Rule-Base"
- The Fuzzification Interface, in this stage, crisp input signals are converted into fuzzified one, these signals are identified by level of membership functions
- The Decision Logic
- The Defuzzification Interface

The DVR has two PI controllers in the grid side converter as it is shown in Fig. 4, in order to have more performed results, two FLC replace the PI controllers as it is shown in Fig. 6.

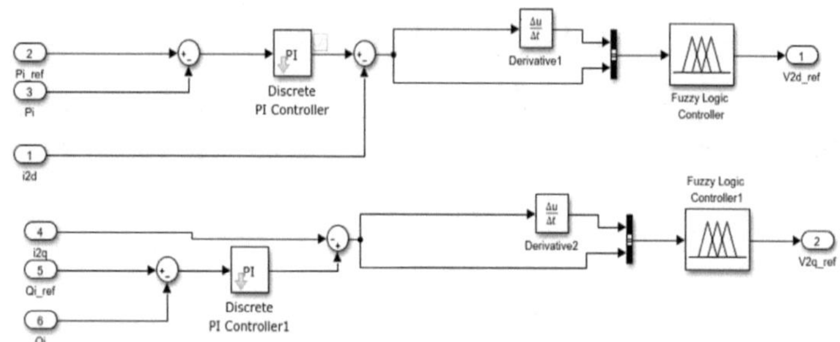

Fig. 6. Schematic of the Fuzzy logic controller for the voltage reference calculation

The proposed membership functions have a triangular shape with 50% overlap for the control precision, Fig. 7 shows the error and the change in error, input and output variables.

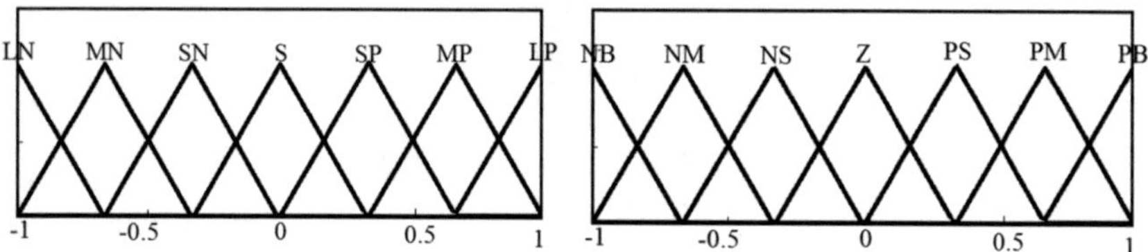

Fig. 7. Membership function used for input variable "on the Left "change in error & error", on the right "output"

The proposed rule base for the FLC is illustrated in Table 1.

Table 1. Rule base representation

Change in error	*Error*						
	NB	*NM*	*NS*	*Z*	*PS*	*PM*	*PB*
NB	PB	PB	PB	PM	PM	PS	Z
NM	PB	PB	PM	PM	PS	Z	NS
NS	PB	PM	PM	PS	Z	NS	NM
Z	PM	PM	PS	Z	NS	NM	NM
PS	PM	PS	Z	NS	NM	NM	NB
PM	PS	Z	NS	NM	NM	NB	NB
PB	Z	NS	NM	NM	NB	NB	NB

4 Simulation Results

The system is simulated for DFIG based WT connected to the grid, the simulation is carried out in Matlab/Simulink environment.

The simulations parameters are:

- DFIG of 1.5 MW, the base wind speed is 9 m/s, the pitch angle controller gain is [kp = 5 ki = 25], the stator volage is 575 V, the frequency is f = 50 Hz
- The DC link voltage Vdc = 300 V
- DVR filter inductance is L = 0.3 mH
- DVR filter capacitance is C = 1 μF
- Series transformer ration 1:1

The DVR performance is evaluated for unbalanced sag of 50%, the sag lasts between instant t = 0.5 s and t = 0.7 s, for 10 cycles harmonics and fault are presented.

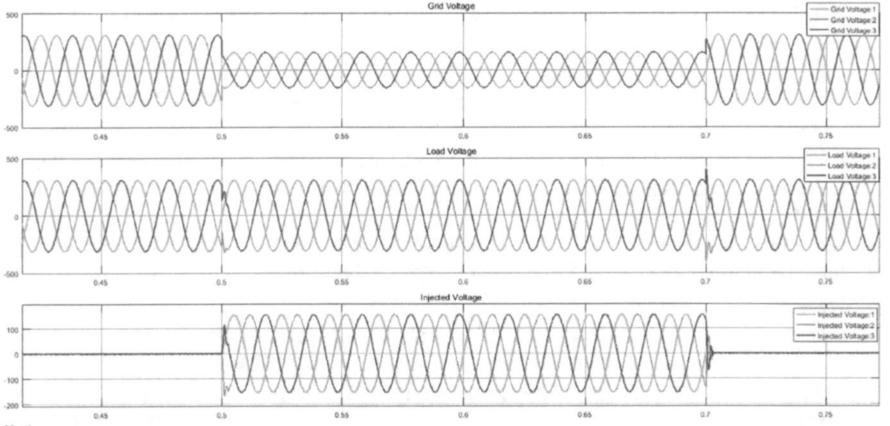

Fig. 8. DVR from up to bottom: supply voltage, DFIG voltage, injected voltage

Under application of sag condition to the system, the performance of the DFIG of the supply voltage is shown in Fig. 8. Between instant 0.5 and 0.7 s; 10 cycles are disturbed. The recovery of the DFIG voltage is observed when the injected voltage by the DVR drop, it reacts rapidly to inject an appropriate voltage with a negative magnitude, this injection corrects the supply voltage from the DFIG, and maintain the stator voltage during the fault using the DVR, without using it, the DC-link of DFIG produce an overvoltage, Fig. 8 shows the control of the DC-link voltage.

The aim of the proposed scheme is to protect and isolate the wind turbine from any transient currents or voltages and asymmetrical faults. The performance of the DVR under sag condition is efficient, it corrects the voltage wave form of the DFIG.

The use of the FLC as a DVR controller allows a rapid response of the proposed strategy, it takes only 0.2 s for the recovery of the DFIG voltage to the appropriate value and form.

5 Conclusion

This paper proposes DVR based FLC connected to the DFIG under sag condition. The design of the DVR with the PLL, FLC, and PI controller mitigates the voltage sags and swells, the simulation of the proposed control strategy using Simulink/Matlab has been presented, the FLC applies the error signal and the PLL synchronize the grid phase angle, for the switching signals of the voltage source converter, the technique of pulse width modulation is used. Added proposed work includes comparison of actual work with an experimental setup.

References

1. Amalorpavaraj, R.A.J., Kaliannan, P., Subramaniam, U.: Improved fault ride through capability of DFIG based wind turbines using synchronous reference frame control based dynamic voltage restorer. ISA Trans. **70**(1), 465–474 (2017)
2. Wu, X., Guan, Y., Ning, W.: Reactive power control strategy of DFIG-based wind farm to mitigate SSO. J. Eng. **2017**(13), 1290–1294 (2017)
3. Gaillard, A., Poure, P., Saadate, S., Machmoum, M.: Variable speed DFIG wind energy system for power generation and harmonic mitigation. Renew. Energy **34**(6), 1545–1553 (2009)
4. Abobkr, A.H., El-Hawary, M.E.: Fault ride-through capability of doubly-fed induction generators based wind turbines. In: Electrical Power and Energy Conference (EPEC). IEEE (2015)
5. Sitharthan, R., Sundarabalan, C.K., Devabalaji, K.R., Nataraj, S.K., Karthikeyan, M.: Improved fault ride through capability of DFIG-wind turbines using customized dynamic voltage restorer. Sustain. Cities Soc. **39**, 114–125 (2018)
6. Farhadi-Kangarlua, M., Babaei, E., Blaabjerg, F.: A comprehensive review of dynamic voltage restorers. Electr. Power Energy Syst. **92**, 136–155 (2017)
7. Rolán, A., Pedra, J., Córcoles, F.: Detailed study of DFIG-based wind turbines to overcome the most severe grid faults. Int. J. Electr. Power Energy Syst. **62**, 868–878 (2014)
8. Leon, A.E., Farias, M.F., Battaiotto, P.E., Solsona, J.A., Valla, M.I.: Control strategy of a DVR to improve stability in wind farms using squirrel-cage induction generators. IEEE Trans. Power Syst. **26**(3), 1609–1617 (2011)
9. González, M., Cárdenas, V., Morán, L., Espinoza, J.: Selecting between linear and nonlinear control in a dynamic voltage restorer. In: Proceedings of IEEE Power Electronics Specialists Conference (PESC), pp. 3867–3872, June 2008
10. Ying, H., Ding, Y., Li, S., Shao, S.: Comparison of necessary conditions for typical Takagi-Sugeno and Mamdani fuzzy systems as universal approximators. IEEE Trans. Syst. Man Cybern. Part A Syst. Hum. **29**(5), 508–514 (1999)
11. Rocha, M.M., da Silva, J.P., das Chagas Barbosa de Sena, F.: Simulation of a fuzzy control applied to a variable speed wind system connected to the electrical network. IEEE Lat. Am. Trans. **16**(2), 512–526 (2018)
12. Ofoli, A.R.: Fuzzy-logic applications in electric drives and power electronics. In: Power Electronics Handbook, 4th edn., pp. 1221–1243 (2018)

Energy Performance and Environmental Impact of an Earth-Air Heat Exchanger for Heating and Cooling a Poultry House

Azzeddine Laknizi[1,2(✉)], Mustapha Mahdaoui[3], Abdelatif Ben Abdellah[1,2], Kamal Anoune[1,2], and Mohsine Bouya[2]

[1] Faculty of Science and Technology, Abdelmalek Esaadi University, Tangier, Morocco
eng.azzeddinelaknizi@gmail.com
[2] Department of Valorization and Transfer, International University of Rabat Parc Technopolis, Sala al Jadida, 11100 Rabat, Morocco
[3] Equipe de Recherche en Transferts Thermiques & Énergétique - UAE/E14FST Département de Physique FST, Université Abdelmalek Essaâdi, Tangier, Maroc

Abstract. The poultry industry in Morocco is an important sector that contributes to the growth of national economy and food security, but this sector faces many problems, one of those problems is the climatic conditions, heat stress wave in summer and cold wave in winter that causes mortality and drops in performance (weight drop). Another problem is the relatively high consumption of conventional energy and corresponding environmental impact of the released greenhouse gas emissions.

In this paper, a proposed system for heating and cooling a poultry house is modeled and simulated for the weather conditions of Tangier city-Morocco. The system consists of an earth-air heat exchanger, a mixing box, and an air-air heat recovery, which are combined to control the temperature inside the poultry house efficiently. The results showed that the proposed system has the potential to save between 34.7% and 96% of the energy demand in the heating mode and in the cooling mode respectively. The required number of tubes was found to be around 80 parallel tubes of 0.2 m in diameter and 30 m in length.

Keywords: Earth-air heat exchanger · Heat recovery · Poultry house · Cooling · Heating · Environmental impact

1 Introduction

Food security has become a global concern because of the actual growth of the population that will reach a peak of 9.2 billion by 2050 [1]. One sector that makes and expects to continue to make a major contribution to global food security and nutrition is the poultry sector. This contribution is limited by many challenges related to climatic conditions and energy consumption. The poultry sector consumes energy for ventilation, lighting, heating, and cooling, which lead to a high energy bill [2] and an increase

M. Ezziyyani (Ed.): AI2SD 2018, AISC 912, pp. 149–157, 2019.
https://doi.org/10.1007/978-3-030-12065-8_15

of the environmental impact of released greenhouses gases [3]. This issue of the environment is addressed by many researchers [4–6].

In Morocco, according to the poultry federation of Morocco [7], the Moroccan poultry industry produced 560 tons of meat, 4.5 milliards of eggs, employs 360 000 people 110 000 directly and 250 000 indirectly with a total investment of 854.5 Million €. The heat wave and the cold wave that strikes the country produce great losses in the sector. For the heat wave, the losses for summer 2012 are estimated at 12% in terms of mortality and 25% in terms of a performance drop (weight loss). For the cold wave an increase in feed efficiency (quantity of food consumed/kg body weight produced) and high heating load for the building, cause a sharp increase in production costs.

To the authors' best knowledge, there is no known case in the Moroccan poultry sector where the earth-air heat exchanger is employed; it is only used in few residential buildings [8, 9]. Some poultry houses only use pad cooling and ventilation during hot periods. During cold periods, they use fuel boiler for heating and they improve insulation by adding plastic. This absent is due to that the energy in Morocco is subsidies by a compensation fund which, in the economic viewpoint, makes conventional solutions strong competitors to the renewable and passive solution. This subsidy will be cut because of Morocco imports around 96% of its energy need, and this represents a large burden to the budget of the country [10]. To reduce this energy dependence, Moroccan government implements many programs to reduce the energy consumption in the agriculture sector by promoting energy efficiency and renewable energy systems production in farms [11].

The aim of this study is to evaluate a system for heating and cooling a typical poultry Moroccan poultry house in reducing the energy consumption. This proposed system combining an earth-air heat exchanger, mixing box, and an air-air heat recovery. The evaluation is conducted under the weather conditions of the Tangier-city in Morocco. In addition, greenhouse gas emissions mitigations by using the proposed system are also evaluated.

2 System Description and Modeling

Figure 1 illustrates the proposed system model. The components used in this model are:

- EAHE is a buried horizontal tube. The length and diameter of the tube are 30 m and 0.2 m respectively.
- Mixing box.
- Heat recovery system is an air-air heat recovery system it consists of the heat exchanger with an efficiency of 0.7.

Bypass valves are used to assure the change between heating and cooling mode. The mixing box is bypassed during heating mode. The Air-air heat recovery is bypassed during cooling mode. The earth air heat exchanger is bypassed in two situations: the first is when the outdoor temperature is equal to the soil temperature, which means that the air will not be cooled or heated as it passes through the tubes. The second situation is when the outdoor temperature is between the soil temperature and

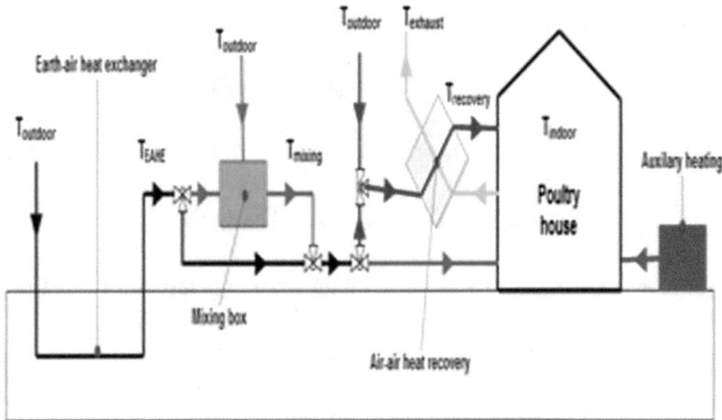

Fig. 1. System model

the indoor desired temperature which means that the use of the earth air heat exchanger has a negative effect, it will lead to cooling instead of heating.

2.1 Earth-Air Heat Exchanger Modeling

The earth-air heat exchanger is a geothermal system, which uses the energy of the ground to heat or cools the fresh air. Depending on weather conditions, day and season, outside air temperature undergo strong variations. In contrast, the ground temperature, a few meters below the surface, varied slightly due to its high thermal inertia of the ground.

The air temperature at the outlet of the earth air heat exchanger is calculated by using the following equation [12].

$$T_o = T_s + (T_i - T_s)e^{-\left(\frac{\pi DLh}{\dot{m}c_p}\right)} \quad (1)$$

Where T_s is the soil temperature, T_i is the inlet temperature which represent the outdoor temperature, D is the inner diameter of the tube (m), L is the length of the tube (m), h is the convection heat transfer coefficient (W/m^2.K), $\dot{m}$ is the air flow rate (kg/s), and c_p is the specific heat capacity (J/kg. K).

2.2 The Energy Gain

The energy gain can be calculated by using the following equations:

Energy gain by EAHE in heating mode:

$$E_{EAHE} = \dot{m}C_p(T_{EAHE} - T_{out})N_h \quad (2)$$

Energy gain by EAHE in cooling mode:

$$E_{EAHE} = \dot{m}C_p(T_{out} - T_{EAHE})N_h \quad (3)$$

Energy gain by heat recovery system when operates in series with EAHE:

$$E_{HR} = \dot{m}C_p(T_{recovery} - T_{EAHE})N_h \quad (4)$$

Energy gain by heat recovery system when operates separately to the EAHE:

$$E_{HR} = \dot{m}C_p(T_{recovery} - T_{out})N_h \quad (5)$$

Where, N_h is the number of the operating hours.

3 Methodology

To evaluate the potential of heating and cooling the dimension of the earth-air heat exchanger was fixed to be D = 0.2 m and L = 30 m. The agriculture year was divided into seven full growing cycles of 42 days, between two successive cycles there is a period for cleaning and disinfection of 7 days. The operating conditions are taken to be the weather conditions of the said city.

The earth-air heat exchanger will be designed to a typical poultry house in Morocco. The said poultry house has as dimension a 1000 m^2 area and 3 m in height with a capacity of 10000 chickens.

The indoor design conditions are giving by ASHRAE [13] (American Society of Heating, Refrigerating and Air-Conditioning Engineers) recommendations Table 1.

Table 1. The required indoor temperature during the growing period

Week	Indoor temperature
1st	33 °C
2nd	30 °C
3rd	27 °C
4th	24 °C
5th	21 °C
6th	18 °C

4 Results of the Energy Performance

For a velocity of 2 m/s and air change rate of 6 h − 1, the modeling equations of the system are solved hourly under the climate conditions of six Moroccan cities. In the first section, the results obtained in each growing cycle are discussed in detail for the case of Tangier city. In the second section, the results of the other cities are presented and discussed.

4.1 Hourly Temperatures Changes

Taking the weather data of Tangier as an example, Fig. 2 shows the outdoor temperature, the required indoor temperature, the soil temperature and the EAHE's outlet temperature during the first cycle. The energy demand is determined by comparison between these temperatures. When the outdoor temperature is lower than the required temperature the energy demand is in heating and is in cooling when the reverse. For this first cycle, it can be seen from Fig. 2 that the outdoor temperature is lower than the required temperature. Therefore, the energy demand is for heating. The ability of the system to provide this energy demand depends on these temperatures.

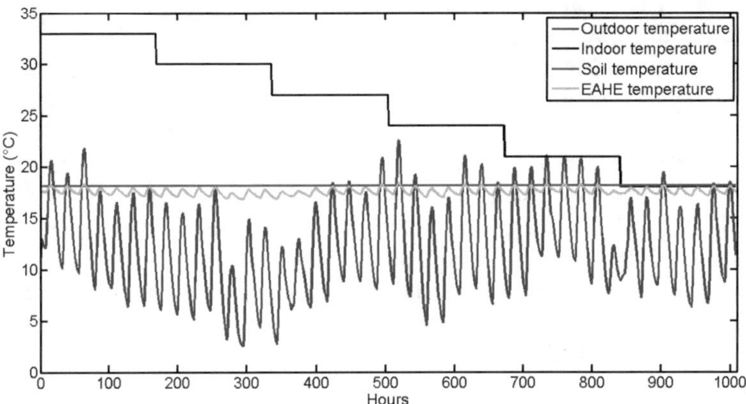

Fig. 2. Hourly temperature changes during the first cycle

4.2 Energy Performance for Tangier City

Figure 3 shows the energy performance of the proposed system in one agriculture year of seven full cycles. The total energy demand in heating and cooling modes, the energy gain in heating and cooling modes and the heat recovery are calculated weekly by integration of the governing equations. One can see from this figure that the heating demand is dominant in the three first cycles and the last cycle, while the cooling demand begin to appear in the 6th week of cycle 3 and become more required in the 3 last weeks of the cycle 4, cycle 5, and cycle 6 which their period coincide with the summer season. The energy performance in each cycle is presented in the following paragraphs:

In the first cycle, the poultry house needs 1008 h of heating with a total heating load of 80.7 MW which represents an energy demand of 81.33 GWh. By using an EAHE coupled to the heat recovery system this energy demand can be reduced by 53% with an energy gain by EAHE of 31.17 GWh in 952 operating hours while the energy gain by heat recovery system is 11.96 GWh in 840 operating hours.

In the second cycle, the energy demand is for cooling and heating. For heating, the poultry house needs 948 h of heating with a total heating load of 66.64 MW with a heating energy demand of 63.17 GWh. This energy demand can be covered by 45% by using an EAHE and heat recovery system. The energy gain by EAHE is 16.99 GWh in 828 operating hours while the energy gain by heat recovery system is 11.46 GWh in 832 operating hours. For cooling, the poultry house needs 57 h of cooling with a total load of 709.33 kW; the EAHE can cover 92.7% of this cooling energy demand with an energy gain of 37.48 MWh in 57 operating hours.

In the third cycle, the energy demand is for cooling and heating. For heating, the poultry house needs 915 h of heating with a total heating load of 55.9 MW; this corresponds to a heating energy demand of 51.15 GWh. This energy demand can be covered by 37.14% by using an EAHE and heat recovery system. The energy gain by EAHE is 8.27 GWh in 649 operating hours while the energy gain by heat recovery system is 10.73 GWh in 831 operating hours.

In the fourth cycle, the energy demand is for cooling and heating. For heating, the poultry house needs 634 h of heating with a total heating load of 28.918 MW with a heating energy demand of 18.33 GWh. This energy demand can be covered by 29% by using an EAHE and heat recovery system, the energy gain by EAHE is 158.983 MWh for 149 operating hours and 5.16 GWh for 618 operating hours by heat recovery system. For cooling, the poultry house needs 371 h of cooling with a total load of 9.528 MW; the EAHE can cover 96.47% of this cooling energy demand with an energy gain of 3.41 GWh for 371 operating hours.

In the fifth cycle, the energy demand is for cooling and heating. For heating, the poultry house needs 548 h of heating with a total heating load of 17.830 MW with a heating energy demand of 9.77 GWh. This energy demand can be covered by 29.92% by using an EAHE and heat recovery system, the energy gain by EAHE is 657.979 kWh for 14 operating hours and 2.92 kWh for 548 operating hours by heat recovery system. For cooling, the poultry house needs 455 h of cooling with a total load of 14.502 MW; the EAHE can cover 96.29% of this cooling energy demand with an energy gain of 6.35 GWh for 455 operating hours.

In the sixth cycle, the energy demand is for cooling and heating. For heating, the poultry house needs 738 h of heating with a total heating load of 28.142 MW with a heating energy demand of 20.768 GWh. This energy demand can be covered by 29.18% by using an EAHE and heat recovery system, the energy gain by EAHE is 132.328 MWh for 134 operating hours and 5.93 GWh for 728 operating hours by heat recovery system. For cooling, the poultry house needs 264 h of cooling with a total load of 5.678 MW; the EAHE can cover 94% of this cooling energy demand with an energy gain of 1.41 GWh for 264 operating hours.

In the seventh cycle, the energy demand is for cooling and heating. For heating, the poultry house needs 992 h of heating with a total heating load of 58.542 MW with a heating energy demand of 58.07 GWh. This energy demand can be covered by 37.66% by using an EAHE and heat recovery system, the energy gain by EAHE is 11.296 GWh for 733 operating hours and 10.57 GWh for 825 operating hours by heat recovery system. For cooling, the poultry house needs 15 h of cooling with a total load of 102 kW; the EAHE can cover 96% of this cooling energy demand with an energy gain of 101 kW for 15 operating hours.

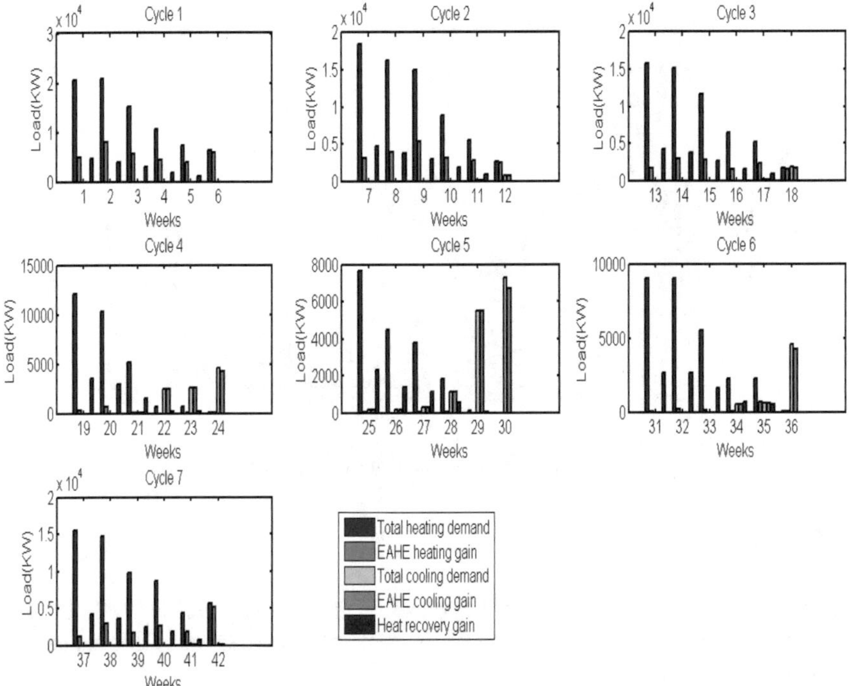

Fig. 3. Weekly load transfer during the seven full cycle

5 CO2 Mitigation

The greenhouse gas emissions mitigation from energy saving due to the use of earth-air heat exchanger in a poultry house is calculated by the following equation:

$$M_{\text{emission mitigated}} = C \times \text{E} \quad (6)$$

Where C is the emission factor (Emissions per kWh of electricity generated). For Morocco, this factor is equal to 0.731211458 kgCO2/kWh for dioxide of carbon, 0.00001301900 kgCH4/kWh for methane 0.00000945179 kgN2O/kWh for dioxide of nitrogen [14]. The values of CH4 and N2O are transformed to their equivalent value of CO2 by using the Global Warming Potential (GWP) factor which equals to 25 times for CH4 and 298 times for N2O [15].

Based on the simulation results, the gas emission mitigation is 101093 Tonnes. Figure 4 shows the amount of CO2 that can be mitigated in each cycle by using the proposed system.

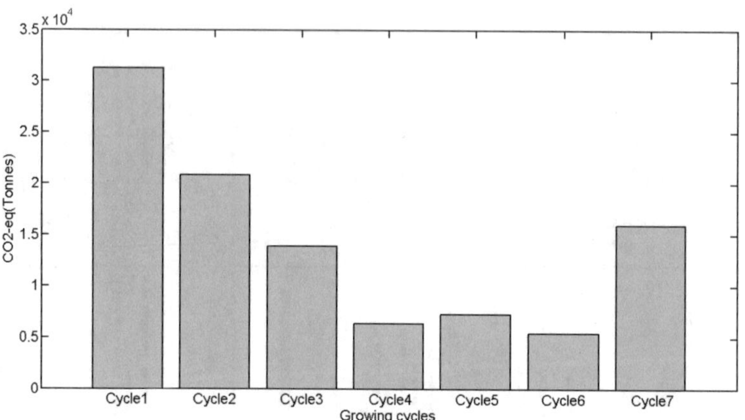

Fig. 4. CO2 mitigation by using EAHE

6 Conclusion

In this paper, a proposed system for application in a poultry house was studied; an hourly simulation analysis was conducted to evaluate the thermal performance of the proposed system. The results show:

- The energy demand varies from a cycle to another. In heating mode, the annual energy demand is 302.609 GWh, while the annual cooling energy demand is 11.9 GWh. This result is important for investors in the field of poultry farming for the choice of zones where the risk of climate conditions and the demand for energy are less.
- The proposed system can cool the air in case of heat stress and heat the air in case of cold stress. Therefore, its implementation in poultry house can attenuate the effect the heat/cold wave and the corresponding losses.
- The proposed system can reduce the energy demand of the poultry house. The coverage in cooling mode is 96% while the coverage in heating mode is from 34.7%.
- Another benefit of the proposed system is the environmental issue; the use of the proposed system results in reduced CO2 emission.

From this study, we can conclude that the application of the system in the Moroccan poultry sector is promising and will reduce the energy bill of the agriculture sector which represents 20.3% of the national consumption [11]. The results of this study can be applicable to the residential sector.

Future works should be conducted to evaluate the economic aspect of this system.

Acknowledgment. The authors would like to express their appreciation to "IRESEN" for providing financial support to carry out this research.

References

1. FAO: How to Feed the World in 2050. http://www.fao.org/fileadmin/templates/wsfs/docs/expert_paper/How_to_Feed_the_World_in_2050.pdf
2. Kilic, I.: Analysis of the energy efficiency of poultry houses in the Bursa region of Turkey. J. Appl. Anim. Res. **44**(1), 165–172 (2016)
3. Fidaros, D., et al.: Numerical study of mechanically ventilated broiler house equipped with evaporative pads. Comput. Electron. Agric. (2017)
4. Mainali, B., Emran, S.B., Silveira, S.: Greenhouse gas mitigation using poultry litter management techniques in Bangladesh. Energy **127**, 155–166 (2017)
5. Yue, Q., et al.: Mitigating greenhouse gas emissions in agriculture: from farm production to food consumption. J. Clean. Prod. **149**, 1011–1019 (2017)
6. Jørgensen, A., Bikker, P., Herrmann, I.T.: Assessing the greenhouse gas emissions from poultry fat biodiesel. J. Clean. Prod. **24**, 85–91 (2012)
7. FISA: http://www.fisamaroc.org.ma/
8. Hollmuller, P.: Etude expérimentale d'un échangeur de chaleur air-sol (puits canadien) pour le rafraichissement d'un bâtiment résidentiel à Marrakech, April 2014
9. Khabbaz, M., et al.: Experimental and numerical study of an earth-to-air heat exchanger for air cooling in a residential building in hot semi-arid climate. Energy Build. **125**, 109–121 (2016)
10. Moroccan Ministry of Energy, Mines, Water & Environment. http://www.mem.gov.ma
11. Al-Badri, A.R., Al-Waaly, A.A.Y.: The influence of chilled water on the performance of direct evaporative cooling. Energy Build. **155**(Supplement C), 143–150 (2017)
12. De Paepe, M., Janssens, A.: Thermo-hydraulic design of earth-air heat exchangers. Energy Build. **35**(4), 389–397 (2003)
13. ASHRAE: Environmental Control for Animals and Plants. HVAC Applications. ASHRAE Inc., Atlanta, GA (2011)
14. Brander, M., et al.: Electricity-specific emission factors for grid electricity. Ecometrica (2011). Technical Paper. Emissionfactors.com
15. IPCC: https://www.ipcc.ch/publications_and_data/ar4/wg1/en/ch2s2-10-2.html

Raman Analysis of Graphene/PANI Nanocomposites for Photovoltaic

Mourad Boutahir[1(✉)], Jamal Chenouf[1], Oussama Boutahir[1], Abdelhai Rahmani[1], Hassane Chadli[1,2], and Abdelali Rahmani[1]

[1] Advanced Material and Applications Laboratory (LEM2A), University Moulay Ismail, FSM-ESTM-FPE, BP 11201, Zitoune, 50000 Meknes, Morocco
mourad.boutahir@gmail.com

[2] Equipe Physique Informatique et Modélisation des Procédés, Ecole Superieur de Technologie, BP 170, 54000 Khènifra, Morocco

Abstract. The nanocomposites based on graphene are single- or few-layer platelets that can be produced in bulk quantities by the in-situ polymerization method. Indeed, graphene is firstly added to a solution of the monomer. The polymerization is initiated either by heat or radiation. With this technique, a variety of polymeric nanocomposites have been prepared using the different types of graphene based nanofillers. Solution mixing techniques have been shown to represent the most effective for the dispersion of graphene nanosheets in polymers, in order to manufacture new high nanocomposite systems performance. In this experimental work, we report the synthesis and characterisation using Raman spectrometer of PANI and graphene/PANI. We study the vibrational properties of polyaniline (PANI) and graphene/PANI. We find a Charges/Energy Transfer in these promising materials, the graphene/PANI nanocomposites.

Keywords: PANI · Graphene · Raman spectrometer

1 Introduction

Organic solar cells are a promising low-cost alternative to silicon solar cells [1–3], but the main handicap is the low power conversion efficiency of these devices [3–5]. One of the alternatives to solve this handicap is the introduction of carbon nanotubes (NTs) or graphene to form an interpenetrating blend with the polymer. Since the discovery of photoinduced charge transfer between organic conjugated polymers (as donor) and nanotubes or graphene (as acceptor), carbon nanotubes and graphene have been used to fabricate photovoltaic devices in combination with different polymers to the aim of increasing the power conversion efficiency [6–8].

The discovery of intrinsically conducting polymers in 1960 has led researchers to study this attractive subject of research because of the interesting properties and numerous application possibilities of conductive polymers [9–11]. Among the most interesting available conductive polymers, polyaniline (PANI) is found to be the most promising : having a controllable electrical conductivity and, varying between an

M. Ezziyyani (Ed.): AI2SD 2018, AISC 912, pp. 158–163, 2019.
https://doi.org/10.1007/978-3-030-12065-8_16

insulator, a semiconductor and a metal [12, 13]. It can be easily synthesized chemically and electrochemically. It is stable chemically, with a strong absorption in the visible spectrum and a high mobility of charge carriers [14]. Its physical properties are controlled by both oxidation and protonation states. Reversible doping is a unique property of polyaniline, it is relatively simple by the addition of acid-base, which makes it possible to control the electrical and optical properties of the PANI.

The electrical conductivity of PANI varies between that of plastic polymers or insulators and that of metals. This conductivity is controllable by the type of polymerization used to form the PANI, and by the amount and nature of the dopant. In addition, PANI can be mixed with plastic polymers using economical and simple methods in solution form or by fusion processing techniques. To form these mixtures, several plastic polymers are used, such as polyethylene terephthalate) (PET) [15] and poly (methyl methacrylate) (PMMA) [16]. On the other hand, aniline can be polymerized in situ in polyvinyl alcohol either [17, 18] or in metal oxides [19]. Due to its properties, PANI is used in many applications, including anti-corrosion treatment in non-acidic environments [20] and also in hybrid solar cells [21, 22].

In this paper we will focus on the vibrational properties of graphene/polyaniline. The synthesis and characterization of polyanilines, using Raman spectrometer.

2 Experimental Methods

2.1 PANI (Polyaniline)

Analytical grade of aniline (C_6H_7N), 37% hydrochloric acid (HCl), and ammonium persulfate (APS, $((NH_4)_2S_2O_8)$) were used as received without further purification. The chemical route for realization of Emeraldine Salt PANI involves an oxidative polymerization of anilinium ion in inorganic acid by APS. The chemical bath deposition (CBD) and stoichiometric conditions was described elsewhere [23]. Briefly, the beaker constitutes of solutions of 2 mL aniline, 1 M 30 mL HCl, and 0.25 M 20 mL APS dissolved in 1 M HCl. Quartz substrates were immersed vertically in the mixture at room temperature. After 2 h, PANI films were removed and rinsed with distilled water. Preparation of thicker films may be carried out by reinserting the initially deposited films into a fresh bath.

2.2 Graphene/PANI

The graphene/PANI composites were synthesized by an in-situ polymerization of aniline in the presence of graphene (GN). The weight percent of GN to aniline was varied from 1% to 5%. The solution of 0.2 M H_2SO_4 in 50 ml of demonized water was divided into two parts. In one part 0.2 M aniline and functionalized GN was added and the mixture was ultrasonicated for 30 min. After the ultrasonication the mixture was kept for stirring for about 5 h at 5 °C to get the better yield. To another part 0.2 M ammonium persulphate (APS) was mixed and added drop by drop to the stirring monomer solution. After mixing the reactants, the solution starts showing greenish tint and afterward it turns violet. The black precipitate was obtained after 6 to 7 h. This precipitate was kept for overnight and diluted with deionised water until the filtrate became colourless. Finally, it was washed with ethanol and dried overnight in oven at 80 °C.

3 Results

Raman spectroscopy is the most useful technique for characterizing the material composition. The Raman spectrum of PANI is shown in Fig. 1. Several distinct bands, at approximately 1596, 1501, 1338, 1247, 1163 and 866 cm^{-1}, have been identified as organic contents of PANI. The peak around 1596 cm^{-1} indicates the stretching modes of C = N of the quinoid ring. The peak around 1501 cm^{-1} is attributed to the C = C stretching deformation of benzenoid rings. The peaks at 1338 cm^{-1} is attributed to the characteristic C-N stretching mode in the benzenoid ring. While the peaks at 1163 cm^{-1} represents an aromatic CH in-plane bending. 1247 cm^{-1} is attributable to CN + stretching and vibration in the polaron structure of PANI. The peak around 866 cm^{-1} correspond to off-plane flexion of C = H in the benzene rings, which again confirms the formation successful PANI during polymerization. In Table 1 we compare our measurements with the previous experimental frequencies measured in Ref [24]. This values are closed.

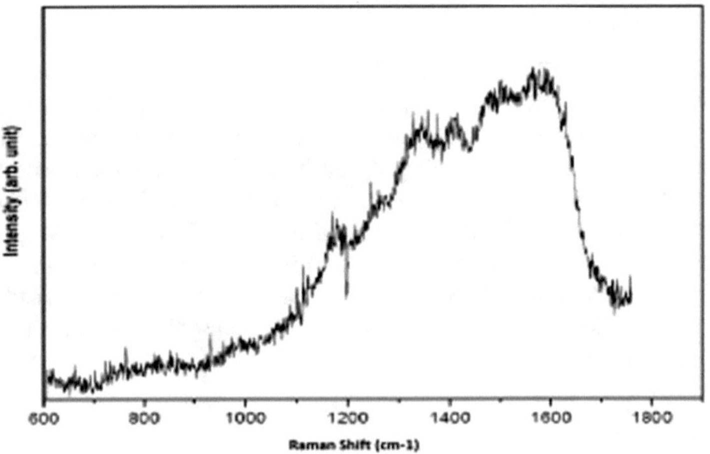

Fig. 1. Raman spectrum of PANI.

The Raman spectrum of the composite, obtained at the 514 nm excitation wavelength, is shown in Fig. 2. The Raman spectrum of graphene Fig. 2-a shows the band D at 1360 cm^{-1} and the band G at 1584 cm^{-1}. The G band is characteristic of all hybridization carbon networks sp^2, which comes from the first order of diffusion of degenerate doubled modes of graphene in the center of the Brillouin zone E2g. Figure 2-b shows that PANI is dominant in all composites with various ratios. This result con_rms the good covering of composites with PANI. As we can see all band

Table 1. Comparison of frequencies obtained with a previous experimental data from Ref [24].

Our work	Work [24]	Font size and stylemlkkmBand characteristics
1596	1593	C = N stretching deformation of the quinoid rings
1501	1503	C = C stretching deformation of benzenoid rings
1338	1328	CN stretching of secondary aromatic amine
1247	1235	CN + stretching and vibration in the polaron structure of PANI
1163	1165	The aromatic CH in-plane bending
866	–	CH bond out-of-plane deformation

positions were shifted to higher frequencies and the intensity of the peaks at 1350 and 1590 cm^{-1} was increased because they are very close to the D and G bands of graphene, which interprets a strong interaction between graphene and PANI due to electron interaction between graphene and PANI. These bands can be attributed to carbon-carbon stretching; however, band D remains due to structural imperfections created by PANI on the carbon plane. The ratio of the intensities (ID = IG) of the two bands D and G is generally used to provide additional information on the quality of nanostructured carbon-based materials. Knowing that, the disorder increases when the ratio ID = IG increases. However, the change of the shape and position of the G band of graphene in graphene nanocomposites (G@PANI) give arise a charge transfer in the nanocomposite between graphene and Polyaniline, which means that this nanocomposite with high structural, thermal and mechanical performances is a good candidate to integrate it in the manufacture of photovoltaic Devices.

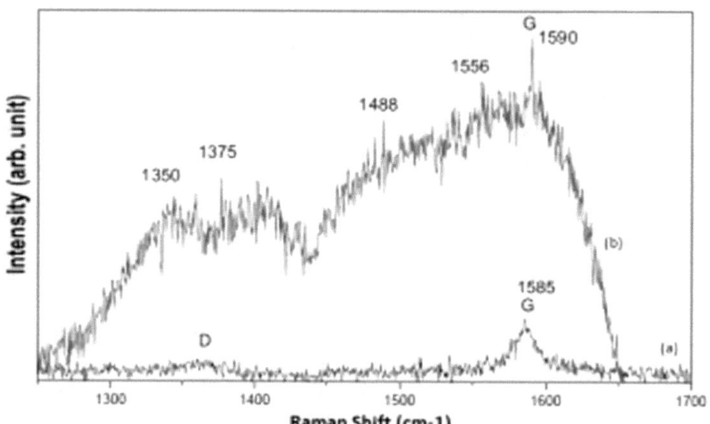

Fig. 2. Raman spectrum of graphene (a) and graphene /PANI (b).

4 Conclusion

In this article the synthesis and characterization polyanilines (PANI) and Graphene/PANI were devoted. The goal is to integrate them in the manufacture of photovoltaic devices based on nanocomposites (graphene/PANI). We have carried out experimentalmeasurements of Raman spectroscopy, of graphene, PANI and graphene/PANI. We discussed the production of new polymeric nanocomposites with high structural, thermal and mechanical performances (by the use of graphene nanosheets). Detailed analysis of the Raman G-band before and after the PANI insertion inside graphene states a charge transfer.

References

1. Sariciftci, N.S.: Plastic photovoltaic devices. Mater. Today **7**(9), 36–40 (2004)
2. Günes, S., Neugebauer, H., Sariciftci, N.S.: Conjugated polymer based organic solar cells. Chem. Rev. **107**, 1324–1338 (2007)
3. Yu, G., Gao, J., Hummelen, J.C., Wudl, F., Heeger, A.J.: Polymer photovoltaic cells: enhanced efficiencies via a network of internal donor-acceptor heterojunctions. Science **270**, 1789–1791 (1995)
4. Zhang, S., Ye, L., Hou, J.: Breaking the 10% efficiency barrier in organic photovoltaics: morphology and device optimization of well-known PBDTTT polymers. Adv. Energy Mater. **6**, 1502529–1502549 (2016)
5. Zhao, W., Li, S., Yao, H., Zhang, S., Zhang, Y., Yang, B., Hou, J.: Molecular optimization enables over 13% efficiency in organic solar cells. J. Am. Chem. Soc. **139**(21), 7148–7151 (2017)
6. Kymakis, E., Amaratunga, G.A.J.: Carbon nanotubes as electron acceptors in polymeric photovoltaics. Rev. Adv. Mater. Sci. **10**, 300–305 (2005)
7. Hyunwoo, K., Abdala, A.A., Macosko, C.W.: Graphene/polymer nanocomposites. Macromolecules **43**(16), 6515–6530 (2010)
8. Wang, Y., Yang, S., Wang, Q., Ta, T., Shi, Y., Hu, J., Wang, H., Zou, B.: The role of surfactant-treated graphene oxide in polymer solar cells: Mobility study. Org. Electron. **53**, 303–307 (2018)
9. Bhadra, S., Khastgir, D., Singha, N.K., Lee, J.H.: Progress in preparation, processing and applications of polyaniline. Prog. Polym. Sci. **34**(8), 783–810 (2009)
10. Unsworth, J., Lunn, B.A., Innis, P.C., Jin, Z., Kaynak, A., Booth, N.G.: Technical review: conducting polymer electronics. J. Intel. Mat. Syst. Str. **3**, 380–395 (1992)
11. Schoch Jr., K.F.: Update on electrically conductive polymers and their applications. IEEE Electr. Insulat. Mag. **10**, 29–32 (1994)
12. Bhadra, S., Chattopadhyay, S., Singha, N.K., Khastgir, D.: Improvement of conductivity of electrochemically synthesized polyaniline. J. Appl. Polym. Sci. **108**, 57–64 (2008)
13. Bhadra, S., Singha, N.K., Khastgir, D.: Polyaniline by new miniemulsion polymerization and the effect of reducing agent on conductivity. Synth. Met. **156**, 1148–1154 (2006)
14. Roe, M., Ginder, J., Wigen, P., Epstein, A., Angelopoulos, M., MacDiarmid, A.: 1988, Photoexcitation of polarons and molecular excitons in emeraldine base. Phys. Rev. Lett. **60** (26), 2789 (1988)

15. Kutanis, S., Karaksla, M., Akbulut, U., Sacak, M.: The Conductive polyaniline/poly (ethylene terephthalate) composite fabrics. Compos. Part A Appl. Sci. Manuf. **38**(2), 609–614 (2007)
16. Juvin, P., Hasik, M., Fraysse, J., Plans, J., Pron, A.., Kulszewicz-Bajer, I.: Conductive blends of polyaniline with plasticized poly (methyl methacrylate). J. Appl. Polym. Sci. **74**(3), 471–479 (1999)
17. Falco, E.H.L., Petrov, D.V., De Azevdo, W.M.: Polyaniline-poly (vinylalco-hol) composite: spectroscopic characterization and diffraction grating recording. Mol. Cryst. Liq. Cryst. **374** (1), 173–178 (2002)
18. Anand, J., Palaniappan, S., Sathyanarayana, D.: Conducting polyaniline blends and composites. Prog. Polym. Sci. **23**(6), 993–1018 (1998)
19. Mo, T.C., Wang, H.W., Chen, S.Y., Yeh, Y.C.: Synthesis and dielectric properties of polyaniline/titanium dioxide nanocomposites. Ceram. Int. **34**(7), 1767–1771 (2008)
20. Mirmohseni, A., Oladegaragoze, A.: Anti-corrosive properties of polyaniline coating on iron. Synth. Met. **114**(2), 105–108 (2000)
21. Liu, Z., Zhou, J., Xue, H., Shen, L., Zang, H., Chen, W.: Polyaniline/TiO_2 solar cells. Synth. Met. **156**(9), 721–723 (2006)
22. Liu, Z., Guo, W., Fu, D., Chen, W.: p-n heterojunction diodes made by assembly of ITO/nano-crystalline TiO_2/polyaniline/ITO. Synth. Met. **156**(5), 414–416 (2006)
23. Amaechi, C., Asogwa, P., Ekwealor, A., Osuji, R., Maaza, M., Ezema, F.: Fabrication and capacitive characteristics of conjugated polymer composite p-polyaniline/n-WO_2 heterojunction. Appl. Phys. A **117**(3), 1589–1598 (2014)
24. Elnaggar, E.M., Kabel, K.I., Farag, A.A., Abdalrhman, G., Al-Gamal, A.G.: Comparative study on doping of polyaniline with graphene and multi-walled carbon nanotubes. J. Nanostruct. Chem. **7**, 75–83 (2017)

Optimization of PV Panel Using P&O and Incremental Conductance Algorithms for Desalination Mobile Unit

Bachar Meryem(✉), Naddami Ahmed, Hayani Sanaa, and Fahli Ahmed

Laboratoire Mathématiques Appliquées Technologies d'Information et de Communication, Univ Hassan 1, 26000 Settat, Marocco
meryem.bachar@gmail.com

Abstract. The increasing demand for water and the depleting fossil fuels for its treatment made renewable energies a better alternative source for feeding water desalination units. Photovoltaic (PV) energy is an important source of renewable energy that could be an alternative to satisfy the broad energy needs in the future.

Our project consists in the realization of a desalination mobile unit of brackish water based on solar energy which will serve as prototype for scientific research to develop many research axes. This prototype consists of different parts such as: The production of electrical energy by photovoltaic panels, DC/DC conversion, DC/AC conversion and water treatment.

PV system produces maximum output power in only one point on Power-Voltage (P-V) curve called Maximum Power Point (MPP). When the weather conditions change (such as temperature and irradiation), the voltage and current in the circuit change. In this case, a new MPP must be found based on Maximum Power Point Tracking algorithms (MPPT) to optimize the power generated by PV. Hence, many methods have been developed to determine MPP.

In this work, a comparison between two MPPT algorithms namely Perturb and Observe (P&O) and Incremental Conductance (InC) is presented. The simulations are accomplished by using a DC/DC Buck converter, a PV array and a load under MATLAB/Simulink environment. The obtained results, in different climatic conditions, reveal that the InC controller is more effective than P&O controller.

Keywords: MPPT · Buck converter · Perturb and Observe · Incremental Conductance · MATLAB/Simulink

1 Introduction

1.1 General Review

Morocco is a country located in the north-west of Africa, on the southern shore of the Mediterranean. It covers an area of 710,000 km^2 with a population of 34,891,915 inhabitants, 40% of them lives in rural areas.

M. Ezziyyani (Ed.): AI2SD 2018, AISC 912, pp. 164–184, 2019.
https://doi.org/10.1007/978-3-030-12065-8_17

In 2014, according to the High Commission for the Plan of Morocco, nearly 1,605,000 people had income below the poverty level. Almost 80% of them lives in rural areas.

The Moroccan House of Representatives announced that 28% of the rural population does not have access to drinking water. The water stress not only affects the rural world, but it will increase throughout all the country, according to the American Water Resources Institute.

Faced to this critical situation, desalination of brackish water remains an effective solution to produce drinking water, especially since Morocco has an important surface water resources such as lakes and runoff.

Therefore, we propose a conception and realization of a hybrid mobile desalination unit of brackish water based on solar energy (thermal and photovoltaic), dragged by a vehicle to make his displacement easy, as illustrated in Fig. 1. This unit will serve as a prototype for scientific research to study its energy consumption, the parameters that influence it and how to reduce the price. It will also serve to compare the Maximum Power Point Tracking (MPPT) algorithms which will be implemented in the DC/DC converter with variable duty cycle, to select the best choice to optimize the energy produced by the photovoltaic panels.

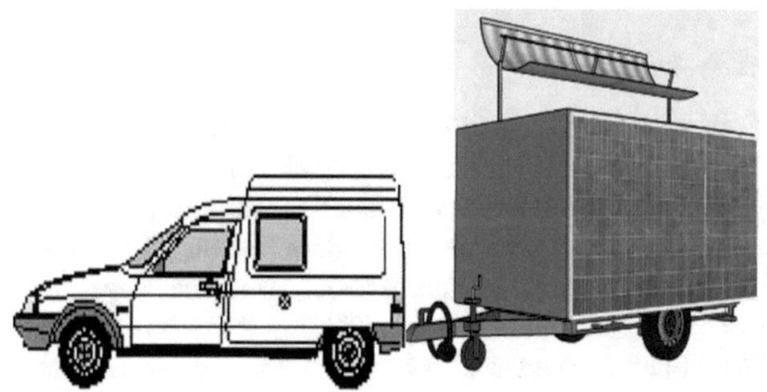

Fig. 1. Proposed desalination pilot

1.2 Desalination

A bibliographical study of the most replicated desalination techniques reveals that the Reverse Osmosis (RO) technique is the most optimal technique of desalination [1–3].

Reverse Osmosis is a semi-permeable membrane technique used for water purification [4]. In this technique, the water comes across high pressure through a very thin membrane which only allows the molecules of water (H_2O) and holds up the particles, the salts and the organic molecules with a size of 7 to 10 mm.

Osmosis is the water flow from the diluted solution to the concentrated solution under the action of a concentration gradient.

To reverse this flow, a higher pressure is applied. The osmotic pressure can be calculated by Van't Hoff's law:

$$P_{Osmotic} = iCRT \tag{1}$$

Where I is the number of dissociated ions, C is the salt concentration in mol.m^{-3}, R is the gas constant R = 8.314 J.mol^{-1}.k^{-1} and T is the absolute temperature of the solution in Kelvin.

The Reverse Osmosis system is influenced by various variables such as: temperature, pressure and salt concentration of water supplying the system.

Equation 1 shows the relation between temperature and pressure: the temperature increases when the osmotic pressure increases.

The increase in temperature provokes the decrease in the viscosity of the water. This decrease brings on the increase in the permeate flow. Consequently, a drop of net energy consumption [5].

According to this, we propose to increase the temperature of salt water before the treatment. A parabolic trough will be used to increase the temperature to 45 °C. This choice is limited by the condition of the membrane used on the desalination by Reverse Osmosis; it can not support a temperature more than 45 °C.

In the next part, we will present our desalination system.

1.3 Proposed System

The proposed system is a mobile and hybrid desalination research prototype driven by solar energy, equipped by a PV panels to supply electrical energy to the equipment for desalination by Reverse Osmosis and a parabolic trough to produce the heat.

Figure 2 shows the global schema of our desalination pilot. The quantity of fresh water produced by our proposed system is about 10 m^3 per day.

It consists of six photovoltaic panels (TS-255-P156-60) with a total power of 1530 W, a DC/DC buck converter with a variable duty cycle controlled by an MPPT algorithm by taking the PV voltage and PV current to extract the MPP using CompactRIO and LabVIEW software. 4 batteries in series (M15-12V-165Ah) to store the electrical energy produced by the PV panel to ensure the power supply in all circumstances. A DC/AC converter (SP5000 Efecto 5 KVA/4 KW) connected to a variable speed drive (ATV312 240 V/50 Hz) to supply the pump motor (TC 150T) to extract the surface water.

The extracted water passes through the parabolic trough to increase its temperature to 45 °C before being stored in a tank of 1 m^3. Chlorine (NaOCI) is added to this water to remove natural microorganisms and oxidize iron and Manganese.

The water is pushed to a pre-filtration stage using a low-pressure pump (4 bar). This stage consists of a multistage filter to remove sediments, suspended solids, dust, etc., and an activated carbon filter to remove organisms that cause color, taste and odor.

A high-pressure pump (10 bars) pushes the water to the RO system which contains an anti-scale dosing unit to prevent deposition of particles on the membranes, filters with sensitivity of 5 μ to 1 μ and finally two RO membranes.

Water will be pumped to a re-mineralization unit before being stored in a drinking water tank with a capacity of 1 m^3.

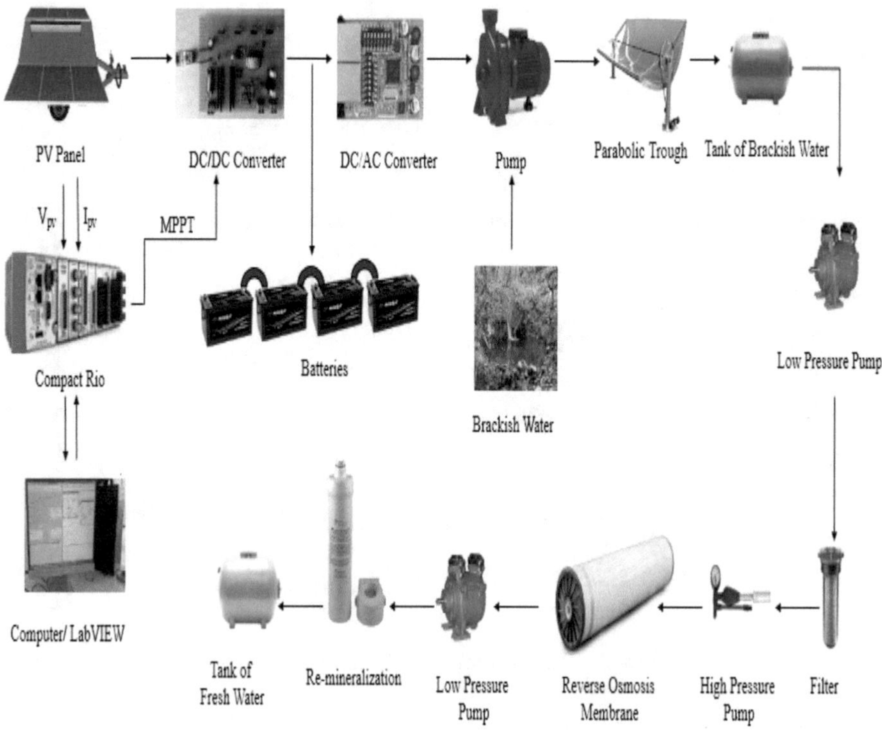

Fig. 2. Global schema of desalination pilot

Figure 3 shows the electrical circuit of desalination unit. From this fig we can see that this project contains 3 fundamental axes:

- DC/DC conversion;
- DC/AC conversion;
- Reverse Osmosis water treatment;

In this work, we attend to optimize the energy produced by the PV panel by using the Perturb and Observe (P&O) MPPT Algorithm, and Incremental Conductance (InC) MPPT Algorithm, which are used to track the Maximum Power Point (MPP) by adjusting the duty cycle of a DC-DC Buck converter. The converter acts as an interface between the load and the PV. The P&O and InC algorithms are tested and compared with MATLAB/Simulink software under a constant and sudden variation in irradiation.

This paper is organized as follows: after this introduction, the Sect. 2 will describe the PV system modelling. The Sect. 3 is dedicated to the DC/DC Buck converter. The Sect. 4 presents the MPPT techniques and focuses on the P&O and InC MPPT algorithms. In the Sect. 5, the simulation results obtained are presented and discussed, and the Sect. 6 concludes the paper followed by the references.

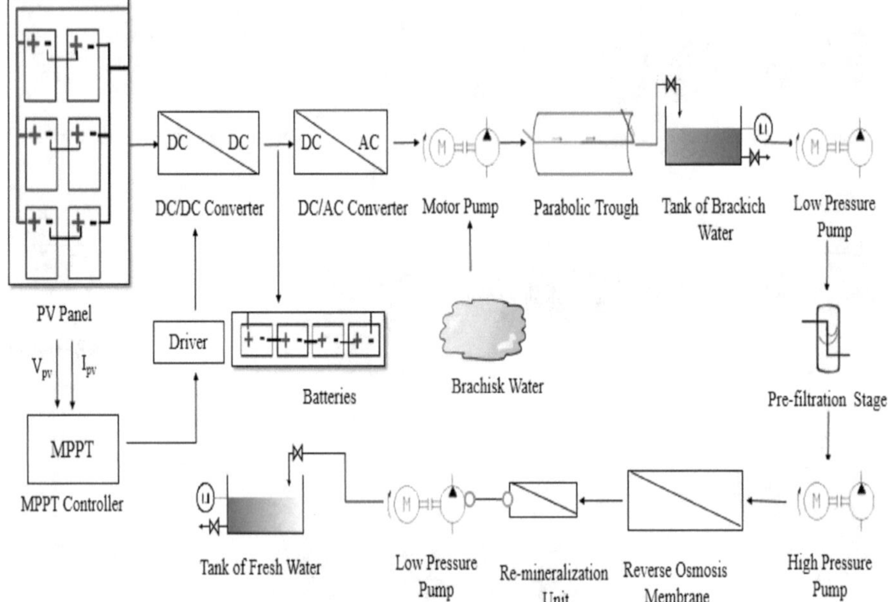

Fig. 3. Electrical schema of desalination unit

2 PV Cell Modelling and Characteristics

The basic component of PV cell is the PN junction diode [6]. It converts solar radiation to electricity by photovoltaic effect. PV cells are mostly made of Silicon (14–17% of cell efficiency) or Gallium Indium Phosphide (30–35% of cell efficiency) as pointed out by [7]. A PV cell can be modeled by the electrical circuit shown in Fig. 4.

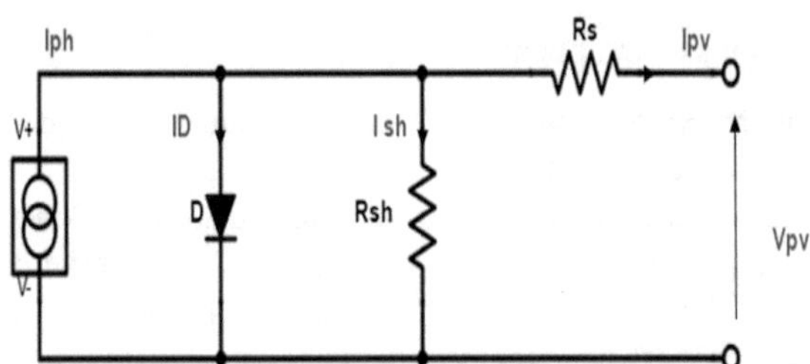

Fig. 4. The equivalent circuit of a solar cell

The five parameters Iph, ID, Ish, Ipv and Vpv allow the analysis and the evaluation of the performances of the photovoltaic model. The PV cell is represented by a current source, which models the conversion of solar radiation into electrical energy. The Rs resistance represents the contact resistance. The resistance in parallel Rsh called the shunt resistance represents the leakage current. The diode D represents the PN junction [8, 9].

The solar cell output current is given by:

$$I_{pv} = I_{ph} - I_D - I_{sh} \tag{2}$$

We replace ID and Ish by their expressions, we obtain the result in (3):

$$I_{pv} = I_{ph} - I_0\left\{e^{q\left(\frac{Vpv+IpvRs}{AKT}\right)-1}\right\} - \frac{Vpv + IpvRs}{Rsh} \tag{3}$$

Where Iph is the photodiode current, I_0 is the inverse saturation current, q is the electron charge $q = 1.6 \times 10^{-19}$C, A is the ideality factor of the PN junction A = 2.8 when T = 289 K, K is the Boltzmann constant $K = 1.38 \times 10^{-23}$ J/K and T is the temperature for PV cell in Kelvin.

The power produced by a photovoltaic cell is low. Hence, to increase the voltage and the current, photovoltaic cells are connected in series or/and parallel to getting desired maximum voltage and current to meet the power requirement of the load [10].

When the temperature increases, the output PV Voltage decreases. The increase of the ambient temperature presents in this case a real inconvenient which many scientists are currently faced. They propose cooling techniques for PV panel to maintain his temperature [11]. When the irradiance increases, the output PV Current increases. The power produced by the PV panel strongly depends on the irradiance values. Knowing the intermittent nature of PV panels throughout a day, it is necessary to use MPPT algorithms to track the MPP under different climatic conditions by using a DC/DC converter [12]. We will present, in the following section, the role of the DC/DC converter.

3 DC/DC Converter

To ensure a proper functioning of a photovoltaic installation supplying a continuous load, several technical problems must be taken into consideration such as:

- The impedance matching between the load and the PV panel [13];
- The losses in the conversion chain;
- The relativity of solar energy to climatic conditions; irradiation and temperature.

Therefore, the use of a DC/DC converter becomes necessary. Nowadays, there are several topologies of converters. In this part, we will use a Buck converter.

The Buck converter converts a higher DC input voltage to lower DC output voltage [14].

Figure 5 shows the circuit of Buck converter; where S1 is the main switch, L1 is the filter inductor, C1 is the filter capacitor and R1 is the load resistor.

The switch S1 operates periodically. T is the period and f is the corresponding frequency. The switch S1 is closed from t = 0 to t = αT and open from t = αT to T. (α is the duty cycle).

When the switch S1 is closed, the input is directly connected to the output. When opened, the input and output work independently. The input is open; the output is short-circuited due to the flyback diode, which ensures the continuity of the inductor current.

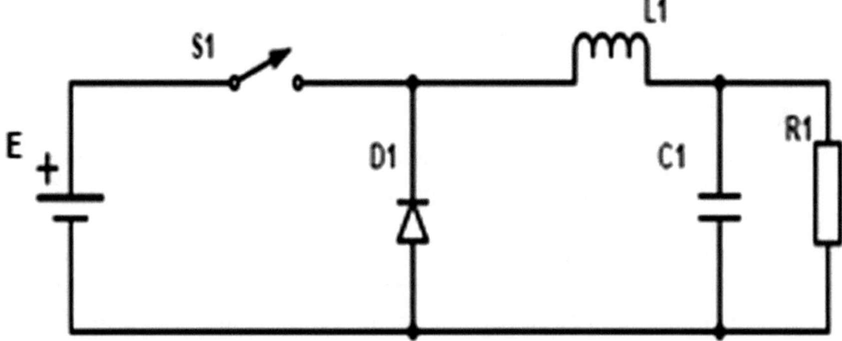

Fig. 5. Circuit of buck converter

The voltage release Vs can be determined by (4):

$$Vs = \alpha E \tag{4}$$

4 Maximum Power Point Tracking (MPPT)

The intermittency of solar energy requires a tracker system to track the maximum power of the PV panel at any environmental conditions. As previously said, MPPT algorithms are used to increase the efficiency of photovoltaic panels by continuously varying the duty cycle of the DC/DC converter [15, 16]. Up today, several MPPT techniques have been developed.

4.1 Perturb and Observe MPPT Algorithm

Perturb and Observe (P&O) algorithm is the most widely used in the world. The commercialized photovoltaic energy applications use P&O MPPT algorithm due to its simplicity as mentioned by [17]. It is based on the disturbance of the system voltage and the observation of its effect on the delivered power.

The flowchart of the P&O algorithm, exposed in the Fig. 6, shows the steps that the algorithm follows to reach the MPP. The controller measures the voltage V (n) and the current I (n) of the photovoltaic panel at the instant "n" to calculate the power P (n) of the photovoltaic panel and then compare it to the power P (n-1) at time (n-1).

When the power increases, the duty cycle is disturbed in the same direction; otherwise, it is disturbed in the opposite direction. The duty cycle is disrupted in each MPPT cycle. Once the power point is found, V oscillates around the optimal voltage [18, 19].

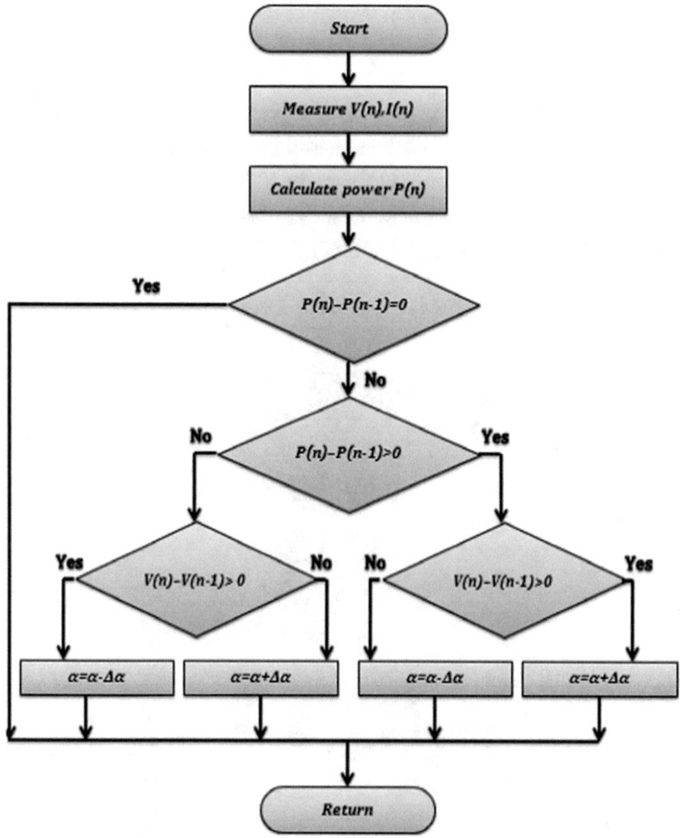

Fig. 6. Flowchart of P&O MPPT algorithm

4.2 Incremental Conductance MPPT Algorithm

The Incremental Conductance algorithm is based on the incremental $\frac{dI}{dV}$ and instantaneous $\frac{I}{V}$ conductance of the PV array. This algorithm uses an expression derived from the condition that at the MPP, $\frac{dP}{dV} = 0$, where $P = VI$.

The equation then become:

$$\frac{dP}{dV} = I + V.\frac{dI}{dV} \tag{5}$$

With these conditions:

$$\frac{dI}{dV} = -\frac{I}{V} \text{ at MPP.} \tag{6}$$

$$\frac{dI}{dV} > -\frac{I}{V} \text{ left of MPP.} \tag{7}$$

$$\frac{dI}{dV} < -\frac{I}{V} \text{ right of MPP.} \tag{8}$$

The flowchart for tracking of V_{ref} is shown in Fig. 7. The algorithm increments or decrements V_{ref} to track the MPP. The controller measures the voltage V (k) and the current I (k) of the photovoltaic panel at the instant "k" to calculate ΔV and ΔI and then compare $\frac{dI}{dV}$ with $-\frac{I}{V}$ to decrease or increase V_{ref} based on Eqs. (7), (8) and (9). The InC can track rapidly increasing and decreasing irradiance conditions compared to the P&O [13].

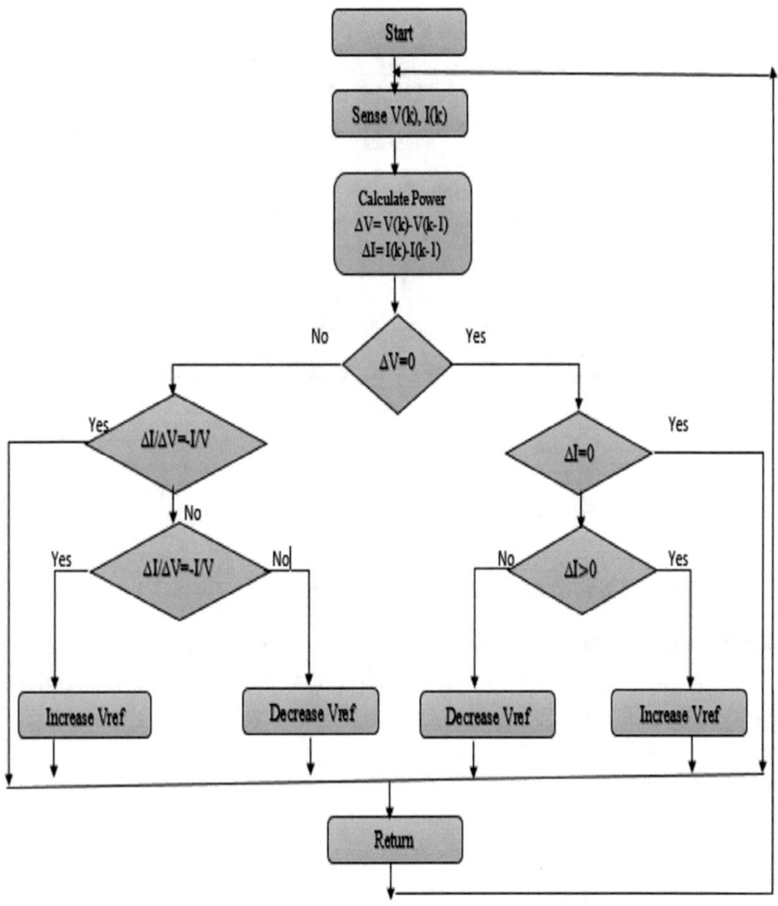

Fig. 7. Flowchart of InC MPPT algorithm

5 Simulations Results

The first part of simulation is dedicated to a PV system without MPPT on MATLAB/Simulink to see the role of the MPPT algorithm.

The PV array is designed based on six Tesla Solar modules (Solar TS255-P150-60). Table 1 shows the characteristics of one PV module.

Table 1. Specifications of PV panel

Parameter	Value
Pm (W)	255 W
Imp (A)	8.32 A
Vmp (V)	30.6 V
Isc (A)	8.75 A
Vsc (V)	37.6 V
Module efficiency	15.58%

PV array is connected directly to batteries and load as presented in Fig. 8.

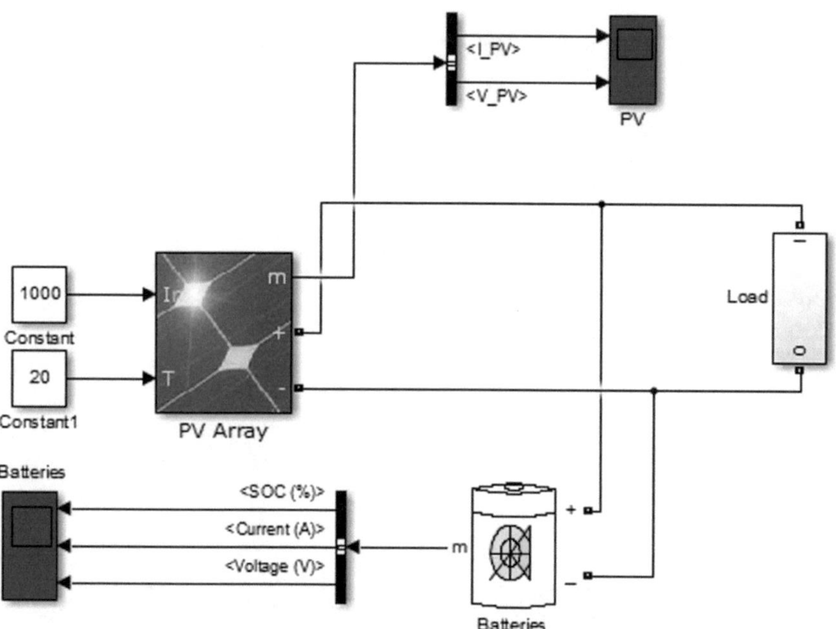

Fig. 8. System without MPPT under MATLAB/Simulink

Figures 9, 10 and 11 show the simulation results of PV system without MPPT for 1000 W/m^2 at 25 °C. The Output PV array Voltage is 51.81 V (Fig. 9), the Output PV array Current is 25.90 (Fig. 10) and the Output PV array Power is 1341.87 W, as shown in Fig. 11.

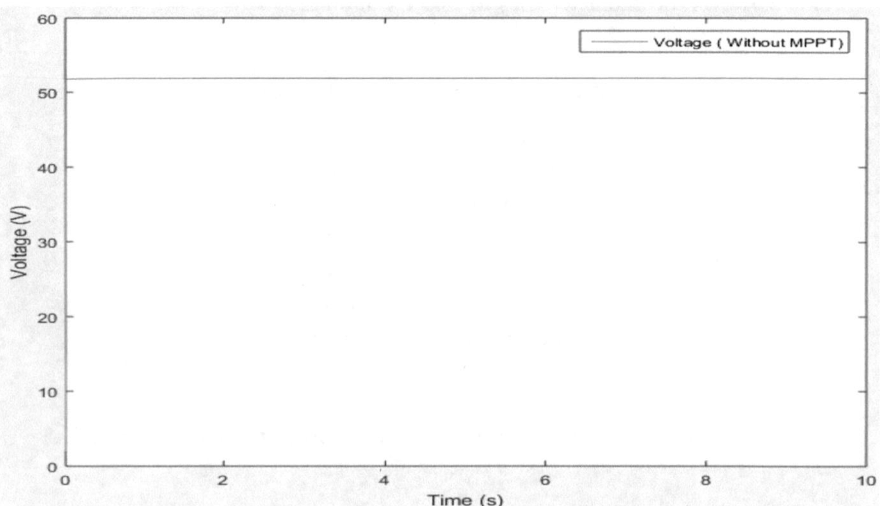

Fig. 9. Output PV Voltage without MPPT

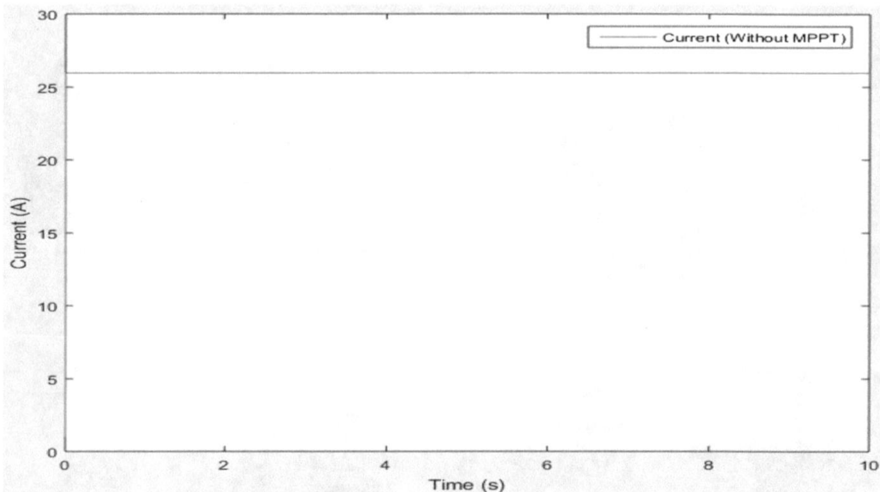

Fig. 10. Output PV Current without MPPT

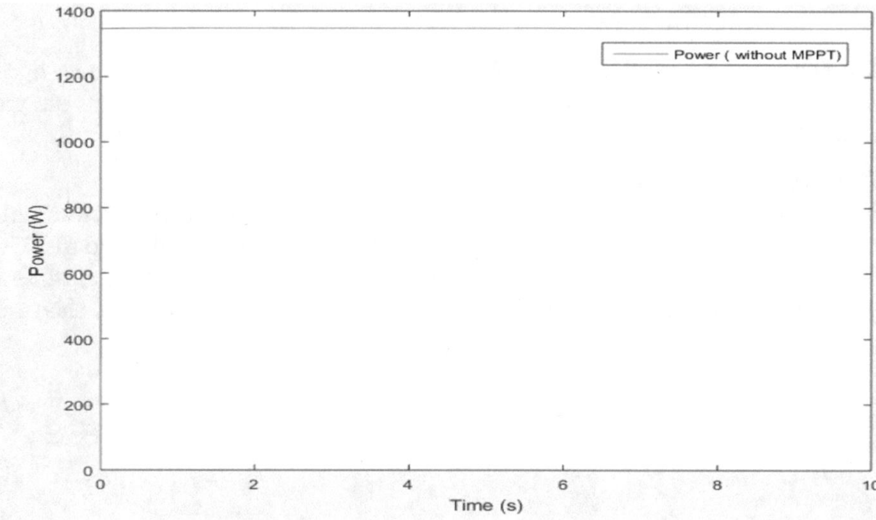

Fig. 11. Output PV array Power without MPPT

The second part of simulations uses MPPT algorithms. To compare the two MPPT algorithms (P&O and InC), the global system was implemented in MATLAB/Simulink environment. The system consists of a PV array that charges batteries and feeds load through a DC/DC Buck Converter, which is controlled by an MPPT controller, as shown in the Fig. 12.

The DC/DC Buck Converter is controlled by P&O and InC MPPT algorithms. The current, voltage and power of the PV array are the main characteristics to take into consideration for the comparison.

The comparative performance of MPPT algorithms is analyzed for constant irradiation and sudden variation in irradiation.

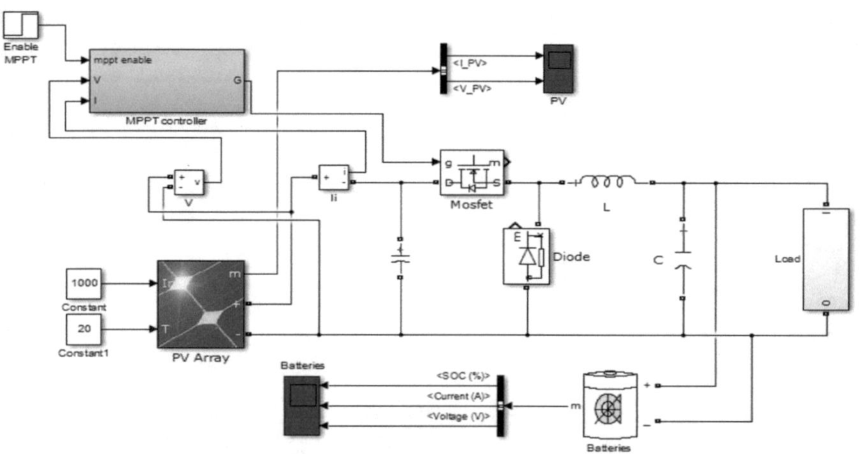

Fig. 12. Global system under MATLAB/Simulink

5.1 Constant Irradiation

The simulations of the PV system with the P&O and the InC algorithms are done under constant irradiation (1000 W/m^2) and fixed temperature (25 °C). The results are presented in the figs below.

P&O MPPT Algorithm

As we can see from Figs. 13, 14 and 15, by using the P&O algorithm, MPP is reached at t = 0.35 s. In Fig. 13, the output Voltage of PV array varies from 59.94 V to 61.37 V. Figure 14 shows the output Current of PV array which varies from 22.45 A to 24.32 A and the output PV array Power reaches 1494 W with clear oscillations as shown in Fig. 15.

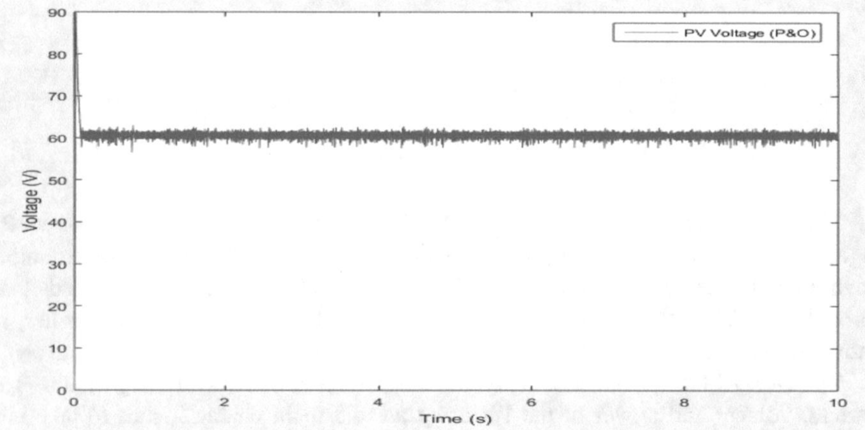

Fig. 13. Output Voltage of PV array with P&O MPPT algorithm for 1000 W/m^2 at 25 °C

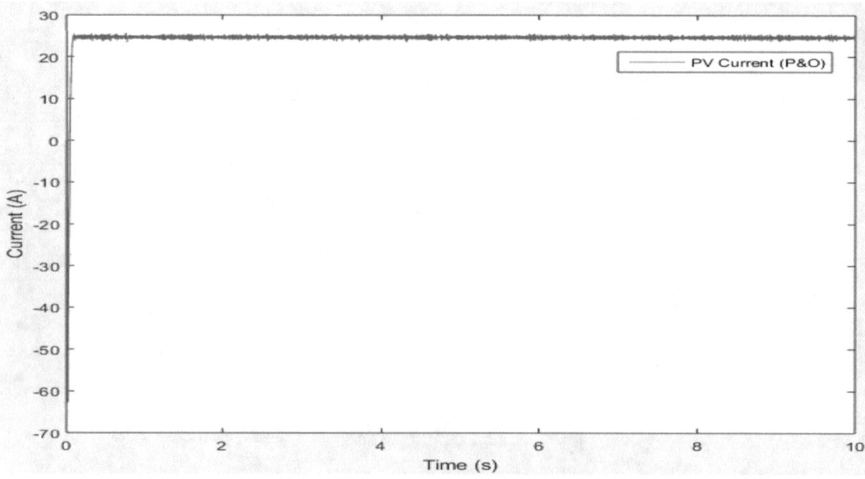

Fig. 14. Output Current of PV array with P&O MPPT algorithm for 1000 W/m^2 at 25 °C

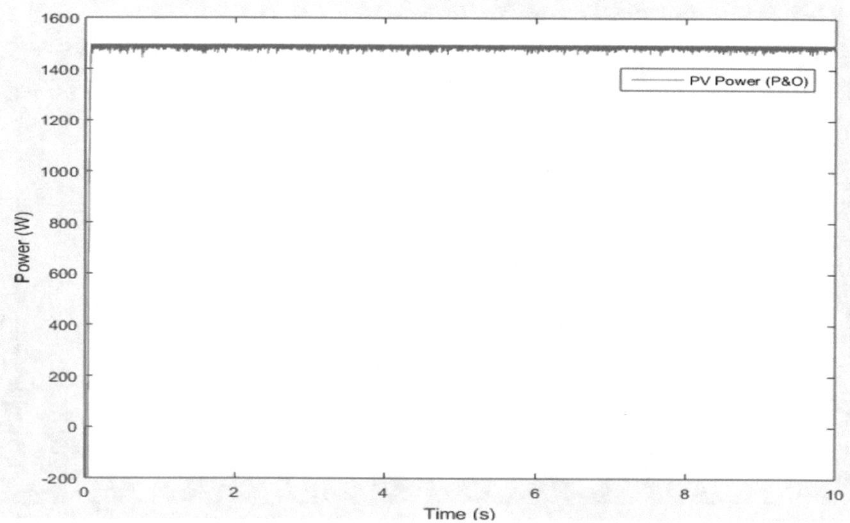

Fig. 15. Output Power of PV array with P&O MPPT algorithm for 1000 W/m^2 at 25 °C

The above simulation results show that the P&O algorithm tracks the MPP under constant irradiation with clear oscillations. Compared the power obtained by the photovoltaic panels with and without MPPT controller, we can clearly see an increase from 1341,8 W to 1494 W. The use of P&O algorithm can have a power increase of about 10.18%.

Incremental Conductance MPPT Algorithm

From Figs. 16, 17 and 18 we can see that the MPP is reached at t = 0.25 s with InC MPPT Algorithm. In Fig. 16 the Output Voltage of PV array is 61.91 V, the

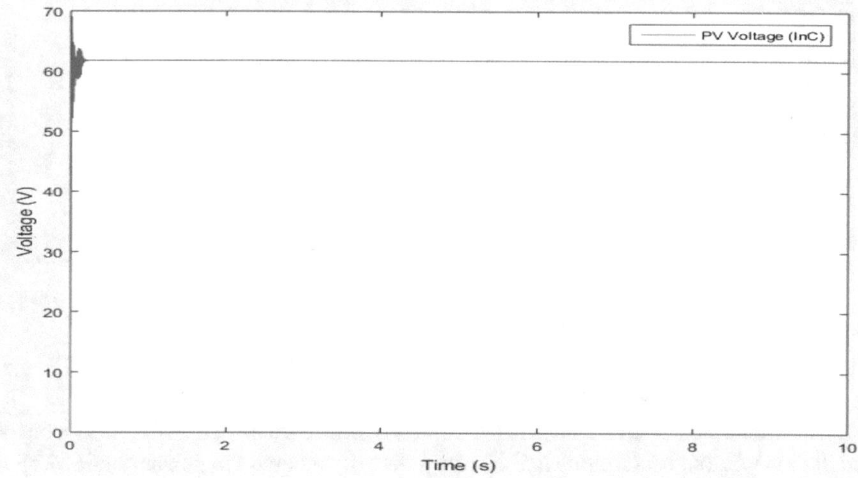

Fig. 16. Output Voltage of PV array with InC MPPT algorithm for 1000 W/m^2 at 25 °C

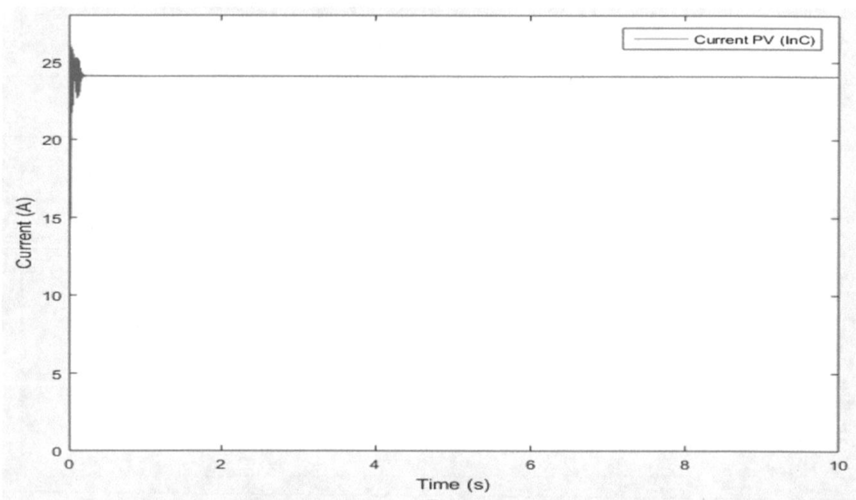

Fig. 17. Output Current of PV array with InC MPPT algorithm for 1000 W/m^2 at 25 °C

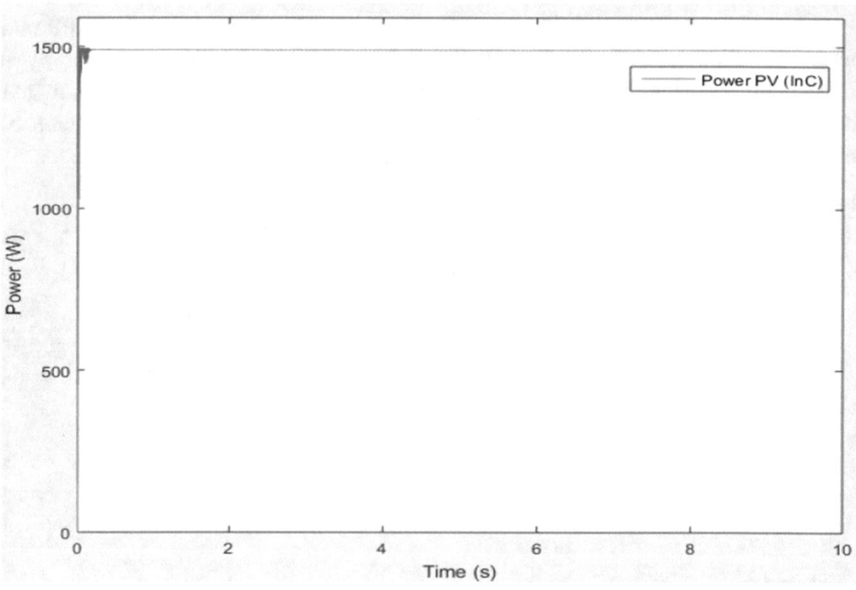

Fig. 18. Output Power of PV array with InC MPPT algorithm for 1000 W/m^2 at 25 °C

Output Current reaches 24.18A as shown in Fig. 17 and the Output Power of PV array is stabilized at 1496 W as presented in Fig. 18.

The simulation results show that the InC algorithm tracks the MPP under constant irradiation without oscillations in the steady state. Compared the power obtained by the photovoltaic panels with and without MPPT controller, we can clearly see an increase from 1341,8 W to 1496 W. The use of InC algorithm can have a power increase of about 10.38%

5.2 Sudden Variation in Irradiation

To test the system operation in different conditions, the PV array are subjected to sudden variation in irradiation as shown in Fig. 19. The temperature is constant at 25 °C, and the irradiation level is varying between two levels. The first irradiation level is 1000 W/m^2; at t = 0.7 s, the irradiation level suddenly changes to 100 W/m^2 and then back to 1000 W/m^2 at t = 1.7 s.

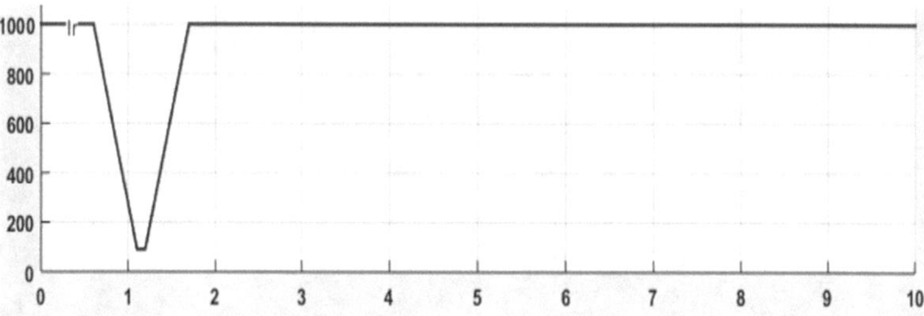

Fig. 19. Change in solar irradiation

P&O MPPT Algorithm

Figures 20, 21 and 22 show, respectively, the outputs Voltage, Current and Power of the PV array with P&O Algorithm under sudden variation in irradiation. At t = 0.7 s, we decrease the solar irradiation from 1000 W/m^2 to 100 W/m^2. Thus, the Output PV array Voltage decreases until 55 V, the output PV array Current decreases until 2.19 A in the first peak and 1.95 A in the second peak. Consequently, the Output PV array Power decrease until 150 W with two peaks.

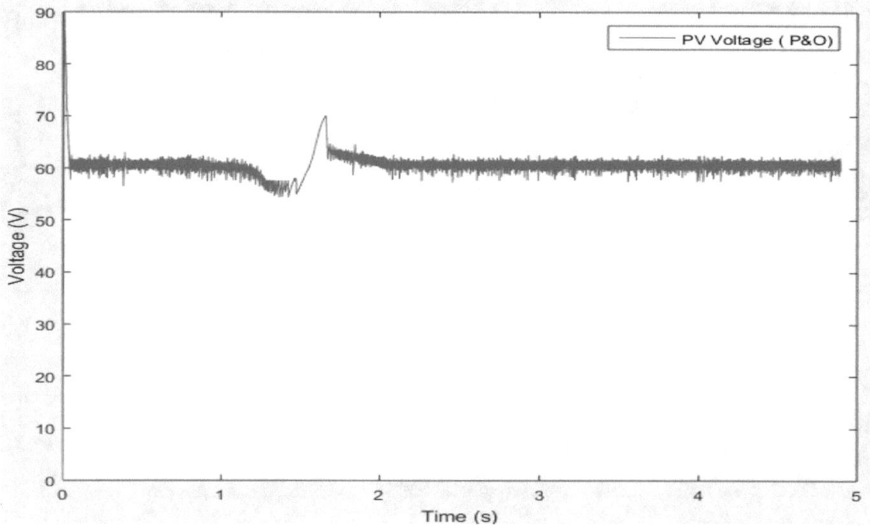

Fig. 20. Output Voltage of PV array with P&O for variable irradiation at 25 °C

At t = 1.87 s, the Output PV array Voltage starts to increase, reaches the V_{MMP} = 61.37 V and oscillates around it at t = 2.25 s. The Output PV array Current increases and oscillates around 24.34 A. The Output PV array Power increases to reach 1494 W with clear oscillations.

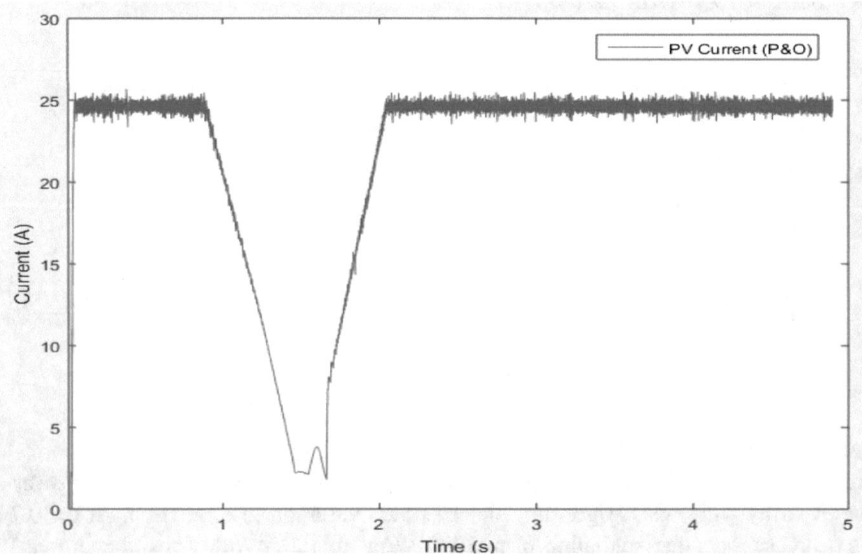

Fig. 21. Output Current of PV array with P&O for variable irradiation at 25 °C

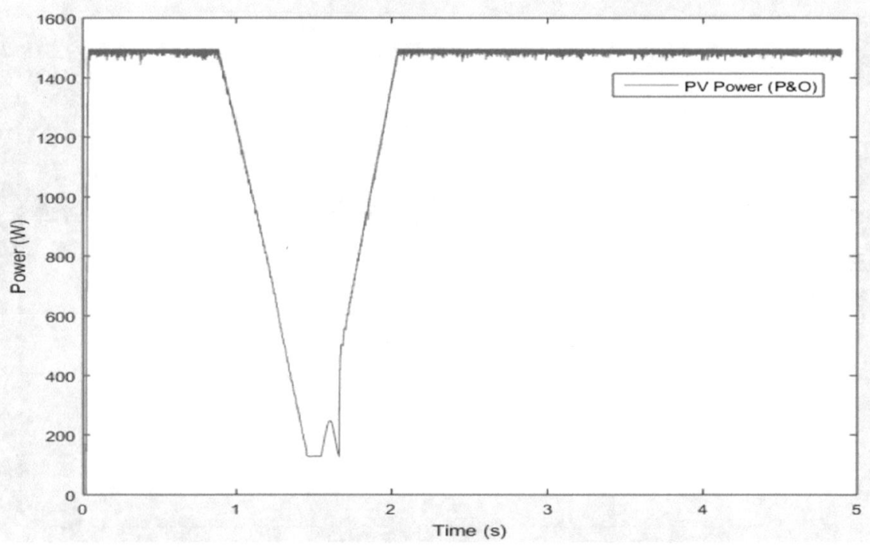

Fig. 22. Output Power of PV array with P&O for variable irradiation at 25 °C

Incremental Conductance MPPT Algorithm

Figures 23, 24 and 25 show, respectively, the outputs Voltage, Current and Power of the PV array with InC Algorithm under sudden variation in irradiation. When the solar irradiation decreases from 1000 W/m^2 to 100 W/m^2, the Output PV array Voltage decreases until 52 V and the output PV array Current decreases until 2.49 A. Consequently, the Output PV array Power decrease until 150 W.

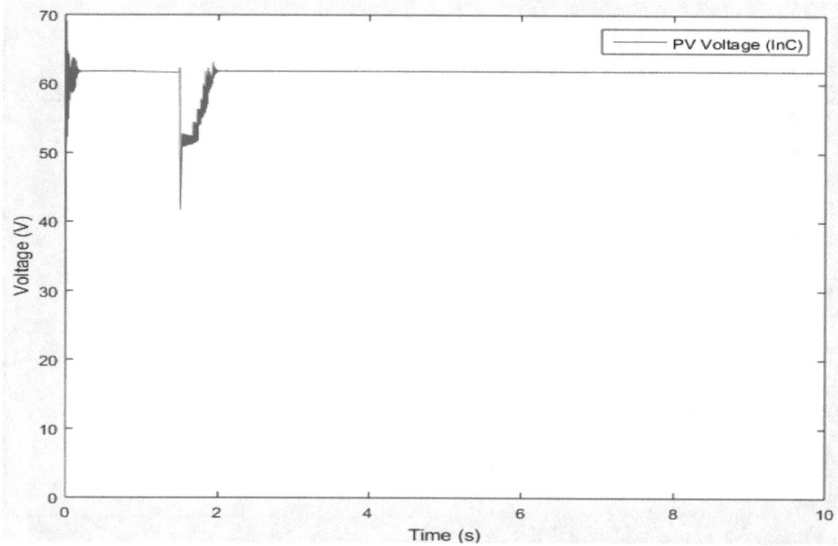

Fig. 23. Output Voltage of PV array with InC for variable irradiation at 25 °C

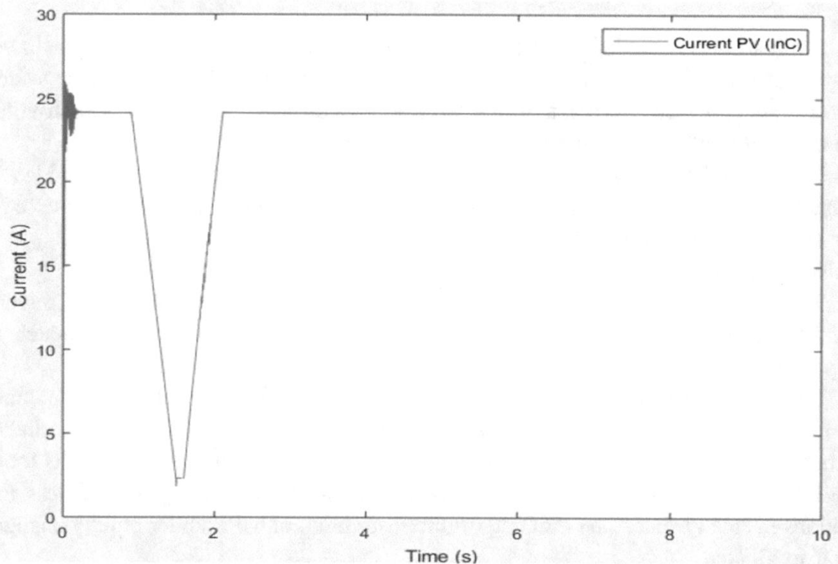

Fig. 24. Output Current of PV array with InC for variable irradiation at 25 °C

At t = 1.77 s, the Output PV array Voltage increases to reach the V_{MPP} = 61.91 V at t = 2.09 s, the Output PV array Current increases to attain I_{MPP} = 24.18 A and the Output PV array Power increases to reach 1496 W.

We can study the robustness of our system by applying different perturbations functions to the irradiation and test the sensitivity of photovoltaic panels in the future work.

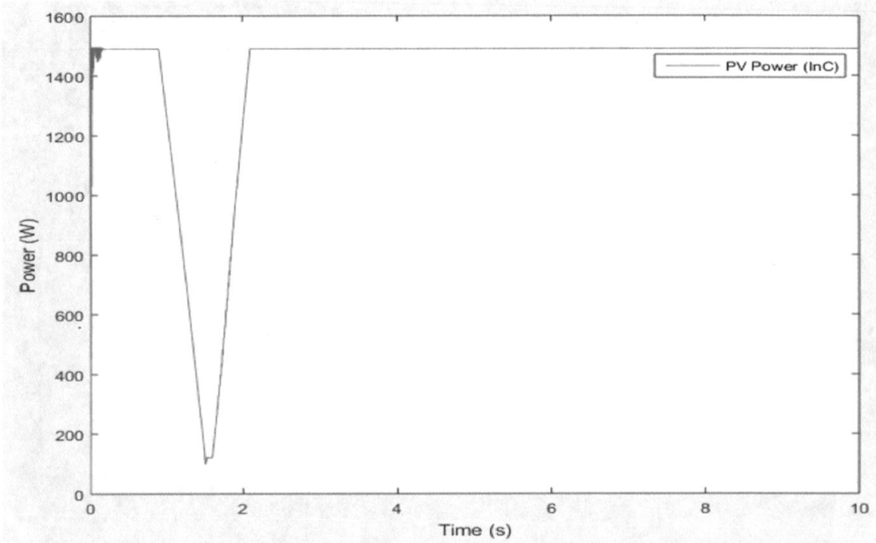

Fig. 25. Output Power of PV array with InC for variable irradiation at 25 °C

From all these simulations it can be deduced that the MPPT algorithms play a very important role in the optimization of the energy produced by the photovoltaic system. The results show that MPPT algorithms increase the power produced by photovoltaic panels by around 11% compared to a system without MPPT controller.

When the PV array is connected directly to the load, it does not produce the maximum power. It operates at the intersection of its V-I curve with the load curve that has a slope of $1/R_{bat}$ which does not always correspond to the MPP, (where $1/R_{bat}$ is the resistance of load).

The implementation of an adaptation stage composed of a DC/DC Buck Converter with a variable duty cycle controlled by an MPPT algorithm allows to match the impedance of the load with the impedance of PV array to extract the MPP.

The comparison between the two algorithms studied (P&O and InC) is done under two different climatic conditions; constant irradiation and fast variation in irradiation. In both cases, P&O and InC techniques reach the MPP. According to the results found: the InC algorithm is faster than P&O algorithm and presents a signal with less fluctuations in steady state. The P&O algorithm loses track of MPP under rapidly changing solar irradiance.

In order to evaluate the performance of the three cases studied (without MPPT, P&O and Inc), we can calculate the efficiency criteria defined by:

$$\eta_{MPPT} = \frac{P_{pv}}{P_{mpp}} \times 100 \quad (9)$$

Where:

- P_{pv} the measured output power of PV panels.
- P_{mpp} the maximum expected output PV power in the same conditions of temperature and irradiation.

Comparison between the three simulations for various parameters is given in Table 2.

Table 2. Simulation results

Techniques	PV Voltage (V)	PV Current (A)	PV Power (W)	Response time (s)	$\eta_{MPPT}(\%)$	Stability
Without MPPT	51.81	25.90	1341	0.01	87.84	+
P&O	59.94–61.37	22.45–24.32	1345–1494	0.35	97.77	–
InC	61.91	24.18	1496	0.25	97.93	+

6 Conclusion

This work is part of the energy optimization of a desalination unit for brackish water. In this context, a deep analysis focuses on the optimization of the energy produced by the photovoltaic panel, especially the Maximum Power Point Tracking (MPPT).

Several simulations have been done to show the importance of MPPT algorithms and compare between the two MPPT algorithms studied P&O and InC by using MATLAB/Simulink.

The results show that the use of a MPPT controller makes it possible to reach a Maximum Output PV array Power that is about 11% higher than the power obtained without MPPT.

Obviously, both techniques have been able to track the MPP and produce an approximately identical power. The InC shows a better performance in constant and fast variation in irradiation.

To verify the simulation results obtained with MATLAB/Simulink, an experimental study will be made by using the CompactRIO and the LabVIEW software from National Instrument.

References

1. Garg, M.C., Joshi, H.: A review on PV-RO process: solution to drinking water scarcity due to high salinity in non-electrified rural areas. Sep. Sci. Technol. **50**(8), 1270–1283 (2015)
2. Miller, S., Shemer, H., Semiat, R.: Energy and environmental issues in desalination. Desalination **366**, 2–8 (2015)
3. Bourouni, K., Ben M'Barek, T., Al Taee, A.: Design and optimization of desalination reverse osmosis plants driven by renewable energies using genetic algorithms. Renew. Energy **36**(3), 936–950 (2011)
4. Chung, K., Yeo, I.-H., Lee, W., Oh, Y.K., Kim, D., Park, Y.-G.: Investigation into design parameters in seawater reverse osmosis (SWRO) and pressure retarded osmosis (PRO) hybrid desalination process: a semi-pilot scale study. Desalin. Water Treat. **57**(51), 24636–24644 (2016)
5. Agashichev, S., Lootahb, K.N.: Influence of temperature and permeate recovery on energy consumption of a reverse osmosis system. Desalination **3**, 253–266 (2003)
6. Choudhary, D., Saxena, A.R.: DC-DC buck-converter for MPPT of PV system. Int. J. Emerg. Technol. Adv. Eng. **4**(7), 813–821 (2014)
7. Srinivas, P., Lakshmi, K.V., Ch, R.: Simulation of incremental conductance MPPT algorithm for PV systems using LabVIEW. IJIREEICE **4**, 34–38 (2016)
8. Agrawal, J.H., Aware, M.V.: Photovoltaic simulator developed in LabVIEW for evalution of MPPT techniques. In: International Conference on Electrical, Electronics, and Optimization Techniques (ICEEOT), pp. 1142–1147 (2016)
9. Selmi, T., Belghouthi, R.: A novel widespread Matlab/Simulink based modeling of InGaN double hetero-junction p-i-n solar cell. Int. J. Energy Environ. Eng. **8**(4), 273–281 (2017)
10. Desai, H.P., Maheshwari, R.: Synchronized pulsed dc-dc converter as maximum power position tracker with wide load and insolation variation for stand alone PV system. Appl. Solar Energy **47**(4), 271–280 (2011)
11. Moharram, K.A., Abd-Elhady, M.S., Kandil, H.A., El-Sherif, H.: Enhancing the performance of photovoltaic panels by water cooling. Ain Shams Eng. J. **4**(4), 869–877 (2013)
12. Karami, N., Moubayed, N., Outbib, R.: general review and classification of different MPPT Techniques. Renew. Sustain. Energy Rev. **68**, 1–18 (2017)
13. Nath-Naidu, B.A.: Voltage based P&O algorithm for maximum power point tracking using labview. Innov. Syst. Des. Eng. **7**(5), 12–16 (2016)
14. Yadav, A.P.K., Thirumaliah, S., Haritha, G., Scholar, P.G.: Comparison of MPPT algorithms for dc-dc converters based PV systems. Int. J. Adv. Res. Electr. Electron. Instrum. Eng. **1**(1), 18–23 (2012)
15. Kermadi, M., Berkouk, E.M.: Artificial intelligence-based maximum power point tracking controllers for Photovoltaic systems: comparative study. Renew. Sustain. Energy Rev. **69**(C), 369–386 (2017)
16. Babaa, S.E., Armstrong, M., Pickert, V.: Overview of maximum power point tracking control methods for PV systems. J. Power Energy Eng. **20**, 59 (2014)
17. Na, W., Carley, T., Ketcham, L., Zimmer, B., Chen, P.: Simple DSP implementation of maximum power pointer tracking and inverter control for solar energy applications. J. Power Energy Eng. **04**(09), 61 (2016)
18. Francis, W.K., Mathew, P.J.: MATLAB/Simulink PV module model of P&O and DC link CDC MPPT algorithms with lab view real time monitoring and control over P&O technique. Int. J. Adv. Res. Electr. Electron. Instrum. Energy **3**(5), 92–101 (1970)
19. Kumar, R., Choudhary, A., Koundal, G., Yadav, A.S.A.: Modelling/simulation of MPPT techniques for photovoltaic systems using Matlab. Int. J. **7**(4), 178–187 (2017)

Enhanced Predictive Model Control Based DMPPT for Standalone Solar Photovoltaic System

Halima Ikaouassen[1](✉), Kawtar Moutaki[1], Abderraouf Raddaoui[1], and Miloud Rezkallah[2]

[1] MEAT, EST-Salé, Mohammed 5 University, B.P. 11000, Rabat, Morocco
halimaikaoaussen@research.emi.ac.ma
[2] Electrical Engineering Department, ETS-Montréal, Notre Dame, QC 1100, Canada

Abstract. This paper discussed an enhanced predictive model control (PMC) strategy based distributed maximum power point tracking DMPPT with a prediction horizon of one sampling time in order to achieve high performances from standalone solar photovoltaic system in the presence of dynamic weather variations and partial shading. In this paper, three PV modules are interfaced to the DC-BUS through three cascaded DC-DC boost power converters used with the enhanced PMC based DMPPT algorithm, the proposed technique calculates all possible switching states before applying to the three converters, and the adequate switching state is selected by minimization of a defined cost function, to regulate the duty cycle of the power converters independently, and to supervise maximum power point of the three cascaded PV modules, in order to avoid mismatching phenomena between modules which is considered the main cause for performance degradation and efficiency drop. The performances of the proposed system and control strategy are verified and confirmed when comparing with other conventional MPPT methods such Perturb and Observe (P&O) algorithm based DMPPT using MATLAB/Simulink interface.

Keywords: Enhanced · PMC · DMPPT

1 Introduction

The solar photovoltaic (PV) systems considered as subdued natural energy resource, have become, the fastest developed renewable energy among the alternative electrical energy source and the most used in several applications [1], especially in remote areas [2] since it's free, no maintenance cost, clean and available over the world, hence the solar PV systems contribute as a first power generation system over all other processes of energy production on earth duo to the high availability of solar irradiation in most geographical locations on earth.

As known, the performance of the PV module is highly dependent on solar irradiation, temperature and load impedance [3, 4]. To achieve maximum power point (MPP) from the PV module, maximum power point tracking system (MPPT) is required to ensure the extraction of the maximum power under any environmental

M. Ezziyyani (Ed.): AI2SD 2018, AISC 912, pp. 185–196, 2019.
https://doi.org/10.1007/978-3-030-12065-8_18

conditions. Several MPPT methods are widely adapted to achieve high performance presented in [5, 6], among existing MPPT methods, Perturb and Observe (P&O) method [7, 8], Hill Climbing (HC) and Incremental Inductance (INC) methods [9], Artificial Neural Network (ANN) method [10], Fuzzy Logic Control (FLC) method [11]. These methods are the most commonly used in practice duo to their simplicity and low-cost implementation. Nevertheless, in all these strategies, the operating point oscillates around MPP at steady state, which leads to low accuracy and efficiency [12]. Recently, model predictive control (MPC) method has been adopted by many authors and becomes broadly used in PV systems applications. MPC using sensor less-current based MPPT has been reported in [13], new MPPT method based on finite control set MPC has been proposed in [14]. This new concept uses equivalent system model to predict system parameters and to calculate the optimal cost function representing the desired behavior of the system in the next sampling time [15]. The MPPT algorithms investigated in [16, 17] suggest Model Predictive Control based MPPT for multi-string PV arrays in smart distribution dc micro-grid systems. The algorithm allows high performances and maximum output power drawn from PV modules in identical characteristics conditions. However, the algorithm meets its weakness under partial shading and mismatching problems between PV modules [18].

In this paper, an enhanced PMC with equivalent model estimation based distributed MPPT is used in order to achieve high performances from solar photovoltaic modules. This strategy extends conventional PMC method with different control parameters. Unlike classical MPC based on MPPT strategies discussed before in literature, the proposed enhanced PMC based DMPPT maximizes the total output power drawn from the three PV modules simultaneously and generates the optimum value of the switching duty cycle of three DC-DC boost power converters by predicting directly the behavior of all PV modules in the future steps using a new estimator to estimate the equivalent system resistance and output PV voltage. The principal contribution is the extraction of maximum power from the proposed standalone power generation system based solar PV modules using the enhanced PMC based DMPPT and to avoid the mismatching problems, thus achieving high performances under dynamic weather variations and partial shading when comparing to the conventional Perturb and Observe (P&O) based DMPPT. Simulation results validate the enhanced proposed control strategy using MATLAB/Simulink interface.

2 System Description and Modeling

As it is described in the Fig. 1. Three identical PV modules are considered as case study using identical cascaded DC-DC power converters to maximize the output power drawn from three PV modules by implementing enhanced PMC based DMPPT control, These converters are connected in parallel to supply a resistive load across the DC-BUS, and to compensate output PV power fluctuations under partial shading and dynamic weather condition variations.

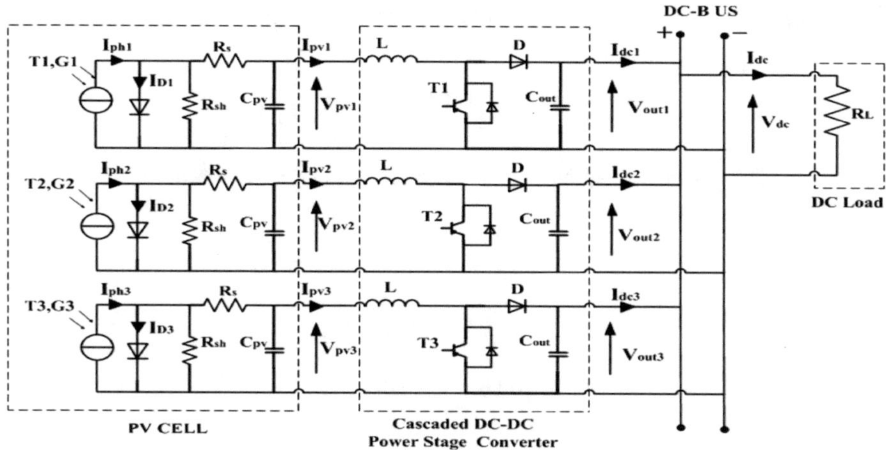

Fig. 1. PV system with cascaded DC-DC power stage converters

2.1 PV Module

The PV system directly converts sunlight into electricity. The basic device of a PV system is the PV cells, which are grouped in parallel and/or in series to form arrays or modules. The PV solar cell contains a light generated source, diode, series and parallel resistances. Its mathematical model is given in Fig. 1. The voltage-current characteristic equation of a solar PV module is given as [19]:

$$I_{pv} = N_p I_{ph} - N_p I_s \left[\exp\left(\frac{q\left(\frac{V_{pv}}{N_s} + \frac{I_{pv} R_s}{N_p}\right)}{N_s AKT} \right) - 1 \right] - \frac{\frac{V_{pv}}{N_s} + \frac{I_{pv} R_s}{N_p}}{R_{sh}} \tag{1}$$

Where I_{ph}, I_S, T, R_S, R_{sh}, q, K, A, N_S and N_p denote light generated-current, cell saturation of dark current, cell's operating temperature, equivalent series resistance of the PV array, equivalent parallel resistance, charge of an electron, Boltzmann's constant, ideal factor, number of cells in series and in parallel respectively.

The light generated current of the PV cell depends linearly on the solar irradiance and the temperature according to the equation given below:

$$I_{ph} = \frac{G}{G_n}\left[I_{sc} + K_i\left(T - T_{ref}\right)\right] \tag{2}$$

Where the I_{SC} is the short circuit current ($V = 0$), G is the irradiance on the device surface, $G_n = 1000\,\mathrm{W/m^2}$ is the solar irradiance in the standard Conditions, T is the operating temperature and $T_{ref} = 24\,°\mathrm{C}$ is the reference temperature, K_i is a cell's short circuit current temperature coefficient.

The diode saturation current varies with cell's temperature and it is described as:

$$I_s = I_{rs}(\frac{T}{T_{ref}})^3 \left[exp \frac{qE_g}{KA} \left(\frac{1}{T_{ref}} - \frac{1}{T} \right) \right] \tag{3}$$

Where E_g is the band-gap energy of the material, and I_{rs} is the nominal saturation current:

$$I_{rs} = \frac{I_{sc}}{\exp\left(\frac{qV_{oc}}{N_s AKT}\right) - 1} \tag{4}$$

Where V_{oc} is the open circuit voltage ($I = 0$).

2.2 Cascaded DC-DC Power Stage Converters

The DC-DC boost converters are operating at continuous conduction mode (CCM). The dynamic of the converter is described based on the position function of the switch and illustrated by the following equations:

When the position function of the switch is set to $T_{n=\{1,3\}} = 1$, we obtained the following equations:

$$V_{pv,n} = L_n \frac{dI_{L,n}}{dt} \tag{5}$$

$$V_{dc,n} = -RC_{out,n} \frac{dV_{dc,n}}{dt} \tag{6}$$

When the position function of the switch is set to $T_{n=\{1,3\}} = 0$, we obtained the following equations:

$$V_{pv,n} = V_{dc,n} - L_n \frac{dI_{L,n}}{dt} \tag{7}$$

$$V_{dc,n} = RI_{L,n} - RC_{out,n} \frac{dV_{dc,n}}{dt} \tag{8}$$

Where n, $C_{out,n}$, L_n, R and $I_{L,n}$ denote number of PV modules, output capacitor and inductor of three DC-DC boost converters, load resistance, current through the inductor which equal to the PV current $I_{pv,n}$, respectively.

Since the three PV modules supply a common DC-Bus and have different power delivery capacities which vary in time, the enhanced PMC method is used to predict the maximum voltage at the next sampling time k + 1 corresponding to the maximum power drawn from the three PV modules.

3 Enhanced PMC Based DMPPT Controller Design

The enhanced PMC based DMPPT controller forces the standalone DC system to operate at its optimal point; it maximizes the total output power drawn from the three PV modules simultaneously, and generates the appropriate value of the switching duty cycle of the three power stage converters by predicting directly the attitude of all PV modules in the future steps using an estimator as illustrated in Fig. 2, which estimates the equivalent system resistance and output voltage for each PV, hence a cost function is defined to select the right response of the three PV modules for the next predicted step. So the algorithm contains three main blocks as detailed in the Fig. 2: (3.1) Estimator block, (3.2) PMC block, and (3.3) Cost function calculation block.

The algorithm calculates firstly the two predicted voltage values for the three PV modules as below:

$$V_{pv,n=\{1,3\}}(k+1)^1 = V_{pv,n=\{1,3\}}(k) + deltaV \tag{9}$$

$$V_{pv,n=\{1,3\}}(k+1)^2 = V_{pv,n=\{1,3\}}(k) - deltaV \tag{10}$$

Where $deltaV$ represents the fixed voltage step size.

3.1 Estimator Block

The estimation of the equivalent resistance and voltage of each PV module using the Thevenin model circuit of the system are expressed as:

$$R_{eq,n=\{1,3\}}(k) = -\frac{V_{pv,n=\{1,3\}}(k) - V_{pv,n=\{1,3\}}(k-1)}{I_{pv,n=\{1,3\}}(k) - I_{pv,n=\{1,3\}}(k-1)} \tag{11}$$

$$V_{eq,n=\{1,3\}}(k) = V_{pv,n=\{1,3\}}(k) + R_{eq,n=\{1,3\}}(k) \times I_{pv,n=\{1,3\}}(k) \tag{12}$$

3.2 Predictive Model Control Block

The estimated value of the equivalent resistance is used to calculate the power provided by each PV module. As given bellow, the predictive block controller predicts three power values of three PV modules depending upon the two possible voltage values already predicted by the algorithm in the next sampling time k + 1 as:

$$P_{pv,n=\{1,3\}}(k+1)^{1,2} = I_{pv,n=\{1,3\}}(k+1)^{1,2} \times V_{pv,n=\{1,3\}}(k+1)^{1,2} \tag{13}$$

And

$$I_{pv,n=\{1,3\}}(k+1)^{1,2} = \frac{V_{pv,n=\{1,3\}}(k) - V_{pv,n=\{1,3\}}(k+1)^{1,2}}{R_{eq,n=\{1,3\}}(k)} \tag{14}$$

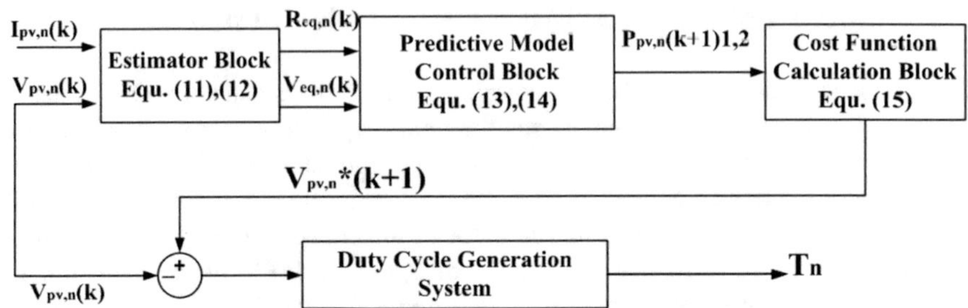

Fig. 2. Enhanced PMC based DMPPT algorithm.

3.3 Cost Function Calculation Block

The cost function is calculated to select the right response of the three PV modules in the next step:

$$\min(g)_{n=\{1,3\}}^{1,2} = P_{pv,n=\{1,3\}}(k+1)^{1,2} - P_{pv,n=\{1,3\}}(k) \tag{15}$$

If the $g_{n=\{1,3\}}^{1}$ is higher than $g_{n=\{1,3\}}^{2}$, then the algorithm selects the cost function $g_{n=\{1,3\}}^{1}$, which correspondingly means that the PV system will balance to $V_{pv,n=\{1,3\}}(k+1)^{1}$ at the next sampling time, and represents the maximum output voltage for each PV module, then the algorithm selects the maximum output voltage and considered as a reference voltage at the next sampling time denoted as $V_{pv}^{*}(k+1)$, then it is compared to the PV output voltage, and the difference is used as input to the duty cycle generation system containing PWM block which generates the desired response.

4 Simulation Analysis

The proposed enhanced controller is simulated in MATLAB/Simulink interface. The performances of the enhanced PMC based DMPPT algorithm are verified and compared to the conventional P&O based DMPPT algorithm. Sampling prediction time T_S of 30 μs is used. The system parameters employed in this work are detailed in Table 1.

As already mentioned, the main contribution of this work is to extract maximum power from the proposed system with the proposed enhanced algorithm in order to avoid the mismatching problems between PV modules, and to acquire high performances in presence of dynamic weather variations and under partial shading.

To evaluate the performances of the enhanced PMC based DMPPT algorithm, two case studies are considered: response to step change in solar irradiance level and response to step change in temperature degree in random way. Thus, three identical PV modules are operating at different solar irradiance levels and temperature degrees changes.

Table 1. Standalone solar PV system parameters.

Item	Symbol	Value
Rated PV power	P_{pv}	5000 W
Open Circuit Voltage	V_{oc}	21.06 V
Short-Circuit Current	I_{sc}	3.75 A
Series Resistance	R_s	2000 Ω
Parallel Resistance	R_{sh}	0.69 Ω
Input Capacitor	C_{pv}	200 μF
Inductor	L	15 mH
Output Capacitor	C_{out}	200 μF
DC-BUS Voltage	V_{dc}	450 V
DC Load Resistance	R_L	20 Ω

To start the analysis and performance evaluation, one applies the solar irradiance levels of G1 to the PV1, G2 to the PV2 and G3 to the PV3 as it is detailed in the Fig. 3, while the temperature is kept constant at the reference temperature value T = 24 °C. The three PV powers extracted from each PV module are given in the Fig. 4; one can clearly remarked that each PV power varies with the solar irradiation level variations, which validate the feasibility of the enhanced PMC as a MPPT algorithm. The step change responses of the P&O based DMPPT and the enhanced PMC based DMPPT algorithms are illustrated in the Fig. 5.

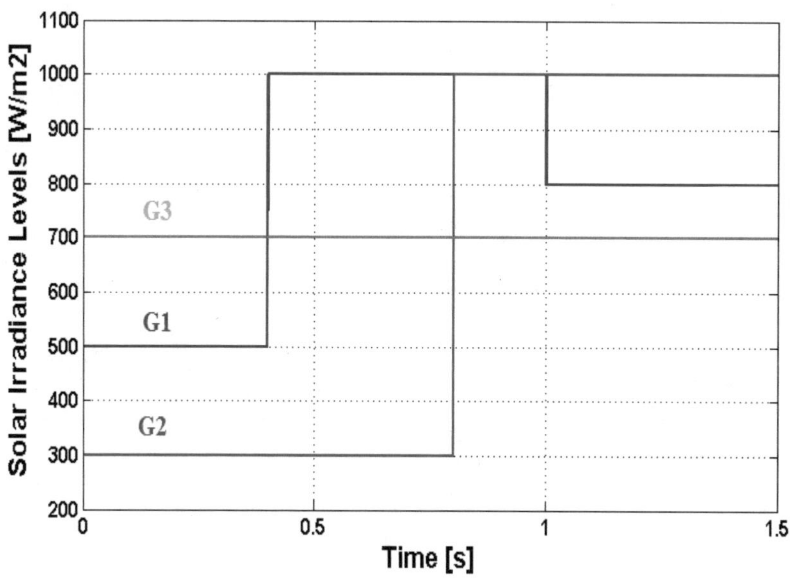

Fig. 3. Solar irradiance levels applied to each of the three PV.

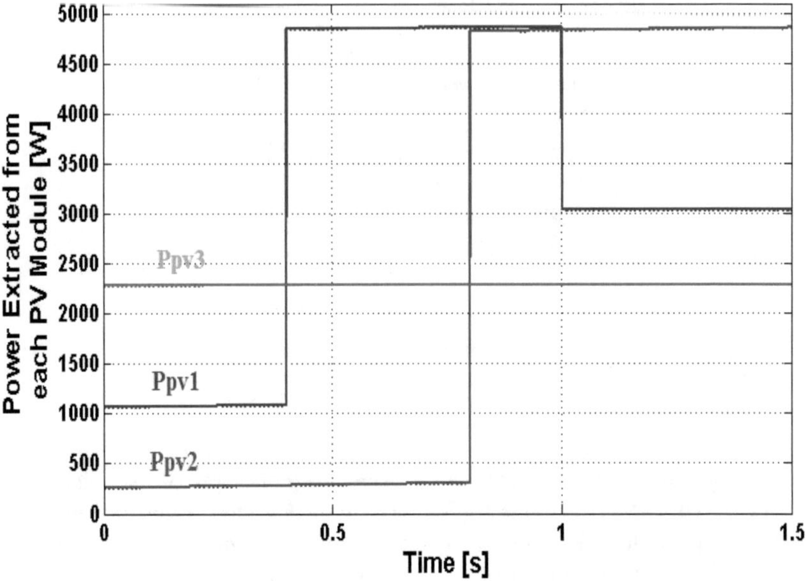

Fig. 4. Power extracted from each PV module using enhanced PMC based DMPPT algorithm under solar irradiance level changes.

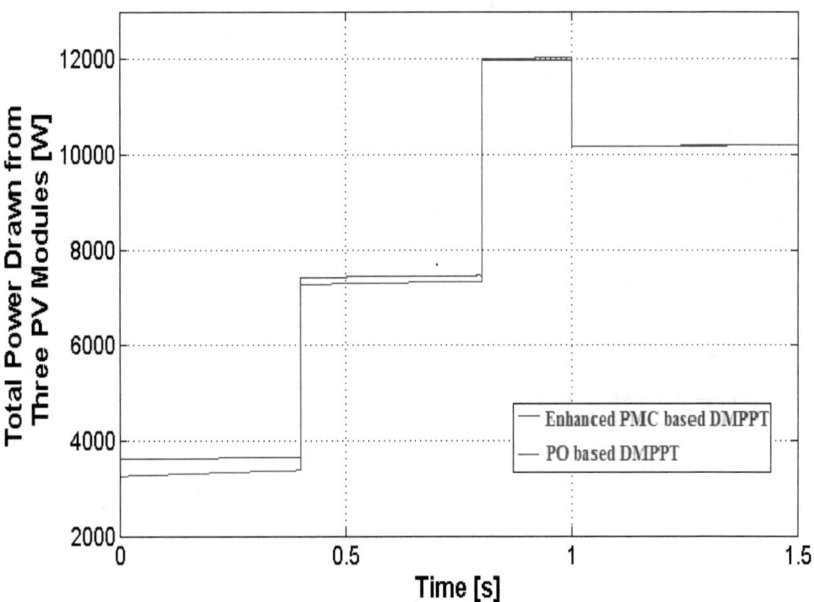

Fig. 5. Total Power drawn from three PV modules using both algorithms under solar irradiance level changes.

In summary, the results demonstrate that the enhanced PMC based DMPPT algorithm has faster dynamic response to step changes with less power ripple at steady state as it is depicted in Fig. 5, thus, the extracted power from the three PV modules using enhanced PMC based DMPPT is more obvious when solar irradiance level is low between time t = 0 s and t = 0.4 s and time t = 0.4 s and t = 0.7 s. Therefore, the enhanced PMC based DMPPT presents efficacy under low solar irradiance levels and partial shading against P&O based DMPPT algorithm.

The descriptive results of the second case study are as follows: One applies the solar temperatures of T1 to the PV1, T2 to the PV2 and T3 to the PV3, while the irradiance is kept constant at the nominal value $Gn = 1000\ \mathrm{W/m^2}$ as it is shown in Fig. 6. The three PV powers extracted from each PV module are displayed in the Fig. 7. It is clearly remarked that each PV power is higher when the corresponding temperature degree is lower, and vice versa; the power is lower when the corresponding temperature degree is higher, which confirm and validate the feasibility of the enhanced PMC as a MPPT algorithm.

According to the Fig. 8, the enhanced PMC based DMPPT can accurately track the MPP with less power ripples and keeps the power extracted from three PV modules constant and higher when the temperature is high between time t = 1 s and t = 1.5 s, comparing to the P&O based DMPPT algorithm Therefore, the enhanced algorithm is more accurate and presents high efficacy under high temperature.

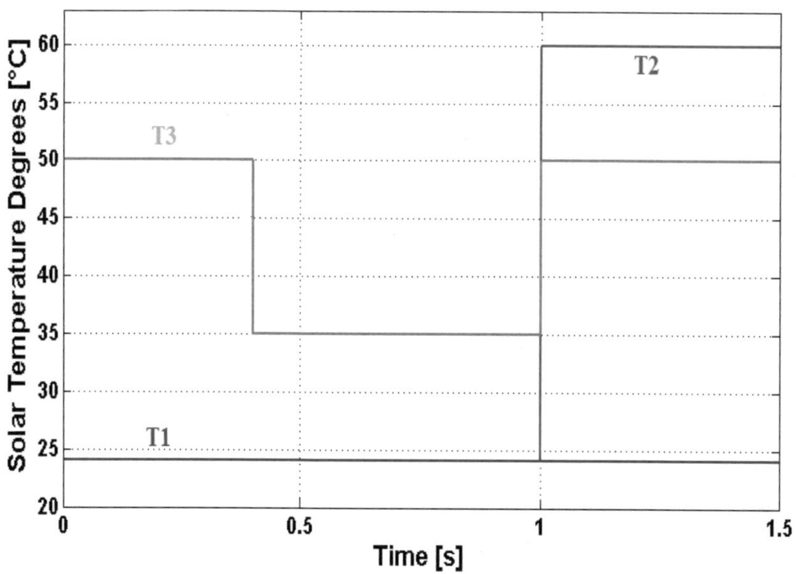

Fig. 6. Solar Temperature degrees applied to each of the three PV modules.

Both of case studies confirm the effectiveness and robustness of the enhanced PMC based DMPPT algorithm against dynamic weather condition variations and partial shading, and presents pronounced performances under low solar irradiance levels and

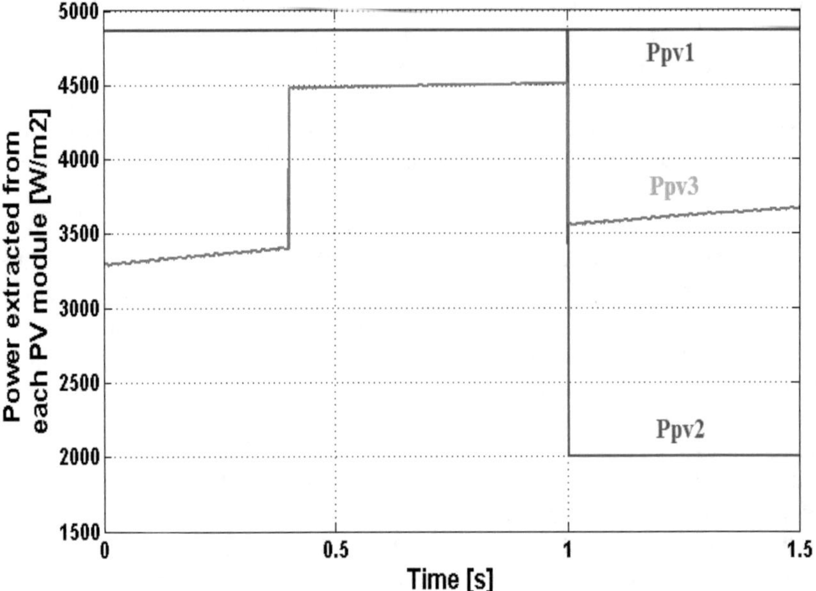

Fig. 7. Power extracted from each PV module using enhanced PMC based DMPPT algorithm under temperature degrees changes.

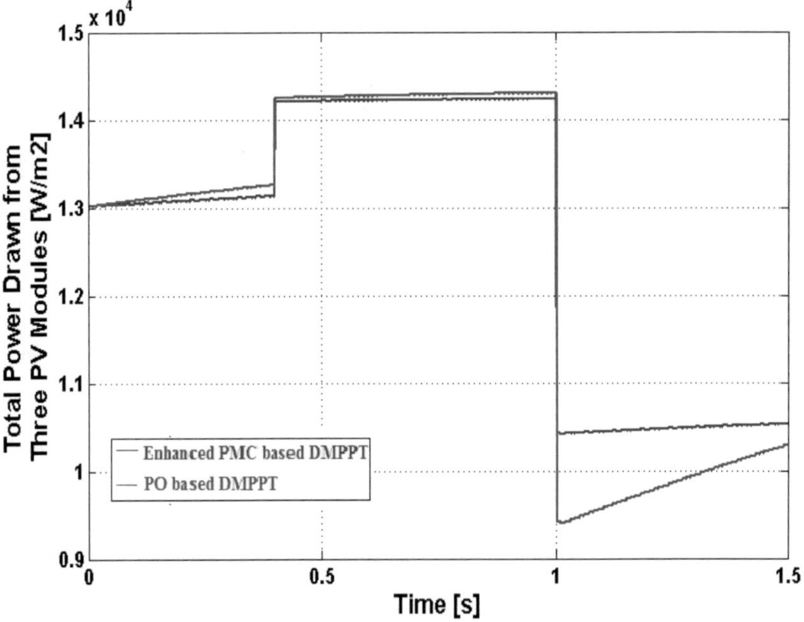

Fig. 8. Total Power drawn from three PV modules using both algorithms under solar temperature degree changes.

high temperature degrees when comparing to the well-known P&O based DMPPT algorithm. Therefore, the proposed enhanced PMC based DMPPT algorithm can be considered as a relevant technique and the results of the simulation analysis show that the proposed method can have interesting applications.

5 Conclusion

In this work, an enhanced PMC framework based DMPPT has been investigated and applied to the proposed PV system. A highly performance predictive model for cascaded PV modules architecture is proposed with cascaded DC-DC power stage converters, The performance of the proposed enhanced controller is compared to the conventional P&O based DMPPT technique. The simulation results demonstrate that the proposed enhanced PMC base DMPPT has faster dynamic response to step changes in solar irradiance levels comparing to the well-known P&O technique based DMPPT and the algorithm present robustness and high efficiency. In addition, the results demonstrate that more power is captured from the PV system when comparing to P&O based DMPPT under low solar irradiances and high temperatures. Hence, less PV modules are required for generation of same amount of energy demanded.

References

1. Ardashir, J.F., Sabahi, M., Hosseini, S.H., Blaabjerg, F., Babaei, E., Gharehpetian, G.B.: A single-phase transformerless inverter with charge pump circuit concept for grid-tied PV applications. IEEE Trans. Ind. Electron. **64**, 5403–5415 (2017)
2. Debnath, D., Chatterjee, K.: Two-stage solar photovoltaic-based stand-alone scheme having battery as energy storage element for rural deployment. IEEE Trans. Ind. Electron. **62**, 4148–4157 (2015)
3. Das, M., Agarwal, V.: Novel high-performance stand alone solar PV system with high gain, high efficiency DC-DC converter power stages. IEEE Trans. Ind. Appl. **51**, 4718–4728 (2015)
4. Heidari, N., Gwamuri, J., Townsend, T., Pearce, J.M.: Impact of snow and ground interference on photovoltaic electric system performance. IEEE J. Photovoltaics **5**, 1680–1685 (2015)
5. El-Helw, H.M., Magdy, A., Marei, M.I.: A hybrid maximum power point tracking technique for partially shaded photovoltaic arrays. IEEE Access **5**, 11900–11908 (2017)
6. Ramyar, A., Iman-Eini, H., Farhangi, S.: Global maximum power point tracking method for photovoltaic arrays under partial shading conditions. IEEE Trans. Ind. Electron. **64**, 2855–2864 (2017)
7. Killi, M., Samanta, S.: Modified perturb and observe MPPT algorithm for drift avoidance in photovoltaic systems. IEEE Trans. Ind. Electron. **62**, 5549–5559 (2015)
8. Ahmed, J., Salam, Z.: A Modified P&O maximum power point tracking method with reduced steady state oscillation and improved tracking efficiency. IEEE Trans. Sustain. Energy **7**, 1506–1515 (2016)
9. Kjær, S.B.: Evaluation of the hill climbing; and the incremental conductance; maximum power point trackers for photovoltaic power systems. IEEE Trans. Energy Convers. **27**, 922–929 (2012)

10. Agha, H.S., Koreshi, Z.-U., Khan, M.B.: Artificial neural network based maximum power point tracking for solar photovoltaics. In: International Conference on Information and Communication Technologies (ICICT), pp. 150–155 (2017)
11. Tang, S., Sun, Y., Chen, Y., Zhao, Y., Yang, Y., Szeto, W.: An enhanced MPPT method combining fractional order and fuzzy logic control. IEEE J. Photovoltaics **7**, 640–650 (2017)
12. Moré, J.J., Puleston, P.F., Kunusch, C., Fantova, M.A.: Development and implementation of a supervisor strategy and sliding mode control setup for fuel cell based hybrid generation systems. IEEE Trans. Energy Convers. **30**, 218–225 (2015)
13. Metry, M., Shadmand, M.B., Balog, R.S., Abu-Rub, H.: MPPT of photovoltaic systems using sensorless current based model predictive control. IEEE Trans. Ind. Appl. **53**, 1157–1167 (2017)
14. Mahmoudi, H., Moamaei, P., Aleenejad, M., Ahmadi, R.: A new maximum power point tracking method for photovoltaic applications based on finite control set model predictive control. In: Applied Power Electronics Conference and Exposition (APEC), pp. 1111–1115. IEEE (2017)
15. Abushaiba, A.A., Eshtaiwi, S.M.M., Ahmadi, R.: A new model predictive based maximum power point tracking method for photovoltaic applications. In: International Conference on Electro Information Technology (EIT), pp. 0571–0575. IEEE (2016)
16. Shadmand, M.B., Balog, R.S., Abu-Rub, H.: Model predictive control of PV sources in a smart DC distribution system: maximum power point tracking and droop control. IEEE Trans. Energy Convers. **29**, 913–921 (2014)
17. Xiao, S., Shadmand, M.B., Balog, R.S.: Model predictive control of multi-string PV systems with battery back-up in a community dc microgrid. In: Applied Power Electronics Conference and Exposition (APEC), pp. 1284–1290. IEEE (2017)
18. Sajadian, S., Ahmadi, R.: Distributed maximum power point tracking using model predictive control for solar photovoltaic applications. In: Applied Power Electronics Conference and Exposition (APEC), pp. 1319–1325. IEEE (2017)
19. Zhou, J., Li, H., Qiao, Y., Gao, Q., Liu, Y., Liu, Z.: A comprehensive method to modeling and simulation of photovoltaic module under natural environment. In: IEEE 40th Photovoltaic Specialist Conference (PVSC), pp. 1353–1357 (2014)

Energy Efficiency Regulation and Requirements: Comparison Between Morocco and Spain

I. Merini[1(✉)], A. Molina-García[1], Ma. S. García-Cascales[2], and M. Ahachad[3]

[1] Department of Electrical Engineering, Universidad Politécnica de Cartagena, 30202 Cartagena, Spain
ikramalmerini@hotmail.com, angel.molina@upct.es
[2] Department of Electronics, Computer Architecture and Project Engineering, Universidad Politécnica de Cartagena, 30202 Cartagena, Spain
socorro.garcia@upct.es
[3] Equipe de Recherche en Transferts Thermiques & Énergétique, Université Abdelmalek Essaadi, FST de Tanger, Tanger, Morocco
ahachad_med@yahoo.fr

Abstract. With the aim of reducing energy dependence and promoting economic development, Morocco has recently issued an energy plan to integrate both sustainable development and environmental protection. This strategy gives a novel thermal construction regulation in Morocco, provided by the Moroccan Energy Efficiency Agency. In terms of energy demand, building sector is currently the second most energy-intensive economic sector in Morocco; after the transport sector, accounting for 33% of the total energy consumption in the country (25% for residential and 8% for tertiary). This paper compares two countries geographically very close, such as Spain and Morocco, in terms of energy efficiency regulation, norms and requirements, discussing the parameters taken into account in both countries to determine their building thermal comfort levels. In this way, BINAYATE package software, proposed by Morocco for implementation and control of Moroccan thermal regulations is used to estimate the energy requirement of an ordinary building and the gain obtained by the application of the Moroccan and Spanish regulations. In addition, economic and environmental improvements of the Moroccan situation from the Spanish experience are also discussed and included in the paper.

Keywords: Building thermal requirements · Thermal comfort · Energy demand · Regulation comparison

1 Introduction

The increase in energy consumption worldwide, as well as the effects it has on the environment has led several countries to develop programs and actions to mitigate those effects and also work on the legislative framework to regulate this energy consumption depending on the sector in question. Despite its low emissions of greenhouse gases, in relation to global emissions worldwide, Morocco has developed ambitious reforms in

M. Ezziyyani (Ed.): AI2SD 2018, AISC 912, pp. 197–209, 2019.
https://doi.org/10.1007/978-3-030-12065-8_19

various sectors to ensure a transition to a green economy. The main challenges in the electricity sector, which Morocco faces, are the constant increase in demand (Fig. 1), the strong dependence on foreign countries, the upward trend and the accentuation of the volatility of world fuel prices, the protection of the environment and the preservation of the purchasing power of the citizen and the strengthening of the competitiveness of national economic operators [1]. The new national energy strategy refers to four fundamental objectives and defines five strategic orientations [2]. The objectives are security of supply and availability of energy; the widespread access to energy at competitive prices; the control of the demand and the conservation of the Environment. In terms of strategic orientations, we have a diversified and optimized mix around reliable and competitive technological options; a mobilization of national resources for the increase in potential of renewable energies; an energy efficiency considered a national priority; a strengthening of regional integration and sustainable development.

In terms of energy efficiency, the country is developing programs and actions that are the following: energy efficiency measures in public administrations; promote energy efficiency in public lighting; generalization of energy saving lamps; price system that incentivize; communication, education and awareness; and sustainable urban development of cities, both new and old, integrating the energy efficiency factor. The objectives set for these programs and actions, which are also developed in Law 47-09 on energy efficiency, consist of making an energy saving of 12% by 2020 and 15% by 2030 [2, 3]. All economic sectors have had an increase in energy demand. In 2013, buildings have had an energy consumption of 33%, of which 25% in the residential sector and 8% in the tertiary sector, see Fig. 2.

At European level, the conclusions of the Council of the European Union of June 2011 on the 2011 energy efficiency plan, stresses that buildings represent 40% of the final energy consumption of the European Union [4]. The measures adopted to reduce energy consumption in the Union, together with the greater share of energy from renewable sources, will allow the Union to achieve the triple objective for 2020 [5], which consists in reducing total gas emissions by 20%. Greenhouse effect, increase energy efficiency by 20% and have 20% of the total energy consumption from renewable sources. According to article 24.2 of Directive 2012/27/EU of the European Parliament and of the Council, of October 25, 2012, regarding energy efficiency, Spain has drawn up the National Action Plan for Energy Efficiency 2017–2020, required of all the Member States of the European Union the presentation of these plans, the first of them no later than April 30, 2014 and then every three years [6].

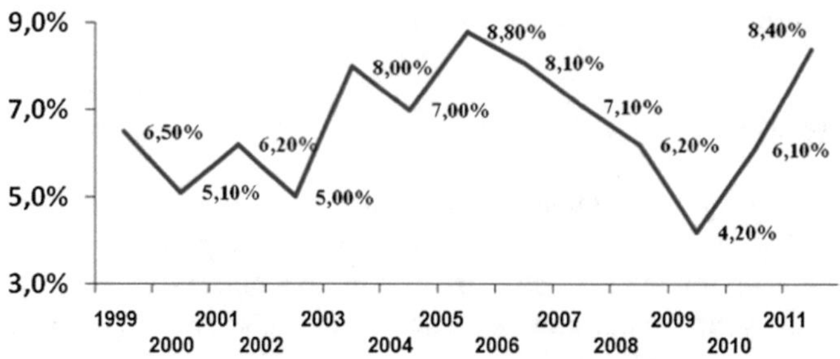

Fig. 1. Evolution of electricity demand

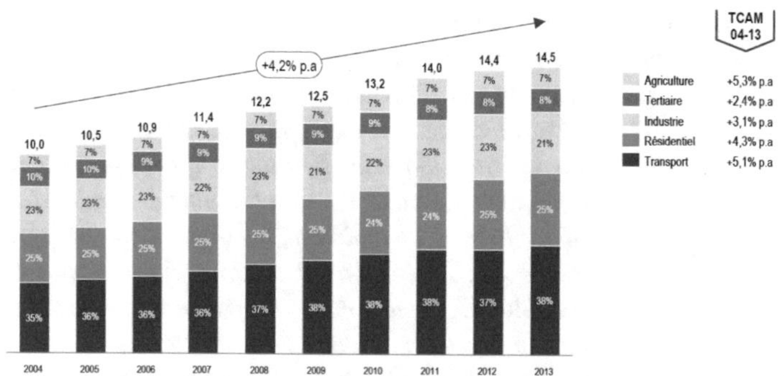

Fig. 2. Final consumption per sector in Morocco

According to the same plan, the preponderance of oil and natural gas in final consumption in Morocco is linked to the sectoral structure of the demand. Indeed, the transport sector was responsible for 41.8% of the total energy consumption in 2015. This sector, which is very dependent on petroleum products, greatly determines the needs and characteristics of energy demand. The industrial sector, with 23.5% of demand, contributes significantly to the fossil fuel consumption and emissions. However, it maintains a progressive loss of weight in the structure of demand. The opposite way has been followed by both residential and tertiary sectors, where their shares in demand remain up against the industry, reaching, as a whole, 31% of demand in 2015.

This paper presents a comparison at a normative level between Spain and Morocco, its application in the real situation of the country, as well as proposes improvements for the case of Morocco.

2 Characterization of Climatic Zones

The identification of climatic zones for energy efficiency proposals focused on buildings is a key element in many programs and policies in order to improve the thermal behavior of buildings. Despite its importance, there is no consensus on the appropriate methodology for climate zoning. Previous studies indicate a wide variety of methods and parameters that are used for this distribution: day-grades, conglomerate analysis and administrative divisions are the most widely used [7]. Therefore, the first objective of this work is to provide a comparison between methods proposed by Spain and Morocco to determine and identificate different climate zones.

Climatic zones in Morocco have been determined through different studies carried out by the National Meteorology Directorate (DMN) and the National Agency for the Development of Renewable Energies and Energy Efficiency (ADEREE), now converted into the Moroccan Energy Efficiency Agency (AMEE), with the support of different international experts. Subsequently, the Moroccan territory is subdivided into homogeneous climatic zones based on the analysis of climatic data collected by 37 meteorological stations for 10 years, from 1999 to 2008. The estimation and identification of the zones has been carried out according to the number of the daily grades of winter as well as of summer. The DMN has established two types of zoning:

Climatic zones considering days with heating requirements based on 18 °C. *It is a measure of the differences between the average temperature of a given day and the setpoint temperature linked to the domestic heating requirements. The reference temperature is 18 °C since, on average values, when the outside temperature is below this threshold, indoor must be heated to maintain a comfortable temperature level. When the outdoor temperature is 18 °C, internal loads and gains can increase the indoor temperature above 20 °C, and thus, it is not required to supply energy/electricity to heat and increase the indoor temperature.*

Climates zones considering days with air conditioning based on 21 °C: *It is a measure of the differences between the average temperature of a given day and the setpoint temperature linked to the domestic climate control during the months of summer. When the outside temperature is equal to the reference temperature, 21 °C, the internal loads and gains can increase the indoor temperature above 24 °C–26 °C, which implies the needs of air conditioning* [8].

Morocco has thus stated the determination of the climatic zoning by the degrees-days method, which is adopted in more than 20 countries [7]. However, and according to the AMEE, it is not currently possible to adopt, for practical thermal regulation, two different seasonal zoning. Thereafter, AMEE has proposed a unique climatic zoning for thermal regulation requirements. This unique zone is based on hourly annual climate files and the results of simulations carried out by the building annual thermal requirements of heating and air conditioning with TRNSYS software, in twelve representative Moroccan cities. For this purpose, seven reference buildings were chosen, obtaining the results shown in Fig. 3. Finally, the final zoning map includes six climatic zones, which is in line with current administrative limits, for an friendly and effective application of the new regulation in Morocco [8], see Fig. 4.

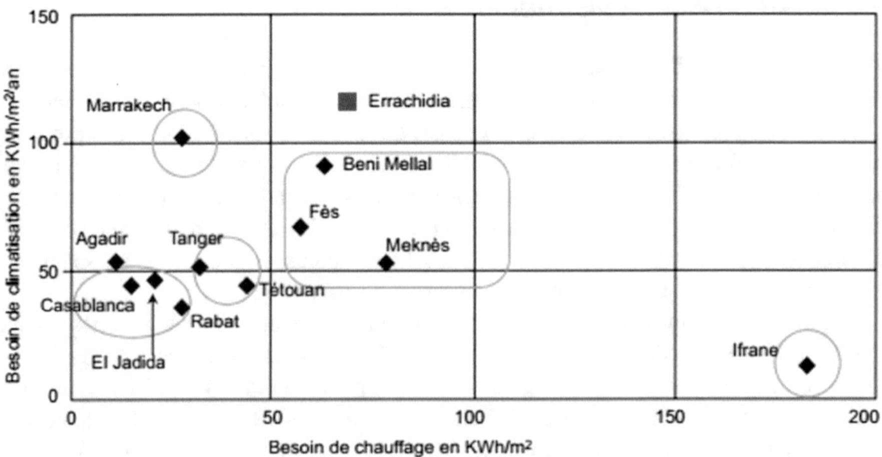

Fig. 3. Heating and cooling requirements for the representative Moroccan cities: definition of reference buildings

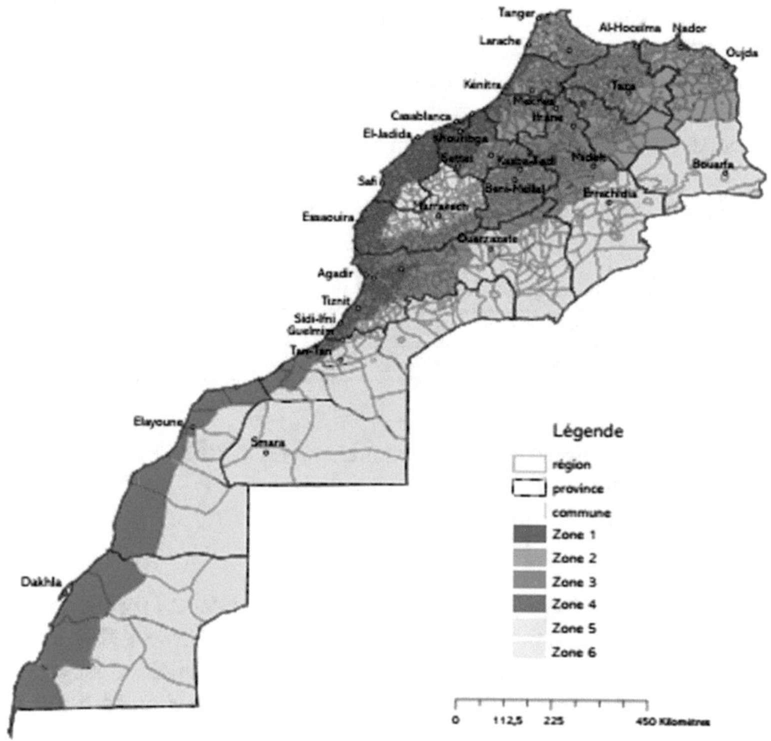

Fig. 4. Climatic zoning for thermal building regulation in Morocco

In Spain, for the purpose of determining the energy demand of a building, a set of climatic zones is established to define the external demands in terms of temperature and solar radiation [9]. The reference climate defines the external demands for a typical year through a series of parameters (temperature, humidity, solar radiation, …), representative of a climatic zone [10]. A set of 12 climatic zones mentioned in the Spanish norms are defined, based on winter and summer climatic severity, for certain Spanish localities. According to this document, climate severity is defined as the ratio between the energy demand of a building in a locality and the energy demanded by the same building in a reference location. In this regulation, the capital of Spain (Madrid) was taken as reference location, and therefore, its climate severity is unity.

The climatic severity combines both degrees-day and solar radiation for a specific location. It is then possible to demonstrate that when two locations present the same winter climate severity (SCI), the heating energy demanded by a same building located in both localities is basically similar. This can also be applied to climate severity in summer (MCS) [11]. The expression of the winter climate severity is obtained as follows (1),

$$SCI = a \cdot GD + b \cdot (n/N) + c \cdot GD^2 + d \cdot (n/N)^2 + e \tag{1}$$

where *GD*, is the sum of winter degrees-day in base 20, from October to May; *n/N* is the ratio between the number of sunshine hours and the maximum number of sunshine hours from October to May; and *a, b, c, d, e*, are the regression coefficients, see Table 1 [10].

The winter climatic zone is then determined according to the winter climate severity (SCI). Each Spanish winter climatic zone of the DB-HE (α, A, B, C, D and E) corresponding to the intervals summarized in Table 2.

Table 1. Regression coefficients for winter climate severity

a	b	c	d	e
3,456E−04	−4,043E−01	8,394E−08	−7,325E−02	−1,137E−01

With regard to the Spanish summer climate severity. It is determined by the following expression (2):

$$SCV = a \cdot GD + b \cdot GD^2 + c \quad (2)$$

where *GD* is the sum of the degrees-day of summer in base 20 for the months from June to September, *a, b, c* are the regression coefficients, see Table 3 [10]. The summer climatic zone is then determined according to the summer climatic severity (MCS). Each summer climatic zone of DB-HE (1, 2, 3 and 4) corresponding to the intervals summarized in Table 4.

Therefore, and according to this comparison between Spain and Morocco in terms of climatic zone definition, it can be affirmed that Morocco provides only a dynamic simulation based on annual hourly data through the TRNSYS software. However, the Spanish methodology establishes two zoning: winter and summer, based on the climatic severity of each of them, and according to the degrees-day and solar radiation.

3 Legislative Framework

3.1 General Overview

Within the framework of the Moroccan energy strategy, the country has developed an energy efficiency policy that aims to clarify the relationships between administrations and operators, establishing a system of institutionalized public management in energy efficiency, an adequate legislative and regulatory framework and appropriate standards and standards. In this sense, Morocco has drafted law 47-09, on energy efficiency, which requires, in Article 3, complete the legislation for urban planning, "*the general construction regulations*", setting the rules of energy performance of buildings in order to guarantee a better energy balance of the buildings by climatic zones, dealing mainly with orientation, lighting, insulation and thermal flows, as well as contributions in renewable energy in order to improve the performance levels of buildings to build or modify. As a result, in Decree No. 2-13-874, the Construction Thermal Regulation in

Table 2. Intervals for winter zoning

α	A	B	C	D	E
SCI $\leq$ 0	0 < SCI SCI $\leq$ 0,23	0,23 < SCI SCI $\leq$ 0,5	0,5 < SCI SCI $\leq$ 0,93	0,94 < SCI SCI $\leq$ 1,51	SCI > 1,51

Table 3. Regression coefficients for summer climate severity

a	b	c
2,990E−3	−1,1597E−07	−1,713E−1

Table 4. Intervals for summer zoning

1	2	3	4
SCV $\leq$ 0,5	0,5 < SCV $\leq$ 0,83	0,83 < SCV $\leq$ 1,38	SCV < 1,38

Morocco (RTCM) was developed and entered into force in November 2015. This regulation expresses the technical specifications of thermal performance, for each climatic zone and type of building (residential or tertiary), in two ways:

(1) "Performancielle" approach, *being the annual energy demand of a building related to thermal comfort are evaluated.*
(2) "Prescriptive" approach, *where the thermal characteristics of the building envelope are evaluated, depending on the type of building, the climatic zone and the overall window rate of the heated and/or cooled spaces.*

According to the RTCM, the project manager has to fill a technical identification card for the project. This information includes the thermal performance of the building, according to the chosen approach. Nowadays, these files are not required by the goverment to obtain a building licence, and they are not yet required in the specifications. According to the Directorate of Renewable Energy and Energy Efficiency, of the Ministry of Energy, Mines and Sustainable Development (Morocco), a technical agreement has to be signed as a mandatory stage to include this technical sheet in the bidding documents.

In terms of European requirements, the Energy Performance of Buildings Directive (EPBD: Energy Performance of Buildings Directive) is the main European standard aimed at guaranteeing compliance with the EU's objectives regarding building, in terms of containment of gas emissions of greenhouse effect, energy consumption and energy efficiency and the generation of energy from renewable sources [12]. The Directive 2002/91/EC establishes some requirements in relation to the general framework to determine the integrated energy efficiency of buildings; the application of minimum energy efficiency requirements for new buildings; the application of minimum energy efficiency requirements for large existing buildings (total useful area over 1000 m^2) that are subject to major reforms; the energy certification of the buildings and the periodic inspection of boilers and air conditioning systems of buildings and, in addition, the evaluation of the state of the heating installation with boilers over 15 years old [13]. This Directive was recast by Directive 2010/31/EU (recast EPBD), which is much more demanding and gives more importance to the quality and impact of energy certificates, including inspections [12]. The EPBD Directive defines the '*almost null building*' as a building with a very high level of energy efficiency. The almost zero or very low amount of energy required should be covered, to a very large extent, by energy from renewable sources, including energy from renewables generated on site or near the building. The Directive also establishes the dates of December 31, 2018 for its application to all buildings that are owned or occupied by public administrations, and December 31, 2020 for all new buildings [12].

In Spain, the regulation on energy matters has progressed in its requirements since the basic building standard, NBE CT-79, which regulated the thermal conditions of buildings. The approval of the Technical Building Code implies a qualitative and quantitative improvement of the thermal performance of the building, establishing new requirements for the building envelope [14]. Subsequently, RD 314/2006 is issued, which approves the Technical Building Code (CTE) including the considerations of the European Directive 2002/91/EC relating to the energy efficiency of buildings.

3.2 Comparison of Spanish and Moroccan Regulations

Nowadays, building regulations are mandatory standards for the design and construction of buildings to ensure the safety and health of people, as well as ensuring energy efficiency [15]. In Spain, the regulations are more demanding since they include energy efficiency measures in buildings not only in the enclosure but also in the facilities. While in Morocco, although article 2 of Law 47-09 requires labeling of appliances and equipment connected to the grid, natural gas, liquid and gaseous petroleum products, coal as well as renewable energies, there is no Decree or norms that develops this law and then, it can be applied on buildings. At this moment, there is the draft Moroccan standard PNM 14.2.300 (2017), concerning the energy labeling of electrical products and household appliances, but it is not yet approved.

On the other hand, the breach of the rules established by the RTCM is not penalized, unlike Spain with the CTE, for lack of a decree that develops chapter 7, in its various articles, regarding sanctions. The integration of photovoltaic installations into buildings, through systems for capturing and transforming solar energy into electrical energy for its own use or supply to the grid, apart from other energy efficiency measures established in the different sections of the DB-HE, allows the Spainsh buildings to achieve one of the objectives of Directive 2010/31/EU, which is an almost '*zero energy*' building.

4 Case Study

4.1 General Overview

The BINAYATE software has been designed by the AMEE (Morocco) within the framework of the energy efficiency building code project, in collaboration with the Global Environment Facility (GEF) and the United Nations Development Program

(UNDP). CYPE, a well-known company developing technical software packages for architecture, engineering and construction professionals, has developed this *ad hoc* software. The BINAYATE application uses the BIM (Building Information Modeling) technology. This technology involves the creation and use of an intelligent 3D modeling to take the best decisions related to a project. BIM solutions make it possible to design, visualize, simulate and collaborate more easily throughout the life cycle of the project, so that it allows to simply reach the objectives of a project [13]. The software is based on a series of rules: NM ISO 6946: Elements and building components. Resistance and thermal transmittance. Calculation method; NM ISO 13370: Thermal performance of buildings - Transmission of heat through the ground; NM ISO 13789: Thermal performance of buildings. Coefficients of heat transfer by transmission and ventilation; NM ISO 13786: Thermal performance of building components. Dynamic thermal characteristics.

The present case study consists of determining and analyzing the thermal energy demanded by a new building under construction. The building is located in the city of Tangier, in northern Morocco, consisting of two floors (ground floor and first floor) using the Moroccan Software BINAYATE The four facades of the building are in contact with the exterior and it is designed for residential purposes. The climatic zone corresponding to the study area is Z2, which is equivalent to the Spanish climatic zone A3.

4.2 Results

Different layers of materials have been introduced in the BINAYATE software, considering the different construction elements of the building: exterior walls, flooring, internal slab, partitions, doors and windows. Figure 5 depicts the 3D modeling without including any type of insulation. As shown in the results obtained from the energy demand, the building is "*Not Conforming*"; that is, it does not comply with the requirements established by the RTCM regarding the limit value of the annual energy demand for heating and air conditioning. Buildings in Morocco for climatic zone Z2, in a residential building, corresponds to 46 kWh/m^2·year. For the equivalent Spanish climatic zone A3 is 30 kWh/m^2·year, while the value obtained in this simulation is 70.52 kWh/m^2·year. To reach the rate of energy demand, required by RTCM, thermal insulation of 3 cm using polyurethane must be adapted to interior wall, exterior wall and roof supported by a double glazing. On the other hand, to reach the value of the energy demand required by the Spanish legislation, we have applied thermal insulation of 6 cm with the same material in the exterior wall, and roof. In this case, the limit values of the energy demand for heating and air conditioning established by the RTCM are respected, obtaining a value of 45.22 kWh/m^2·year, being the energy demand reduced by 36%. With these results, we can say that the application of the RTCM through the BINAYATE software allows to improve the energy performance and therefore obtain thermal comfort in the home, as well as obtain energy savings in relation to the common and current energy consumption. Nevertheless, it must be pointed out the these two systems, heating and air conditioning, are not commonly used in Moroccan buildings, as shown in Fig. 6, and as a result, in most cases the only thing that is obtained with the application of the RTCM is thermal comfort of the occupants of the building. In order to achieve the commitments established by Morocco, when developing its RTCM, in terms of greenhouse gas emissions, it is not enough to apply energy efficiency measures in the envelope, but it would have to apply additional measures as well, such as lighting systems and household appliances, which together with the envelope represent a potential saving of 69%, see Fig. 7. Table 5 summarizes a comparison between RTCM and CTE results in terms of energy requirements, energy consumption and emissions. According to these results, CTE provides an important energy saving of around 28%, in comparison to RTM application.

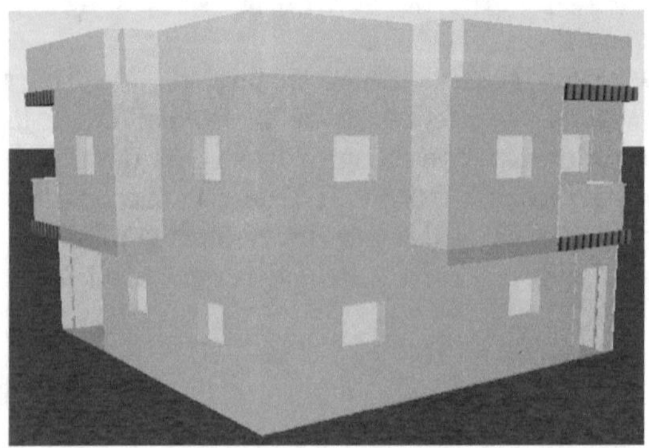

Fig. 5. 3D modeling of the housing

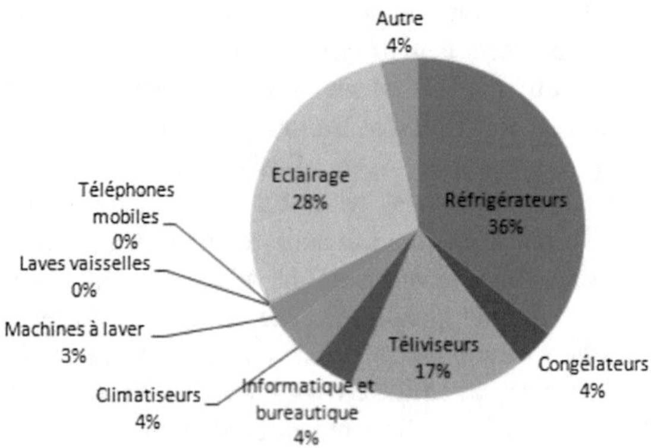

Fig. 6. Distribution of electrical consumption by uses (Morocco)

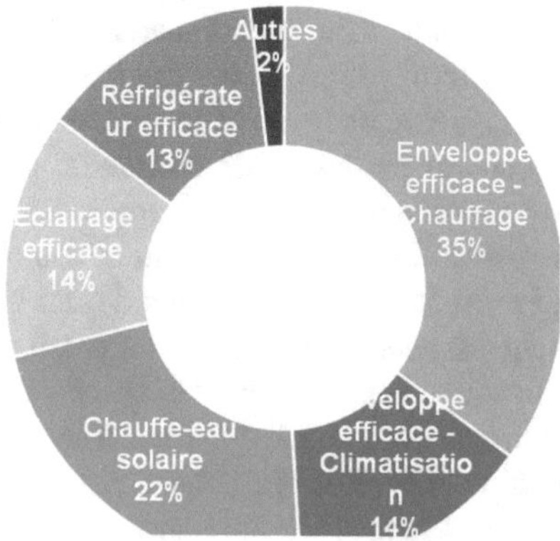

Fig. 7. Potencial of energy efficiency of a building in the south of Mediterranean region in the period 2010–2030

Table 5. Intervals for winter zoning

		Energy needs kWh/year	Consumption kWh/year/year	CO_2 emission Kg CO_2/year
Building	Ordinary	14104	4701.3	3008.8
Building	RTCM	9044	3014	1929.4
Building	CTE	5800	1933.3	1275.7
Gain	RTCM	5060	1687.3	1079.4
Gain	CTE	8304	2768	1733.1

5 Conclusion

The zoning divided by degrees-day allows us to provide a clear comparison between climates and their relationship with the use of energy through heating, ventilation and air conditioning systems, particularly in cold climates [7].

In Spain, the zoning is based on the climatic severity that combines between degrees-day and solar radiation, giving 12 climatic zones. In addition, within each of them, there may be differences among climatic zones depending on their differences in height with the capital of the province. While in Morocco, despite having conducted a study based on degrees-day, it has been decided to establish a zoning that is the result of a dynamic simulation carried out with TRNSYS in only twelve Moroccan localities. As a result, six different climatic zones are defined throughout the country where, according to the comparison of some climatic data, relevant differences can be observed, which considerably affect the estimations of the energy demand of buildings located in different areas of the same climatic zone.

According to the results obtained in the comparison of both legislations, Spanish and Moroccan, it is necessary to develop decrees to be able to apply several articles of law 47-09, relating to energy efficiency in Morocco, in order to have an adequate energy labeling for the electrical products and household appliances and some sanctions applicable in case of non-compliance with the norm. Indeed, the current thermal regulation allows us to calculate the energy demand of a building for heating and air conditioning needs, while most Moroccan buildings do not have such facilities due to its high cost and the consequent energy increase that it entails. For that reason, only the application of energy efficiency measures in the envelope is widely extended and generalized as a common thermal comfort requirements by people. For this reason, the RTCM should also include energy efficiency measures for the facilities that have a building, including lighting and household appliances, since, together with the envelope, energy savings of 68% would be obtained.

On the other hand, about 450 million people live on both shores of the Mediterranean and consume about 1,000 MTEP of primary energy every year, that is, a little more than 8% of the world's demand. The buildings sector is the first consumer of electricity and the second after transport for fossil fuels. In addition, the progressive increase of the global standard of living in the Southern and Eastern Mediterranean countries, should be translated in a tendential way in a recovery of the energy consumption of the SEMCs in relation to the Northern Mediterranean Countries in the 2050 horizon. This rate of development is not sustainable from the environmental point of view, due to the scarcity of fossil resources on the one hand, and the effects of greenhouse gas emissions and their consequences on the climate. another part [16]. This situation compromises the adoption of energy policies and more energy efficient programs, where Morocco should improve its Thermal Regulation for Construction. With this aim, and in line with Spain, the basic document of energy saving should be extended by including the performance of thermal installations, energy efficiency of the lighting installations, the minimum solar contribution of sanitary hot water, as well as the minimum photovoltaic contribution of electrical energy. This additional requirements would be included in the Moroccan regulation, as well as additional research studies to obtain a suitable zoning methodology and optimize thermal requirements, which is currently under study and analysis by the authors.

References

1. Renewable Energy in Morocco: Strategies and action plan. Technical reports, Ministry of Energy, Mines, Water and Environment (2012)
2. Les energies renouvelables au Maroc: Bilan et Perspectives. Technical reports, Ministry of Energy, Mines, Water and Environment (2015). (in French)
3. Hamdaoui, S., Mahdaoui, M., Allouhi, A., Alaiji, R.E., Kousksou, T., Bouardi, A.E.: Energy demand and environmental impact of various construction scenarios of an office building in Morocco. J. Cleaner Prod. **188**, 113–124 (2018)
4. Ballarini, I., Corrado, V., Madonna, F., Paduos, S., Ravasio, F.: Energy refurbishment of the Italian residential building stock: energy and cost analysis through the application of the building typology. Energy Policy **105**, 148–160 (2017)

5. Hernández, R., Velasco, A., San José, E., Martinez, J.R., Javier, F.: Propuesta de la certificación energética, mediante simulación dinámica, como herramienta de gestión energética ISO 50001 Versus auditoria energética e edificios. III CIEEMAT, pp. 343 (2017)
6. National Action Plan for Energy Efficiency 2017–2020. Technical reports, Ministry of Energy, Tourism and Digital Agenda (2017)
7. Walsh, A., Cóstola, D., Labaki, L.C.: Comparison of three climatic zoning methodologies for building energy efficiency applications. Energy Buildings **146**, 111–121 (2017)
8. ADEREE: Guide de la Reglementation Thérmique de Construction au Maroc (RTCM in French). Technical reports, ADEREE (2013)
9. Basic Document HE Technical Report (2017). (in Spanish)
10. Ministry of Public Works, DG: Descriptive document climates of reference. Technical reports, Ministry of Development (2015). (in Spanish)
11. Lrulegi, Serra, Hernández, Torres, R.-P.: Ventilated facades active to reduce the demand for heating in office buildings. The case of Spain. Construction reports (2012). (in Spanish)
12. Casal, A.G.: Comparative analysis of energy efficiency in existing buildings with different energy simulation tools. Ph.D. dissertation, School of Industrial Engineering, University of Valladolid, Spain (2015)
13. Asrih, N., Boudra, A.: Evaluation eco-énergétique des bâtiments construit avec des materiaux a base de chanvre « Etude de cas: Nouvelle ville de Chrafate ». ENSAH (2016). (in French)
14. Suárez, R., Fragoso, J.: Passive strategies for energy optimization of social housing in Mediterranean climate. Reports of the Construction, no. 68, pp. 136 (2016)
15. Chandel, S.S., Sharma, A., Marwaha, B.M.: Review of energy efficiency initiatives and regulations for residential buildings in India. Renew. Sustain. Energy Rev. **54**, 1443–1458 (2016)
16. Thibault, H.-L., Andaloussi, E.A.: L'efficacité énergétique dans le bâtiment en Méditerranée. Futuribles **376**, 47–59 (2011)

Lyapunov Function Based Control for Grid-Interfacing Solar Photovoltaic System with Constant Voltage MPPT Technique

Kawtar Moutaki[1(✉)], Halima Ikaouassen[1], Abderraouf Raddaoui[1], and Miloud Rezkallah[2]

[1] University Mohammed V in Rabat, MEAT, Ecole Supérieure de Technologie Salé, Salé, Morocco
moutaki.kawtar@gmail.com
[2] Electrical Engineering Department, ETS, Montréal, QC, Canada

Abstract. In this paper nonlinear control design to obtain high performance from solar PV system using three-phase grid-connected LCL-filtered voltage source inverter is presented. The grid-connected system is modeled in the synchronously rotating frame. For perfect synchronization of solar photovoltaic and clean power injection to the grid new control based on Lyapunov function is used. Lyapunov function based control is derived from the Lyapunov's direct method which guarantees the global stability of the closed-loop system. The output PV voltage is used as DC link voltage which will be maintained at its reference value using constant voltage MPPT tracking method. In the proposed control strategy, the measurement of inverter currents, capacitor voltages and grid currents are essential. The generation of reference functions in the d- and q-components can be achieved by using the reference d-component grid current. The performance of the proposed scheme and its developed control strategy, are validate using MATLAB Simulink.

Keywords: Single-stage inverter · Photvoltaic system · Lyapunov function based control

1 Introduction

Recently, interest in renewable energy has grown in response to increased concern for the environment. Solar energy is the cleanest and most abundant renewable energy source. Moreover, Due to a number of social and economic factors, photovoltaic (PV) technology has become a favored form of renewable energy technology [1]. The most of distributed generator (DG) resources are interfaced to the utility grid by way of Single-stage voltage source inverter (VSI) [2] systems because it's more efficient than the double-stage and multi-stage topologies [3]. Most control schemes of CC-VSI are established by making linear the nonlinear representation around a nominal operating point [4, 5]. Designers examine each part separately to establish the scheme to harmonize a chosen set of large signal model which forms an approach of the integral dynamical behavior of the global system seeing that, the PV source is expressed by a nonlinear electrical behavior that influences the variables of the system, by

M. Ezziyyani (Ed.): AI2SD 2018, AISC 912, pp. 210–219, 2019.
https://doi.org/10.1007/978-3-030-12065-8_20

consequence the system is not linearizable about a single operating point. This work is based mainly on the Lyapunov functions. This technique has been proved efficient for DC-DC converters [6, 7] and AC-DC converters [8–11].

In addition, the quality of energy at the PCC are highly important in a power distribution system, and since current harmonics raised from pulse width modulation nature of the converter, passive LC or LCL filters are used at the converter AC terminals to regulate the output voltages [12–16]. Commonly, LCL filter is the more appropriate in grid-connected converters because they outcome in good attenuation, smaller ripple in the injected current, and smaller inductance values. The resonance must be strictly damped in the LCL filter, otherwise the filters can lead in important power quality and stability problems, therefore, both passive damping as well as active damping techniques have been studied in [17, 18]. Different authors have closely examined these methods in the case of grid-connected converters [16, 19–21]. The synchronization of the inverter with the grid is a serious problem that has been discussed in [5, 22–24], demonstrating how the output of the inverter should be synchronized with phase-locked loops (PLLs).

The power produced by the photovoltaic modules is intermittent in nature and depends on the intensity of solar radiation and temperature of solar cells therefore the exploitation gains of PV panels depend on the ability of the inverter to draw out the maximum profitable power that can be instantly taken from the PV farm. In order to acquire this objective, different authors have discussed these problem by means of different maximum power point tracking techniques [25–27].

A study of sliding mode and Lyapunov function methods used for the control of DC-DC boost converter and the DC-AC inverter have been presented in a previous research [28].

In this research work the main contributions are focused on achieving MPPT only by maintain the DC link voltage constant, and injecting clean and stable power quality in the grid. The remainder of this paper is structured as following. The system description and modeling is presented in Sect. 2. The proposed control scheme is presented in Sect. 3. Section 4 presents simulation results focused on the validation of the suggested controller. Conclusion is drawn in Sect. 5.

2 System Description and Modelling

As it is described in Fig. 1, the proposed system configuration consists of a PV array connected through a DC bus to a three-phase VSI that is connected to an ideal 380 V AC grid via an LCL filter that leads to better switching harmonic attenuation. The DC link capacitor C_{dc} serve to filter the voltage fluctuations at the output of the PV. As results, we get a uniform flow of current from the DC bus to the utility grid.

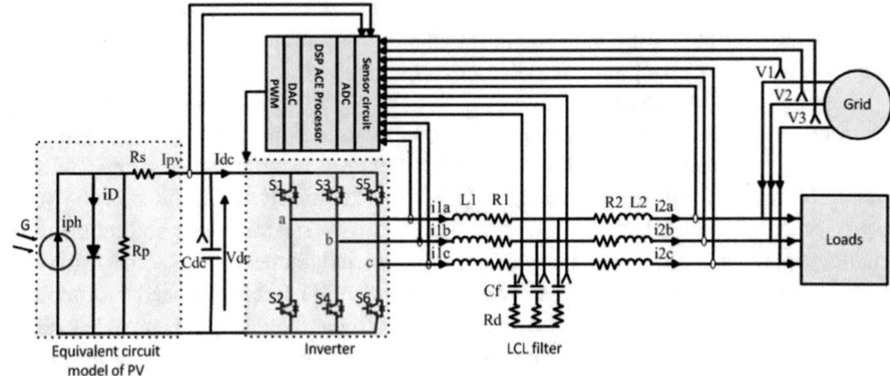

Fig. 1. System configuration

2.1 Model of the Solar PV Array

The solar photovoltaic module is composed by an assemblage of PV cells, which are actually a p-n junction. The output PV power depends on the irradiation and temperature of the PV-array and load resistance R. If one neglect both shunt and series resistances, then the output PV power can be described by the following equation:

$$P_{pv} = V_{pv}.I_{pv} = N_p I_{ph} V_{pv} - N_p I_s V_{pv} \left[e^{\frac{1}{\gamma V_T}\left(\frac{V_{pv}}{N_s}\right)} - 1 \right] \quad (1)$$

where, I_{ph}, I_s, N_p, N_s, V_{pv}, V_T, γ denote the photo-generated current, the saturation current, number of modules which are connected in parallel, total number of cells, which are connected in series, the output voltage of PV farm, thermal voltage, ideality factor of the p-n junction respectively.

According to [7], the constant voltage MPPT technique utilizes empirical results, indicating that the voltage at the MPP V_{mpp} is about 70–80% of the solar cell open-circuit voltage. This provides a reference V^*_{dc} to which the output voltage can be forced to track.

2.2 Model of Three-Phase Grid-Connected PV System

To examine the controller with no actual system, a mathematical model is needed to be extracted from Fig. 1, which represents the dynamic behavior of the global system. The equations describing the operation of the global system in d-q axis are given as:

$$\dot{v}_{dc} = \frac{i_{pv}}{C_{dc}} - \frac{3\alpha_d}{2C_{dc}} i_{1d} - \frac{3\alpha_q}{2C_{dc}} i_{1q} \quad (2)$$

$$\dot{i}_{1d} = -\frac{R_1}{L_1} i_{1d} + \omega i_{1q} - \frac{v_{cd}}{L_1} + \frac{\alpha_d V_{dc}}{2L_1} \quad (3)$$

$$\dot{i}_{1q} = -\frac{R_1}{L_1} i_{1q} - \omega i_{1d} - \frac{v_{cq}}{L_1} + \frac{\alpha_q V_{dc}}{2L_1} \tag{4}$$

$$\dot{v}_{cd} = \frac{i_{1d}}{C_f} - \frac{i_{2d}}{C_f} + \omega v_{cq} \tag{5}$$

$$\dot{v}_{cq} = \frac{i_{1q}}{C_f} - \frac{i_{2q}}{C_f} - \omega v_{cd} \tag{6}$$

$$\dot{i}_{2d} = -\frac{R_2}{L_2} i_{2d} + \omega i_{2q} + \frac{v_{cd}}{L_2} - \frac{V_d}{L_2} \tag{7}$$

$$\dot{i}_{2q} = -\frac{R_2}{L_2} i_{2q} - \omega i_{2d} + \frac{v_{cq}}{L_2} - \frac{V_q}{L_2} \tag{8}$$

where, C_{dc}, L_1, R_1, L_2, R_2, C_f, v_{cd}, v_{cq}, i_{1d}, i_{1q}, i_{2d}, i_{2q}, v_d, v_q, ω, α_d and α_q, denote DC link capacitor, equivalent resistance and inductance at the inverter side, equivalent resistance and inductance at the grid side, filter capacitor, voltage across filter capacitor, inverter currents, filter current, equivalent grid voltage, angular frequency of the grid and the control laws, in dq0 reference, respectively.

3 Control Design

As already discussed the main objective of the improved Lyapunov-based control is to achieve MPPT by controlling the DC-link voltage, as well as injecting clean power into the grid by controlling the active currents.

Let us define the state variables errors as:

$$\varepsilon_1 = i_{1d} - i_{1d}^* \tag{9}$$

$$\varepsilon_2 = i_{1q} - i_{1q}^* \tag{10}$$

$$\varepsilon_3 = i_{2d} - i_{2d}^* \tag{11}$$

$$\varepsilon_4 = i_{2q} - i_{2q}^* \tag{12}$$

$$\varepsilon_5 = v_{cd} - v_{cd}^* \tag{13}$$

$$\varepsilon_6 = v_{cq} - v_{cq}^* \tag{14}$$

$$\varepsilon_7 = v_{dc} - v_{dc}^* \tag{15}$$

In synchronous dq0 reference frame, $v_q = 0$. Knowing that $Q = -3/2 v_d . i_q$, with v_d is fixed, thus i_q can control Q. Thence, the objective is to regulate directly the current i_q at the reference current $i_{2q}^* = 0$.

To decrease the complexity of the system, as well as, the number of sensed parameters without affecting the global stability improved Lyapunov based control is used as detailed and shown in Fig. 2.

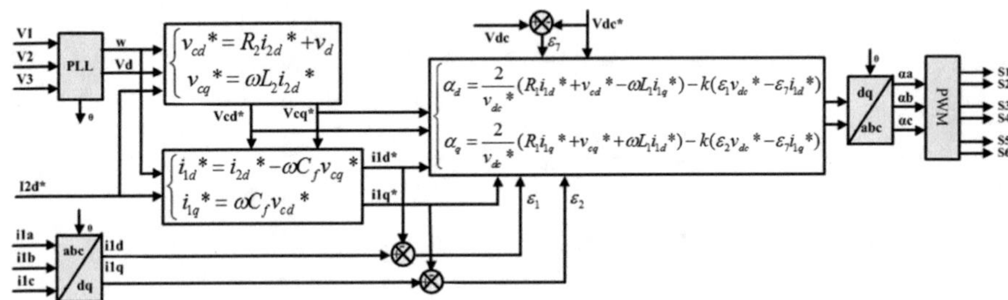

Fig. 2. Lypunov-based control for single-stage inverter

The objective of the proposed controller is to extend zero state variable errors in presence of dynamic changes. The control laws are defined as:

$$\alpha_d = \alpha_{d0} + \Delta\alpha_d \tag{16}$$

$$\alpha_q = \alpha_{q0} + \Delta\alpha_q \tag{17}$$

Where, α_{d0} and α_{q0} denote the d- and q-components of the switching function in the steady-state, and $\Delta\alpha_d$ and $\Delta\alpha_q$ are the perturbation of the switching function. Supposing that the inverter currents, capacitor voltages and grid currents track their references in the steady-state, one can get upon calculation the steady-state parts of the inverter switching functions as:

$$\alpha_{d0} = \frac{2}{v_{dc}^*}(R_1 i_{1d}^* + v_{cd}^* - \omega L_1 i_{1q}^*) \tag{18}$$

$$\alpha_{q0} = \frac{2}{v_{dc}^*}(R_1 i_{1q}^* + v_{cq}^* + \omega L_1 i_{1d}^*) \tag{19}$$

The expression of the total saved energy of system is defined as:

$$V = \frac{3}{2}(L_1\varepsilon_1^2 + L_1\varepsilon_2^2 + L_2\varepsilon_3^2 + L_2\varepsilon_4^2 + C_f\varepsilon_5^2 + C_f\varepsilon_6^2) + \frac{C_{dc}\varepsilon_7^2}{2} \tag{20}$$

The time derivative of V can be written as:

$$\dot{V} = 3(L_1\varepsilon_1\dot{\varepsilon}_1 + L_1\varepsilon_2\dot{\varepsilon}_2 + L_2\varepsilon_3\dot{\varepsilon}_3 + L_2\varepsilon_4\dot{\varepsilon}_4 + C_f\varepsilon_5\dot{\varepsilon}_5 + C_f\varepsilon_6\dot{\varepsilon}_6) + C_{dc}\varepsilon_7\dot{\varepsilon}_7 \tag{21}$$

The errors dynamics of the state variables errors are given as:

$$\overset{\bullet}{\varepsilon}_1 = -\frac{R_1}{L_1}\varepsilon_1 + \omega\varepsilon_2 - \frac{\varepsilon_5}{L_1} + \frac{\alpha_{d0}}{2L_1}\varepsilon_7 + \frac{(\varepsilon_7 + V_{dc}^*)}{2L_1}\Delta\alpha_d \tag{22}$$

$$\overset{\bullet}{\varepsilon}_1 = -\frac{R_1}{L_1}\varepsilon_2 - \omega\varepsilon_1 - \frac{\varepsilon_6}{L_1} + \frac{\alpha_{d0}}{2L_1}\varepsilon_7 + \frac{(\varepsilon_7 + V_{dc}^*)}{2L_1}\Delta\alpha_q \tag{23}$$

$$\overset{\bullet}{\varepsilon}_3 = -\frac{R_2}{L_2}\varepsilon_3 + \omega\varepsilon_4 + \frac{\varepsilon_5}{L_2} \tag{24}$$

$$\overset{\bullet}{\varepsilon}_4 = -\frac{R_2}{L_2}\varepsilon_4 - \omega\varepsilon_3 + \frac{\varepsilon_6}{L_2} \tag{25}$$

$$\overset{\bullet}{\varepsilon}_5 = \frac{\varepsilon_1}{C_f} - \frac{\varepsilon_3}{C_f} + \omega\varepsilon_6 \tag{26}$$

$$\overset{\bullet}{\varepsilon}_6 = \frac{\varepsilon_2}{C_f} - \frac{\varepsilon_4}{C_f} - \omega\varepsilon_5 \tag{27}$$

$$\overset{\bullet}{\varepsilon}_7 = -\frac{3\alpha_{d0}}{2C_{dc}}\varepsilon_1 - \frac{3\alpha_{q0}}{2C_{dc}}\varepsilon_2 - \frac{3(\varepsilon_1 + i_{1d}^*)}{2C_{dc}}\Delta\alpha_d - \frac{3(\varepsilon_2 + i_{1q}^*)}{2C_{dc}}\Delta\alpha_q \tag{28}$$

Substituting the errors dynamics of the state variables errors in (21), leads to:

$$\begin{aligned}\overset{\bullet}{V} = & -3R_1\varepsilon_1^2 - 3R_1\varepsilon_2^2 - 3R_2\varepsilon_3^2 - 3R_2\varepsilon_4^2 + \frac{3}{2}(\varepsilon_1 v_{dc}^* - \varepsilon_7 i_{1d}^*)\Delta\alpha_d \\ & + \frac{3}{2}(\varepsilon_2 v_{dc}^* - \varepsilon_7 i_{1q}^*)\Delta\alpha_q\end{aligned} \tag{29}$$

The global stability of the system is guaranteed only when the time derivative of Lyapunov function is strictly less than zero. The choice of $\Delta\alpha_d$ and $\Delta\alpha_q$, where $k > 0$ leads us to a system that is globally asymptotically stable.

$$\Delta\alpha_d = -k(\varepsilon_1 v_{dc}^* - \varepsilon_7 i_{1d}^*) \tag{30}$$

$$\Delta\alpha_q = -k(\varepsilon_2 v_{dc}^* - \varepsilon_7 i_{1q}^*) \tag{31}$$

Finally, from (18), (19), (30) and (31) one obtains the control laws that yield the tracking v_{dc}^* reference and to achieve the unit power factor, which are expressed as follows:

$$\alpha_d = \frac{2}{v_{dc}^*}(R_1 i_{1d}^* + v_{cd}^* - \omega L_1 i_{1q}^*) - k(\varepsilon_1 v_{dc}^* - \varepsilon_7 i_{1d}^*) \tag{32}$$

$$\alpha_q = \frac{2}{v_{dc}^*}(R_1 i_{1q}^* + v_{cq}^* + \omega L_1 i_{1d}^*) - k(\varepsilon_2 v_{dc}^* - \varepsilon_7 i_{1q}^*) \tag{33}$$

4 Simulation Results and Analysis

In order to validate the design and robustness of the improved Lyapunov based control, numerical simulations have been executed using Matlab/Simulink. Two cases are considered in this validation. In the first test performance for a 5 kW-rated inverter under irradiation changes are validated. Whereas in the second test the performance under fixed irradiation are evaluated. The system parameters used in this work are detailed in the appendix.

Figure 3 shows the simulation results for the PV system under irradiation changes with no-load condition. In Fig. 3a, it is noticed that, at 0.4 s, when the system was experiencing increases of the solar irradiation, the PV current i_{pv} and the grid current i_g have been augmented. Whereas, in Fig. 3b when the solar irradiation decrease at 1.1 s, i_{pv} and i_g have been diminished simultaneously. The DC voltage follows rapidly the desired value. It is remarkable that the DC voltage is regulated even in presence of variations in PV power; also the steady state error is very small. The controller proved excellent robustness in front of irradiation changes.

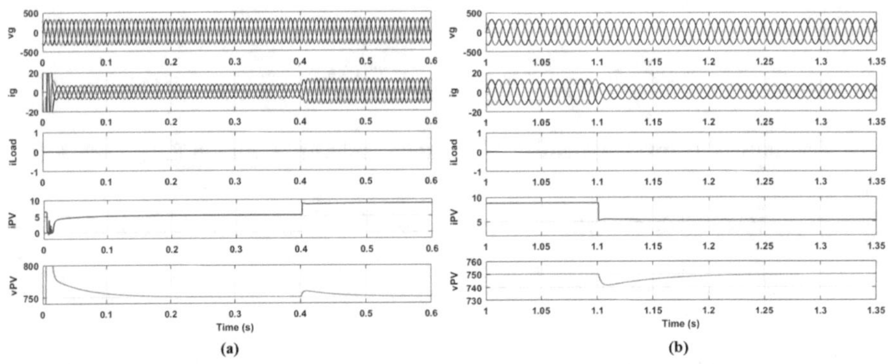

Fig. 3. Dynamic reaction of the inverter being the subject of irradiation changes

Figure 4 shows the simulation results of the system under study when a nonlinear load composed of a three-phase rectifier feeding resistive-inductive load of $R_L = 20\ \Omega$ and $L_L = 10$ mH is connected at 0.3 s and a linear load has been disconnected, furthermore the load on phase ‘b’ is switched off between 0.4 s and 0.5 s. It is noticed that, the grid currents are balanced and the DC voltage follows the desired value even with load disturbances. The obtained results prove the robustness of the improved Lyapunov based control under load changes.

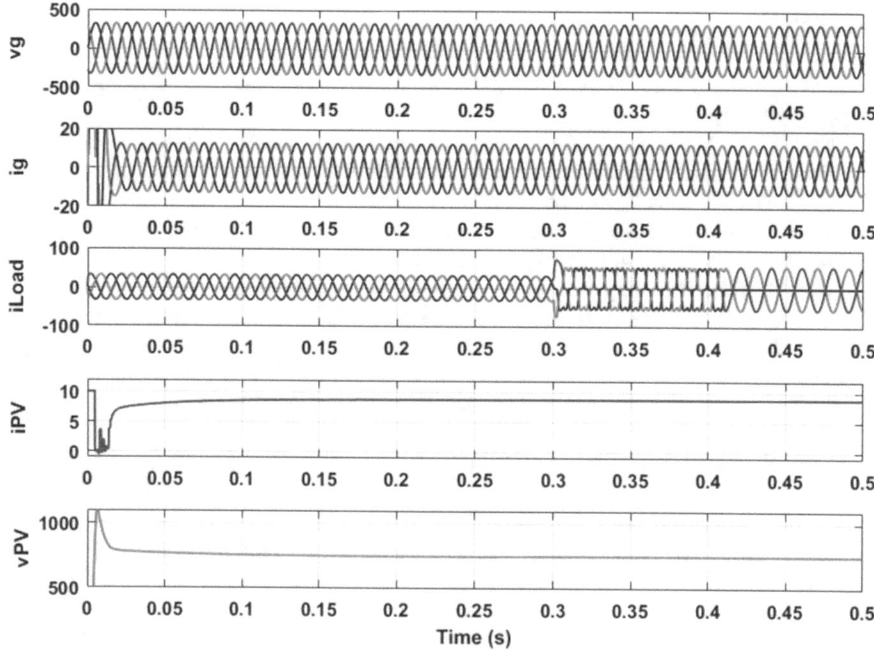

Fig. 4. Dynamic reaction of the inverter being the subject of different load disturbances

5 Conclusion

In this paper Lyapunov based control for single stage inverter for PV system interfacing grid, is presented. To carry out the unity power factor and to follow the optimal voltage under various perturbations, a nonlinear model of the system based on Lyapunov functions has been discussed. The obtained results show satisfactory performance under different conditions in terms of stability, accuracy and time response.

Appendix

See Table 1.

Table 1. System parameters.

Parameter	Value
PV array rated power	5 kW
DC-link voltage	530 V
DC-link capacitor	1500 μF
Capacitance C_f	4.3 μF
Inductance L_2	0.4 mH
Inductance L_1	1.4 mH
Controller gains	0.003
Grid frequency	50 Hz
Switching frequency	20 kHz

References

1. Lena, G.: Rural Electrification with PV Hybrid Systems. International Energy Agency, pp. 7–10, July 2013
2. Darwish, A., Holliday, D., Ahmed, S., Massoud, A.M., Williams, B.W.: A single-stage three-phase inverter based on Cuk converters for PV applications. IEEE J. Emerg. Sel. Top. Power Electron. **2**, 797–807 (2014)
3. Agarwal, R.K., Hussain, I., Singh, B.: Three-phase single-stage grid tied solar PV ECS using PLL-less fast CTF control technique. IET Power Electron. **10**, 178–188 (2017)
4. Teodorescu, R., Blaabjerg, F., Liserre, M., Loh, P.C.: Proportional-resonant controllers and filters for grid-connected voltage-source converters. IEE Proc. Electr. Power Appl. **153**, 750–762 (2006)
5. Blaabjerg, F., Teodorescu, R., Liserre, M., Timbus, A.V.: Overview of control and grid synchronization for distributed power generation systems. IEEE Trans. Industr. Electron. **53**, 1398–1409 (2006)
6. Leyva, R., Cid-Pastor, A., Alonso, C., Queinnec, I., Tarbouriech, S., Martinez-Salamero, L.: Passivity-based integral control of a boost converter for large-signal stability. IEE Proc. Control Theory Appl. **153**, 139–146 (2006)
7. Sanders, S.R., Verghese, G.C.: Lyapunov-based control for switched power converters. In: 21st Annual IEEE Conference on Power Electronics Specialists, pp. 51–58 (1990)
8. Komurcugil, H., Kukrer, O.: A new control strategy for single-phase shunt active power filters using a Lyapunov function. IEEE Trans. Industr. Electron. **53**, 305–312 (2005)
9. Komurcugil, H., Kukrer, O.: Lyapunov-based control for three-phase PWM AC/DC voltage-source converters. IEEE Trans. Power Electron. **13**, 801–813 (1998)
10. Meza, C., Biel, D., Jeltsema, D., Scherpen, J.M.A.: Lyapunov-based control scheme for single-phase grid-connected PV central inverters. IEEE Trans. Control Syst. Technol. **20**, 520–529 (2012)
11. Komurcugil, H., Altin, N., Ozdemir, S., Sefa, I.: Lyapunov-function and proportional-resonant-based control strategy for single-phase grid-connected VSI with LCL filter. IEEE Trans. Industr. Electron. **63**, 2838–2849 (2016)
12. Campanhol, L.B.G., da Silva, S.A.O., Goedtel, A.: Application of shunt active power filter for harmonic reduction and reactive power compensation in three-phase four-wire systems. IET Power Electron. **7**, 2825–2836 (2014)
13. Ahmed, K.H., Finney, S.J., Williams, B.W.: Passive filter design for three-phase inverter interfacing in distributed generation. In: 2007 Compatibility in Power Electronics, pp. 1–9 (2007)
14. Zheng, X., Xiao, L., Lei, Y., Wang, Z.: Optimization of LCL filter based on closed-loop total harmonic distortion calculation model of the grid-connected inverter. IET Power Electron. **8**, 860–868 (2015)
15. Xu, J., Xie, S., Tang, T.: Improved control strategy with grid-voltage feed forward for LCL-filter-based inverter connected to weak grid. IET Power Electron. **7**, 2660–2671 (2014)
16. Balasubramanian, A.K., John, V.: Analysis and design of split-capacitor resistive inductive passive damping for LCL filters in grid-connected inverters. IET Power Electron. **6**, 1822–1832 (2013)
17. Liu, Q., Peng, L., Kang, Y., Tang, S., Wu, D., Qi, Y.: A novel design and optimization method of an LCL filter for a shunt active power filter. IEEE Trans. Industr. Electron. **61**, 4000–4010 (2014)

18. Hatua, K., Jain, A.K., Banerjee, D., Ranganathan, V.T.: Active damping of output LC filter resonance for vector-controlled VSI-fed AC motor drives. IEEE Trans. Industr. Electron. **59**, 334–342 (2012)
19. Reznik, A., Simões, M.G., Al-Durra, A., Muyeen, S.M.: LCL filter design and performance analysis for grid-interconnected systems. IEEE Trans. Ind. Appl. **50**, 1225–1232 (2014)
20. Sen, S., Yenduri, K., Sensarma, P.: Step-by-step design and control of LCL filter based three phase grid-connected inverter. In: 2014 IEEE International Conference on Industrial Technology (ICIT), pp. 503–508 (2014)
21. Tang, Y., Yao, W., Loh, P.C., Blaabjerg, F.: Design of LCL filters with LCL resonance frequencies beyond the Nyquist frequency for grid-connected converters. IEEE J. Emerg. Sel. Top. Power Electron. **4**, 3–14 (2016)
22. Dong, D., Wen, B., Boroyevich, D., Mattavelli, P., Xue, Y.: Analysis of phase-locked loop low-frequency stability in three-phase grid-connected power converters considering impedance interactions. IEEE Trans. Industr. Electron. **62**, 310–321 (2015)
23. Cheng, P., Nian, H.: Direct power control of voltage source inverter in a virtual synchronous reference frame during frequency variation and network unbalance. IET Power Electron. **9**, 502–511 (2016)
24. Se-Kyo, C.: A phase tracking system for three phase utility interface inverters. IEEE Trans. Power Electron. **15**, 431–438 (2000)
25. Qahouq, J.A.A., Jiang, Y.: Distributed photovoltaic solar system architecture with single-power inductor single-power converter and single-sensor single maximum power point tracking controller. IET Power Electron. **7**, 2600–2609 (2014)
26. Zainuri, M.A.A.M., Radzi, M.A.M., Soh, A.C., Rahim, N.A.: Development of adaptive perturb and observe-fuzzy control maximum power point tracking for photovoltaic boost dc-dc converter. IET Renew. Power Gener. **8**, 183–194 (2014)
27. de Oliveira, F.M., da Silva, S.A.O., Durand, F.R., Sampaio, L.P., Bacon, V.D., Campanhol, L.B.G.: Grid-tied photovoltaic system based on PSO MPPT technique with active power line conditioning. IET Power Electron. **9**, 1180–1191 (2016)
28. Rezkallah, M., Sharma, S.K., Chandra, A., Singh, B., Rousse, D.R.: Lyapunov function and sliding mode control approach for the solar-PV grid interface system. IEEE Trans. Industr. Electron. **64**, 785–795 (2017)

Library Pruning for Power Saving During Timing and Electrical Design Rules Optimization

Mohamed Chentouf[1,3(✉)], Lekbir Cherif[1,2], and Zine El Abidine Alaoui Ismaili[3]

[1] ICDS Department, Mentor Graphics, Rabat, Morocco
{mohamed_chentouf, lekbir_cherif}@mentor.com
[2] Laboratory of Systems Engineering, ENSA, Ibn Tofail University, Kénitra, Morocco
[3] Information, Communication and Embedded Systems (ICES) Team, National Superior School of Computer Science and System Analysis, University Mohammed V, Rabat, Morocco
z.alaoui@um5s.net.ma

Abstract. Timing optimization techniques are widely used to meet the frequency and electrical design rules requirement of integrated circuits, they use logical and physical transformation to speed up the problematic signals and to close the design setup and hold constraints. On the other side, each technique induces a power increase as a cost for signal speed up. In this paper, we propose a standard cell library tuning methodology to reduce the timing optimization impact on power increase. We divide each optimization step of the place and route process into two sub-steps, the first one uses only low power standard library cells and try to correct the maximum number of violations, and the second uses all the available cells in the library to close the remaining violations. Experimental results on 45 industrial designs of different processes show that the proposed methodology provides a leakage power reduction of 5%, a total power reduction of 1.3% and a timing improvement of 55.8% in Total Negative Slack and 37.5% in Worst Negative Slack.

Keywords: Library pruning · Timing optimization · CMOS · Electrical Design Rule Constraints · Electronic design automation · System on Chip · Physical design · Place & route · Power optimization

1 Introduction

In the previous decade, power consumption has been a second order priority in the Integrated Circuit market, since the primary concern of chip designers was to reduce the cost, the area, and the timing. But nowadays, the trend has shifted and the power budget became one of the major constraints of an electronic device development process.

Power density and total power dissipation reduction are major challenges in the design of High Performance Integrated Circuits (HPC) that embed tens of millions of

M. Ezziyyani (Ed.): AI2SD 2018, AISC 912, pp. 220–235, 2019.
https://doi.org/10.1007/978-3-030-12065-8_21

gates in a small die and needs a very sophisticated packaging and cooling infrastructures. Overpassing the power budget can lead to severe consequences on the product, whether it means moving from cheap packaging materials and processes to more expensive ones, or making any necessary power-performance trade-offs to meet the reliability and battery life requirements. This power density has increased to a threshold where it is no longer possible to increase the clock frequency even with technology shrinking, and hence the designers are aiming for multi-processor chips to overcome this blocking obstacle.

On the other hand, battery-powered devices represent a very big segment of the electronics market, and here also the power consumption is a deterministic and a differentiating parameter that pushes the designers to use aggressive and complex approaches at every step of the design development cycle to stay competitive in the field. These approaches include power gating, where blocks are powered on and off depending on the operational modes [1], clock gating, where the clock signal is stopped from propagating to a module or a sub-module that is not needed for a specific operation [2]. Also, designers have moved from single-voltage and frequency implementations to multi-voltage/multi-frequencies architectures, where different blocks run at different voltages and frequencies depending on the amount of computation needed. And in some cases, designers are using dynamic voltage and frequency scaling to adapt the supply voltage and the operational frequency to the workload and hence optimize the power-performance factor.

Although, many techniques have been introduced to reduce the power consumption and density of System on Chips (SoCs). New power reduction techniques are mandatory to continue to add performance and features and grow the IC businesses.

In this paper, we propose a new timing optimization technique based on standard cell libraries pruning to optimize a set of industrial designs of different structural, operational, and technological characteristics. We classified the standard library cells based on their power footprint, and we divided the optimization process into two passes, the first pass aim to optimize the design, using only the lowest power consuming cells and the second pass uses all the available cells in the library.

The remainder of this paper is organized as follows. Section 2 presents background about physical design, timing analysis, delay estimation, power estimation, and optimization methodologies. Section 3 gives a brief description of the setup timing and electrical design rules optimization techniques used in the industry of physical design. Section 4 describes our proposed optimization flow based on the new pruning methodology. The experimental results are illustrated and discussed in Sect. 5. Finally, Sect. 5 draws conclusions from our study.

2 Background

In this section, we begin with a brief background on physical design flow, timing analysis, and power estimation techniques.

2.1 Standard Place and Route Flow

The place-and-route (P&R) flow is the process of generating the layout from a gate level netlist using EDA tools. The netlist contains the gates from a standard cell library. The library is a collection of generic logic gates and different transistor sizes in a specific process technology. The main steps of a standard P&R flow are:

Floor planning
The floor-planning step consists of die size and aspect ratio estimation, I/O pad ordering and placement, block partitioning and placement (control logic, data path, on-chip memory, etc.), and power/ground net routing.

Placement
The goal of standard cells and macro blocks placement is mainly the following:

- Placing cells in the specified regions of the floorplan, and minimizing the routed wire length.
- Meeting the timing requirement of the circuits.

Pre Clock Tree Synthesis
After placement is done, multiple passes of optimization with different objectives (setup timing, max transition, max capacitance, area reduction, and power reduction) are done to meet timing, area, power, and electrical constraints.

Clock Tree Synthesis (CTS)
CTS is the phase where buffers are inserted on a clock network to drive the clock signal to registers, balancing the latency to the clock tree leaves to minimize latency and skew.

Post CTS
More timing violations are exposed at this stage since the real clock timing is now known, and hence further passes of optimization with different objectives (setup, hold, max transition, max capacitance, area, and power) are done to fix the exposed violations.

Routing
The routing stages give the nets a physical representation as wires and vias. There are two routing steps:

- *Global Routing*. This task assigns the topological paths of nets into switch boxes. The floorplan is partitioned into many sub-regions called global routing cells. It provides a simplified view of the wiring, providing estimates of wire capacitance and resistance for timing analysis.
- *Detail Routing*. This step routes the nets assigning different layers to each wire segment with vias connecting between layers and to cell pins. Routing tracks are used based on the design rules for routing layers.

Post Routing
With more accurate timing analysis provided by detail routing, additional timing violations are exposed at this stage. Coupling noise between metal wires is also considered to provide a more accurate timing picture. So another round of optimization is needed to clean the design.

2.2 Basics of Timing Analysis and Delay Estimation

Static timing analysis is an important step in the physical design flow, used to ensure that a design will operate correctly at a specific frequency in a specific environment without any timing violation. Timing violations could be either setup, hold, or Electrical Design Rule Constraints (eDRCs) violations such as max input transition and max load. In our study, we will focus on setup timing violations.

Setup timing is the minimum amount of time before the clock's active edge that the data must be stable for it to be latched correctly [3]. Setup slack is the difference between the required time (RT) and the arrival time (AT):

$$\text{Setup Slack} = \text{RT} - \text{AT} \tag{1}$$

A positive slack means that the design is meeting the setup timing constraints. A negative slack indicates that the design has not satisfied the specified timing constraints, so optimization should be carried out to fix the timing violation. The main contributors to a setup violation are the cell delays and the interconnect delays [3].

Cell Delays
In a standard cell design methodology, cells are pre-characterized in the library and their timing is calculated from a look-up table referred as NLDM (Non-Linear Delay Model) used to estimate delay, output slew, or other timing metrics. For delay, an NLDM model is represented in a two-dimensional matrix, its indices are the input transition time and the output load capacitance. [3] Composite current source (CCS) timing models are used for higher accuracy, at the cost of higher timing analysis run time.

Interconnect Delay
To estimate the interconnect delay, there are many models which are based on interconnect equivalent resistance and capacitance. The resistance is deduced from the interconnect wires in various metal layers and vias in a design's physical implementation. The capacitance is extracted also from the metal wires, cell pins, and includes the coupling capacitance with the ground as well as the coupling capacitance with neighboring routes [3].

2.3 Power Reduction Techniques

Various methods were developed to reduce the power consumption. The most known power optimization approaches are: Multi-VDD used to primarily reduce dynamic power by supplying the high-performance blocks with high Vdd and the low-performance blocks with low Vdd [4]. Power gating to cut off the power supply to unused parts of the circuit in to prevent leakage currents [5]. Clock gating to prevent the clock from propagating into disabled portions of the circuit. Cell sizing used to find the appropriate sizes of the cells to satisfy the trade-off between power, area, and delay [6]. Pin reduces power by swapping connections to logically identical input pins of a cell, by assigning a high switching net to a pin with low capacitance value and the low switching net to a pin with higher capacitance [6]. Dynamic voltage and frequency scaling (DVFS) focus on the adjustment of power and frequency of some areas in the chip to optimize the resources needed for each task and minimize the power consumption when these resources are not used. Cell voltage threshold swap to reduce leakage power by substituting low voltage threshold (LVt) cells that are in a non-critical path by equivalent high voltage threshold (HVt) cells [7–9]. Using higher transistor threshold voltages also reduces short-circuit power by reducing the portion of time that both PMOS and NMOS transistors in the cell are conducting. For a specific technology node, up to 20% reduction in internal power is possible by changing to an equivalent higher Vt cell.

3 Timing Optimization Techniques

The setup timing optimization is a very critical task of ASIC's designers, especially in the physical design phase. Many techniques are available to help to close and to fix the setup timing violations. All these techniques have the same objective which is to resolve the existing violations, but each one has a different power footprint. This section presents the most used setup timing optimization techniques in the physical design phase.

3.1 Buffer Insertion

In advanced technology nodes, wire RC delay which is proportional to the wire equivalent resistance (R_{eq}) multiplied by the equivalent capacitance (C_{eq}) is no longer negligible compared to gate delay. Careful attention should be taken when routing a design to reduce wire delay.

Buffer insertion is one of the techniques used to balance the wire load and linearize the quadratic dependence between delay and wire length. Long interconnects impact the signal's transition since the delay is quadratically proportional to the wire length [10]. Buffer insertion on long nets is used to resolve this limitation by linearizing the delay/wire-length relationship. Its key principle is to cut long wires into small wires connected by buffers. This helps to reduce the path delay, and thus reduces setup violations [9].

Let's assume that we have a timing violation caused by a long interconnect. In order to solve the violations, buffer insertion on the long net is used as shown in Fig. 1a and 1b. It splits the net by a chain of buffers, where each buffer drives a part of the net's load. The resulting timing will be less than the original, and the violation will be reduced.

As demonstrated in [4], after a certain wire length a long interconnect consumes more delay than a buffer with two small interconnects.

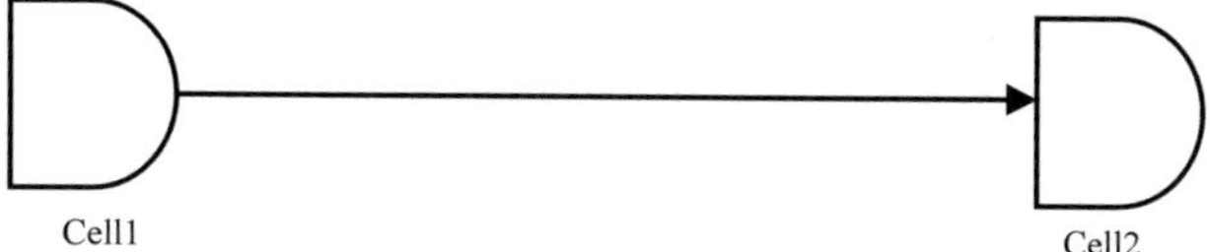

Fig. 1a. Long net before optimization

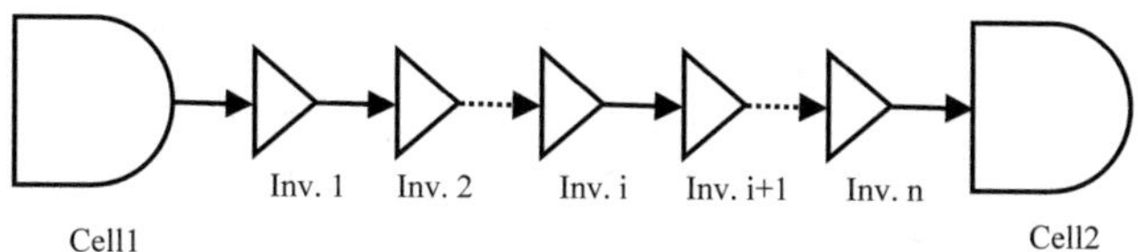

Fig. 1b. After buffering optimization

Buffering is used by both synthesis and P&R tools. The synthesis tools optimize the design's setup timing before generating the final netlist which is provided to the P&R stage.

Since the synthesis tools don't take into account the physical constraints, sometimes the buffers inserted in that stage are suboptimal. For those nets which still present setup violations in the physical design stage, the P&R tool can re-buffer it in an optimal way to close the violation.

Also, buffering helps drive loads from weak drivers. As stated in [9], the delay is optimal if the ratio of each output capacitance load divided by the gate input capacitance is equal to a constant stage ratio K (C_i/C_{i-1} = e = 2.71). To reduce the delay of a timing path with different gates, the optimizer tries to upsize the drivers or downsize the sinks, so the ratio K is respected. If a gate is not present in the library with a specific size, a buffer may be inserted to tackle the problem.

3.2 Cell Sizing

Timing optimization by gate sizing is a widely used technique in standard-cell based design methodology, each logic gate is available in the library in multiple sizes, and each size corresponds to a different drive strength. In the case of setup timing optimization, the sizing approach consists of substituting the weak drivers in timing critical paths by their equivalent library cells with higher drive strength [9, 12].

Increasing a cell's drive strength means increasing the cell width, which leads to a decrease of its equivalent resistance (2) and an increase of the delivered current, and by consequence, reduces the gate delay (3).

$$R = \frac{1}{k.\frac{W}{L}.(Vdd - Vth)} \tag{2}$$

$$CellDelay = R.C \tag{3}$$

Where R_{drive} is the drive resistance of the cell, C_{load} is the pin and wire load driven by the cell, and $D_{internal}$ is the internal delay of the cell. The drive resistance of the cell can also be reduced by decreasing the threshold voltage V_{th} or channel length, or increasing the supply voltage Vdd.

In physical design, upsizing a driver consists of swapping a weak driver on timing violated path with a better driver from the same library (Fig. 2).

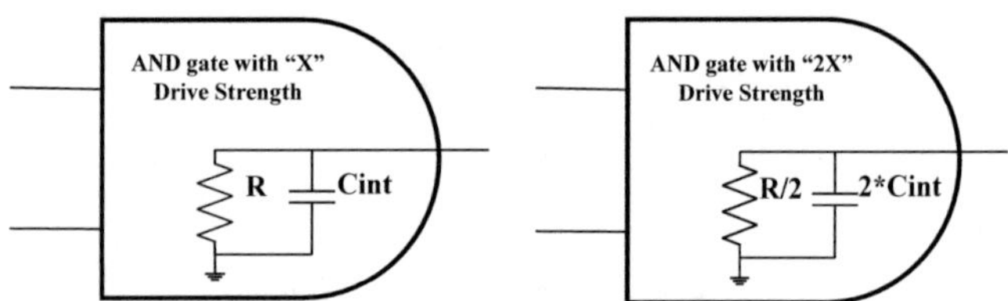

Fig. 2. AND gate drive strength

Standard cell libraries have a number of equivalent cells with different drive strengths. Substituting a cell with an equivalent cell with higher drive strength will decrease setup violations but will increase area and power consumption.

Cell upsizing doesn't focus only on increasing the drive strength of combinational gates but can also upsize a flip-flop to reduce the Clk->D setup constraint [3].

3.3 Load Capacitance Optimization

Load capacitance reduction is also a timing optimization technique. It deals with finding the optimal cell position to reduce interconnect wire length and to balance its load, to speed up the setup path. This technique consists of performing local cell movements to reduce or balance the net capacitance and resistance. It takes into consideration the criticality of pins, which is the ratio of the worst negative slack of the pin divided by the worst negative slack of the design (4), and the centrality of a pin which determines how many critical endpoints are affected by that pin [13, 14].

$$\text{Criticality} = (\text{WNS}(\text{pin}))/(\text{WNS}(\text{circuit})) \tag{4}$$

In order to reduce setup timing violations, load capacitance optimization transforms targets to decrease wire load capacitance and resistance by buffer balancing, wire sizing, or pin swapping.

Buffer balancing is used for the drivers that drive only one buffer. While the driver and the sink are fixed, the buffer can freely move between them (see Fig. 3a and 3b). The best position for the buffer is determined accounting for the drive strengths and the interconnect delay. The same approach is extended to cell balancing to operate on cells with multiple pins and sinks, accounting for the impact on each timing arc.

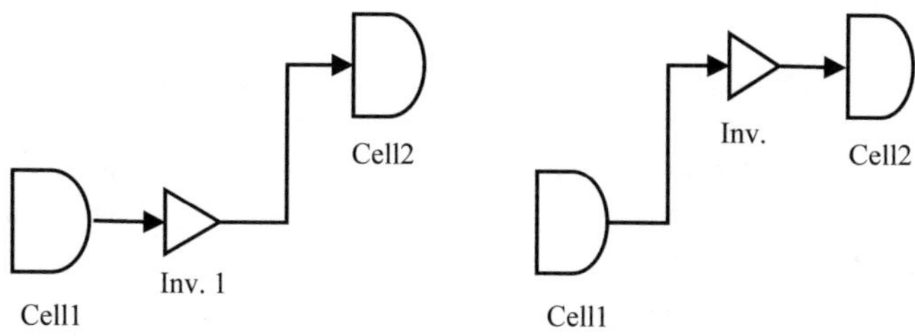

Fig. 3a. Original buffer placement **Fig. 3b.** Buffer moved

Buffer balancing alone may not solve all the setup violations, or it may impact the routability of the design. As a remedy of such limitations, routing layer assignment may be used. It takes into consideration several metrics (delay, congestion, area) before assigning any wire segment to a specific layer. As shown in Fig. 4a and 4b, the wire can be routed in different layers while taking into consideration the wire/vias delay [15].

Fig. 4a. Metal layer thickness

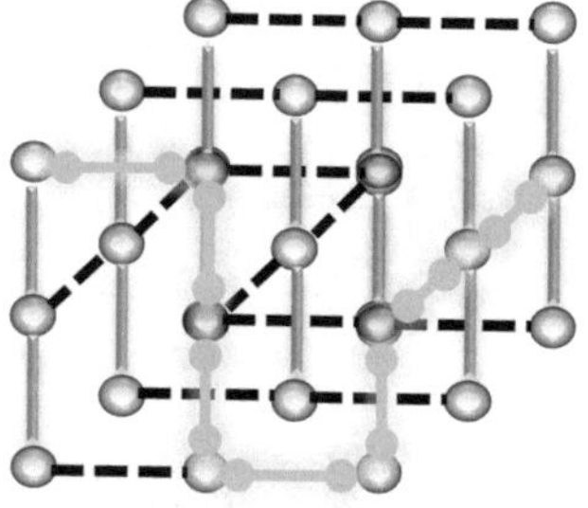

Fig. 4b. Multi-layer model for net routing

Pin swapping surveys the slack on each gate's input on a violated timing path. It switches the pin with the least amount of slack to the fastest propagating path through the gate, and the pin with a large amount of slack to the slowest propagating path to balance the timing without impacting the logical function (Fig. 5).

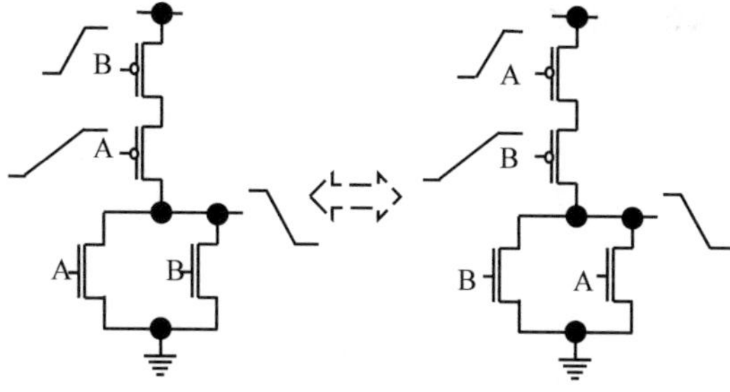

Fig. 5. Swap approach.

3.4 Cell Duplication

Gate duplication is used to improve delay by adding and connecting an identical gate between the critical node and the fan-out (Fig. 6a and 6b). It leads to an improvement in the overall negative slack. Duplicating a gate leads to a reduction in the fan-out load causing a gate delay reduction. Although the fan-out of the duplicated gate's drivers increases, the gain associated with duplicating a gate should be able to compensate for this increase and improve the circuit slack [16].

Each time this technique is applied, the network needs to be locally updated by modifying the fan-outs related to nodes. Then timing is updated at all the nodes of the network, to know if there are more improvements that can be made.

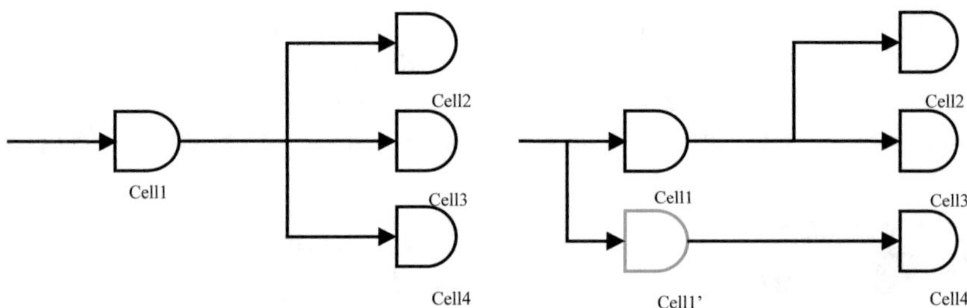

Fig. 6a. Before cloning

Fig. 6b. After cloning

3.5 Remapping

The remapping technique applies the technology mapping algorithms to replace slow gates with equivalent faster logical structures. Many logically equivalent decompositions for a given circuit can be found depending on the objective (delay, power, or area minimization). Remapping consists mainly of decomposing a gate into several gates or combining several gates into a single one. A technology mapping with the binning technique was presented in [17].

This technique uses Boolean transformations such as DeMorgan's law to restructure the netlist. The equivalent new structure will have less delay due to its higher drive strength (fewer stacked PMOS transistors), which makes the transition faster within those gates and helps to reduce the setup timing violations. Figure 7 gives a set of equivalent logic structures that could be swapped with each other.

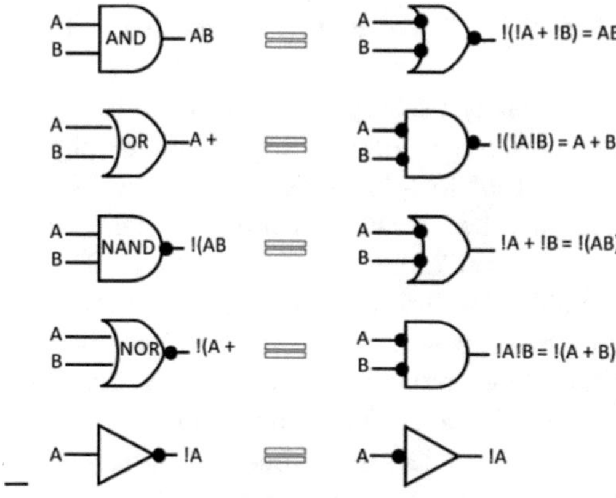

Fig. 7. Equivalent logic gates

3.6 Useful Skew

Skew is the difference in arrival time between two or more signals. If a clock tree has a skew of 10 ps, it means that the difference in latency between the longest and the shortest clock paths is 10 ps.

The goal of the useful skew technique is to ensure that only the necessary skew buffer stages are added to meet timing. An intentional skew may be added between specific leaf nodes (that require longer timing paths) to help close timing. Instead of balancing the clock at all leaf nodes, the clock at some leaf nodes is pushed out or alternately brought forward [18, 19].

As shown in Fig. 8. The clock period is 5 ns, adding a useful skew of 2 ns allows this path to meet the setup timing requirement. If another optimization technique was used instead of useful skew, additional buffers would have to be inserted which will lead to power consumption increase.

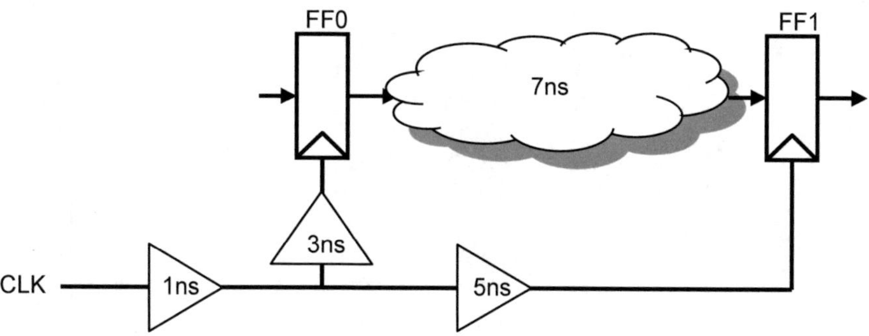

Fig. 8. Useful skew to meet timing

4 New Power Friendly Timing Optimization Flow

As described in Sect. 3, many techniques are available to fix the timing and eDRC violations. Each technique uses a different logical and/or physical transformation to speed-up the data signal or to slow-down the clock propagation. The optimizer which embeds all the above-presented techniques uses the standard cells available in the library to implement the right timing solution (flow 1). Depending on the library structure and the available standard cells the optimizer may optimize the timing and power concurrently or may degrade the power while optimizing the timing. In this section, we present the library pruning and the optimization flow used to reduce the power overhead generated due to timing optimization.

Our new approach (flow 2) divides the optimization process into two steps. In the first step, the optimizer tries to apply the logical transformation using only low power cells, which results in more cells' downsizing and load reduction and fewer cells' upsizing and duplication, and hence reduces the timing with a minimum power impact. The second pass of optimization uses all the available standard cells to close the design timing.

Table 1. Designs characteristics

TestCase	# Modes	# Corners	# Clocks	Max Freq (Hz)	# Std Cells	# Macro Cells	Area (mm^2)	Node (nm)	Stage
TC1	1	2	86	184.0 M	607644	253	108.20	180	PreCTS
TC2	2	5	7	251.0 M	795766	238	17.74	32	PreCTS
TC3	1	5	6	250.0 M	259973	71	3.11	90	PreCTS
TC4	2	14	68	500.0 M	3290622	178	148.08	180	PreCTS
TC5	1	2	3	500.0 M	1184163	61	17.87	90	PreCTS
TC6	1	4	114	1.0 G	218994	87	6.42	90	PreCTS
TC7	1	5	75	541.0 M	348247	70	3.35	90	PreCTS
TC8	2	10	1	1.0 G	46807	0	0.14	45	PreCTS
TC9	1	2	84	833.0 M	2474385	152	72.15	90	PreCTS
TC10	2	6	11	515.0 M	1132771	181	27.05	90	PreCTS
TC11	4	4	358	498.0 M	318618	46	3.32	90	PreCTS
TC12	3	4	33	363.0 M	98564	11	1.11	90	PreCTS
TC13	2	3	20	533.0 M	1919389	160	38.12	90	PreCTS
TC14	1	5	3	300.0 M	212337	4	2.37	90	PreCTS
TC15	1	2	24	752.0 M	185731	60	6.11	180	PreCTS
TC16	1	5	12	405.0 M	64288	5	1.22	90	PreCTS
TC17	2	6	5	400.0 M	944829	16	7.02	90	PreCTS
TC18	2	2	126	619.0 M	768014	88	18.92	90	PreCTS
TC19	4	5	895	833.0 M	911073	179	52.49	32	PreCTS
TC20	1	2	9	313.0 M	862311	144	65.63	180	PreCTS
TC21	2	3	6	600.0 M	647190	71	7.17	90	PreCTS
TC22	2	3	20	533.0 M	1844279	160	38.12	90	PreCTS
TC23	2	6	7	400.0 M	1015464	143	18.99	90	PreCTS
TC24	1	2	2	500.0 M	277005	8	9.25	180	PreCTS
TC25	1	6	4	1.0 G	1326408	57	3.13	32	PostCTS
TC26	1	4	5	602.0 M	292795	32	4.24	90	PostCTS
TC27	4	7	158	556.0 M	1555269	47	15.16	40	PostCTS
TC28	2	4	2	250.0 M	128606	0	0.21	28	PostCTS

(*continued*)

Table 1. (*continued*)

TestCase	# Modes	# Corners	# Clocks	Max Freq (Hz)	# Std Cells	# Macro Cells	Area (mm^2)	Node (nm)	Stage
TC29	2	4	1	556.0 M	52221	0	0.13	45	PostCTS
TC30	4	7	300	400.0 M	1607258	107	15.58	40	PostCTS
TC31	4	5	10	500.0 M	840997	82	1.86	32	PostCTS
TC32	2	18	3	833.0 M	724107	40	1.97	28	PostCTS
TC33	10	22	101	1.0 G	1037368	48	1.69	28	PostCTS
TC34	1	4	5	909.0 M	1272643	75	5.63	40	PostCTS
TC35	1	4	5	602.0 M	278348	32	4.24	90	PreCTS
TC36	2	4	70	671.0 M	570946	0	0.57	14	PreCTS
TC37	2	4	2	250.0 M	127132	0	0.21	28	PreCTS
TC38	2	4	1	556.0 M	52007	0	0.13	45	PreCTS
TC39	2	3	2	1.0 G	347106	35	2.66	40	PreCTS
TC40	2	6	3	225.0 M	636125	60	3.58	40	PreCTS
TC41	2	8	178	3.0 G	1136421	105	10.78	28	PreCTS
TC42	6	7	678	2.0 G	328120	5199	8.84	32	PreCTS
TC43	2	18	3	833.0 M	708548	40	1.97	28	PreCTS
TC44	3	14	41	500.0 M	846726	164	1.90	28	PreCTS
TC45	10	22	101	1.0 G	1024869	48	1.69	28	PreCTS

Flow 1: ***Default Optimization Approach***

1: Read Design Database
2: Report initial timing and power metrics
3: Optimize timing and eDRC
4: Detail Placement
5: Global Routing
6: Quality of Results assessment

Flow 2: ***New Optimization Approach***

1: Read Design Database
2: Report initial timing and power metrics
3: Filter Low Power STD Cells
4: Optimize Timing and eDRC ***using Low Power Lib Cells***
5: Optimize Timing and eDRC ***using All Lib Cells***
6: Detail Placement
7: Global Routing
8: Quality of Results assessment

To measure the power gain of the new approach compared to the default one, we developed the proposed algorithm (Flow 2) in TCL programming language and integrated it in the PnR flow, and we compared the generated results with a commercial

EDA tool (Nitro-SoC™). The experiments were conducted on a set of 45 designs from different technologies, sizes, and complexities (Table 1). We applied both the default optimization (Flow 1) and the newly proposed flow (Flow 2) on the sample of testcases (Table 1) and we compared the timing improvement and the power consumption increase engendered. As summarized in Fig. 9, the new flow -compared to the default flow- achieves an average setup timing gain of 37.5% and 55.8% in WNS and TNS respectively, with an average power decrease of 5% in leakage and 0.12% in dynamic power, which represented a gain of 1.3% in total power consumption (Figs. 10, 11 and 12). The timing gain is caused by the two optimization passes in the new algorithm, which means that each target is optimized twice with different techniques dictated by the possibilities offered by the available lib cells. While the power improvement is the results of imposing the usage of more low power cells in the first optimization pass.

Since the new approach emphasizes the usage of low power cells for timing optimization, these cells are in general the smallest ones in the library, and hence their usage lead to total load and total parasitic capacitance reduction, which gave a gain of 95% of Max capacitance violations and 45% in Max transition violations (Fig. 13). Since we are

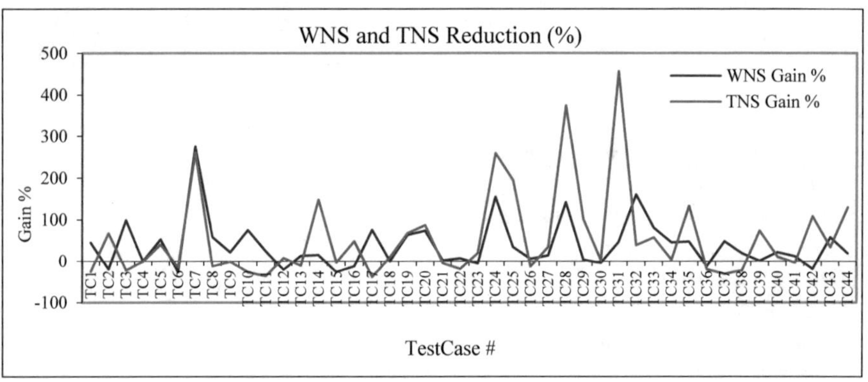

Fig. 9. Worst and total negative slacks reductions

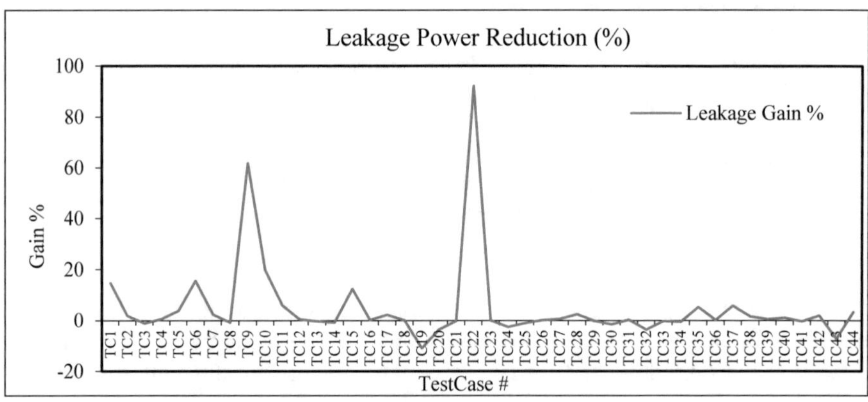

Fig. 10. Leakage power reduction

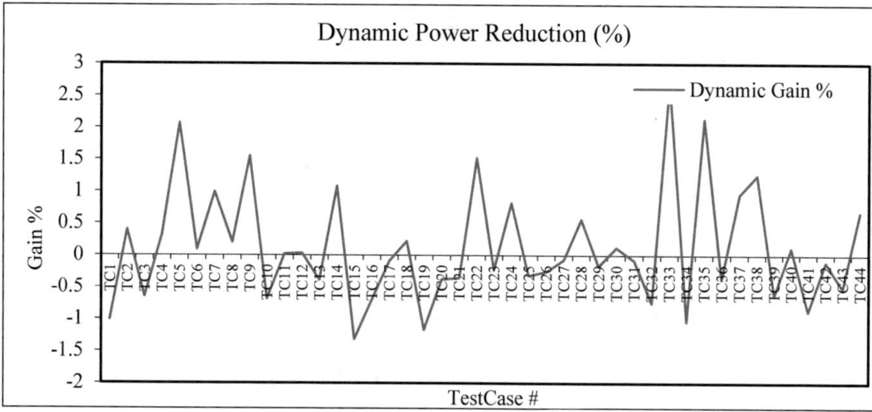

Fig. 11. Dynamic power reduction

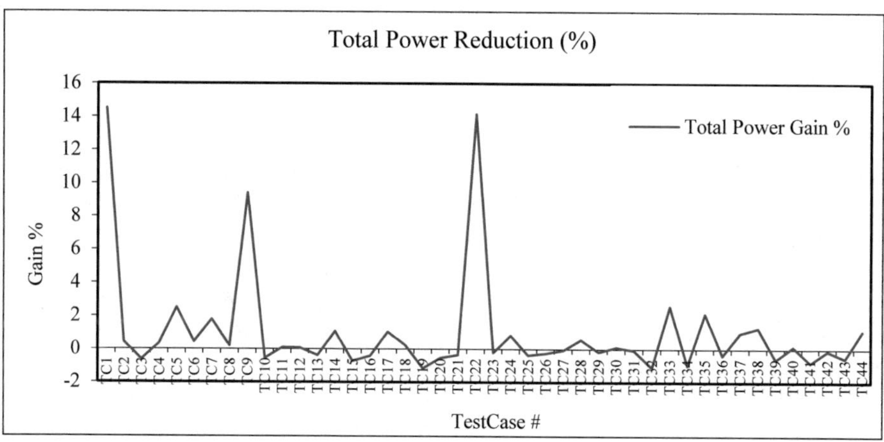

Fig. 12. Total power gains

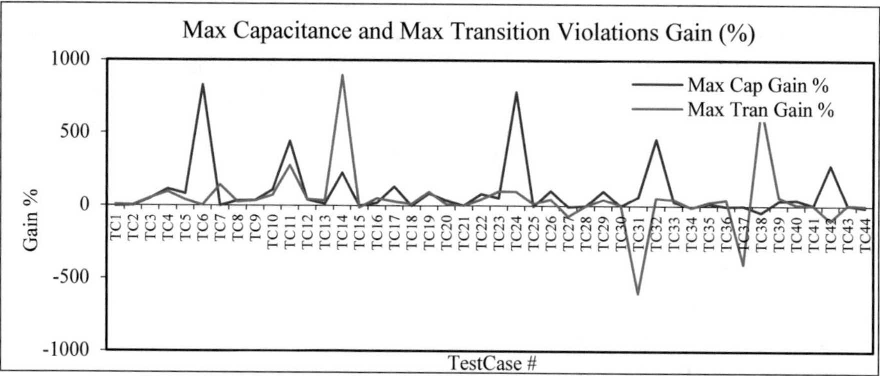

Fig. 13. Max capacitance and transition violations gains

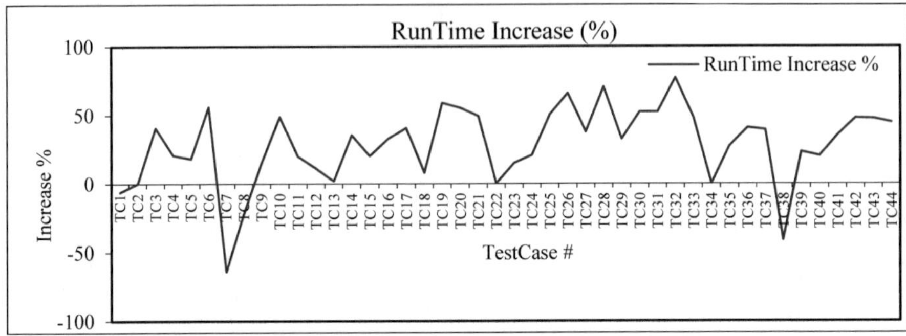

Fig. 14. Runtime increase

doing two passes of optimization, the runtime has increased by 28%, which is acceptable compared to the other achieved benefit in power and timing (Fig. 14).

5 Conclusion

In this paper, we propose a power efficient methodology to optimize the timing and the eDRC using library pruning techniques. We found an opportunity to filter the STD cells supplied to the optimizer to perform the logical and physical transformation with minimum power impact. We showed evidence that in order to reduce the power consumption of a System on Chip (SoC) optimally, one should tune the library cell before starting any optimization. The flow presented in Sect. 4 was applied to more than 40 industrial designs, and we were able to achieve an eDRC average gain of 95% in Max capacitance violations and 45% in Max transition violations, a timing gain of 37.5% and 55.8% in WNS and TNS respectively, with an average power decrease of 5% in leakage and 0.12% in dynamic power, which have generated a gain of 1.3% in total power consumption.

Acknowledgment. This research was supported by Mentor Graphics Corporation. We thank our colleagues from CDS division who provided insight and expertise that greatly assisted the research, although they may not agree with all of the interpretations/conclusions of this paper.

We thank Dr. Hazem El Tahawy (Mentor Graphics, Managing Director MENA Region) for initiating and supporting this work, Chinnery David (Architect, CSD Nitro R&D Optimization) for assistance, help and guidelines through the research, and Bhardwaj Sarvesh (Group Architect, ICDS P&R Solutions Optimization) for the opportunity to work on such advanced topic.

References

1. Jeon, B.-K., Hong, S.-K., Kwon, O.-K.: A low-power 10-bit single-slope ADC using power gating and multi-clocks for CMOS image sensors. In: 2016 International SoC Design Conference (ISOCC) (2016)

2. Melikyan, V., Babayan, E., Melikyan, A., Babayan, D., Petrosyan, P., Mkrtchyan, E.: Clock gating and multi-VTH low power design methods based on 32/28 nm ORCA processor. In: 2015 IEEE East-West Design & Test Symposium (EWDTS) (2015)
3. Bhasker, J., Rakesh, C.: Static Timing Analysis for Nanometer Designs A Practical Approach. Springer, New York (2009)
4. Shin, Y., Kim, H.-O.: Analysis of power consumption in VLSI global interconnects. IEEE (2015)
5. Arkadiy, M.: Short-circuit power reduction by using high-threshold transistors. J. Low Power Electron. Appl. **2**, 69–78 (2012)
6. Benjamin, C., Ivailo, N.: Power compiler: a gate-level power optimization and synthesis system. IEEE (1997)
7. Srivastava, A., Sylvester, D., Blaauw, D.: Statistical Analysis and Optimization for VLSI: Timing and Power. Springer, New York (2005)
8. Konstantin, M., Avinoam, K., Shmuel, W.: Multi-Net Optimization of VLSI Interconnect. Springer, New York (2015)
9. Kahng, A.B., Igor, J.L., Markov, L., Hu, J.: VLSI Physical Design: From Graph Partitioning to Timing Closure. Springer, Dordrecht (2011)
10. War, K.M., Rosdi, B.A.B., Wee, C.E.: CAD automation module based on cell moving algorithm for ECO timing optimization, pp. 284–288. IEEE (2011)
11. Alpert, C., Chu, C., Gandham, G., Hrkic, M., Hu, J., Kashyap, C., Quay, S.: Simultaneous driver sizing and buffer insertion using a delay penalty estimation technique. IEEE Trans. Comput. Aided Des. Integr. Circ. Syst. **23**(1), 136–141 (2004)
12. Chen, C., Tsui, C., Ahmadi, M.: A gate duplication technique for timing optimization. Can. J. Electr. Comput. Eng. **28**(1), 37–40 (2003)
13. Guilherme, F., Mateus, F., Jucemar, M., Marcelo, J., Ricardo, R.: Drive strength aware cell movement techniques for timing driven placement. ACM (2016)
14. Liu, D., Yu, B., Chowdhury, S., Pan, D.Z.: Incremental layer Assignment for critical path timing. In Proceedings of the 53rd Annual Design Automation Conference, DAC 2016, June 2016
15. Nitro-SoC™ and Olympus-SoC™ User's Manual, Mentor Graphics, November 2016
16. Ankur, S., Ryan, K., Majid, S.: On the complexity of gate duplication. IEEE Trans. Comput. Aided Des. Integr. Circ. Syst. **20**(9), 1170–1176 (2001)
17. Mohammed, M., Siva, Y.: Study and implementation of multi-VDD power reduction technique. In: 2015 International Conference on Computer Communication and Informatics (ICCCI), pp. 1–4 (2015)
18. Hima Bindu, K., Hamid, M.: ASIC design flow tutorial using synopsys tools. Nano-Electronics & Computing Research Lab School of Engineering San Francisco State University, San Francisco, Spring 2009
19. Flach, G., Fogaça, M., Monteiro, J., Johann, M., Reis, R.: Drive strength aware cell movement techniques for timing driven placement. In: Proceedings of 2016 on International Symposium on Physical Design, ISPD 2016, pp. 73–80 (2016)

Modelling Wind Speed Using Mixture Distributions in the Tangier Region

Fatima Bahraoui[1(✉)], Hind Sefian[1], and Zuhair Bahraoui[2]

[1] Faculty of Sciences and Techniques, Tangier, Morocco
fbahraoui@gmail.com

[2] University of Chouaib Doukali, EST, El Jadida, Morocco
bahraouizuhair@gmail.com

Abstract. The main objective of this research is to improve the predictability of wind generation, these include to propose a probabilistic prediction approach of the moment of appearance of these variations. In this perspective, we will compare the adjustment of wind speed distribution by Weibull distribution and two mixture distribution function as a solid alternative model to the eolian energy models. First with the mixture of the Weibull and Pareto distribution, and second with Lognormal and Pareto distribution. Our aim is to capture the outlier if there exists in the data and gives a most precise predictive estimation of the power density energy, evaluate the wind potential and predict the electrical energy produced in the site in order to size a wind farm on a site in Tangier while based on a judicious choice of wind turbines.

Keywords: Wind speed · Wind power density · Wind variation · Statistical analysis · Weibull · Lognormal · Pareto · Probability density function

1 Introduction

In the past few years in order to reduce global warming, air pollution and the depletion of unconventional fossils such as fuels many developed and developing countries have adopted policies for the use of renewable energy sources such as solar, hydroelectric, geothermal and wind energy because they are clean and inexhaustible, consequently, about 18% of energy needs are covered by renewable energies. It has been increasing rapidly but at the same time, energy demand will continue to increase significantly in the coming decades, according to some predictions it will reach 700–830 EJ by 2040 [1].

Renewable energy systems such as wind power have less of a negative impact on the environment compared to other conventional systems of power generation. Although wind energy is abundantly available everywhere, the wind characteristics and wind potential to create electricity are different. Therefore it is important to study the wind characteristics and make an estimation of wind potential at a particular location to determine the capacity of wind energy for the installation of wind turbines. During the analysis of the wind energy potential of a region, knowledge of wind characteristics helps to define site requirements and choose a proper turbine. In addition, it is necessary to determine the hourly average, the daily average, the monthly average, the

M. Ezziyyani (Ed.): AI2SD 2018, AISC 912, pp. 236–245, 2019.
https://doi.org/10.1007/978-3-030-12065-8_22

seasonal average of the wind speed and the density of the wind energy, In addition in order to optimize the profits from the wind energy production sometimes it is necessary to determine the hourly average, the daily average, the monthly average, the seasonal average of the wind speed and the density of energy. It is very important to analyze the wind characteristics of a region in order to acquire the maximum wind potential from the site, since wind speed fluctuates rapidly over time, some critical errors might occur between the estimated and he actual energy output and using statistical methods leads to a sophisticated estimation of the energy. Many researchers have been using several methods to predict the wind potential of a specific location, Weibull probability density function is the most commonly used tool to estimate the wind power density, Ozay and Celiktas [2] used Weibull and Rayleigh distributions to estimate the wind potential in the Alacati region of Turkey. They conducted their study keeping in view of the rapidly increasing population of Turkey (8% annually). [3] Ali, Lee and Jang analyzed wind characteristics using two variables Weibull and Rayleigh PDF in Deokjeok-do which is a small island situated in the west of South Korea. [4] Qing used two-parameter of Weibull distribution they first applied it to model the wind speeds on various timescales and to determine wind energy potential in Santiago island. [5] Wais compared the two and three-parameter Weibull distributions directly for wind energy calculations, and he checked whether or not the three-parameter Weibull distribution can take advantages comparing to the typical Weibull distribution for high percentages of null wind speeds. [6] Soulouknga, Doka, Revanna, Djongyang and Kofane, analyzed wind speed for a period of 18 years (1960–1978) and they used the two-parameter Weibull distribution.

[7] Pishgar-Komleh and Akram examine the wind energy potential by finding the Weibull and Rayleigh distribution parameters, [8] MertKantar and Usta used upper-truncated Weibull distribution, in modelling wind speed data and also in estimating wind power density. [9] Usta, Arik, Yenilmez and MertKantar introduced a new estimation approach that could be used in calculating the Weibull parameters for the estimation of wind power. This new approach called the method of multi-objective moments (MUOM). [10] GülAkgül, Şenoğlu and Arslan used the Inverse Weibull (IW) distribution to model the wind speed.

In order to determine the available power and energy density, but mixing two statistical distribution to get a new distribution carrying the properties of its components can be very effective to arrive at an accurate estimate of power. The main purpose of the present study is to analyze the wind characteristics of Tangier using three methods: Weibull, a mixture of Weibull-Pareto and Lognormal-Pareto. Finally, we will carry out a comparison, in order to select an accurate estimation of wind potential of a wind farm, Tangier (latitude: 35.583333, Longitude: −5.75) is known as a windy city situated in the north of Morocco. The parameters of the mixture wind speed distribution will be estimated by the maximum likelihood method.

2 Data Analysis

The data was collected from the meteorological center of Tangier, Fig. 1 shows a satellite view of the center, the height of the anemometer is 10 m from ground level and the wind data, including wind speed and wind direction, were recorded for five years

(Jan. 2013 to Dec. 2017) with a 1 h time interval. According to international standards, wind data should be collected for more than one year to increase confidence in the results. Table 1 shows the brief details about recorded data. For the density's distribution, we will use the daily maximum.

Figure 2 present the hourly (24 h) variation of average wind speed of each month and season. During all the months, wind speed first stabilizes until 8 AM (8 h in the figure) and starts to increase until it reaches the maximum value of the day at approximately 14 h (6.58 m/s). After 14 h it starts to decrease and stabilize again. In Fig. 2, it is observed that most of the wind speeds are between 4–6 m/s during all seasons. Wind speed variation during daytime is faster than during night time. This is due to the difference in temperature between land and sea. This temperature difference creates a density difference between the sea air and earth air which make the wind to move by natural convection. If the temperature difference increase, the value of mean wind speed will increase.

Fig. 1. Location of the Meteorological centre.

The wind rose diagram is presented in Fig. 4. The wind rose diagram summarizes the wind characteristics for a specific time period to find out the dominating wind directions at Tangier so that we can determine the optimal position of wind turbines. As the figure shows that most of the winds are in the range of 7 m/s–11 m/s and it comes mostly from the east side. The observed wind directions and wind speed are due to the climate characteristics of Tangier since Tangier is a windy city such as seasonal monsoons (especially in the summer). Monsoons are the result of temperature differences between land and sea, this temperature difference influence the wind characteristics such as mean wind speed and fluctuations in wind speed. Figure 3 Show the variation of monthly mean wind speed in 2017, the highest average of wind speed was observed in April and that's due to the easterly wind.

A correct statistical analysis of wind data is an important phase in a wind resource evaluation, Table 2 presents the statistical parameters of the data. The choice of the wind speed distribution function influences the estimation of the available wind energy or the wind turbine performance and gives the significant impact on the investment profitability.

Table 1. Detail collected data position

Location	Longitude	Latitude	Data period	Time-interval	Height (m)
Meteorological center of Tangier	−5,75°	35,583°	2013–2017	1 h	10 m

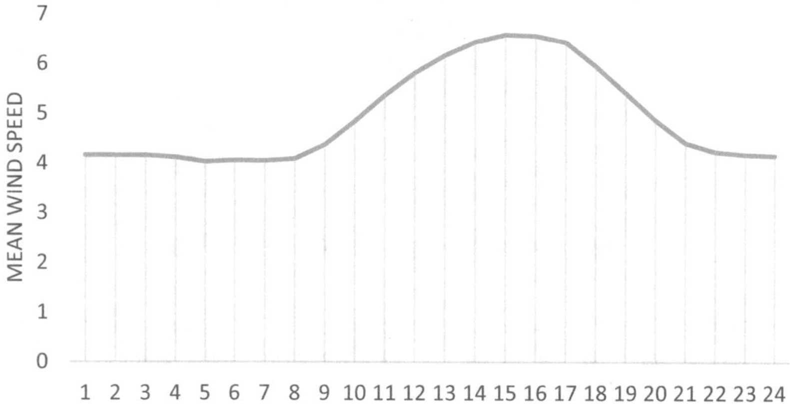

Fig. 2. The hourly variation of average wind speed (2013–2017).

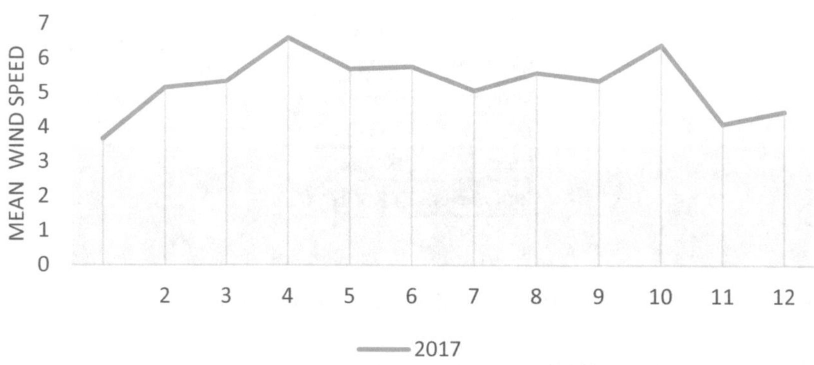

Fig. 3. Variation of average wind speed (2017).

The Weibull distribution model often allows a good approximation of the distribution of the wind speed, because of its ability to assume the character the Wind speed. Many studies have shown that the two-parameter Weibull distribution provides a good prediction of the wind energy potential. In this study, we will compare the Weibull distribution function and new probability distributions obtained from mixing Pareto (two parameters α *and* β) with Weibull (two parameters k *and* c) and Lognormal with Pareto. The parameters will be estimated by maximum likelihood method (MLE) and the simulation is done by **R** software.

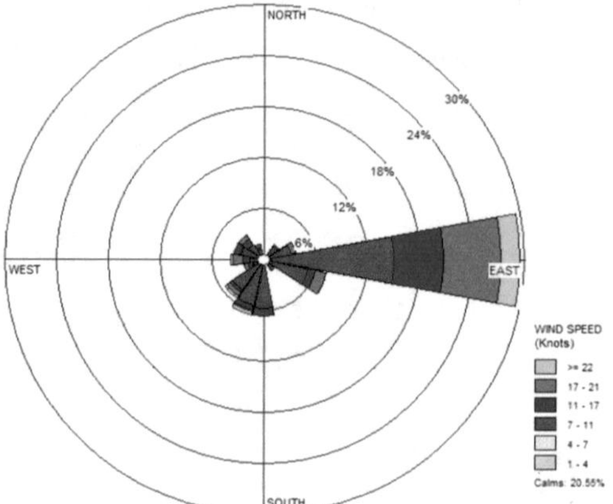

Fig. 4. Windrose diagram by WRPLOT (2017).

Jarque-Bera statistic test is used to tests the normality hypothesis. The P-value of Jarque Beta test confirms that the data is not Normal distributed and the skewness and kurtosis shows the asymmetric of the data.

Table 2. Descriptive statistics

Average	7.977 m/s
Sd	2.730
Skewness	0.681
kurtosis	0.200
Min	3 m/s
Max	19 m/s
Jarque-Bera	2.2×10^{-16}

3 Alternative Probability Models for Wind Speed

The first alternative probability density function (pdf) is a mixture of the Weibull and Pareto distribution function. The Pareto distribution is a heavier tailed distribution and it can capture the extremes value if they exist. The probability density (pdf) of a mixture is:

$$f(v) = pf_1(v) + (1 - p)f_2(v) \tag{1}$$

$f_1(v)$: *the Weibull probability density function*
$f_2(v)$: *the Pareto probability density function,*

With: $0<p<1$ and,

$$f_1(v) = \left(\frac{k}{c}\right)\left(\frac{v}{c}\right)^{(k-1)} e^{\left(\frac{-v}{c}\right)^k} \tag{2}$$

$$f_2(v) = \frac{\alpha}{\beta}\left(\frac{v}{\beta}\right)^{-(\alpha+1)} \tag{3}$$

$$f(v) = p\left(\left(\frac{k}{c}\right)\left(\left(\frac{v}{c}\right)^{(k-1)} e^{\left(\frac{-v}{c}\right)^k}\right) + (1-p)\left(\frac{\alpha}{\beta}\right)\left(\frac{v}{\beta}\right)^{-(\alpha+1)}\right) \tag{4}$$

v: wind speed $\left(\frac{m}{s}\right)$
(*k*, *c* > 0: *Weibull parameters*)
($\alpha, \beta > 0$: *Pareto parameters*)

The second alternative is the Lognormal-Pareto mixture. The Lognormal is asymmetric and can give a better adjustment in the centre and for low values. The probability density function: (p.d.f) of their mixture with Pareto distribution is:

$$f(v) = pf_1(v) + (1-p)f_2(v) \tag{5}$$

$f_1(v)$: *the probability density function of Lognormal*
$f_2(v)$: *the probability density function of* Pareto:

Follows:

$$f_1(v) = \frac{1}{v\sigma\sqrt{2\pi}} e^{\left(\frac{-1}{2}\left(\frac{\log(v)-\mu}{\sigma}\right)^2\right)} \tag{6}$$

$$f_2(v) = \left(\frac{v}{\beta}\right)^{-(\alpha+1)} \tag{7}$$

$$f(v) = p\left(\frac{1}{v\sigma\sqrt{2\pi}} e^{\left(\frac{-1}{2}\left(\frac{\log(v)-\mu}{\sigma}\right)^2\right)}\right) + (1-p)\left(\frac{\alpha}{\beta}\left(\frac{v}{\beta}\right)^{-(\alpha+1)}\right) \tag{8}$$

With: ($\mu, \sigma > 0$: *Lognormal parameters*).

4 The Goodness of Fit Models

Table 3 summarizes the parameters estimations of each method and its AIC criteria values. Generally, Akaike's information criterion (AIC) compares the quality of a set of statistical models to each other. It is a fined technique based on in-sample fit to estimate the likelihood of a model to estimate the future values. A good model is the one that has minimum AIC among all the other models.

Table 3. Statistical estimation of the parameters

Model	Parameters	AIC
Weibull	$k = 3.093$	8818.25
	$c = 8.923$	
Weibull-Pareto	$k = 3.093$	8824.25
	$c = 8.921$	
	$\alpha = 4.979$	
	$\beta = 0.903$	
	$p = 0.936$	
Lognormal-Pareto	$\sigma = 2.018$	8668.78
	$\mu = 0.345$	
	$\alpha = 2.880$	
	$\beta = 1.741$	
	$p = 0.908$	

The parameters were estimated using maximum likelihood minimization method to estimate the parameters (Cohen) [11]. This method is easy simple tools and consists to maximize the product density:

$$f(x) = \prod_{i=1}^{n} f(x_i, \theta)$$

Where θ consists the set parameters to be estimated.

We note that the Lognormal-Pareto model has the lowest value of the AIC, therefore is the most accurate distribution for the regions wind characteristics. Figure 5 shows the histogram of the wind speed data and the fitted probability density function. We observe that the Lognormal adjust well the empirical value in the centre and differ to the other distribution and is most heavier tailed. The Weibull and their mixture with Pareto distribution function give almost the same adjustment. We can say that the most accurate distribution function for estimating wind speed is Lognormal-Pareto because it takes into account the extreme values and the values centre and even the low values as shown in Fig. 5.

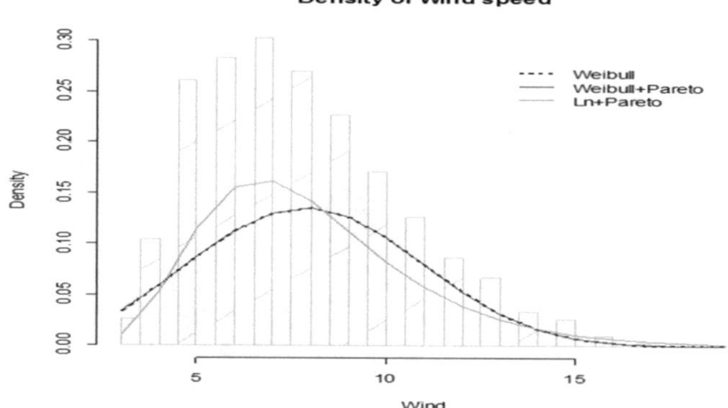

Fig. 5. The distribution 's density of three methods

5 Wind Power Energy

Estimation of the available energy in the wind at a site, in which the wind turbine will be, is one of the essentials steps in the planning of a wind energy project. The available power of the wind is:

$$P = \frac{1}{2}\rho A v^3 \tag{9}$$

A is the cross sectional area.

Mean wind power:

The mean wind power at the observation vc of the wind speed is:

$$MP(vc) = \frac{1}{2}\rho A \int_0^{vc} v^3 f(v)dv \tag{10}$$

$f(v)$: *the density of the chosen model.*

When $vc \to \infty$, we have the average of the power energy.

Equations (9) and (10) enable to calculate the available wind power directly from the measurements. The simulation is for the maximum wind speed of each day for one year. The available wind power and energy can be estimated by adding up the energy corresponding to all wind speeds. Figure 6. Presents the estimated power using the three methods from the Eq. (9). It can be noticed that the Lognormal-Pareto mixture gives the highest estimation of wind power. Figure 7 shows the mean wind power, Eq. (10) at the simulate daily observation of the wind speed. We note that using the Weibull pdf function the average of the density power energy is equal 429.792 and from Mixture Weibull Pareto is 429.599, while the average of the density power of Mixture of the Lognormal and Weibull is 446.052 stabilize in higher value.

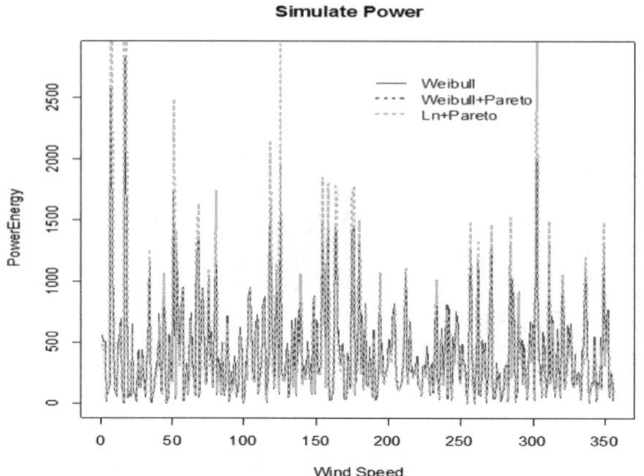

Fig. 6. Density power energy of three methods

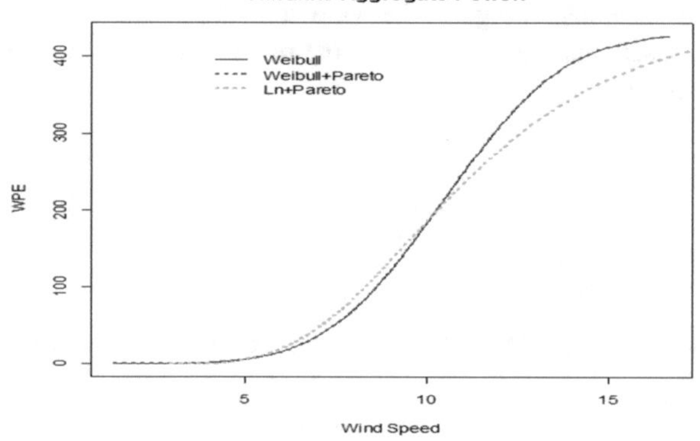

Fig. 7. Estimated power using the three methods

6 Conclusion

The goal of this paper is to propose a procedure to estimate accurately possible the wind energy. Wind speed and wind direction data from 2013 to 2017 for the Tangier region have been statistically analyzed. The paper presents that the two-parameter Weibull distribution is not always effective to evaluate the wind speed distribution and to estimate the available wind power. In addition, it is noticed that the Lognormal-Pareto distribution can offer a better evaluation of the wind speed data. The energy gained from Lognormal-Pareto distribution is higher and this mixture model can help the analysts, or the wind farms companies to obtain more accurate predictive values of wind speed energy. Future work will focus on using the proposed distribution in order to evaluate wind in farm sites.

References

1. WEC: World energy scenarios. Technical report, World Energy Council (2016)
2. Ozay, C., Celiktas, M.S.: Statistical analysis of wind speed using two-parameter Weibull distribution in Alaçatı region. Energy Convers. Manag. **121**, 49–54 (2016)
3. Ali, S., Lee, S.-M., Jang, C.-M.: Statistical analysis of wind characteristics using Weibull and Rayleigh distributions in Deokjeok-do Island e Incheon, South Korea. Renew. Energy **123**, 652–663 (2018)
4. Xiangyun, Q.: Statistical analysis of wind energy characteristics in Santiago Island, Cape Verde. Renew. Energy **115**, 448–461 (2018)
5. Wais, P.: Two and three-parameter Weibull distribution in available wind Power analysis. Renew. Energy **103**, 15–29 (2017)
6. Soulouknga, M.H., Doka, S.Y., Revanna, N., Djongyang, N., Kofane, T.C.: Analysis of wind speed data and wind energy potential in Faya-Largeau, Chad, using Weibull distribution. Renew. Energy **121**, 1–8 (2018)
7. Pishgar-Komleh, S.H., Akram, A.: Evaluation of wind energy potential for different turbine models based on the wind speed data of Zabol region, Iran. Sustain. Energy Technol. Assess. **22**, 34–40 (2017)
8. MertKantar, Y., Usta, U.: Analysis of the upper-truncated Weibull distribution for wind speed. Energy Convers. Manag. **96**(15), 81–88 (2015)
9. Usta, U., Arik, I., Yenilmez, I., MertKantar, Y.: A new estimation approach based on moments for estimating Weibull parameters in wind power applications. Energy Convers. Manag. **164**, 570–578 (2018)
10. GülAkgül, F., Şenoğlu, B., Arslan, T.: An alternative distribution to Weibull for modelling the wind speed data: inverse Weibull distribution. Energy Convers. Manag. **114**, 234–240 (2016)
11. Cohen, A.C.: Maximum likelihood estimation in the Weibull distribution based on complete and on censored samples. Technometrics **7**, 579–588 (1965)

Prediction of Time Series of Photovoltaic Energy Production Using Artificial Neural Networks

A. Elamim[1(✉)], B. Hartiti[1,2], A. Barhdadi[3], A. Haibaoui[1,4], A. Lfakir[5], and P. Thevenin[6]

[1] ERDYS Laboratory, MEEM & DD Group, Hassan II University of Casablanca, FSTM BP 146, Mohammedia 20650, Morocco
elamimabderrazzak7@gmail.com
[2] ICTP, UNESCO, Trieste, Italy
[3] Energy Research Centre, Ecole Normale Supérieure (ENS) Mohammed V University, Rabat, Morocco
[4] LIMAT Laboratory, Department of Physics, University Hassan II FSB, Casablanca, Morocco
[5] University Sultan Moulay Slimane FSTB, BP 523 Beni Melall, Morocco
[6] LMOPS Laboratory, Department of Physics, University of Lorraine Metz, Lorraine, France

Abstract. An artificial neural network (ANN) model is used for forecasting the power provided by photovoltaic solar panels using feed forward neural network (FFNN) of a photovoltaic installation located in the city of Mohammedia (Morocco). One year of hourly data on solar irradiance, ambient temperature and output PV power were available for this study. For this, different combinations of inputs with different numbers of hidden neurons were considered. To evaluate this model several statistic parameters were used such, as the coefficient of determination (R^2), the Root Mean Squared Error (RMSE) and the Mean Absolute Error (MAE). The results of this model, tested on unknown data, showed that the model works well, with determination coefficients lying between 0.98 and 0.998 for sunny days and between 0.82 and 0.96 for cloudy days.

Keywords: Photovoltaic installation · Feed forward neural network · Artificial neural network

1 Introduction

PV production depends on climatic factors, namely the temperature and the amount of global solar radiation incident on PV modules [1], but the latter presents a source of uncertainty for the designers of PV generators. This is mainly due to the lack of long series of data, poor quality data, the quality of measuring instruments and the non-uniform nature of solar radiation over time, which can create anomalies in the design and sizing photovoltaic of systems [2, 3]. In addition, one of the main challenges of the

M. Ezziyyani (Ed.): AI2SD 2018, AISC 912, pp. 246–257, 2019.
https://doi.org/10.1007/978-3-030-12065-8_23

massive integration of photovoltaic is the ability to accurately predict the electrical energy produced by PV systems [4].

Several approaches to production forecasting can be found in the literature. We cite direct and indirect methods and they can be divided into three categories: physical, statistical and soft-computing [5].

Physical methods attempt to create analytical models based on geographic and meteorological parameters [6, 7]. Statistical approaches attempt to establish a mapping link between historical data and the real power produced by the PV system to minimize the errors [8, 9], but these approaches are somewhat limited because they are unable to produce short-term and real-time predictions [10].

Thus, several studies have indicated that soft-computing methods perform more competitively than the other methods cited earlier [11–13].

Among the most useful methods for predicting the power produced by a PV system, we quote the method of artificial neural networks (ANN). It seems to be a very promising alternative to solve this kind of difficulties, especially for complex, non-linear or multi-variable problems and which require knowledge that is difficult to specify but for which there is an enormous number of examples, artificial neural Networks (ANN) is an approach that does not need to know the information about the process or system that generates the data. ANN techniques have been widely used for modelling, simulation, prediction and optimization of complex systems, for several energy systems, including photovoltaic systems. [13]. Recently, ANN models have been used in the prediction of electrical power produced in many locations with different climates. Relevant research has been conducted in different countries [14–18], but no work has been done in Morocco.

The aim of this article is the modelling and prediction of the power generated by a photovoltaic installation of 2.04 kWp on the roof of the Faculty of Science and Technology of Mohammedia, Morocco using the ANN models. The developed neural model can be used to predict and model the power produced by the photovoltaic system and to analyze the performances in terms of electrical energy delivered to the network.

2 Materials and Methods

2.1 Site Information and System Description

The PV system under study is a 2.04 kWp rated power, facing South; it is composed of 8 polycrystalline silicon photovoltaic panels, 30° tilted, fixed on the roof of the research building in the Faculty of Science and Technology of Mohammedia. The electrical diagram of the system is shown in Fig. 1.

The PV generator injects electricity into the grid throughout a Sunny boy inverter 2000 HF SMA with a nominal efficiency of 96.6%. The specifications of the studied module and the inverter are represented in Tables 1 and 2, respectively.

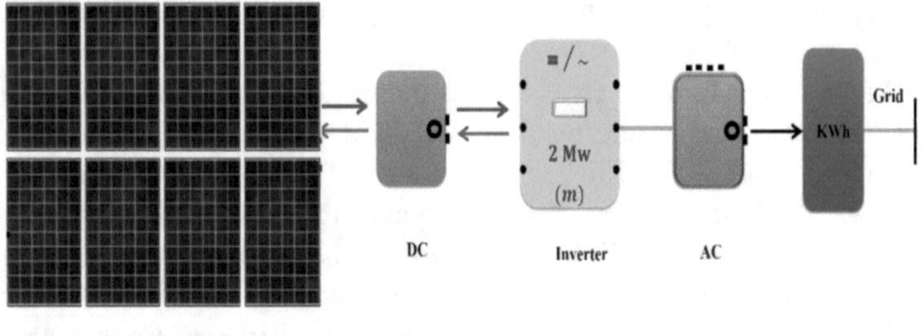

Fig. 1. Schematic diagram of solar photovoltaic plant at Mohammedia, Morocco

Table 1. PV panel properties

Trademark	Solarworld
Model	Sunmodule
Solar cell	pc-Si
Maximum power at STC	255 Wp
Optimum operating voltage	30.9 V
Optimum operating current	8.32 A
Open circuit voltage	38 V
Short circuit current	8.88 A
Temperature coefficient of maximum power	–0.41%/°C
Module efficiency	15.2%

2.2 Data Base Description

The photovoltaic plant and its weather station are equipped with several sensors that record and provide physical measurements and meteorological parameters related to climatic conditions (Fig. 2).

Table 2. Inverter specifications

Inverter model	SB 2000HF30
Max PV power	2100 W
Max voltage	700 V
Nominal voltage	220 V/230 V/240 V
Voltage range	180 V–280 V
Grid frequency; range	50/60 Hz
Max. output current	8.3 A
Max input current	12.0 A
Maximum efficiency	96%

Fig. 2. Meteorological data instrumentation.

The data collected from the PV power plant site correspond to: the average of power output (PAC), the ambient temperature, the horizontal and inclined irradiation, wind speed and direction.

The collected data are saved in CSV files every five-minute and then converted to hourly averages.

2.3 Artificial Neural Network Approach

The artificial neural network method is an information processing system, non-algorithmic and massive parallel learning technique inspired by biological neurons. ANNs are used to link inputs and outputs using a historical database to generate outputs when they are missing [3–19].

The ANN architecture is composed of three parts: an input layer that receives the data, an output layer where the supposedly calculated data is sent, and one or more hidden layers that connect the input to the output data layers, which can be translated by the following Eq. (1):

$$Z_i = -f(\sum_{j=1}^{N} w_{ij}xj) \tag{1}$$

The function f is called the activation function of the neuron. It models the behavioral nature of the system or the physical approach modelled by the ANNs, which can be in the form of a hyperbolic, sigmoid, exponential or radial basic function.

These layers are generally classified according to their architecture. There are two main categories: feedback neural networks and feed forward neural networks [19]. The multilayer Perceptron with the back-propagation (BP) training algorithm is the most widely used type of FFNN in the literature; the diagram of this type of network is shown in Fig. 3.

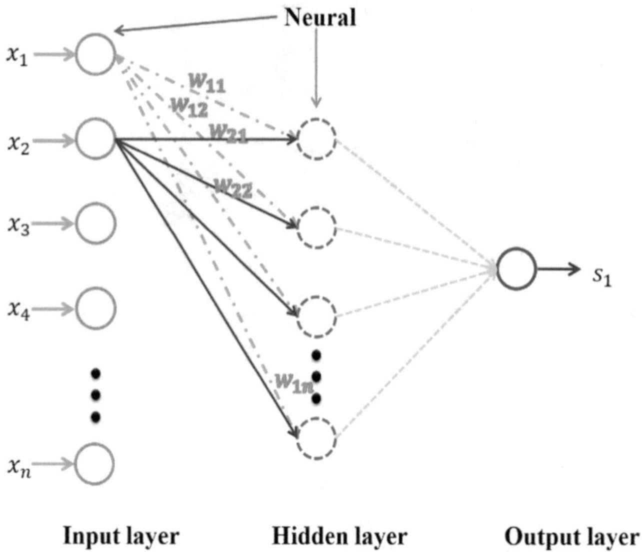

Fig. 3. Feed forward neural network

2.4 Methodology

To forecast the power output of the PV power plant, the Feed Forward Neural Network has been designed and formed using the MATLAB software and the neural network toolbox implemented on MATLAB. A simplified diagram of this network is shown in Fig. 4, The main parameters for this model are mentioned as follows:

- There are two input neurons (the ambient temperature (T_{am}) and the solar irradiance (G_I)), one output neuron (the power output P_{AC}) and one hidden layer.
- The activation function is sigmoidal for the hidden layer and linear for the output layer.
- The FFNN algorithm is back-propagation based on the Levenberg-Marquardt (LM) minimization method to adjust the weights. (see Fig. 5).

In addition to the factors pointed above, a database of 304 days from 1/10/2016 to 31/10/2017 have been selected. The other days were discarded due to a variety of reasons such as system malfunctions, updates, etc.

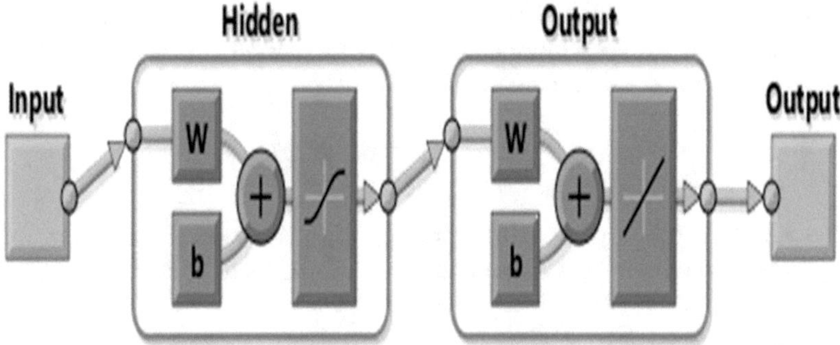

Fig. 4. Diagram of the used feed forward neural network in MATLAB

Overall, the selected data has been divided into two parts, 80% for the training and the rest for testing and validating the neural network.

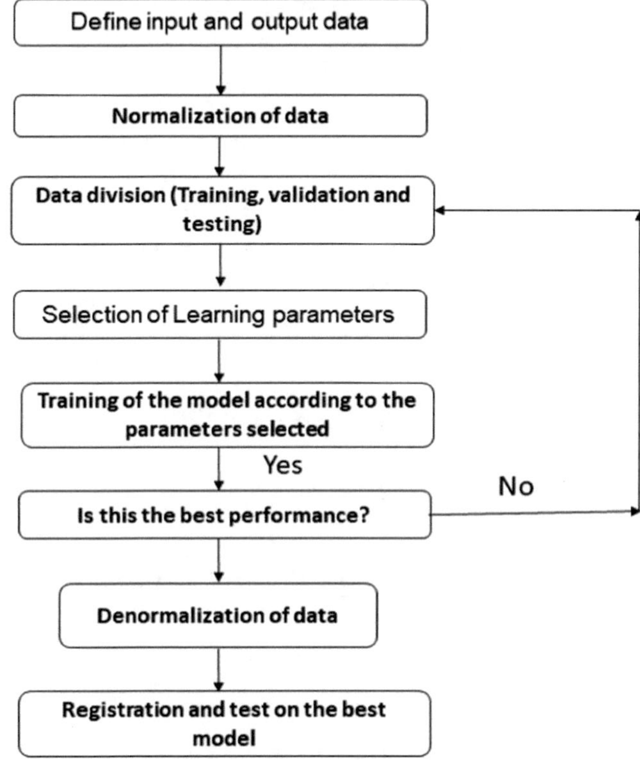

Fig. 5. The flowchart of output PV power forecasting

Before proceeding with the network formation process, a pre-treatment of the network training data was established, through a normalization of the input and output data between –1 and 1 using the equation n°2.

$$I_{Normalized} = \frac{2 * (I_{actual} - I_{\max} - I_{\min})}{I_{\max} - I_{\min}} \tag{2}$$

Where I_{actual} is the original data value; $I_{Normalized}$ is the normalized data; and I_{min} and I_{max} are the minimum and maximum values of the input data, respectively.

2.5 Model Evaluation

To predict the PV power produced, many methods exist to verify the effectiveness of a predictor. In this paper, the method used is called "cross-validation". Thus, we used statistical parameters for comparison. The tools used are described below:

- Coefficient of determination (R^2):

$$R^2 = 1 - \frac{\sum_{t=1}^{m} (P_{AC,m} - P_{AC,f})^2}{\sum_{1}^{m} (P_{AC,m} - \bar{P})^2} \tag{3}$$

- Root mean square error (RMSE):

$$RMSE = \sqrt{\frac{1}{m} \sum_{t=1}^{m} (P_{AC,m} - P_{AC,f})} \tag{4}$$

- Mean Absolute error (MAE):

$$MAE = \frac{1}{m} \sum_{t=1}^{m} \left| P_{AC,m} - P_{AC,f} \right| \tag{5}$$

- Normalized mean Absolute error (nMAE):

$$\text{nMAE} = \sum_{t=1}^{m} (P_{AC,m} - P_{AC,f}) \tag{6}$$

- Normalized Root mean square error (nRMSE):

$$nRMSE\% = \frac{\sqrt{\frac{1}{m} \sum_{t=1}^{m} (P_{AC,m} - P_{AC,f})^2}}{\max(P_{AC,m})} .100 \tag{7}$$

- Mean Bias Error (MBE):

$$MAE = \frac{1}{m}\sum_{t=1}^{m}(P_{AC,m} - P_{AC,f}) \tag{8}$$

Where $P_{AC,f}$ is the forecasting output PV power, $P_{AC,m}$ is the actual measured PV power. $\bar{P}$ is the rated output PV power. m represents the hours of the forecasting horizon.

3 Results and Discussion

This section treats the results of the prediction of PV power by artificial neural networks. For this purpose, a network of the FFNN type has been formed to generate a model for predicting the PV power produced by the photovoltaic system described above.

The construction of a prediction model by ANN involves the formation of its network using input and output data from a real system, and then testing it with different data. After several trials, the best configuration corresponds to the model with two nodes in the input layer (incident solar irradiation and ambient temperature) and 5 neurons in the hidden layer. In the learning process, the training was repeated several times, and for several numbers of hidden neurons.

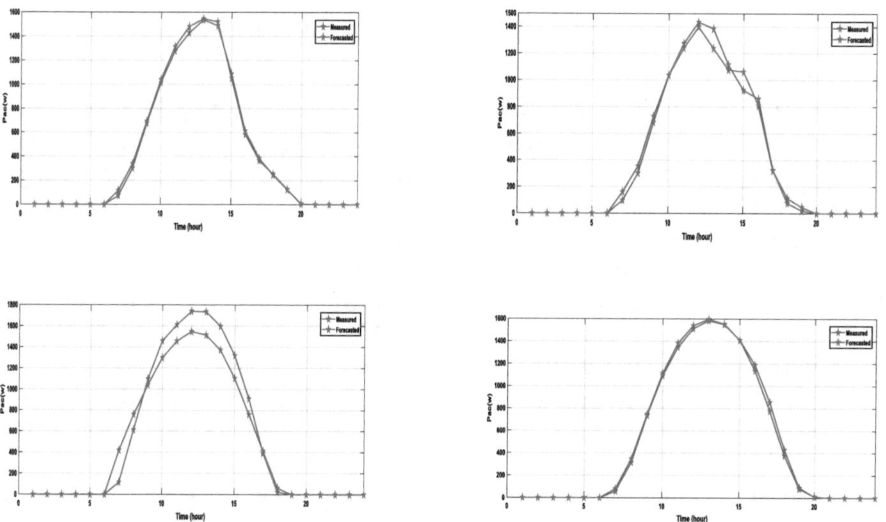

Fig. 6. Predicted and measured output PV power (PAC) as a function of time for sunny days

To evaluate the precision of prediction by FFNN, a comparison between the predicted values of the generated PV power produced for 24 h and the actual values for some typical days (sunny and cloudy) is shown.

On the Fig. 6, a comparison in the case of sunny days is represented. It appears on this figure that the measured PV power is totally in agreement with the predicted values, which can be justified by the regression Fig. 7 which show that most of the points fall along the diagonal, with coefficients of determination (R^2) between (0.98 and 0.998).

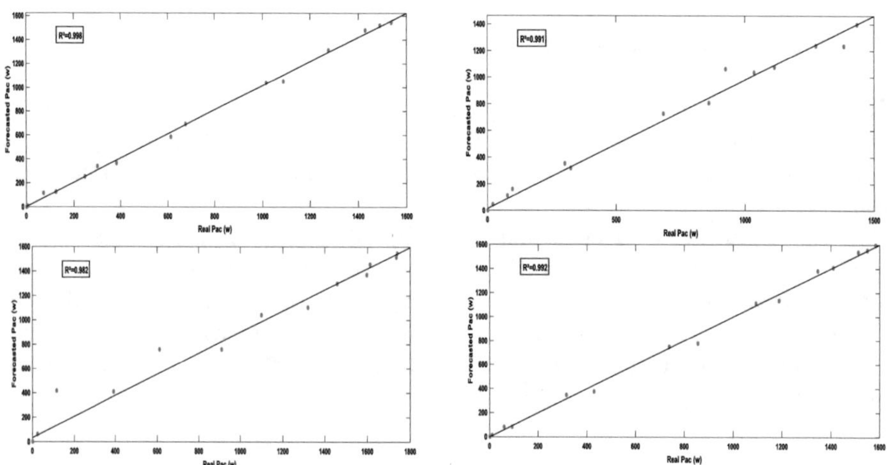

Fig. 7. Scatter plot 1 day ahead forecasting during sunny days

In the second case of cloudy days, it is clear that the results between the measured PV power and the predicted PV power values (Fig. 8) are acceptable. Concerning the dispersion Fig. 9, some fluctuations can be found with coefficients of (R^2) arranged between (0.82 and 0.96).

To discuss the validity of these results, a thorough evaluation of the two cases using statistical parameters (RMSE, nRMSE MAE, nMAE) is presented in Table 3. According to the results presented in Table 3, we note that the RMSE varies between 5.54 Wh and 22.18 Wh, the nRMSE is between 1.69% and 5.17%, the MAE between 1.13 Wh and 36.15 Wh and the nMAE is arranged between 1.5% and 7%, for sunny days.

For cloudy days, the RMSE varies between 55.19 Wh and 97.14 Wh, the nRMSE is between 1.75% and 96%, the MAE between 4.35 Wh and 27.35 Wh, the nMAE is arranged at 2% and 65%.

In the presented tow cases, we can conclude that the most important errors occur mainly during cloudy days and the accuracy is very interesting for sunny days.

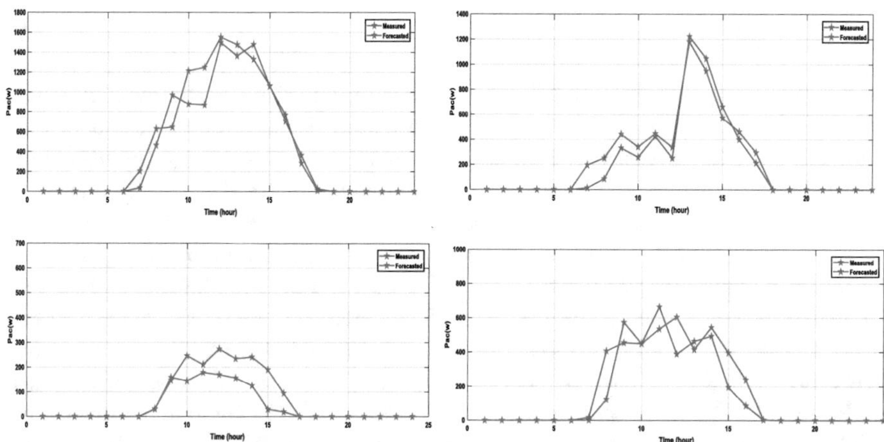

Fig. 8. Predicted and measured output PV power (PAC) as a function of time for cloudy days

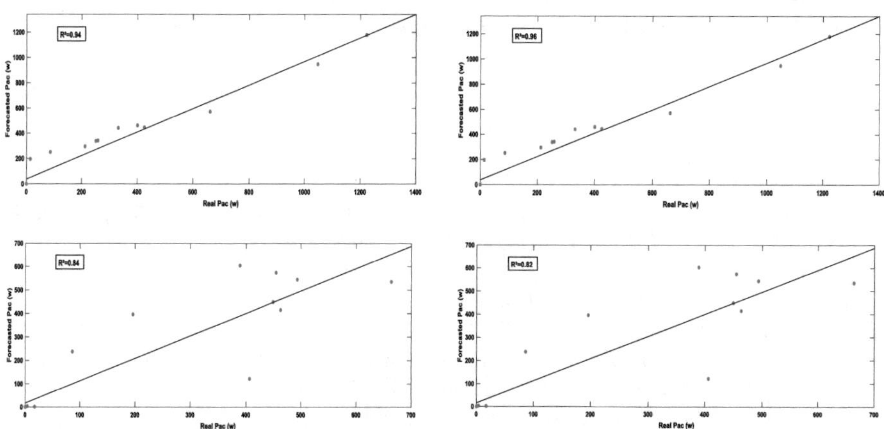

Fig. 9. Scatter plot 1 day ahead forecasting during cloudy days

Table 3. Error metrics in testing days

Days	Weather condition	RMSE	nRMSE	nMAE	MAE
day1	sunny	22.18	5.17	2%	7.59
day2	sunny	10.18	2.59	1.5%	2.1
day3	sunny	16.25	3.34	7%	36.15
day4	sunny	5.54	1.69	3%	1.13
day5	cloudy	70.25	1.75	2%	4.35
day6	cloudy	69.9	34.16	11%	23.35
day7	cloudy	55.19	30.96	65%	27.35
day8	cloudy	97.14	64.33	8%	10.84

4 Conclusion

In this paper, a prediction model of PV power produced by a photovoltaic system installed on the roof of the research building of the Faculty of Sciences and Technologies of Mohammedia, based on artificial neural networks of type (FFNN) is presented. The input parameters of the proposed neural model are the solar irradiance and the ambient temperature, while the output is represented by the 24 h ahead of the PV AC power output. After several trials, the best configuration corresponds to a network of a single layer with two neurons (G_I, T_{am}) and a hidden layer of five neurons. The model was evaluated using several statistic parameters such as (RMSE, nRMSE, MAE, R^2). Results during unknown days of the year showed that the accuracy over sunny days is higher than the one over cloudy days, with determination coefficients of (0.99 and 0.96) for sunny and cloudy days, respectively. We plan to apply this method in other studies using data from other places and with other PV technologies in order to develop a model that represents all of Morocco.

Acknowledgements. The authors would like to thank "Institute for Research in Solar Energy and New Energies (IRESEN)" for the financing of the project PROPRE.MA.

References

1. Taşçıoğlu, A., Taşkın, O., Vardar, A.: A power case study for monocrystalline and polycrystalline solar panels in bursa city, Turkey. Int. J. Photoenergy **2016**, Article ID 7324138, 7 pages (2016). https://doi.org/10.1155/2016/7324138
2. Atwater, M.A., Ball, J.T.: A numerical solar radiation model based on standard meteorological observations. SolarEnergy **21**, 163–170 (1978)
3. Loutfi, H., Bernatchou, A., Raoui, Y., Tadili, R.: Learning processes to predict the hourly global, direct, and diffuse solar irradiance from daily global radiation with artificial neural networks. Int. J. Photoenergy **2017**, Article ID 4025283, 13 pages (2017)
4. Sozen, A., Arcaklıoglu, E.: Solar potential in Turkey. Appl. Energy **80**(1), 35–45 (2005)
5. Ulbricht, R., Fischer, U., Lehner, W., Donker, H.: First steps towards a systematical optimized strategy for solar energy supply forecasting. In: European Conference on Machine Learning and Principles and Practice of Knowledge Discovery in Databases, ECMLPKDD (2013)
6. Mellit, A., Pavan, A.M.: A 24-h forecast of solar irradiance using artificial neural network: application for performance prediction of a grid connected PV plant at Trieste, Italy. Sol. Energy **84**(5), 807–821 (2010)
7. Monteiro, C., Fernandez-Jimenez, L.A., Ramirez-Rosado, I.J., Munoz-Jimenez, A., Lara-Santillan, P.M.: Short-term forecasting models for photovoltaic plants: analytical versus soft-computing techniques. Math. Probl. Eng. **9** (2013)
8. Cornaro, C., Pierro, M., Bucci, F.: Master optimization process based on neural networks ensemble for 24-h solar irradiance forecast. Sol. Energy **111**, 297–312 (2015)
9. Antonanzas, J., Osorio, N., Escobar, R., Urraca, R., Martinez-de-Pison, F., AntonanzasTorres, F.: Review of photovoltaic power forecasting. Sol. Energy **136**, 78–111 (2016)
10. Dolara, A., Lazaroiu, G.C., Leva, S., Manzolini, G.: Experimental investigation of partial shading scenarios on PV (photovoltaic) modules. Energy **55**, 466–475 (2013)

11. Kalogirou, S.A., Sencan, A.: Artificial intelligence techniques in solar energy applications. In: Manyala, R.I. (Ed.) Theory and Applications, p. 444, Book (2010)
12. Rehman, S., Mohandes, M.: Artificial neural network estimation of global solar radiation using air temperature and relative humidity. Energy Policy **36**, 571–576 (2008)
13. Mellit, A., Kalogirou, S.A.: Artificial intelligence techniques for photovoltaic application: a review. Prog. Energy Combust. Sci. **34**, 574–632 (2008)
14. Simonov, M., Mussetta, M., Grimaccia, F., Leva, S., Zich, R.: Artificial intelligence forecast of PV plant production for integration in smart energy systems Int. Rev. Electr. Eng. **7**(1), 3454–3460 (2012)
15. Dolara, A., Grimaccia, F., Leva, S., Mussetta, M., Ogliari, E.: A physical hybrid artificial neural network for short term forecasting of PV plant power output Energies **8**(2), 1138–1153 (2015)
16. Leva, S., Dolara, A., Grimaccia, F., Mussetta, M., Ogliari, E.: Analysis and validation of 24 hours ahead neural network forecasting of photovoltaic output power. Math. Comput. Simul. **131**, 88–100 (2017)
17. Mellit, A., Pavan, A.M.: Energy conversion and management **51**, 2431–2441 (2010)
18. The MathWorks, "MATLAB" (2017). https://www.mathworks.com/products/neural-network.html
19. Sfetsos, A., Coonick, A.H.: Univariate and multivariate forecasting of hourly solar radiation with artificial intelligence techniques. Sol. Energy **68**(2), 169–178 (2000)
20. http://dx.doi.org/10.1016/j.rser.2017.08.017

Adaline and Instantaneous Power Algorithms for Offshore and Onshore Wind Farms Based VSC-HVDC for Oil Gas Station Application

Seghir Benhalima[1(✉)], Ambrish Chandra[1], Rezkallah Miloud[1], and Abdderaouf Raddaoui[2]

[1] École de Technologie Supérieure, 1100 Notre-Dame, Montreal, QC H3C1K3, Canada
Seghir.benhalima.1@ens.etsmtl.ca

[2] University Mohammed V in Rabat, MEAT, École de Technologie Salé, B.P277, Salé médina, Morocco

Abstract. In this paper High Voltage Direct Current system based on Voltage Source Converter (VSC-HVDC) connecting large–scale Offshore Wind Farm (LSOWF) for gas station, is presented. The three level Neutral Point diode Clamped (NPC) converter is used in VSC-HVDC system to improve the power reliability and transmission capacity. Exploitation of a platform Oil and gas industry has been used in the last few years, two options to run its machines either using a gas turbine or to convert the energy from coast by submarine cables. This issue is solved using a cleaner and less expensive renewable energy, such as offshore wind turbine. However, efficiency and safe operation of the gas platform in case of defect or system disturbance should be solved. To achieve these objectives, Photovoltaic system is employed to ensure stable operation of sensible elements of the platform during fault. Furthermore, crowbar protection topology, instantaneous power and Adaline algorithms are also employed to solve the problem caused by positive DC fault and to control the offshore and onshore VSCs. MATLAB/SIMULINK is employed to demonstrate and validate of the proposed concept. The proposed scheme may enable optimal integration of highly distributed of LSOWF and demonstrate the good performance of VSC-HVDC.

Keywords: Wind turbine · VSC-HVDC · Three level converter (NPC) · Photovoltaic · Adaline control · Instantaneous power · Crowbar protection

1 Introduction

Recently, wind energy has attracted interest due to the increased on sustainable energy resources and advanced research in the field of wind turbines. As the distance between OWF and earthly is long, the system HVDC transmission based VSC has more advantages in comparison to the conventional HVDC system [1]. The converter (VSC-NPC) is recognized as the most used new generation of the VSC-HVDC, its can support high voltage and carry more power than two-level converter [2]. The NPC-based three-level VSC-HVDC system has been classified as the best in conversion and power generation in the last few years and has a good prospect using the conversion energy.

M. Ezziyyani (Ed.): AI2SD 2018, AISC 912, pp. 258–271, 2019.
https://doi.org/10.1007/978-3-030-12065-8_24

The fast progress application of wind energy especially offshore is a new key clean renewable energy [3]. In the last year, several companies and countries in the world are interested in offshore renewable industry, and especially to supply the oil and gas development platform [4, 5], among companies can be found who enjoys a solid reputation of power converter.

For a very long time, the offshore gas plant uses power generators and large compressors that are driven by gas turbines or diesel engines. This technology has large amounts of pollution and low efficiency [21]. The CO_2 gas emissions are penalized by Kyoto, which supports the control and exchange of greenhouse gas emissions. The penalization of gas emissions, the maintenance of equipment and the cost of gas and oil have led to an increase in prices for the offshore production platform.

The first transmission link was commissioned in 1954. It was an undersea transmission in the Baltic Sea. The choice of HVDC over HVAC is generally motivated by the benefits of HVDC links [6–20]:

- Low transmission losses over long distances.
- Access submarine cables over long distances. A special case is the connection of offshore wind farms.
- Connection of asynchronous grid.

The magnetic fields of the HVDC cable are negligible related to HVAC lines.

However, for high power HVAC transmission, three cables are required with high losses and more maintenance. In addition, HVDC cables have smaller diameters, lower losses and high efficiency [7, 20].

This research investigated the transmission VSC-HVDC of offshore wind farm energy based PMSG to the terrestrial grid using power converters topologies, thus adding an Oil and gas platform with a study of different power options knowing that each platform needs an emergency power supply which brings us back to using offshore photovoltaic energy as a secondary source to use in case of failure of the link with the main power supply or a fault on the system. The control of the system is studied in case of DC fault applied to the DC transmission with analysis of their effects on the different systems and on the oil and gas platform.

However, the crowbar protection for offshore wind farm is used when the system is affected with severe DC fault voltage. Thus, for DC fault proposed in [8], the system is affected when the crowbar has not been activated.

The rest of the main paragraphs are reproduced as follows. In Sect. 2, different elements of wind farm and platform gas are introduced. The Sect. 3 includes the description of the system. In Sect. 4, modeling of the system and control are recommended. In Sect. 5, case study results are presented and illustrated. At last conclusions are drawn in Sect. 6.

2 Wind Offshore and Platform Oil and Gas

After the global energy crisis in 1970 during the huge increase in prices of such fuels as oil and natural gas [9], the appearance of the law of protection of the environment, so the high risk nuclear energy forces the world to seek other energy based on the wind and the sun [10], like the tidal and wave energy are also called supplies infinite or green energy. Renewable energy flows are present in the environment of different forms and geographical locations. The energy supplied by the offshore wind farm is transported by the VSC-HVDC system to the onshore station.

2.1 Offshore Wind Farm

The energy conversion system is to increases the power capacity of wind farm and to reduce the purchase price (kW/h). Numerous research efforts have been done for broad-based systems, targeting 5–10 MW level for broad-based systems and offshore applications. Depending on the capacity of the grid and the size of the wind farms, the use of HVDC is applicable provided that the distance to the connection AC grid is greater than 50 km [11]. Another important advantage is that the AC breakdowns will not be propagated back and forth by both onshore and offshore stations.

2.2 Sun Converters

The solar irradiation resource provides an energy density of approximately 1000 W/m^2 [12]. These technologies should be considered intermittent. Solar photovoltaic system is widely used for small and large generation, for our case; we used this technology to generate electricity to the offshore platform in case of failure.

However, the technical and economic feasibility depends on the available solar resource, to determine the daily average of the global irradiance and the photovoltaic installation should not be detrimental to the operational safety and the legal requirements of the maritime codes [13]. The advantage of the solar photovoltaic is the low maintenance and easy to manipulate and control. Some disadvantage in offshore photovoltaic production is marine corrosion also with a large installation area and intermittent power generation, but with an energy storage system we will cover all the basic operation in the offshore platform station. Solar energy shall be used in the event of an electrical failure and the safety of the station.

2.3 HVDC Cable

The world aims to develop and integrate renewable energy into the electricity grid and feed the gas platform. The use of submarine cables and more cost effective and economical in [14], and another type of flat cable filled with oil has been installed from 172 km which transform the power between Sweden, Denmark and Germany [15]. It is considered the first cable in the world to provide a transmission capacity of 600 MW continuously; it maintains an efficient electrical system with a high level of reliability. The connection with this cable reduces the greenhouse gases while respecting the environment submarine transmission for the continental interconnection, in the offshore park and for supplying the platform gas and oil offshore. Underwater cables mass impregnated HVDC is the most economical, can withstand very high mechanical stresses.

Then, the increasing use of VSC technology encourages using the synthetic insulated cables in long distance underwater.

In addition, practical experience with 225 kV Cu XLPE cables has been successfully achieved [16].

For applications with voltages below 320 kV, the XLPE underwater HVDC Cable is complex and consists of several concentric layers and is the most preferred. In VSC-HVDC schemes, insulation is typically XLPE because it is less expensive and more robust than mass-impregnated cables [22].

3 System Description

Offshore wind farm and gas platform is shown in Fig. 1. The VSC-HVDC station is connected to the offshore station which provided 200 MW. The crowbar is installed between offshore farm and offshore VSC-HVDC station. VSCs are three-phase, twelve-pulse, and three level bridges using IGBT semiconductors. The electrical system of a production platform is connected to the offshore VSC-HVDC. Electrical charges can be classified as normal, auxiliary, essential and emergency. The emergency loads are electrical equipment that supports critical systems. These systems are necessary to protect life on board, for the operational safety of wells and critical equipment. These loads are powered by photovoltaic system (this power supply is replaced the diesel group). The proposed electrical system is shown in Fig. 1.

Those main generators are connected in the main bus at 13.8 kV and the feeders distribute power at 4.16 kV and 400 V, as industrial load requirements. The normal loads are allocated in load distribution centers (LDC) and motor control center (MCC), at voltages of 220 V, 400 V, 4.16 kV and 13.8 kV. These loads are mostly motors below 75 kW, which are used to pump fluids and fans large normal loads at 13.8 kV are compressors (11 MW) and water injection pumps (5.9 MW). The cargo pumps of 590 kW are supplied with 4.16 kV. One of the largest loads fed at 400 V is the 630 kW heater of sulfate removal unit. The auxiliary loads are allocated in LDCs and CCMs, at voltage of 220 V and 400 V. The essential loads are allocated in LDCs and CCMs, at voltage of 220 V and 400 V.

Outstanding are compressed air motors at 400 V, for air conditioning systems and essential lights. Some examples of essential loads are:

- Air compressors, instrument and service.
- Engine room ballast pump.
- Emergency lifting pump.
- Essential illumination system.

The emergency loads are allocated in LDCs at voltages of 24, 125 and 220 VDC, and 220 VAC. Outstanding at the emergency stop system and emergency lighting system.

Some of the loads of emergency are:

- Control panel of emergency and auxiliary generation.
- Emergency stop system.
- Emergency illumination.
- Control panel and start panel of the fire pumps.

- Navigation system.
- Control system and process instrumentation (ECOS/ESC).
- Gas/Fire detection system.

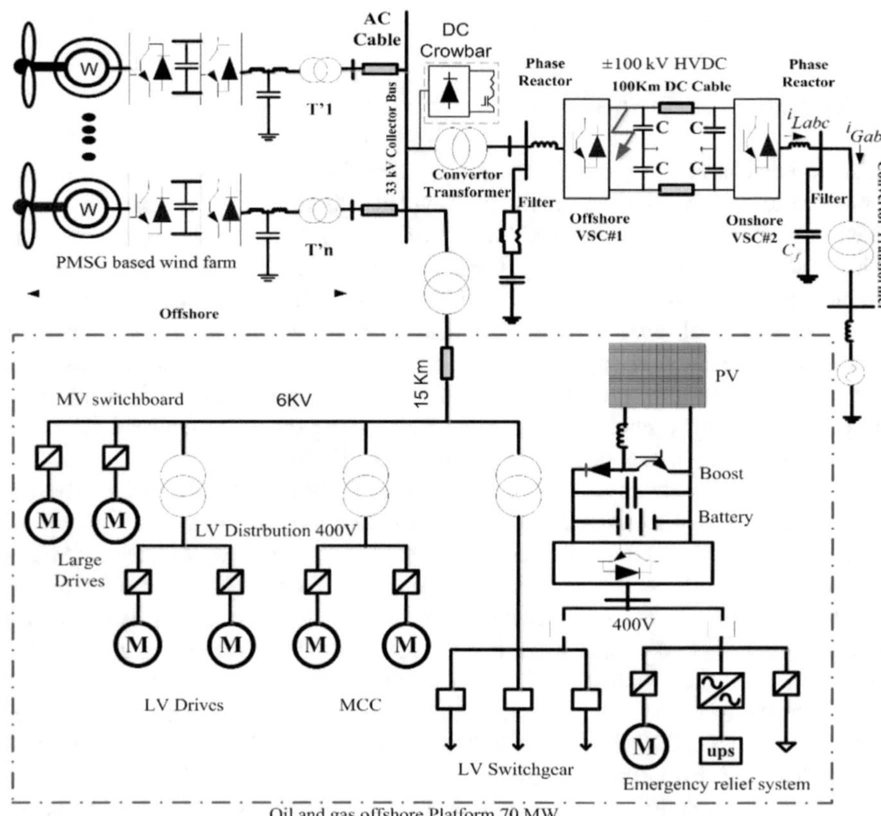

Fig. 1. System under study.

4 System Modelling and Control

Generally, speaking the VSC-HVDC systems consists of following parts: a high voltage DC transmission cables; a rectifier station; and inverter station. Figure 1 shows the principal station VSC-HVDC wind farm and the gas platform.

4.1 Modeling of Offshore Station

The independent regulations of active and reactive powers of each station help to regulate the offshore station by instantaneous power.

(a) PMSG side converter control: The principal equations of PMSG in d-q are given below:

$$\begin{cases} V_d = -R_s i_{Td} - L_{sd}\frac{di_{Td}}{dt} + L_{sq}\omega_r i_{Tq} \\ V_q = -R_s i_{Tq} - L_{sq}\frac{di_{Tq}}{dt} - L_{sd}\omega_r i_{Td} + \omega_r \phi_m \end{cases} \tag{1}$$

where V_{dq} are stator terminal voltages, R_s and L_s are stator resistance and inductance respectively and i_{Tdq} are output currents.

The expression of the maximum power is given by:

$$P_{\omega,max} = 0.5C_{p,max}\rho A\left(\frac{R}{\lambda_{opt}}\omega_{opt}\right)^3 \tag{2}$$

The value of the optimal speed (ω_{opt}) is dependent on the λ_{opt} and $P_{\omega,max}$ as,

$$\omega_{opt} = \frac{\lambda_{opt}}{R}\left(\frac{P_{\omega,max}}{0.5C_{p,max}\rho A}\right)^{1/3} \tag{3}$$

The reference current the q axis component is,

$$i^*_{Tq} = 2\left(\frac{2}{3}\right)\left(\frac{T^*_e}{p\phi_m}\right) \tag{4}$$

(b) Control strategy for offshore VSC#1 station: The developed control for VSC-HVDC offshore station shown in Fig. 2 is based on the instantaneous power calculation [17]. However, the calculation of different components, are presented follows.

$$\begin{cases} P = \sum u_{sj} i_{sj} \\ Q = \frac{[(u_{sb}-u_{sc})i_{sa} + (u_{sc}-u_{sa})i_{sb} + (u_{sa}-u_{sb})i_{sc}]}{\sqrt{3}} \end{cases} \tag{5}$$

where P, Q, represent the measured active and reactive power controllers from offshore farm.

The reference active and reactive current components i^*_d and i^*_q, are calculated as follows,

$$\begin{cases} i^*_d = (k_{p1} + k_{i1}/s)(P_s - P) \\ i^*_q = (k_{p3} + k_{i3}/s)(Q_{ref} - Q) \end{cases} \tag{6}$$

where k_{p1}, k_{i1}, k_{p3} and k_{i3}, P_s and Q_{ref}, represent the proportional and integral gains of the active and reactive power controllers, total real generated power from offshore farm, and reference reactive power offshore farm (Fig. 3).

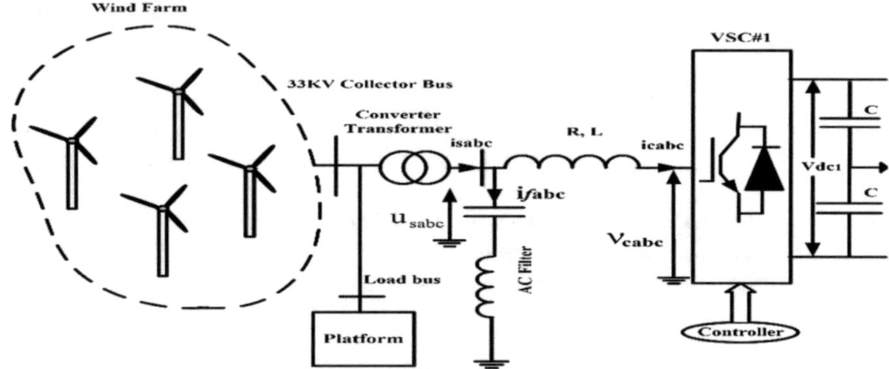

Fig. 2. Wind farm and platform oil and gas offshore station.

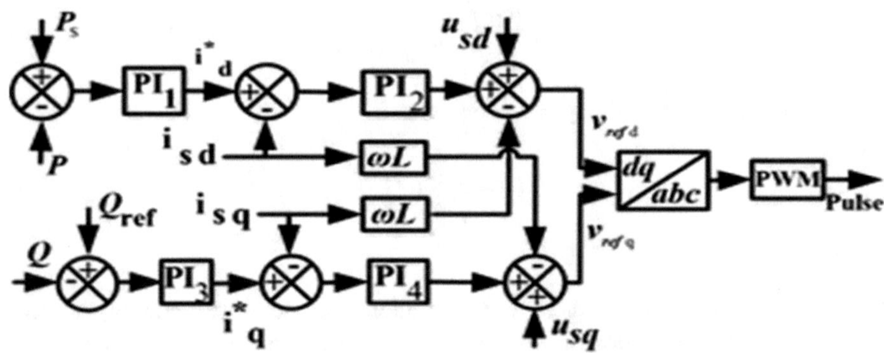

Fig. 3. Control of VSC-HVDC offshore station.

4.2 Modeling of Onshore Station

(a) Adaline control method for onshore VSC#2 station: The proposed Adaline control strategy for onshore VSC_2 is presented in Fig. 4, to regulate and correct the Vdc to maintain stable system operation under presence of sever DC short-circuit, to inject the total generated offshore power with high power quality and less energy losses [18].

   (1) DC-link voltage regulation: The DC-link voltage is regulated using PI controller. The measured DC-link voltage is compared to its reference, the output of PI controller gives the reference current (i^{*}_{dc}). It's expressed as.

$$i^{*}_{dc} = (k_{p1} + \frac{k_{i1}}{s})(v^{*}_{dc} - v_{dc}) \tag{7}$$

where k_{i1} and k_{p1} are respectively, the integral and proportional gains of the DC voltage controller.

The DC current (i_{dc}) of VSC-HVDC onshore is calculated by dividing the DC power (P_{dc}) by DC-link voltage (V_{dc}) is expressed as,

$$i_{dc} = \frac{P_{dc}}{v_{dc}} \tag{8}$$

(2) AC voltage regulation: As shown in Fig. 4, the AC grid voltage is regulated using PI controller. The sensed amplitude of the AC grid voltage (v_G) is compared to its reference (v_G^*), which is calculated as follows,

$$w_{cq} = (k_{p2} + \frac{k_{i2}}{s})(v_G^* - v_G) \tag{9}$$

where k_{i2} and k_{p2} are respectively, the integral and proportional gains of the AC grid voltage controller.

(3) Estimation of the active and reactive components of the current: The weight vectors for the active and reactive components of current are estimated using the following calculation as,

$$\begin{cases} w_{pa}(k) = \sum \eta \begin{bmatrix} \{ (i_{La}(k) - u_{\text{pqa}}(k) w_{pqa}(k-1)) \} * \\ \{ (u_{\text{pqa}}(k) + K i_{La}(k) w_{pqa}(k-1)) \} \end{bmatrix} \\ w_{pb}(k) = \sum \eta \begin{bmatrix} \{ (i_{Lb}(k) - u_{\text{pq}b}(k) w_{pqb}(k-1)) \} * \\ \{ (u_{\text{pq}b}(k) + K i_{Lb}(k) w_{pqb}(k-1)) \} \end{bmatrix} \\ w_{pc}(k) = \sum \eta \begin{bmatrix} \{ (i_{Lc}(k) - u_{\text{pq}c}(k) w_{pqc}(k-1)) \} * \\ \{ (u_{\text{pq}c}(k) + K i_{Lc}(k) w_{pqc}(k-1)) \} \end{bmatrix} \end{cases} \tag{10}$$

where $w_{pqa}(k)$, $w_{pqb}(k)$ and $w_{pqc}(k)$, $i_{La}(k)$, $i_{Lb}(k)$ and $i_{Lc}(k)$ denote the active and reactive components of the current in phases a, b, and c, and grid current in phases a, b, and c, respectively.

(4) In phase and quadrature unit templates: The division of the grid voltage by the amplitude of phase voltages is gives as,

$$u_{abcp} = \frac{v_{Gabc}}{v_p} \tag{11}$$

And

$$v_p = \sqrt{\frac{2}{3}(v_{Ga}^2 + v_{Gb}^2 + v_{Gc}^2)} \tag{12}$$

Also

$$\begin{cases} u_{aq} = (u_{cp} - u_{bp})/\sqrt{3} \\ u_{bq} = (3u_{ap} + u_{bp} - u_{cp})/2\sqrt{3} \\ u_{cq} = (-3u_{ap} + u_{bp} - u_{cp})/2\sqrt{3} \end{cases} \quad (13)$$

where v_{Gabc}, v_p, u_{abcp}, u_{aq}, u_{bq} and u_{cq} represent the instantaneous grid voltages, amplitude of phase voltages, phase unit templates and the quadrature unit templates in phases a, b, and c, respectively.

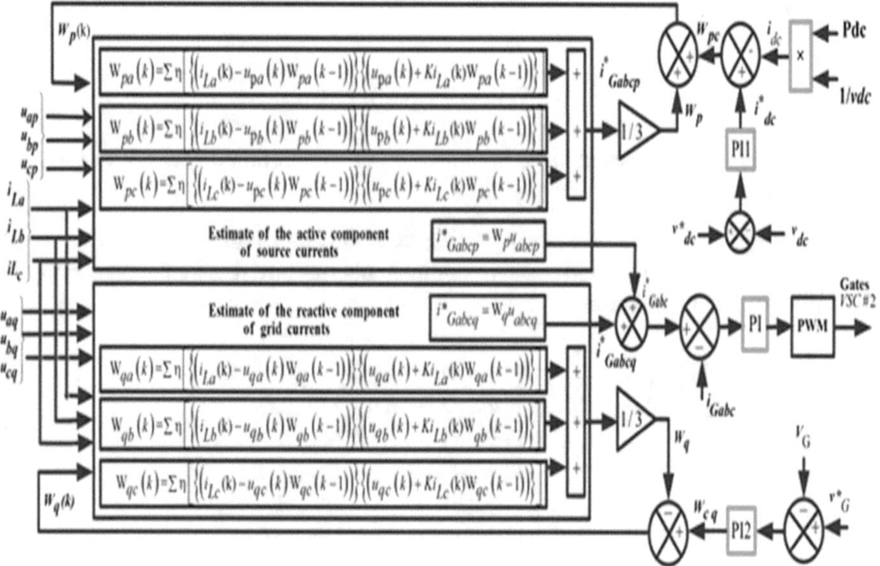

Fig. 4. Control of VSC-HVDC onshore station.

5 Simulation and Experimental Results

The simulation model was developed in Mathlab/Simulink. The simulated case uses 100 of wind based PMSG (200 MW), which is connected to three level NPC VSC-HVDC transmission system via 100 km DC submarine cable and transformer (35 kV/110 kV). The direct current transmission voltage between −100 kV and +100 kV. The IGBT of the VSC#1 and VSC#2 are modeled as ideal switches with antiparallel diodes [19]. They are controlled by PWM at switching frequencies of 1350 kHz.

Figures 5 and 7 shows the waveforms of the AC voltage and current of the wind farm. It can be seen clearly the voltage and current are constant and sinusoidal. But to test the performance of the proposed topology and control approach of all the system, the positive DC fault is applied at 2.5 s. It can be seen in Figs. 6 and 7 minor disturbances in the AC voltage of the wind offshore farm and DC voltage in offshore and onshore. It is observed in Fig. 6 the active and reactive power are influenced by DC fault.

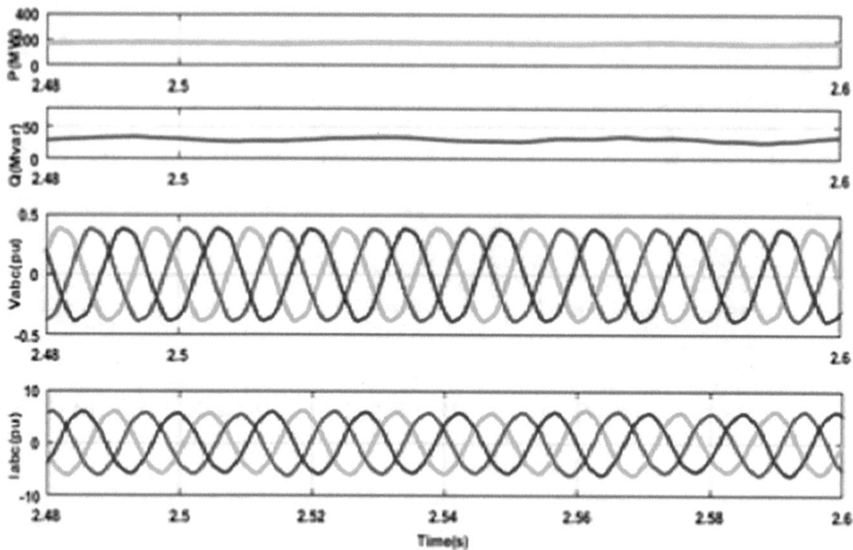

Fig. 5. Waveforms at wind farm offshore station without DC fault: Source voltage, source current, power active and reactive.

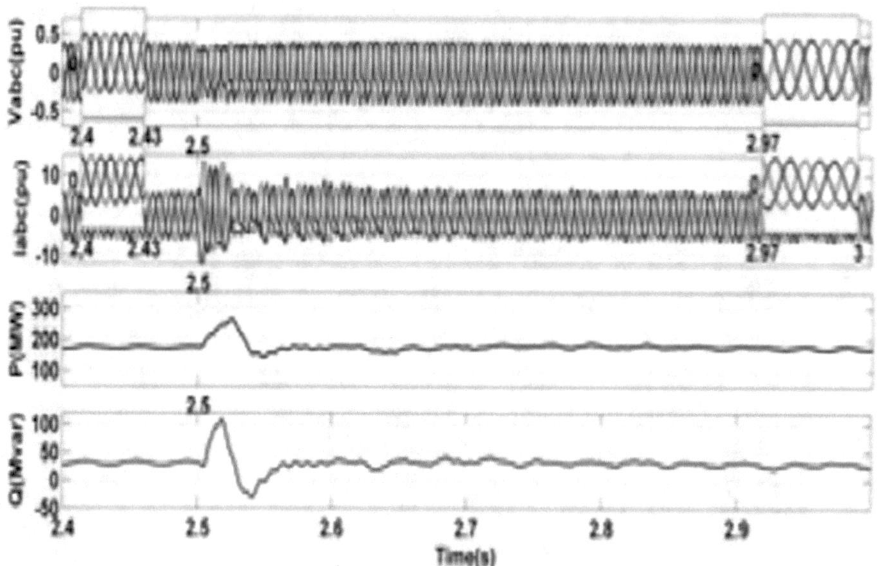

Fig. 6. Waveforms at wind farm offshore station before and after DC fault at 2.5 s. Source voltage, source current, power active and reactive.

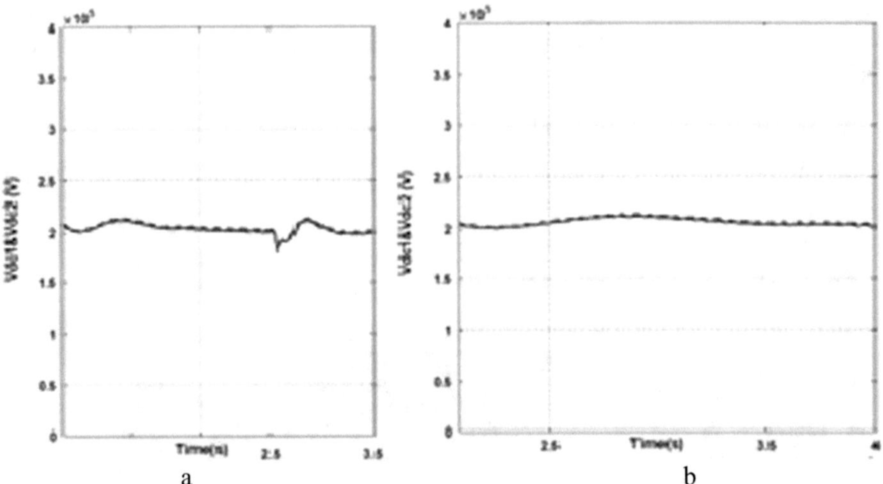

Fig. 7. Vdc response in offshore and onshore stations: (a) with DC fault, (b) without DC fault.

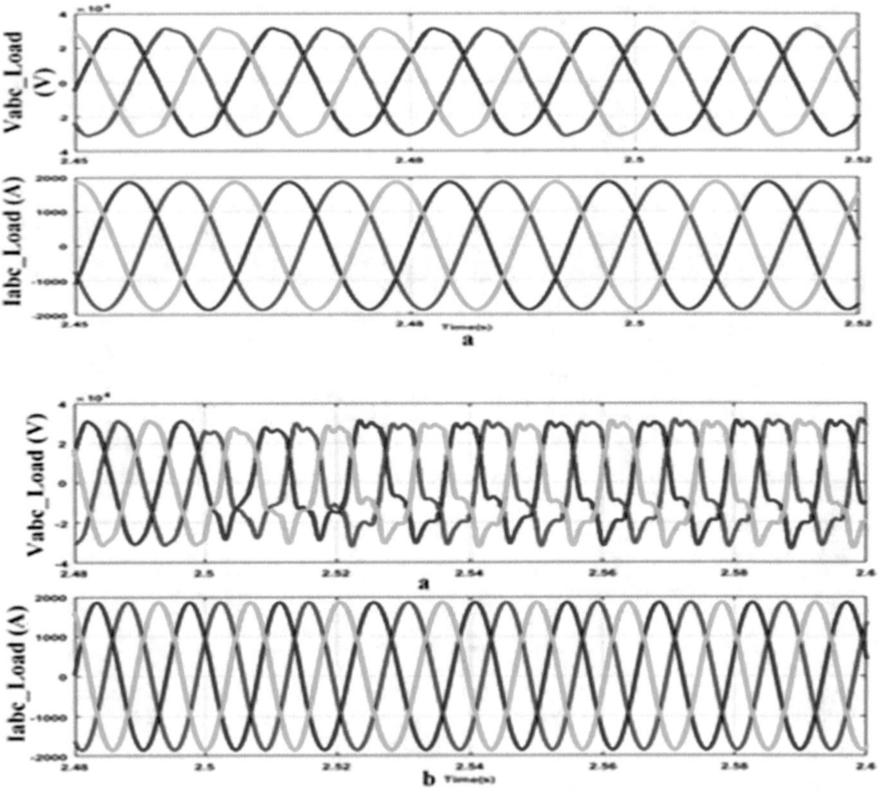

Fig. 8. Vabc and Iabc at Gas Platform offshore station: (a) without DC fault, (b) with DC fault.

Figure 8 shows the waveforms at gas platform for two scenarios as follows, (a) without positive DC fault. And (b) with positive DC fault. It is observed in (a) the load voltage and current are constant and sinusoidal. At 2.5 s one can see clearly the voltage is weakly affected but the current kept constant and sinusoidal. From the simulation one can see clearly the Adeline algorithm based PI controller confirms the high performance and fast response about DC voltage.

The performance of the system is approved in the laboratory. The real time implementation using dSPACE DS1104, the VCS inverter and the load presented in Fig. 11. Figure 9 shows the experimental results at study state, it can see clearly the DC voltage kept constant and the currents having good waveforms.

Figure 10 shows the experimental results of the DC voltage, grid current, load current and inverter current in the variation between absence and presence of the load. It is observed that the DC voltage kept constant which confirm the efficiency of the proposed approach.

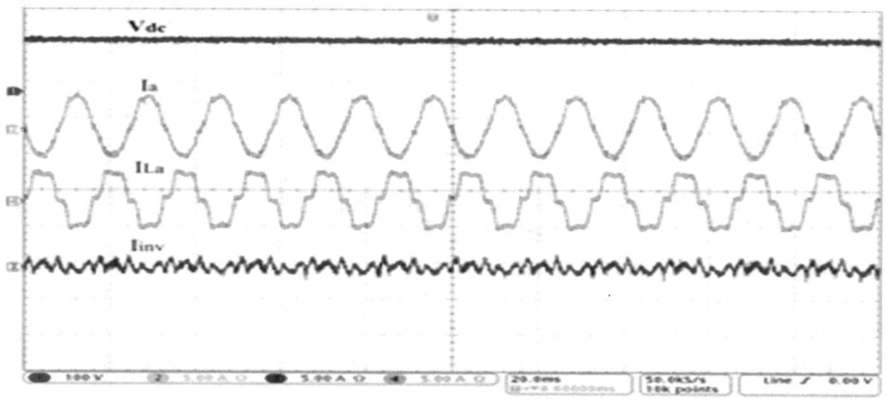

Fig. 9. Waveforms of study state: DC voltage, grid current, load current and inverter current.

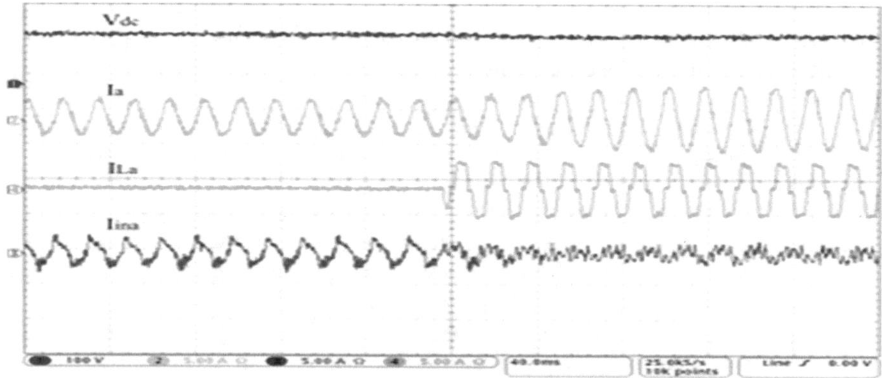

Fig. 10. Experimental results with and without load of DC voltage, grid current, load current and inverter current.

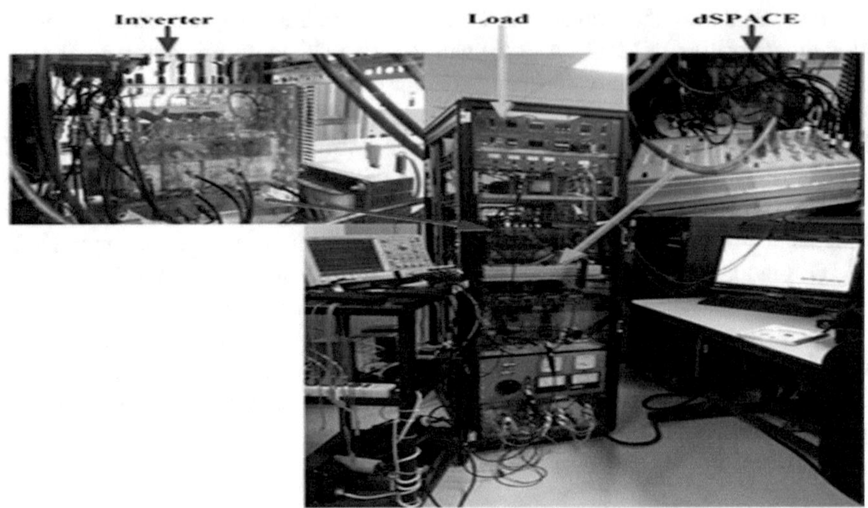

Fig. 11. Experimental setup.

6 Conclusion

In this paper, detailed analysis of different control algorithms to operate effectively and with high safety the oil and gas stations using onshore and offshore wind farm have been presented. It has been demonstrated that the proposed Adaline and instantaneous algorithms for VSC-HVDCs in onshore and offshore sides, as well as the crowbar protection in the offshore station have been improve the reliability of the full system and solve the problem caused by positive DC fault. The obtained results show satisfactory performance.

References

1. Jonathan, R., Ronan, M., Terence, O.: Low frequency AC transmission as an alternative to VSC-HVDC for grid interconnection of offshore wind. In: 2015 IEEE Eindhoven PowerTech, pp. 1–6. IEEE, July 2015
2. Qidi, T., Xinglai, G., Yong-Chao, L.: Improved switching-table-based DTC strategy for the post-fault three-level NPC inverter-fed induction motor drives. IET Electr. Power Appl. **12** (1), 71–80 (2017)
3. Camurça, L.J., Lago, J., Heldwein, M.L.: High efficiency wind energy conversion system based on the three-level delta-switch t-type converter and PMSG model-based loss minimization. In: 2015 IEEE 13th Brazilian Power Electronics Conference and 1st Southern Power Electronics Conference (COBEP/SPEC), pp. 1–6 (2015)
4. Ramya, S., Napolean, A., Manoharan, T.: A novel converter topology for stand-alone hybrid PV/Wind/battery power system using Matlab/Simulink. In: 2013 International Conference on Power, Energy and Control (ICPEC), pp. 17–22 (2013)
5. Jacobs, I.S., Bean, C.P.: Fine particles, thin films and exchange anisotropy. In: Rado, G.T., Suhl, H. (eds.) Magnetism, vol. III, pp. 271–350. Academic, New York (1963)

6. Andrés, H.D., Antonio, H.E., Gallego, R.A.: An MILP model for the static transmission expansion planning problem including HVAC/HVDC links, security constraints and power losses with a reduced search space. Electr. Power Syst. Res. **143**, 611–623 (2017)
7. Joseph, S., Maja Harfman, T., Ranjan, G.: A modular stacked DC transmission and distribution system for long distance subsea applications. IEEE Trans. Ind. Appl. **50**(5), 3512–3524 (2014)
8. Vidal, J., Abad, G., Arza, J., Aurtenechea, S.: Single-phase DC crowbar topologies for low voltage ride through fulfillment of high-power doubly fed induction generator-based wind turbines. IEEE Trans. Energy Convers. **28**(3), 768–781 (2013)
9. Gustafson, M.: "Diversified Utilization" can play a strong hand in dealing with the energy crisis. IEEE Trans. Power Appar. Syst. **PAS-100**, 4661–4664 (1981)
10. Mahyar, K., Saeed, H., Davood, K.: A novel hybrid model-based MPPT algorithm based on artificial neural networks for photovoltaic applications. In: 2017 IEEE Southern Power Electronics Conference (SPEC), pp. 1–6. IEEE (2017)
11. Sandeep, K., Prasanth, T., Rajesh, S.: Subsea power transmission cable modelling: reactive power compensation and transient response studies. In: 2016 IEEE 17th Workshop on Control and Modeling for Power Electronics (COMPEL), pp. 1–6. IEEE, September 2016
12. Sergei, K., Moshe, S., Simon, L.: Solar irradiation independent expression for photovoltaic generator maximum power line. IEEE J. Photovolt. **7**(5), 1416–1420 (2017)
13. Ruther, R., Beyer, H.G., Montenegro, A.: Performance results of the first grid-connected, thin-film PV installation in Brazil: temperature behaviour and performance ratios over six years of continuous operation. In: 19th European Photovoltaic Solar Energy Conference, Paris, France, pp. 3091–3094 (2004)
14. Paola, B., Wil, K., Hendriks, L.: HVDC connection of offshore wind farms to the transmission system. IEEE Trans. Energy Convers. **22**(1), 37–43 (2007)
15. Carsten, W., Elberling, T.: The kontek HVDC link between Denmark and Germany. In: An Implementation of an Adaptive Control Algorithm for a Three-Phase Shunt Active Filter: Power Engineering Society Winter Meeting, pp. 572–574. IEEE (2000)
16. Kiyotaka, U., Osami, T., Shigeo, N.: R&D of a 500 m superconducting cable in Japan. IEEE Trans. Appl. Supercond. **13**(2), 1946–1951 (2003)
17. Hirofumi, A., Edson Hirokazu, W., Mauricio, A.: The Instantaneous Power Theory. Wiley, Hoboken (2007)
18. Bhim, S., Jitendra, S.: An implementation of an adaptive control algorithm for a three-phase shunt active filter. IEEE Trans. Industr. Electron. **56**(8), 2811–2820 (2009)
19. Nguyen, M.C., Rudion, K., Styczynski, Z.A.: Improvement of stability assessment of VSCHVDC transmission systems. In: 5th International Conference on Critical Infrastructure (CRIS) (2010)
20. Saksvik, O.: HVDC technology and smart grid. In: 9th IET International Conference on Advances in Power System Control Operation and Management (APSCOM 2012) (2012)
21. Rodrigues, S., Teixeira Pinto, R., Bauer, P., Wiggelinkhuizen, E., Pierik, J.: Optimal power flow of VSC-based multi-terminal DC networks using genetic algorithm optimization. In: IEEE Energy Conversion Congress and Exposition (ECCE) (2012)
22. Beddard, A., Barnes, M.: HVDC cable modelling for VSC-HVDC applications. In: IEEE PES General Meeting, Conference & Exposition (2014)

Performance Analysis of 4.08 KWp Grid Connected PV System Based on Simulation and Experimental Measurements in Casablanca, Morocco

Amine Haibaoui[1,3](✉), Bouchaib Hartiti[1,2], Abderrazzak Elamim[1], and Abderraouf Ridah[3]

[1] ERDyS Laboratory, MEEM and DD Group, Hassan II University of Casablanca, FSTM, BP 146, 20650 Mohammedia, Morocco
Haibaoui.amine93@gmail.com
[2] ICTP, UNESCO, Trieste, Italy
[3] LIMAT Laboratory, Department of Physics FSB, Hassan II Casablanca University, B.P 7955, Casablanca, Morocco

Abstract. The energy generated from photovoltaic (PV) panels depends usually on the PV cell technology used and meteorological data at a given location. This work presents a comparison study of 2 × 2.04 KWp grid-connected PV module technology systems, constituted by two types of photovoltaic solar panels (Monocrystalline and Polycrystalline)-silicon, installed on the roof of faculty of sciences Ben M'sik Casablanca. Three types of results are presented. The first type is the performance evaluation for one year of exposure under natural outdoor conditions including: System efficiency, reference and final yield as well as the performance ratio. The second type is based on simulation data given by PVsyst 6.4.3, compared to experimental data obtained through the inverters of the installation and meteorological station. The third type is an economic analysis including the most commonly used financial parameters, which are the annual incomes (Ai), the cost of electricity of operating period (LCOE) and the payback period (PB) in order to determine the optimal technology for the city. The investigation of the annual productivity shows that Monocrystalline and Polycrystalline deliver an energy of 3325,711 Kwh/year and 3250,842 Kwh/year respectively. The experimental results show that the monocrystalline-silicon is the best technology for Casablanca city.

Keywords: Performance analysis · Simulation · Economic analysis · PV cell · Grid-connected · Monocrystalline · Polycrystalline

1 Introduction

Renewable energies play a significant role in the sustainable development of countries. Morocco, non-oil producing country, benefits from a great solar energy potential with an average radiation value of 2600 KWh/m^2/year and a significant number of hours of sunshine, about 300 days per year [1]. To meet growing demand (about 7% per year) [2], Morocco has developed a new national energy strategy, to secure energy supply

M. Ezziyyani (Ed.): AI2SD 2018, AISC 912, pp. 272–287, 2019.
https://doi.org/10.1007/978-3-030-12065-8_25

while adopting a sustainable development approach. This strategy aims to increase the share of renewable energies from 42% installed capacity for 2020, to 52% for 2030 [3]. One of the pillars of this strategy is the production of electricity based on photovoltaic solar. Among the projects envisaged an installation with a total capacity of 2000 MW, in solar power connected to the network, spread over five sites by 2020: Ouarzazate, Ain Beni Mathar, Foum El Ouad, Boujdour and Sabkhat Tah [4]. The implementation of this vast integrated project will reduce the country's energy dependence from 97% to 85%, and will save Morocco one million tons of oil equivalent and avoid the emission of more than 3.5 million tons of carbon dioxide per year [4]. Photovoltaic materials include several types of technologies, but silicon remains the most advanced technology [5]. The cells can be made from crystallized (c-Si) silicon wafers in two distinct categories, monocrystalline silicon (mc-Si) and polycrystalline silicon (pc-Si) [6, 7]. In recent years, various studies have been carried out on the performance parameters of photovoltaic power plants installed in different geographical locations, with different climatic conditions [8–21]. Performance analysis can be made on hourly, daily, monthly and yearly basis. Tripathi et al. [22] have investigated monitoring data from two 500 KWp solar PV power plants of amorphous and monocrystalline technologies, for a period of one year, located at the same place in Gujarat, Western India. They have reported that, the PR of a-Si ranged from 53.72% to 87.64%, and for mc-Si, it ranged from 57.1% to 93.14%. The final yield of a-Si power plant ranged from a lower value of 2.62 h/day to a maximum value of 4.84 h/day, and for mc-Si power plant, it ranged from a lower value of 2.79 h/day to a maximum value of 5.14 h/day. Pietruszko and Gradzki [23] conducted a performance investigation for solar PV systems from a 1 kW grid connected PV system of amorphous technologies, located in Poland. Performance analysis was done over an entire year, the PR of a-Si ranged from 60% to 80% and the annual energy yield was about 830 kW h. This study is part of the "PROPRE.MA" project, financed by IRESEN, with a budget of 5 million dirhams; it involves 20 Moroccan public universities for the development of a photovoltaic productivity mapping in Morocco. The result of the project represents a major innovation and differs from the conventional methods, since it allows obtaining maps calibrated to the soils with a high correlation by using three technologies of photovoltaic modules, polycrystalline, monocrystalline and amorphous based on silicon material [24]. This paper presents three types of results: the first is the performance of grid connected PV system consisting of two technologies namely, mono-Si and poly-Si for 12 months period under Casablanca climate conditions. The second is a comparison study of simulation and experimental results, and the third is an economic analysis. The overall idea of this approach is to know exactly the behavior of each technology, which can help us to determine the optimal technology for Casablanca city.

2 Materials and Methods

2.1 Description of Location

The 4.08 KWp grid-connected PV system are installed on the roof of the Faculty of sciences Ben M'Sik Casablanca, with a latitude of 33°33′56.33″ North and a longitude

7°32′29.22″ West (see Fig. 1). Casablanca located in the central western part of the country bordering the Atlantic Ocean, with a latitude of 33°35.2986′ North and a longitude 7°36.6828′ West, at 80 km south of Rabat, the administrative capital. It is the largest city in Morocco with an area of about 149 m^2, one of the most important cities in Africa, both economically and demographically (see Fig. 2).

Fig. 1. Location of roof by google earth

Fig. 2. Location of Casablanca by google earth

2.2 PV System

8 polycrystalline panels and 8 monocrystalline panels, each with 255 Wp PV panel power were placed on the roof of FSBM in Casablanca, mounted facing south with a near to latitude tilt angle (30°), supposed close to the optimal value to provide a maximum annual yield. The appearance of PV systems on the roof surface is given in Fig. 3.

Fig. 3. FSBM PV systems (front: mono, rear: poly)

The specifications of the modules of each technology array are given in the Table 1.

Table 1. The specifications of the modules of each technology array.

	System-1	System-2
Trademark	SOLARWORLD	
Model	SUNMODULE plus SW 255 poly	SUNMODULE Plus SW 255 mono
Solar cell	Poly crystalline	Mono crystalline
Maximum power at STC	255 Wp	255 Wp
Maximum power point voltage (Vmp)	30.9 V	31.4 V
Maximum power point current (Imp)	8.32 A	8.15 A
Open circuit voltage (Voc)	38 V	37.8 V
Short circuit current (Isc)	8.88 A	8.66 A
Module efficiency	15.21%	15.51%
Length	1675 mm	
Width	1001 mm	
Weight	18 kg	21.2 kg
Temperature coefficient (Pmax)	−0.41%/°C	−0.45%/°C

To convert DC to AC, two identical Sunny Boy 2000-HF 30 inverters are used with a maximum AC power of 2100 W. It is possible to observe DC currency and voltage, converted into AC produced from PV panels and to follow instant, daily and total production information by the inverter with monitoring system. The size of the inverters are 348/580/145 mm and it weighs about 17 kg. Technical properties of Sunnyboy inverter are shown in Table 2.

Table 2. Technical properties of Sunnyboy inverter

Technical data		
Input (DC)	Maximum DC power	2100 W
	Max. DC voltage	700 V
	MPP voltage range	175 V–560 V
Output (AC)	AC nominal power	2000 W
	Max. AC apparent power	2000 VA
	Nominal AC voltage	220, 230, 240 V
Efficiency	96.3%/95.0%	
Dimensions (W/H/D)	348/580/145 in mm	
Weight	17 kg	

2.3 Meteorological Station

Meteorological parameters, horizontal and 30° titled solar irradiations, ambient temperature, PV modules temperatures, wind speed and its direction are measured by meteorological station (see Fig. 4) based on PC Duino which is acts like a computer, allows meteorological data to be recorded every 5 min, as well as a recording period between 00 h and 24 h.

Fig. 4. Meteorological station.

Meteorological data are collected by 4 types of sensors:

– Solar irradiance sensors: The sensor for measuring the illumination (irrespective of orientation) is the pyranometer. Although it has good accuracy and can measure the overall illumination with a single sensor, its price remains relatively high. Which justifies our option to use cheaper sensors and which give very satisfactory results. Two modules are used (Phaesun Sun Plus 20), one horizontal and the second with the same angle 30°, it were installed near PV panels, allowing us to have the horizontal and inclined irradiations, by the use of the voltage given by the modules during their exposure to the sun, by the introduction of a resistance of 0.5 Ω in the junction box of the two panels, and by a calibration with the aid of a pyranometer used temporarily, and this by the determination of a sensitivity factor connecting the voltage and the illumination. The formula allowing us to calculate it is:

$$E = V_m/S; \tag{1}$$

With:

E: irradiation (w/m^2).
V_m: the module voltage (V).
S: the sensitivity factor ($V.m^2/w$).

- Wind speed and direction sensors: Benefiting from a very good finish and a connection via RJ11 sockets, the sensors are very easy to operate.

The anemometer uses a simple reed contact whose frequency of switching will allow us to measure the wind speed.

The wind vane uses a potentiometer whose value gives us an indication of the direction of the wind.

- Temperature sensors: To measure the temperature of two photovoltaic technologies and ambient temperature we used three-temperature sensors PT100 module. Which can be used in temperature range between −35 °C to +105 °C, with a weight of 40 g, and with a probe length of 50 mm and a total length of 100 cm.

The station installed in June 2016 allowed us to collect one year of meteorological data, between August 2016 and July 2017. The first month of installation was dedicated to the start-up so that the operating problems do not appear anymore and the results be unfail.

The collection and processing of data was done monthly to eliminate the insignificant data, the results of this work led to the determination of the average daily sunshine, for each month of the year.

Figure 5 shows the solar radiation between August 2016 and July 2017 as well as the sunshine duration of each month. It can be seen that the radiation varies regressively and gradually according to the months of the year and the duration of sunshine. We can also note that the radiation varies from 4.24 $KWh/m^2/day$ in December to 6.335 $KWh/m^2/day$ in April. Regarding the duration of sunshine, it varies from 10 h in December to 14.5 h in June.

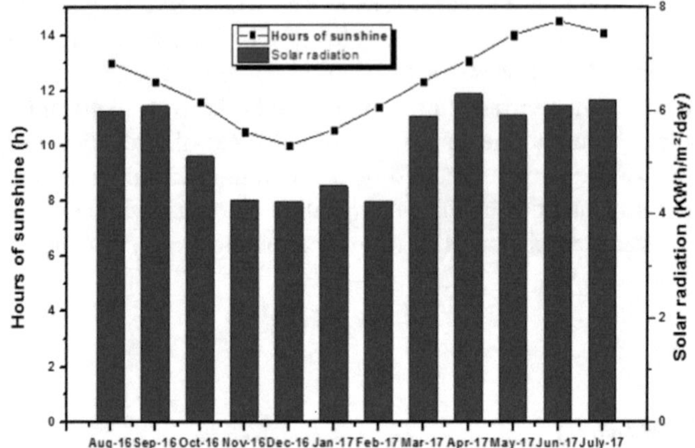

Fig. 5. Average daily radiation energy reflecting onto PV surfaces according to months and hours of sunshine

2.4 PV Characteristic Parameters

The elements used to study and evaluate the performance parameters of the PV systems, include: monthly production ($E_{AC,m}$), system efficiency (η_{sys}), reference and final yield (Y_R, Y_F), as well as the performance ratio (P_R), which are described by IEC 61724:1998 and the international energy agency task II database on photo-voltaic power system [25, 26].

$$E_{AC,m} = \sum_{1}^{N} E_{AC}; \tag{2}$$

$$\eta_{sys}(\%) = \frac{E_{AC}}{S \times G_{opt}}; \tag{3}$$

$$Y_R = \frac{G_{opt}}{G_{STC}}; \tag{4}$$

$$Y_F = \frac{E_{AC}}{P_{STC}}; \tag{5}$$

$$P_R(\%) = \frac{Y_F}{Y_R}; \tag{6}$$

The integration of the instantaneous values of the AC power on 5 min intervals, like it was done by [27] generates the energy E_{AC} [28].

- E_{AC}: (monthly) AC energy amount transferred to power plant by the inverter (kWh).
- G_{opt}: amount of radiation reflecting onto unit area of panel surface (KWh/m^2).
- S: Total surface area of the panels (m^2).

- G_{STC}: total solar radiation under standard test condition (1KW/m2).
- P_{STC}: the rated power of the installed PV array at standard test conditions (STC) (KWp).
- N is the number of days in the month.

3 Results and Discussion

3.1 Performance Analysis

In this part, data produced between August 2016 and July 2017 by two different panel types installed on the roof of the Faculty of Sciences ben m'sik Casablanca, were compared with each other. Monthly production collected through the two inverters used for both monocrystalline and polycrystalline systems, system efficiency, reference and final yield, as well as the performance ratio, have been interpreted in order to determine the behavior of the two systems during a year of operation, which will allow us to know the technology with the best performance for the site location.

Monthly total electrical energy generated by PV systems is shown in Fig. 6. The month with the highest electric power production is March. The electricity values for polycrystalline and monocrystalline panels were 345.0 KWh and 347.1 KWh, respectively. Figure 6 shows also that the system consisting of monocrystalline panel's generated energy more than polycrystalline panels during autumn, winter and spring and polycrystalline panels generated more during summer.

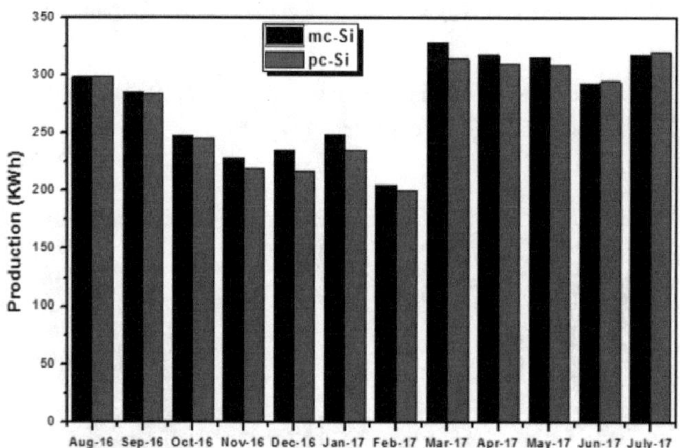

Fig. 6. Total electrical energy produced according to months.

For annual total production, the installation constructed by polycrystalline panels generated 3250,842 KWh, the second installation constructed by monocrystalline panels generated 3325,711 KWh annually. However, we can conclude that the perceived difference between the two technologies of the installation implies that monocrystalline modules have the best performance in terms of the productivity of Casablanca City.

According to the literature, the efficiency of PV energy conversion is much lower in months with high average ambient temperature compared to other months with low average ambient temperature, what is appearing in the Fig. 7. The minimum value of energy conversion efficiency for polycrystalline PV panel was 11.56% in October and the maximum value was 12.81% in March. The maximum value of energy conversion efficiency for monocrystalline PV panels was about 13.39% in March and minimum value was 11.62% in September.

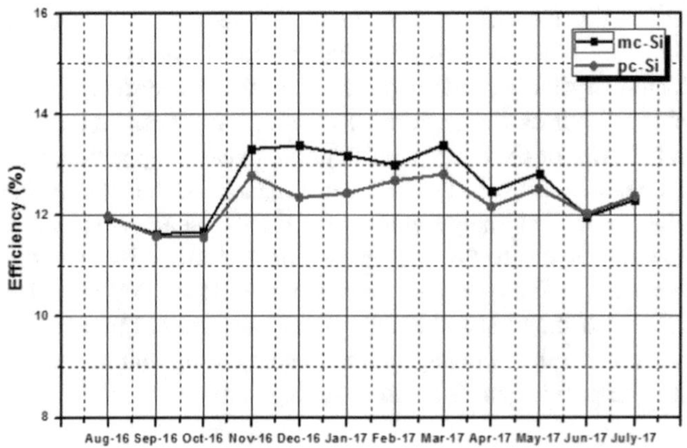

Fig. 7. Monthly energy efficiency of two different PV panel types.

Monthly average energy efficiency values for polycrystalline and monocrystalline panels were 12.26% and 12.6% respectively (Fig. 8).

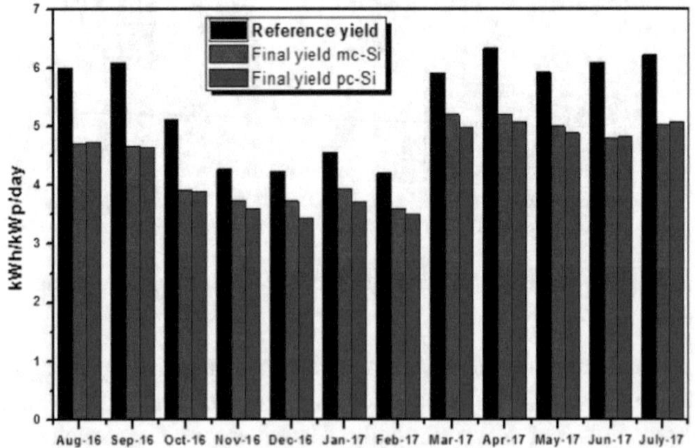

Fig. 8. Reference and final energy yield of each technology measured monthly.

The reference yield has parallel results with radiation rate, it has a minimum value in December and maximum value in March. The measured monthly normalized energy yields of monocrystalline PV panel's values change in range of 3.58–5.20 (KWh/KWp/day) and 3.44–5.06 (KWh/KWp/day) for polycrystalline in December and July respectively, which shows that mc-Si PV modules have performed better in comparison to pc-Si under similar outdoor conditions.

Yearly average value of monocrystalline Y_F is determined by 4.46 (KWh/KWp/day). Although the value of polycrystalline is 4.36 (KWh/KWp/day).

The performance of each PV system is depicted by its performance ratio (P_R) that is calculated and presented in Fig. 9. It is noticeable that P_R system range from 76.45% to 88.04% for (mc-Si module), and from 76.04% to 84.23% for (pc-Si module). Performance ratio above 80% is always desirable, as the highest performance, which constitute an economic gain. Monocrystalline PV technology shows better performance in terms of Performance ratio than Polycrystalline technology. This may be due to the efficiency of this technology, which is superior under standard test conditions.

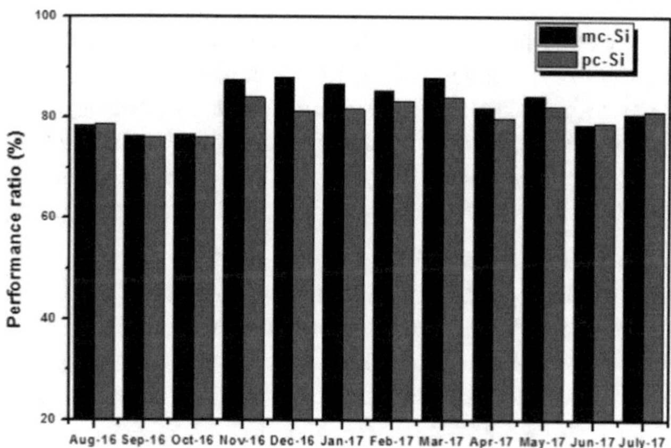

Fig. 9. PR rate of two different PV panel types.

According to the results interpreted before and Fig. 10, the difference in production between monocrystalline and polycrystalline is justified by the relationship between ambient temperature and irradiation at the PV module temperature, because the only major difference between seasons is the average ambient temperature and radiation. For more explanation, the low average ambient temperature and radiation give a low PV modules temperature, which increase the efficiency of solar panels, and the high average ambient temperature and radiation give a high PV modules temperature, which decrease the efficiency. But as regards, the difference in production between technologies, the low temperatures of monocrystalline modules, leaves the efficiency which is of about 15.51% of the monocrystalline performed so as to give more electricity, while the efficiency of the polycrystalline is 15.21%.

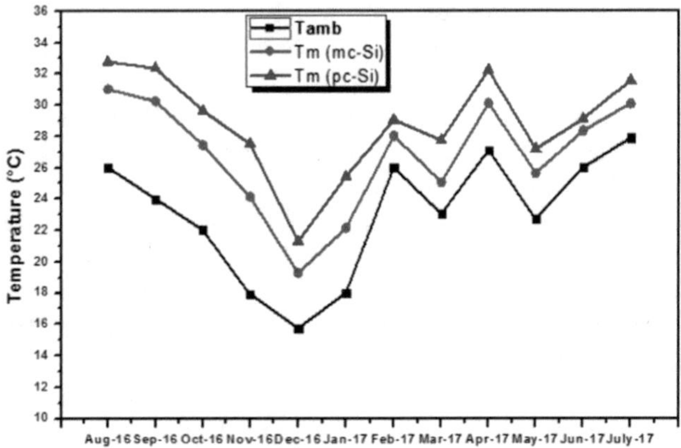

Fig. 10. Monthly average annual ambient temperature

3.2 Simulation

The simulation of energy production requires a large amount of meteorological data and technical parameters [29]. Photovoltaic software is widely used in the design of photovoltaic systems to calculate expected energy yield, in addition to quantifying the disruptive effects in order to identify weaknesses and optimize the entire installation. For our case, we will use the software PVsyst 6.4.3 to have a general idea on the performance produced by a 2.04 KWp installation for two technologies (mc-Si and pc-Si).

Simulation With PVsyst 6.4.3

The simulation methodology for the PVsyst 6.4.3 consists of the introduction of the location of the site to enable it to use the appropriate meteorological data, secondly the introduction of PV installation information, such as the exact type of modules used, inverter and the arrangement of the modules. Following all of these steps, data for Casablanca of PVsyst 6.4.3 are shown in Table 3.

Depending on the simulation result, PVsyst 6.4.3 gave us an idea of all PV characteristic parameters generated by the two systems, and confirms the experimental results obtained.

The data shown in Table 3 show a difference in annual production between experimental and simulated results of 197.29 KWh/year for monocrystalline, and of 253.158 KWh/year for polycrystalline technology.

For the efficiency, the results show a difference of 0.02% of the monocrystalline, whereas it is of 0.23% for the polycrystalline technology, this also manifests in the performance ratio with a difference of 0.25% for the technologies mc-Si while it is 1.75% for pc-Si.

For the meteorological data, which are the radiation and the ambient temperature, the difference is more larger with a difference for reference yield of 0.31 KWh/KWp/day for both technologies and a difference in temperature of 4.7 °C.

Table 3. Results of PV characteristic parameters by experimental and simulation methodology.

	Monocrystalline-Si	
	Experimental results	Simulation results
Yearly production	3325.711 KWh/year	3523 KWh/year
Yearly production/KWc	1630.25 KWh/KWp/year	1727 KWh/KWp/year
System efficiency	12.59%	12.57%
Reference yield	5.41 KWh/KWp/day	5.72 KWh/KWp/day
Final yield	4.46 KWh/KWp/day	4.73 KWh/KWp/day
Performance ratio	82.44%	82.69%
Ambient temperature	23.02 °C	18.32 °C
	Polycrystalline-Si	
	Experimental results	Simulation results
Yearly production	3250.842 KWh/year	3504 KWh/year
Yearly production/KWc	1593 kwh/KWp/year	1718 KWh/KWp/year
System efficiency	12.27%	12.50%
Reference yield	5.41 KWh/KWp/day	5.72 KWh/KWp/day
Final yield	4.36 KWh/KWp/day	4.71 KWh/KWp/day
Performance ratio	80.59%	82.34%
Ambient temperature	23.02 °C	18.32 °C

Therefore, we can draw that the results of PVsyst are significant, which can be used as a basis to dimension and know the behavior of the different technologies under climatic conditions.

To conclude, the simulation and experimental results show that the technology of monocrystalline silicon produces more than polycrystalline silicon in weather conditions of Casablanca city.

3.3 Economic Analysis

Based on the production data presented before, the initial investment cost (N) and the fixed feed in tariff per kWh (c), we have carried out a financial study including the most commonly used financial parameters which are: the annual incomes (Ai), the cost of electricity of operating period (LCOE) and the payback period (PB).

Annual Incomes (Ai)

The Ai represents the amount of money that a person earns in a year from the use of PV electrical energy. This value has been calculated with Eq. (7):

$$Ai = P \times c; \tag{7}$$

Where P is the yearly energy production of the installation [KWh] and c the fixed feed in tariff per kWh [$/KWh].

The Cost of Electricity of Operating Period (LCOE)

The levelized cost of electricity (LCOE) as called in the literature is one of the most frequently used financial parameters to compare technologies. It represents the total cost expressed in dollars including the cost of installing and operating a system per kilowatt hour of electricity generated by the system throughout its lifetime [30, 31]:

$$\mathrm{LCOE} = \frac{\sum_{t=0}^{Lf} \frac{\mathrm{N}_t + \mathrm{O}_t + \mathrm{M}_t}{(1+r)^t}}{\sum_{t=0}^{Lf} \frac{\mathrm{P}_t}{(1+r)^t}}; \tag{8}$$

Where Lf the lifetime, t is the year, and r is the effective annual reduction rate (%), N_t is the initial investment cost per years, O_t and M_t are the operation and maintenance costs for t, respectively. P_t expresses the energy production of the installation for t [KWh].

With:

$$\mathrm{P_t} = \mathrm{S_t}(1 - \mathrm{d})^{\mathrm{t}}; \tag{9}$$

Where S_t expresses the yearly rated energy output [kWh] and d is the degradation rate of the PV panels [%]. The literature gives a value of d between 0.2–0.5 for the crystalline technology [32]. For our case the constructor has estimated a value of d = 0.35% which is between the two values.

Payback Period

For the PB, it represents the time needed to accumulate through economic saving the equivalent of the total initial investment N obtained using Eq. (8) [33]:

$$\mathrm{PB} = \frac{\mathrm{N}}{\mathrm{A}_i}; \tag{10}$$

Where Ai (expressed in $/year) is the Annual incomes.

Considering the net initial investment cost (N) including the final cost of photovoltaic panels, inverters, cabling, assembly structures, labor costs and taxes which is approximately 8335$.

Table 4 show a study of economic performance of two types of solar panels, based on silicon material, the two silicon technologies are monocrystalline and polycrystalline, it presents at the level of the last three line the financial parameters, which will allow us to determine the optimal PV power system for a 12 months period under Casablanca climate conditions.

The first line of the last three shows the annual income of the two technologies. Considering the annual production and the fixed feed in tariff per kWh (c), which is the tariff determined by the electricity distribution company, it shows that the system consisting of monocrystalline type modules is more productive than the polycrystalline in terms of annual incomes.

Table 4. Economic performance of the installed systems.

	System-1	System-2
Components	Polycrystalline-silicon	Monocrystalline-silicon
PV modules	295$ × 8 = 2360$	298$ × 8 = 2384$
Inverter	1397$ × 1 = 1397$	1397$ × 1 = 1397$
Taxes and parallel equipment	398.56$	398.56$
Total	4155.56$	4179.56$
Net initial investment cost (N)	4155.56$	4179.56$
PV costs ($/Wp)	1.155$	1.158$
System costs ($/Wp)	2.035$	2.030$
Production [kWh] per year (P)	3250.842	3325.711
Fixed feed in tariff per kWh (c)	0.1237 $/KWh	0.1237 $/KWh
Annual income (Ai)	402.13$	411.4$
The unit cost of energy for 25-year running (LCOE)	0.0734 $/KWh	0.0724 $/KWh
Payback period (PB)	10.33 years	10.15 years

The second line of the last three shows the unit cost of energy of operating period (LCOE), which shows that the monocrystalline technology is the less expensive between the two technologies.

The final line in Table 4 compares the PB value for the two PV plants, and shows that the nearest payback period among the two-system is 10.15 year and belongs to the system consisting of monocrystalline.

4 Conclusion

In this paper, we described the PV plant and the meteorological station which are installed on the roof of the Faculty of Sciences Ben M'sik Casablanca and their performance were assessed, as a conclusion the following results are obtained:

- Monocrystalline presents the maximum value of system efficiency, final yield, and performance ratio with values of 12.59%, 4.46 KWh/KWp/day and 82.44% respectively.
- Polycrystalline presents the minimum value of system efficiency, final yield, and performance ratio with values of 12.27%, 4.36 KWh/KWp/day and 80.59% respectively.
- The total annual electricity delivered to the grid was found to be 1630.25 KWh/KWp for monocrystalline technology and 1593 KWh/KWp for polycrystalline technology.
- The economic analysis shows that, the annual incomes, the cost of electricity of operating period and the payback period for monocrystalline installation are respectively, 411.4$, 0.0724$ and 10.15 years, while for polycrystalline they are respectively, 402.13$, 0.0734$ and 10.33 years.

This study, allowed us to know the behavior and its reason for the two technologies during the 12 months of the year and to recognize the optimal technologies of the region, which undoubtedly the monocrystalline technology.

Acknowledgment. All the authors as well as all the rest of the "PROPRE.MA" partner are grateful to IRESEN for financing this study and would like to thank all the IRESEN staff for their support.

References

1. Attari, K., Elyaakoubi, A., Asselman, A.: Performance analysis and investigation of a grid-connected photovoltaic installation in Morocco. Energy Rep. **2**, 261–266 (2016)
2. https://www.finances.gov.ma/fr/pages/strat%C3%A9gies/strat%C3%A9gie-dans-le-domaine-de-l%E2%80%99%C3%A9nergie.aspx?m=Investisseur&m2=Investissement
3. http://www.mem.gov.ma/SiteAssets/Monographie/DirectionsCentrales/DEREE.pdf
4. http://taqaway.net/sites/default/files/uploads/documents/doc98.pdf
5. Yilmaz, S., et al.: Renew. Sustain. Energy Rev. **52**, 1015–1024 (2015)
6. Abella, M.A., Chenlo, F., Nofuentes, G., Ramírez, M.T.: Analysis of spectral effects on the energy yield of different PV (photovoltaic) technologies: the case of four specific sites. Energy **67**(1), 435–443 (2014)
7. Raugei, M., Frankl, P.: Life cycle impacts and costs of photovoltaic systems: current state of the art and future outlooks. Energy **34**, 392–399 (2009)
8. Zhou, W., Yang, H., Fang, Z.: A novel model for photovoltaic array performance prediction. Appl. Energy **84**, 1187–1198 (2007)
9. Kim, J.-Y., Jeon, G.-Y., Hong, W.-H.: The performance and economical analysis of grid-connected photovoltaic systems in Daegu. Korea. Appl. Energy **86**, 265–272 (2009)
10. Li, D.H.W., Cheung, K.L., Lam, T.N.T., Chan, W.W.H.: A study of grid-connected photovoltaic (PV) system in Hong Kong. Appl. Energy **90**, 122–127 (2012)
11. Su, Y., Chan, L.-C., Shu, L., Tsui, K.-L.: Real-time prediction models for output power and efficiency of grid-connected solar photovoltaic systems. Appl. Energy **93**, 319–326 (2012)
12. Sharma, V., Kumar, A., Sastry, O.S., Chandel, S.S.: Performance assessment of different solar photovoltaic technologies under similar outdoor conditions. Energy **58**, 511–518 (2013)
13. Padmavathi, K., Daniel, S.A.: Performance analysis of a 3 MWp grid connected solar photovoltaic power plant in India. Energy. Sustain. Dev. **17**, 615–625 (2013)
14. Chemisana, D., Lamnatou, C.: Photovoltaic-green roofs: an experimental evaluation of system performance. Appl. Energy **119**, 246–256 (2014)
15. Khan, F., Baek, S.-H., Kim, J.H.: Intensity dependency of photovoltaic cell parameters under high illumination conditions: an analysis. Appl. Energy **133**, 356–362 (2014)
16. Ferrada, P., Araya, F., Marzo, A., Fuentealba, E.: Performance analysis of photovoltaic systems of two different technologies in a coastal desert climate zone of Chile. Sol. Energy **114**, 356–363 (2015)
17. Wu, X., Liu, Y., Xu, J., Lei, W., Si, X., Du, W., et al.: Monitoring the performance of the building attached photovoltaic (BAPV) system in Shanghai. Energy Build. **88**, 174–182 (2015)
18. Kumar, K.A., Sundareswaran, K., Venkateswaran, P.R.: Performance study on a grid connected 20 KWp solar photovoltaic installation in an industry in Tiruchirappalli (India). Energy. Sustain. Dev. **23**, 294–304 (2014). https://doi.org/10.1016/j.esd.2014.10.002

19. Bakos, G.C.: Distributed power generation: a case study of small scale PV power plant in Greece. Appl. Energy **86**, 1757–1766 (2009). https://doi.org/10.1016/j.apenergy.2008.12.021
20. Khatib, T., Sopian, K., Kazem, H.A.: Actual performance and characteristic of a grid connected photovoltaic power system in the tropics: a short term evaluation. Energy Convers. Manage. **71**, 115–119 (2013). https://doi.org/10.1016/j.enconman.2013.03.030
21. Adaramola, M.S., Vågnes, E.E.T.: Preliminary assessment of a small-scale rooftop PV-grid tied in Norwegian climatic conditions. Energy Convers. Manage. **90**, 458–465 (2015). https://doi.org/10.1016/j.enconman.2014.11.028
22. Tripathi, B., Yadav, P., Rathod, S., Kumar, M.: Performance analysis and comparison of two silicon material based photovoltaic technologies under actual climatic conditions in Western India. Energy Convers. Manage. **80**, 97–102 (2014). https://doi.org/10.1016/j.enconman.2014.01.013
23. Pietruszko, S.M., Gradzki, M.: Performance of a grid connected small PV system in Poland. Appl. Energy **74**, 177–184 (2003)
24. Aarich, N., Erraïssi, N., Akhsassi, M., Lhannaoui, A., Raoufi, M., Bennouna, A.: "Propre. Ma" project: roadmap & preliminary results for gridconnected PV yields maps in Morocco. In: IEEE International Renewable and Sustainable Energy Conference (IRSEC), pp. 774–777 (2014)
25. Bhattacharjee, S., Bhakta, S.: Analysis of system performance indices of PV generator in a cloudburst precinct. Sustain. Energy Technol. Assess. **4**, 62–72 (2013)
26. IEC: Photovoltaic system performance monitoring—guidelines for measurement, data exchange, and analysis IEC Standard 61724, Geneva, Switzerland (1998)
27. Ayompe, L.M., Duffy, A., McCormack, S.J., Conlon, M.: Measured performance of a 1.72 kW rooftop grid connected photovoltaic system in Ireland. Energ. Convers. **52**, 816–825 (2010)
28. Guenounou, A., Malek, A., Aillerie, M.: Comparative performance of PV panels of different technologies over one year of exposure: Application to a coastal Mediterranean region of Algeria. Energy Convers. Manage. **114**, 356–363 (2016)
29. Huld, T., Šúri, M., Dunlop, E.D.: Comparison of potential solar electricity output from fixed-inclined and two-axis tracking photovoltaic modules in Europe. Prog. Photovolt. Res. Appl. **16**, 47–59 (2008)
30. Darling, S.B., You, F., Veselka, T., Velosa, A.: Assumptions and the levelized cost of energy for photovoltaics. Energy Environ. Sci. **4**, 3133–3139 (2011)
31. Branker, K., Pathak, M.J.M., Pearce, J.M.: A review of solar photovoltaic levelized cost of electricity. Renew. Sustain. Energy Rev. **15**(9), 4470–4482 (2011)
32. Vats, K., Tomar, V., Tiwari, G.N.: Effect of packing factor on the performance of a building integrated semitransparent photovoltaic thermal (BISPVT) system with air duct. Energy Build **53**, 159–165 (2012)
33. Allouhi, A., Saadani, R., Kousksou, T., Saidur, R., Jamil, A., Rahmoune, M.: Grid-connected PV system installed on institutional buildings: Technology comparison, energy analysis and economic performance. Energy and Buildings http://dx.doi.org/10.1016/j.enbuild.2016.08.054

Energy Study of Different Solar Water Heating Systems in MENA Region

Mohamed Amine Ben Taher, Abdelmounim Al Jamar, Firdaous Akzoun, Mohammed Ahachad, and Mustapha Mahdaoui(✉)

Equipe de Recherche en Transferts Thermiques & Énergétique - UAE/E14FST
Département de Physique FST, Université Abdelmalek Essaâdi, Tangier, Morocco
mustapha_mahdaoui@yahoo.com

Abstract. This article presents an energy study during the year associated with the results of two solar water heating systems with flat plate collectors (FPC) and evacuated tube collectors (ETC) operating in the Middle East and North Africa region. Annual simulations are performed using the TRNSYS software by considering a typical meteorological year (TMY) for all countries in the MENA region. Energy efficiency between these systems was compared on a daily, monthly and annual basis. It is discovered that high estimations of solar fraction and gatherer efficiency can be come to in studied regions with the preference of utilizing (ETC).

Keywords: MENA · Flat plate collector · Evacuated tube collector · Solar fraction · System efficiency · TRNSYS · TMY2

1 Introduction

The MENA region has about 57% and 41% of the world's proven reserves of natural gas and oil respectively. It also has unique solar energy resources on the planet. The energy context in the MENA region continues to be marked by an unbalanced dependence on non-renewable fossil energy sources which sometimes reach 100% of the energy mix of some of them such as Bahrain, Kuwait, Libya or Oman. The strong population growth and intensive energy consumption in the region will lead to an increase in energy needs of 6 to 7% per year in the future if the consumption patterns remain the same. It is conjectured that will lead to a doubling of energy needs by 2020 [1].

To deal with the problem of dependence, there are various solutions proposed such as the integration of renewable energies, strengthening the societal discussion on energy efficiency and the increase in the consumption of natural gas or the installation of nuclear centers for the production of electricity. Overall, there is great potential at the regional level to improve the efficiency of energy supply, to better conserve energy and to develop renewable energy resources. The MENA region has started to scale up its renewable energy potential, and this initiative has the full support of the World Bank. To date, it is estimated that clean energy projects in the Middle East and North Africa region currently require 200 $ billion in investment with more than 67GW of clean vitalizing projects in different phases of the plan and study arrange as per sustainable power source in the MENA region [2].

M. Ezziyyani (Ed.): AI2SD 2018, AISC 912, pp. 288–296, 2019.
https://doi.org/10.1007/978-3-030-12065-8_26

One of the most promising clean energy sources is solar energy the most multi-functional and abundant sources of renewable energy at our disposal. It can be caught and utilized both directly and indirectly way and can make an essential contribution to the diminishment of carbon emissions from petroleum products. The annual solar thermal power yield in 2016 collected to 375 TWH which associate to savings of 130 million tons of co2 and 40.3 million tons of oil. Properly, different advancements were created so as to tackle the sun's power [3]. The SWH system utilizes natural sun powered technology which is where solar radiation is changed into heat and transmitted into a exchange medium such as water or air. Almost part, the system is simple, since it works when the warmth exchange liquid is acquired contact with a surface that is presented to daylight, which at that situation raise the coolant temperature. On the whole, the SWH is a very universal application today, that why the major challenge facing the development of these systems is to ensure their adaptation to different climates with a better performance. There are numerous types of stationary collectors utilized in SWH systems, flat plate collectors and evacuated tube gatherers are the most broadly deployed for these applications [4].

Different studies have been realized on the efficiency of the sun collectors and the outcomes have been distributed by a few papers. Economic and environmental aspects of their integration were discussed as well. Perers et al. [5] carried out distinction of thermal performance for FPC and ETC installed in Sweden. Their results show that the performance of these collectors has produced near 400 kWh/m^2 and 300 kWh/m^2. Nikoofard et al. [6] presented an economic study with the plan of evaluating the energy saving and the greenhouse gas emissions reduction resulting from the retrofitting of SWH systems to Canadian houses. Morrison et al. [7] evaluated the performance of evacuated tube and flat plate collectors and found that evacuated tube solar collectors have more efficiency than flat-plate collectors when supplying heat at temperatures above 100 °C. Zambolin and Del Col [8] carried out a side by side testing of FPC and ETC in Padova, Italy. They performed steady-state and quasi-dynamic efficiency tests. Druck et al. [9] did a comparison test of different thermal solar systems for domestic hot water. Ayompe et al. [10] presented comparative field performance of FP and ET collectors for domestic SWH under the weather conditions in Ireland. This research, intends to compare the energy and environmental performance of two solar water heating systems with FPC and ETC subjected to similar operating conditions and weather conditions operating in the MENA region.

2 Methodology

Solar water systems are likewise a key component in the spread out of renewable in the region. Today, in MENA regions they account for about 9 million square meters (m^2) of collector area, representing 6.3 gigawatts-thermal of installed capacity [11]. For this study, we considered two complete forced circulation SWH systems with flat plate (FPC) and evacuated tube (ETC) collectors were installed in a same domestic house (4–5 occupants) in the representative cities of all countries in the MENA region. The two waters heating systems each had a 300-L hot water tank and the hot water is assumed to be delivered at a temperature of 45 °C. These two systems are modeled and simulated using TRNSYS-16.

3 System Description

3.1 System Design

A schematic diagram of a forced-circulation SWH system designed by TRNSYS systems illustrated in (Fig. 1).

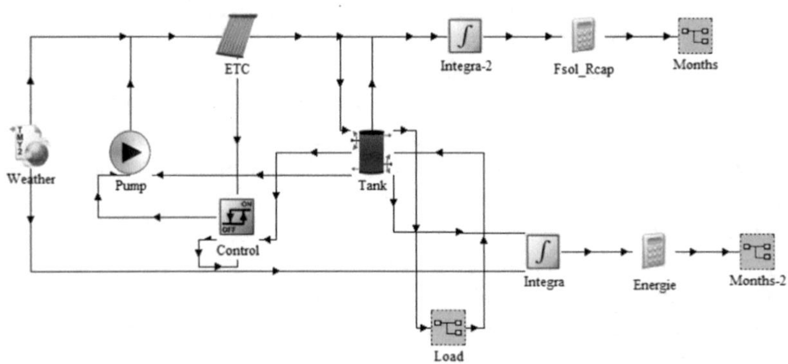

Fig. 1. Diagram of SWH with evacuated tube collector

- **Collectors.** The total area of solar collector's (FPC and ETC) consisted of 2.5 m^2; The collector optical efficiency, heat loss coefficient, and temperature dependence of the heat loss coefficient values are 0.82, 3.31 W/m^2K and 0.0181 $W/m^2K^2$for the ETC system while for the FPC system the respective values 0.821,2.82 W/m^2K and 0.0047 W/m^2K^2.
- **Solar Storage Tank.** The solar storage tank has a direct influence on the performance of the system. For SWH. The thermal storage (stratified storage tank) had a capacity of 300 L. The tank height and diameter were 1600 mm and 650 mm respectively. This storage tank had equipped lower & upper coil heat exchanger of 3.1/2 kW.
- **Pump.** The pump is used to circulate the fluid into the first collector, for this work the pump (single speed) selected with maximum flow rate approximately 40 $kg/hr.m^2$.
- **Control Unit.** The regulation is an automatic system that allows to maintain a set point, temperature of hot water. The on/off differential controller can generate a control function which can have a value of 1 or 0, the controller blocks the system when the temperature exceeds 100 °C.

3.2 Weather Data

The input to the weather data in TRNSYS is the typical meteorological year weather data (TMY). In what follows, each country will be characterized by its representative city (TMY2 for each Capital). The ambient temperature, the global horizontal irradiation and total radiation on titled surface of the studied cities are generated by the Meteonorm database.

3.3 Hot Water Demand Profile

The load profile is the most important parameter that needs to be considered for SWH over a certain period of time. Average daily requirements per person over a year are on average 56 ± 23 L at 45 °C and 35 ± 14 L at 55 °C for an average annual cold-water temperature of 16 °C [12]. In Ghana, a family (of 4 people) will demand 200 L of hot water daily, assuming an average hot water demand of 50 L per day [13].

For this study an averaged hot water profile will be distributed 3 times per days of 240 L/day of water at 45 °C for a Moroccan individual house (composed of 4–5 occupants) is used. This rating based on the result of Allouhi et al. [14].

4 Energy Analysis

The energy effectiveness records assessed in this study incorporate:

- **Energy collected**
  The rate of useful energy gain by the solar collector fluid is given as:

$$Q_c = mc_p(T_{out} - T_{\text{in}}) \tag{1}$$

- **Energy to load**
  The rate at which energy is expelled from the tank to outfit the load is:

$$Q_l = \text{m}_{load}c_p(T_{load} - T_{cold}) \tag{2}$$

- **Energy delivered and total energy losses**
  The useful energy delivered from heat source to the hot water tank is given as:

$$Q_d = \text{m}_{h,source}c_p\left(T_{hot} - T_{h,source}\right) \tag{3}$$

  The rate of thermal energy loss to the environment

$$Q_{loss} = UA_i(T_i - T_{amb}) \tag{4}$$

  T_i, A_i is temperature (°C) and the area (m^2) of the tank, U loss coefficient (W/m^{2}K).
- **Solar fraction**
  Solar fraction is the most important factor for evaluate the performance of the SWH system expressed by the ratio of solar heat yield to the total energy requirement for water heating and is given as:

$$SF = \frac{Q_s}{Q_s + Q_{aux}} \tag{5}$$

- **Collector efficiency**
  The efficiency of the collector (FPC or ETC) is calculated as:

$$\eta = \frac{Q_c}{G_{inc}} \tag{7}$$

5 Results and Discussion

The simulation was carried to study the efficiency of FPC and ETC systems. For a significant analysis, we will detail our study by the case of Rabat-Morocco, this simulation included ambient temperatures (Fig. 2), total radiation on tilted surface and on horizontal (Fig. 3) generate from Meteonorm data. Then we will present the results of the simulation for the MENA region. Using the software TRNSYS we calculated the different energies for daily, monthly and annual solar incident radiation.

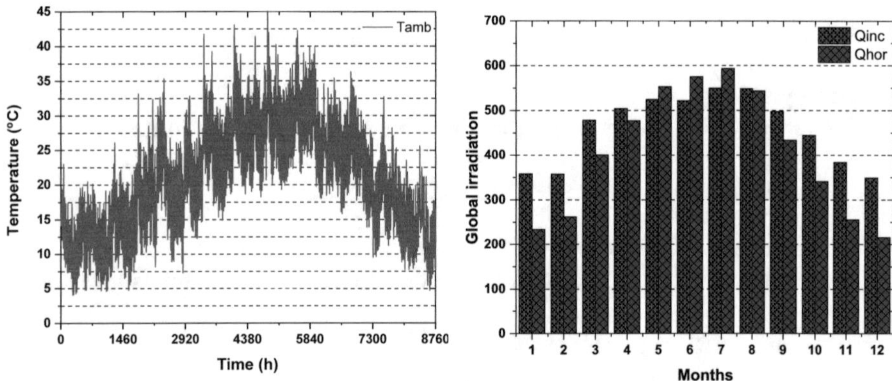

Fig. 2. Ambient temperature (Rabat-Morocco)

Fig. 3. Monthly solar radiation on horizontal and a tilted surface (Rabat-Morocco)

5.1 Energy Study

The results are shown in Table 1 for FPC and ETC installations in Rabat city, as a result, the annual total energy collected by the FPC and ETC was 3103 kWh/y and 4828 kWh/y respectively.

We can see that the ETC system generated 25% more energy than FPC. In order the ETC delivers even less quantity of auxiliary electric energy (1882 kWh/y) than the FPC (3397 kWh/y). The results also show that the annual energy losses for the FPC and ETC had of 893 kWh/y and 1098 kWh/y respectively.

Table 1. Energy quantities of the two SWHs (FPC and ETC)

Months	Qinc (kWh)	Qhor (kWh)	Qcoll (kWh)		Qaux (kWh)		Qloss (kWh)		Qsol (kWh)	
			FPC	ETC	FPC	ETC	FPC	ETC	FPC	ETC
1	359	234	195	306	358	258	71	80	124	225
2	357	262	195	310	300	194	65	75	130	235
3	479	401	262	425	289	147	76	95	186	330
4	504	477	277	445	258	113	74	97	203	349
5	525	553	285	458	266	115	76	99	209	359
6	522	575	291	450	245	108	75	97	216	353
7	550	593	315	478	240	101	79	102	236	375
8	548	543	321	486	235	94	79	103	241	382
9	497	434	293	445	244	112	77	97	216	348
10	445	341	258	393	292	170	76	91,3	186	302
11	385	256	215,1	332,5	316	211	71	81,7	144	251
12	349	216	192,2	296,9	355	257	71	79,3	124	218

5.2 Solar Fraction and Collector Efficiency

The Table 2 indicated the solar fraction and collector efficiency of the both systems. We note that the ETC configuration can achieve a SF around 81% and a collector efficiency of 90%.

Table 2. Solar fraction and collector efficiencies for FPC and ETC systems

Months	Solar fraction (%)		Collector Efficiency (%)	
	FPC	ETC	FPC	ETC
1	25,76	46,60	25,76	46,60
2	30,23	54,81	30,23	54,81
3	39,17	69,21	39,17	69,21
4	44,05	75,48	44,05	75,48
5	43,96	75,68	43,96	75,68
6	46,90	76,63	46,90	76,63
7	49,64	78,72	49,64	78,72
8	50,67	80,32	50,67	80,32
9	47,01	75,60	47,01	75,60
10	38,48	63,88	38,48	63,88
11	31,37	54,31	31,37	54,31
12	25,39	45,86	25,39	45,86

The SF of the FPC system range between 25% and 60% during the months of the year. The range of efficiencies of the overall FPC system varied from 54% in January to 59% in August.

5.3 Performance Evolution in MENA Region

The figures (Fig. 4) represent the solar fractions for the countries of the region, in many cases the solar water heating systems are sized to operate more in winter than in summer. From these figures, it can be seen that the ETC system had higher solar fractions, which makes ETC technology perform better than FPC in almost all

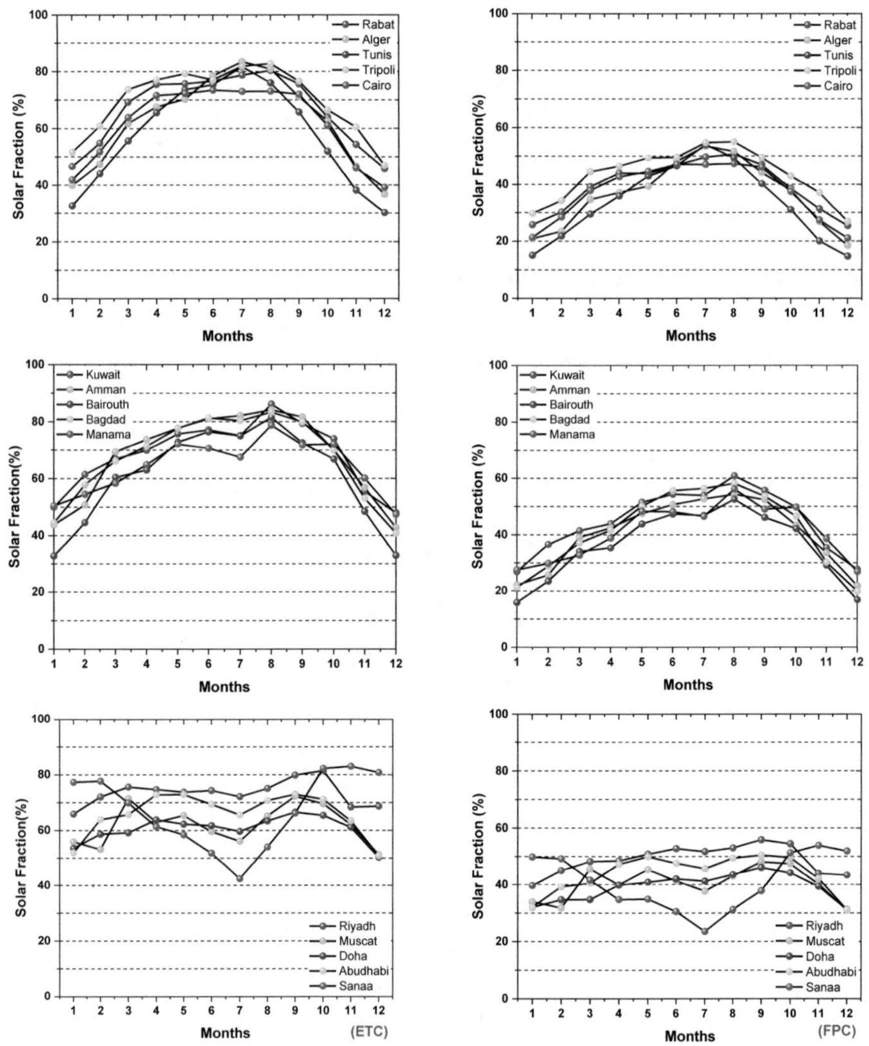

Fig. 4. Monthly solar fractions in Middle East and North Africa (ETC and FPC case)

countries in the region. During the summer season, it can also be noted that the maximum values of the solar fraction were reached by the CES system in the South zone of the Middle East. This can be explained by the huge solar radiation and the hot and dry climate all year long. In addition, there is a difference in the southern region of the Middle East in general and in Yemen in particular compared to other regions. Because of the cloud cover brought by the southwestern monsoon, hence the sun, good all year, decreases a little in summer.

6 Conclusion

In the current project, the performance of two forced circulation solar water heating systems with flat plate collectors (FPC) and evacuated tube collectors (ETC) are simulated and compared for the Middle East and North Africa (MENA) climate conditions by TRNSYS software.

According to the results of the simulation over the year according to Rabat condition, which shows a total annual in-plane solar installation of 5519 kWh, a total of 3103 kWh and 4828 kWh of heat energy were collected by the 2.5 m^2 FPC and ETC systems. For 3397 kWh and 1882 kWh of auxiliary energy to the FPC and ETC systems their annual energy losses were 893 kWh and 1098 kWh respectively. The annual average solar fractions were 39.39% and 66.43% while the collector efficiencies 56.23% and 87.35% respectively for the FPC and ETC respectively.

In this paper, the results of the energy performance analysis show that the 2.5 m^2 ETC system compares quite favorably with the 2.5 m^2 FPC system when connected to a 300-L hot water tank in the Middle East and North Africa region. For all cities MENA region simulated, we found that the solar fraction obtained by the ETC system larger than obtained by the FPC system.

Finally, to make this study relevant, an environmental and economic study is recommended. Through this study we can decide the best technology for each region.

References

1. Anabtawi, Y.: The Future for Renewable Energy in the MENA Region. Squire sanders, Riyadh
2. MEED. Renewable Energy in the MENA region, May 2017. http://www.meed.com/
3. Weiss, W., Spörk-Dür, M., Mauthner, F.: Solar Heat World Wide. IEA Solar Heating & Cooling Program, May 2017
4. Jamar, A., Majidb, Z.A.A., Azmi, W.H., Norhafana, M., Razake, A.A.: A review of water heating system for solar energy applications. Int. Commun. Heat Mass Transfer **76**, 178–187 (2016)
5. Perers, B.: Comparison of thermal performance for flat plate and evacuated tubular collectors. J. Adv. Solar Energy Technol. **1**, 615–619 (1988)
6. Nikoofard, S., Ismet Ugursal, V., Beausoleil-Morrison, I.: An investigation of the technoeconomic feasibility of solar domestic hot water heating for the Canadian housing stock. Sol. Energy **101**, 308–320 (2014)

7. Morrison, G.L., Budihardjo, I., Behnia, M.: Water-in-glass evacuated tube solar water heaters. J. Sol. Energy **76**, 135–140 (2004)
8. Zambolin, E., Del Col, D.: Experimental analysis of thermal performance of flat plate and evacuated tube solar collectors in stationary standard and daily conditions. Sol. Energy **84**(8), 1382–1396 (2010)
9. Druck, H., Heidemann, W., Muller-Steinhagen, H.: Comparison test of thermal solar systems for domestic hot water preparation and space heating. In: Proceedings of EuroSun, Freiburg, Germany (2004)
10. Ayompe, L.M., Duffy, A., Mc Keever, M., Conlon, M., Mc Cormack, S.J.: Comparative field performance study of flat plate and heat pipe evacuated tube collectors (ETCs) for domestic water heating systems in a temperate climate. J. Energy **36**, 3370–3378 (2011)
11. Riahi, L.: (REN21 Secretariat). MENA Renewables Status Report (2013)
12. ADEME, Les besoins d'eau chaude sanitaire en habitat individuel et collectif, July 2016
13. Country Report - Ghana. Status of Solar Heating/Cooling and Solar Buildings (2016). http://www.iea-shc.org/country-report-ghana
14. Allouhi, A., Jamil, A., Kousksou, T., El Rhafiki, T., Mourad, Y., Zeraouli, Y.: Solar domestic heating water systems in Morocco: an energy analysis. Energy Convers. Manage. **92**, 105–113 (2015)

Study and Improvement of the FMECA in a Production Way

Ilyas Mzougui(✉) and Zoubir El Felsoufi

Faculty of Sciences and Technologies,
Route Boukhalef, 90000 Tangier, Morocco
Ilyas.mzougui@gmail.com, Elfelsoufi_zoubir@yahoo.fr

Abstract. Failure mode and effect and criticity analysis is a tool highly used for the identification and the elimination of the failures. He has been used in the first time by the national aeronautics and space agency on 1977. He has been created as a development methodology. Since that, many works have been established to improve it some of them use probabilistic methods. Some others use the Multi criteria decision making. But as we know, there is any works that try to solve the interdepency problem in this tool. This is what we will try to do in this article.

Keywords: FMECA · Interdependency · DSM

1 Introduction

The Failure mode, effects and criticality analysis (FMECA) judge each failure by 3 criteria: Severity, Occurrence and Detection. The factors of this tree are used to have the risk priority number (RPN). More this value is high more the failure is critical to the system. We use to have a critical threshold to classify the failures that need some action to reduce their RPN. However, FMECA present a weakness. Each failure mode is treated separately. In this way, we cannot see if there is a link between two of them and we can have some hidden risk who can escape to this analysis.

Many works tried to eliminate the first weakness and to have a performed FMECA. The first study has been established by Kara-Zaitri [1]. He proposes a methodology of the fuzzy numbers modelization on a simple matrix. This study has been used as a tool for the creation of data bases for the experts. The problem with this method is that she needs many times to be modelized. Martins and Gilson [2] propose a stochastique method based on the monte carlo for the analyses and the simulation of the risks. The advantage with this method is that she covers the all failures that can emanate on a system with her probability despite we need to have a complete understanding of the system with all the information's and the data needed. Facing this constraint, Chen [3] propose a model of linguistic conversion with the comparatives high, medium and low. With this method, we don't need any more to have all the information's before staring a FMECA analysis. The Fuzzy numbers are used to overpass the weakness of this that the first paragraph of a section or subsection is not indented. The first paragraphs that follows a table, figure, equation etc. does not have an indent, either. method. The linguistic terms are then converted to a numerical form.

M. Ezziyyani (Ed.): AI2SD 2018, AISC 912, pp. 297–306, 2019.
https://doi.org/10.1007/978-3-030-12065-8_27

Braglia and Bevilacqua [4] show their doubt about the efficiency of the FMECA. They try to define a set of rules and functions to improve this tool. They propose to use the Analytical hierarchy process (AHP) to obtain a score for the weight of the criteria used. Hua, Hsu, Kuo, and Wua [5] use an approach by using the fuzzy AHP (FAHP) to evaluate the weight of each risk factor to analyse the risk of green components in the European Union.

Kutlu and Ekmekçioğlu [6] used a fuzzy technique for order of preference by similarity to ideal solution (TOPSIS) combined with a fuzzy AHP to obtain a performed FMEA.

Peeters, Basten, Tinga [7] use an approach by combining the fault analysis tree (FTA) and the FMEA.

In this article, we will use a design structure method (DSM), we talk about some works who studied this method. Braglia and Bevilacqua [8] use the analytic hierarchy process (AHP) for selecting the best maintenance strategy. This method was proposed in the first time by Saaty [9].

Kuo and Wua [10] use the Fuzzy AHP to determine the relative weights of the factors and after that the FMEA to analyze the risks of green components. Braglia, Frosolini and Montanari [11] use a multi-attribute decision making approach for prioritizing failures in FMECA. Wang, Liu, Quan [12] propose a new FMEA model which integrates complex proportional assessment (COPRAS) and analytic network process (ANP) method to assess and rank the risk of failure mode.

Yassine et al. [13] use a DSM approach for analyzing the design relationships in product development and for capturing the interrelation among multiple matrices. Tang, Zhu, Tang, Xu, He [14] track with the use of the DSM the interaction on a product design process. It's used to show the hidden risks. Danilovic and Browning [15] use an approach with a domain mapping matrix (DMM) to compare two DSM to improve the decision making among engineers and managers by providing a basis for communication and learning across domains to that of availability which has lately become the biggest challenge especially within the sustainable development objectives and the contradictory industrial context. However, if the most influencing part in production systems can manage its resources sustainably, it will become the most enduring and flexible to face market pressure. Maintenance logistics is more than a simple service but a wider logistics vision which is able to solve availability problems respecting the triptych of the sustainable development in spite of the difficulty to answer the question of which is the practical model to follow.

Browning [16] (2000) highlight the advantage of the use of the DSM. He reviews two type of DSM: Time and statistic DSM and discuss the research directions and new DSM applications.

2 A Performed FMECA

As we have previously explained, FMECA is a linear tool (who don't take in consideration the interdependency between two failure mode). He considers each failure with a dependent and unrelated way. It's not the reality such as in a system each component is linked to the others. Therefore, a failure of a component can lead to the failure of some others. In another way, the combination of multiple component failures will create a greater risk for the main function of the system than if each component fails in a separate manner. With a normal FMECA we cannot take into consideration these cases and certainly the analysis will let escape these risks and we will have a wrong criticality.

The proposed approach start from a normal FMECA then we will try to identify all the relationships of interdependencies between different failure modes. We will obtain a simple DSM matrix. After that, we will try to transform it into a digital DSM matrix using a board of conversion of linguistic terms to numerical number. The result obtained will be produced at risk priority number for each failure part of the interdependency relation in FMECA. In the end, we will obtain a risk priority number that considers the relationships of interdependence and therefore all risks

2.1 The Design Structure Matrix

The DSM is a structural method that represents and visualizes the interactions between the elements of a system. It is widely used as an information flow tracing tool in the product development processes and to represent interactions between features, objectives, and different components of the system that need to be analyzed. She allows a good analysis for these interactions. She is represented by a square matrix that contains the functions, objectives, criteria or components in the rows and columns. The same elements must be in columns and rows. The elements in the diagonal are empty because it's absurd that it's otherwise. The elements outside the diagonal when they are not empty, they represent an interaction between the element in column and in the line.

The Lower triangular matrix represents the upstream information flow and the upper triangular matrix represents the downstream feedback. The Fig. 1 shows an example of a DSM matrix. The interdependencies can be read as follows: In line C, we have two interactions. We can say that C depends on b and that d depends on C.

The DSM was first introduced by Steward. However, it is only recently that this method has begun to attract attention about managing the development of complex products and systems. Robert and Morrow [17] felt that DSM is an effective tool for representing and analyzing interactions. They compared it with other tools to validate its effectiveness. Yassine et al. [13] presented an approach based on DSM. They used connectivity links to analyze design relationships in product development. Tang et al. [14], Eppinger et al. [18] used the DSM in the reengineering process by creating design parameter groupings. Saridakis and Dentsoras [19] used the DSM to structure the parameters of the development hierarchy.

The relevance of the DSM method is based on two factors:

- A good decomposition of the system per the right method based on the right information or expert judgment
- The Identification of all interdependencies.

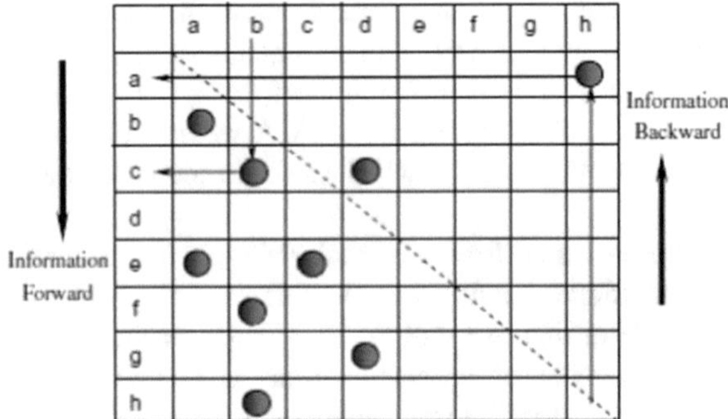

Fig. 1. DSM example

DSMs are effective at visualizing the interdependencies between components. Thanks to its simplicity it is quickly put in place. However, they don't give information on the situation describes. The main weaknesses of this method are:

- The lack of information about the causes of the interdependence
- The lack of information about the importance of the interdependencies.

2.2 The Numerical DSM

The numerical DSM has a big advantage over a simple one. It contains a set of information that can give a detailed overview of the relationships between the different components of the system.

If after the application of a binary DSM, we obtain that the component B depends on the component A. We cannot deduce that it's a strong or weak dependence and whether we will keep this relationship or subsequently delete it. This is the big gap in the simple DSM method. The binary DSM can identify interdependency on a system but it doesn't give any information's about how this relation is. With this tool, we can't have a useful information and its resumed into a graphical tool.

The numerical DSM provides a more detailed description of these interdependencies and subsequently a better interpretation of the system to analysts. Each interdependency is represented by two type of values:

- A value that represents the cause of the interdependency.
- A value that represents the importance of the interdependency.

In our case, we are specifically interested in eradicating the second weakness by transforming a simple or binary DSM to a digital DSM. In our study, we will use a numerical DSM with only value the importance of the interdependency. The level of interdependence is only intended to classify interdependencies for future manipulation on the matrix. In our case, our interest is to use the matrix as it is without manipulations that can lead the values to an unnecessary modification. The only value we are interested in is the importance of interdependencies. It will be directly exploitable with the criticality value obtained will a normal FMECA.

To judge this importance, we opt for the linguistic conversion method of Chen shown in Fig. 2. It offers 9 different levels of judgment which offer more possibilities to experts. The advantage with this method is that it allocates to simple linguistic terms qualitative a value so it can be used easily. The other advantage of this method is that it does not require a total understanding of the system and the analysis can be start immediately.

We will after that transforms the Chen linguistic conversion board into a simple one with a simple column value. The interests to do this transformation is:

- The simplicity of the method
- The time saving opportunity
- The No requirements of advanced knowledge to judge interdependency
- The new method doesn't need to pass by the TOPSIS method. (Technical for Order Preference by Similarity to Ideal Solution) It's already simplified.

Fuzzy evaluation scores for alternatives.

Linguistic terms	Fuzzy score
Very poor (VP)	(0, 0, 1)
Poor (P)	(0, 1, 3)
Medium poor (MP)	(1, 3, 5)
Fair (F)	(3, 5, 7)
Medium good (MG)	(5, 7, 9)
Good (G)	(7, 9, 10)
Very good (VG)	(9, 10, 10)

Fig. 2. The conversion method

We will transform the board on the Fig. 2 to simple one with a simple column. This will avoid to us to pass by the FTOPSIS algorithm and to simplify the method used. The new board is presented on the Fig. 3.

The function of transformation from the Chen board into the simplify board is:

$$Value\,New = 1 + \;(Sum\,of\,value\,old\,/10)$$

The new values are all inferior to 4. The idea is to correct the RPN by taking into consideration the interdependency risks by having an acceptable new value.

So, we will judge each interdependency by her importance we will select a linguistic termed to describe this importance. After that, we will use the Fig. 3 board by selecting the value that come with the term describing this fact.

We will obtain in the final a numerical DSM matrix with values that we can use to calculate the new risqué priority number.

The new board for linguistic conversion	
Linguistic terms	Score
Very Poor (VP)	1,1
Poor (P)	1,4
Medium poor (MP)	1,9
Fair (F)	2,5
Medium Good (MG)	3,1
Good (G)	3,6
Very Good (VG)	3,9

Fig. 3. The new conversion board

The faction that we will use to calculate the new RPN is:

New RPN = RPN X Valor of the interdependency

In the end, we will get a new value of RPN that take into consideration the risk coming from the interdependency.

In Fig. 4, we can see an example of a numerical DSM.

	System Drivers	Stakeholders	Objectives	Functions	Objects	Activities
System Drivers	D X D	S X D	V X D	F X D	O X D	A X D
Stakeholders	D X S	S X S	V X S	F X S	O X S	A X S
Objectives	D X V	S X V	V X V	F X V	O X V	A X V
Functions	D X F	S X F	V X F	F X F	O X F	A X F
Objects	D X O	S X O	V X O	F X O	O X O	A X O
Activities	D X A	S X A	V X A	F X A	O X A	A X A

Fig. 4. Example of numerical DSM

3 The Proposed Methodology

Step 1: This step consists to analyses a system by identifying the set of failures using a n conventional FMECA. This analysis is leaded by a competent and multidisciplinary analysis team of experts. The causes of failures are listed per the different failure modes. Each one is judged using the 3 criteria of FMECA: Severity, Occurrence and detection. We can after that calculates the RPN. For the critical failures, an action plan is done to reduce its criticality.

Step 2: After finishing the analysis of FMECA, the next step consists by using the DSM matrix method, each failure is related to the others to see if there is an interdependencies with other modes. In the end, we will get all interdependency identified.

Step3: After highlighting all the relationships, a simple DSM is built. The elements in the diagonal are empty. the cases outside a diagonal are not empty when there is an interdependency between the element in line and in the column.

Step 4: The simple DSM matrix is transformed into a Numerical one by using the board on the Fig. 3. Each interdependency is evaluating by their importance and per the board of conversion associate to a value. The binary DSM is transformed into a digital one that we can easy used.

Step 5: The value of interdependency importance obtained in step 4 is produced into the risqué priority number of the normal FMECA. We have in the final a new valor of the RPN that take into consideration the interdependency factor.

4 An Example with This New Approach

We will take a simple example with a hydraulic circuit. After performing a FMECA analysis and identifying all failure of the system, we will start the use of the DSM method to identify all interdependency between failures of the system. We dress after that the interdependency matrix. In our system, the interdependency that we identified are:

- The relation between check valve and pump circuit: if there is a failure in the pump circuit and also in the check valve there will be certainly a failure on the pump itself. This relation will be nominated interdependency 1
- The relation between Distributor and pressure regulator: if there is a failure in the distributor and in the pressure regulator, the system will stop. This relation will be nominated interdependency 2.

After identifying all relations, we will highlight the simple DSM. The interdependency is represented by a mark of X.

	Pump circuit	check valve	distributor	pressure regulator
Pump circuit				
check valve	X			
distributor				X
pressure regulator				

Fig. 5. The example DSM

We will transform this simple DSM to a Numerical one by using the Fig. 3 and then each interdependency will have her importance value.

The Interdependency between check valve and pump circuit is judged Medium Good with the valor of 3,1.

The Interdependency between distributor and pressure regulator is judged Medium poor with the valor of 1,9 (Fig. 6).

	Pump circuit	Check valve	Distributor	Pressure regulator
Pump circuit				
Check valve	3,1			
Distributor				1,9
Pressure regulator				

Fig. 6. Numerical DSM

After getting these values of importance, we can transform the binary DSM into a Numerical one. The numerical matrix is shown in Fig. 5.

After that comes the step of calculating the new value of RPN. For this failure mode, the new RPN value is:

- The RPN of the check valve was 12, it will have the valor of $12 \times 3{,}1 = 37.2$
- The RPN of Pump circuit was 6, it will have the valor of 18,6.
- The RPN of the distributor was 8, it will have the valor of 15,2.
- The RPN of the pressure regulator was 12, it will have the valor of 22.8.

We can now change the RPN and have the final form of FMECA. This form is shown in Fig. 7.

failure mode effects and criticity analysis										
CETIM SENLIS	System : system of tooling machine lubrification Sub system : Lubrification pumping			Phase de fonctionnement : MACHINE NORMALE		Analysis Date				AMDEC Machine
						RPN				
Element	Function	Failure mode	failure cause	Failure effect	Detection	Q	D	G	C	correctives actions
Engin	training the pump	No rotation	No alimentation	breakdown machine mano°		1	4	2	8	
			No commande	breakdown machine mano°			4	2	16	
			Engin out of service	breakdown machine mano°		1	4	3	12	
		Inversed rotation	error in harness	breakdown machine mano°		1	2	2	8	
Suction strainer	Filtring the lubrifiant	clogging	Presence of various impurities in the filling	breakdown machine mano°	visual	1	3	3	9	
		bad filtration	deterioration of Suction strainer	Pump attritied		1	3	2	6	
PUMP	Debit the lubrifiant with pression	No debit	Mating failure	breakdown machine mano°		1	4	3	12	
			Internal rupture / blockage	breakdown machine mano°		1	4	3	12	
		Debit not enough	internal attrition	breakdown machine mano°	visual	1	2	4	8	
			Lubrifiant non conform	breakdown machine mano°	visual	1	2	4	8	
Pump cricuit	Establish the hydraulic connection between the pump and the pressure relief valve	clogging	impurities due to attritions	breakdown machine mano°	visual	1	3	2	6/18,6	
		lack	Vibration loosened connections / defective seals	breakdown machine mano°	visual	1	3	2	6/18,6	
Decompression valve	Decompress the primary circuit at the end of the cycle	bloquing	Seizure / Broken spring / Impurity / Wear seal	breakdown machine mano°	visual(Manométre	1	3	4	12	
		clogging	Impureties	breakdown machine mano°	visual(Manométre	1	3	4	12	
Check valve	Avoid overpressure in the primary circuit during the cycle	lack		breakdown machine mano°	visual(Manométre	1	3	4	12/37,2	MPT: verify increasing of pression
			Impuretés	breakdown machine mano°	visual(Manométre	1	3	4	12/37,2	MPT: verify increasing of pression
			breaking	breakdown machine mano°	visual(Manométre	1	3	4	12/37,2	PR: Spring
		Triggering in low pressur	loss of adjustement	breakdown machine mano°	visual(Manométre	1	3	4	12/37,2	
Pressure regulator	permet de régulariser la pression dans le systéme	clogging	impurities due to attritions	No indication of pression	...	1	3	4	12/22,8	NPA:retight the fixations PR: joints, pipe fittings
			attributions due to vibrations	wrong Indication of pression	...	1	3	4	12/22,8	
Manometer	Indicate the pressure in the internal circuits	wrong indicator	loss of adjustement	...	...	1	4	2	4	
Distributor	Establish distribution of the lubrifiant	bad functionment	internal attrition	breakdown machine mano°	visual(Manométre	1	4	2	8/17,1	

Fig. 7. FMECA with new RPN

In Fig. 7, we have the FMECA with the new RPN calculated from interdependency.

In Fig. 8, we have comparison of value between the old RPN calculated by the traditional one and the new RPN of the new method.

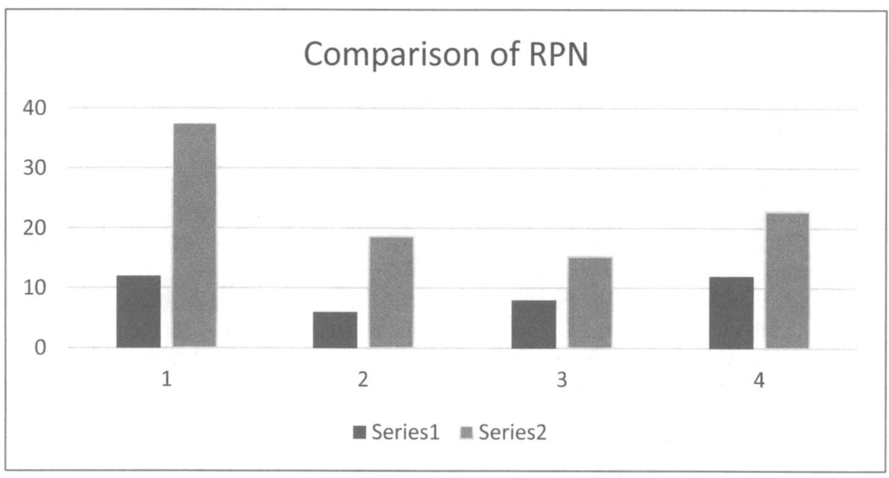

Fig. 8. Comparison of new and Old RPN

5 Conclusion

FMECA is a tool used in all industries. It allows analyzing risks in a system in order to identify all risks. Each failure mode is then listed and associated to a risk priority number value. This valor allows a classification between the failures and subsequently to put an action plan for failure modes exceeding a given threshold. FMECA can also give some recommendations to improve the system.

However, the FMECA treats each failure mode in a separate way. It doesn't allow treating the risks related to the impact of the combination of two or many failures in the same times. This is let escape many hidden risks to this tools.

FMECA need a total understanding of systems and having all information's about the system before starting analysis. We can have a good analysis in the case of a new system that just have been installed.

FMECA have only 3 criterions. It's not suitable for all uses and especially for risky activities such as nuclear, aeronautical or Petrolia industry. In these activity, the most important criteria are security.

This method comes to overcome one of these weaknesses. We can therefore detect all the possible interactions between the various failure modes and make global analysis that considers all the risks of a system. The numeration of the DSM with this method gives a value to each interaction according its importance. We can after that recalculates the RPN for all the failure modes that interact with another's. We produce the value of the DSM by the RPN obtained with a normal FMECA. With this correction, we obtain a RPN more realistic that take into consideration also the hidden risks. For complex systems, we must go through a breakdown into sub-systems, components and sub-components to have a relevant analysis. We propose the use of fault analysis tree that can identify all failures of the systems.

References

1. Author, F.: Article title. Journal **2**(5), 99–110 (2016)
2. Keller, A.Z., Kara-Zaitri, C.: Further applications of fuzzy logic to reliability assessment and safety analysis. Microelectron. Reliab. **29**(3), 399–404 (1989)
3. Sant'Anna, A.P., Martins, E.F., Lima, G.B.A., da Fonseca, R.A.: Beta distributed preferences in the comparison of failure modes. Proc. Comput. Sci. **55**, 862–869 (2015)
4. Jia Chen, C.: Extensions of the TOPSIS for group decision-making under fuzzy environment. Fuzzy Sets Syst. **114**, 1–9 (2000)
5. Braglia, M., Bevilacqua, M.: Fuzzy modeling and analytic hierarchy processing as a means to quantify risk levels associated with failure modes in production systems (2000)
6. Chin, K.-S., Wang, Y.-M., Poon, G.K.K., Yang, J.-B.: Failure mode and effects analysis using a group-based evidential reasoning approach. Comput. Oper. Res. **36**(6), 1768–1779 (2009)
7. Kutlu, A.C., Ekmekçioğlu, M.: Fuzzy failure modes and effects analysis by using fuzzy TOPSIS-based fuzzy AHP. Expert. Syst. Appl. **39**(1), 61–67 (2011)
8. Peeters, J.F.W., Basten, R.J.I., Tinga, T.: Improving failure analysis efficiency by combining FTA and FMEA in a recursive manner. Reliability engineering & system safety **172**, 36–44 (2017)
9. Braglia, M., Bevilacqua, M.: The analytic hierarchy process applied to maintenance strategy selection. Reliab. Eng. Syst. Saf. **70**(1), 71–83 (2000)
10. Saaty, T.L.: A note on the AHP and expected value theory. Socio Econ. Plan. Sci. **20**(6), 397–398 (1986)
11. Hu, A.H., Hsu, C.-W., Kuo, T.-C., Wu, W.-C.: Risk evaluation of green components to hazardous substance using FMEA and FAHP. Expert Syst. Appl. **36**, 7142–7147 (2009)
12. Braglia, M., Frosolini, M., Montanari, R.: Fuzzy TOPSIS approach for failure mode, effects and criticality analysis. Qual. Reliab. Eng. Int. **19**, 425–443 (2003)
13. Wang, L.-E., Liu, H.-C., Quan, M.-Y.: Evaluating the risk of failure modes with a hybrid MCDM model under interval-valued intuitionistic fuzzy environments. Comput. Ind. Eng. **102**, 175–185 (2016)
14. Yassine, A.A.: An introduction to modeling and analyzing complex product development processes using the design structure matrix (DSM) method. Urbana **51**(9), 1–17 (2004)
15. Tang, D., Zhu, R., Tang, J., Xu, R., He, R.: Product design knowledge management based on design structure. Adv. Eng. Inform. **24**(2), 159–166 (2009)
16. Danilovic, M., Browning, T.R.: Managing complex product development projects with design structure matrices and domain mapping matrices. Int. J. Proj. Manage. **25**(3), 300–314 (2006)
17. Browning, T.: Applying the design structure matrix to system decomposition and integration problems: a review and new directions. IEEE Trans. Eng. Manage. **48**(3), 292–306 (2001)
18. Smith, R.P., Morrow, J.A.: Product development process modeling. Des. Stud. **20**(3), 237–261 (1999)
19. Eppinger, S.D., et al.: A model-based method for organizing tasks in product development. J. Eng. Des. **6**, 1–13 (1994)
20. Saridakis, K.M., Dentsoras, A.J.: Integration of fuzzy logic, genetic algorithms and neural networks in collaborative parametric design. Adv. Eng. Inform. **20**(4), 379–399 (2006)

Ship Operational Measures Implementation's Impact on Energy-Saving and GHG Emission

Abdelmoula Ait Allal(✉), Khalifa Mansouri, Mohamed Youssfi, and Mohammed Qbadou

Laboratory: Signals, Distributed Systems and Artificial Intelligence (SSDIA), ENSET Mohammedia, University Hassan II, Casablanca, Morocco
aitallal.abdelmoula67@gmail.com, khmansouri@hotmail.com, med@youssfi.net, qbmeden7@gmail.com,

Abstract. The improvement of ship operation efficiency and the environmental protection are the main pillars for a competitive and sustainable shipping industry. This sustainability depends on the fluctuation of the fuel price market, compulsory international maritime organization environmental regulations and the shipowners policy regarding the energy-saving and their commitment in the reduction of greenhouse gas emission. To ensure a sustainable competitiveness and compliance with environmental requirements of their fleets, the shipping companies have implemented several innovative solutions. Some innovative solutions might be implemented at ship design stage, while others might be implemented at ship operation stage. This paper focuses on the solutions which might be implemented at the operation stage, i.e. ship speed optimization, weather routing optimization, ship trim optimization and hull and propeller condition based maintenance. The effectiveness of these solutions has been demonstrated through study of ship voyages performance reports and simulation of case study.

Keywords: Environment · Energy-saving · GHG emission · Trim optimization · Shipping · Autonomous

1 Introduction

The shipping industry is the backbone of the worldwide trading economy. Over 80% of global trade by volume and more than 70% of its value being carried on board ships and handled by seaports worldwide. United Nations of Trade and Development (UNCTAD) forecasts an increase in world seaborne trade volumes between 2017 and 2022 [1]. This increase in commercial fleet results in an increase of energy consumption. In 2015, the energy used by the maritime transport is estimated at 12EJ and in 2050 at 13EJ [2]. The main source of this energy is fossil marine fuels in particular, heavy fuel oil (HFO). This fuel is widely used in the shipping industry and expected to represent about 40% of fuel use by 2030 [3]. The use of fuel for the ship propulsion results in the production of CO2, SO2, NOx, and pollutant particles. In 2009, the maritime industry GHG emission contribution is estimated at 3% of the global emission [4]. To minimize the shipping global emission share, the international maritime

M. Ezziyyani (Ed.): AI2SD 2018, AISC 912, pp. 307–319, 2019.
https://doi.org/10.1007/978-3-030-12065-8_28

organization (IMO) has set standards for new ships, and associated operational energy efficiency measures for existing ships. These standards consist of the IMO convention Marpol (73/78) annex VI, energy design efficiency design index (EEDI) and the ship energy efficiency management plan (SEEMP). The reduction of ship greenhouse gases (GHG) emission, maximization of energy efficiency and operational expenditure are the key priorities to ensure the sustainability of the shipping industry. This may be met by implementation of technical and operational innovative measures. These measures consist of implementation of solutions, either at ship design stage or at ship operation stage. The solutions that can be implemented at ship design stage are ship hull design optimization, design and operation of ship machinery, structural optimization and light weight construction. Whereas at ship operation stage, the shipowners can implement operational measures, such as speed optimization, weather routing optimization, ship trim optimization, hull and propeller condition based maintenance.

This paper focuses on the implementation's impact of ship operation measures. For an accurate assessment of this impact an in-situ study has been conducted on board of ships of type containers. This study includes relevant documents study, ship performance reports analysis and case study simulation. The result of this study shows considerable impact of these measures on energy-saving and GHG emission.

Several studies have been conducted to assess the impact of the implementation of the operational measures on energy-saving GHG emission. Lu et al. [5] had proposed an empirical fuel consumption prediction approach based on Kwon's added resistance modeling (Kwon, Y.J. 2008) with a specific application to Suez-Max oil tanker. The results of the two case studies indicate that the modified empirical approach for the Suez-Max oil tanker can predict the fuel consumption reasonably well considering the uncertainty factors in the ship actual onboard data recording process and that is more accurate considering the uncertainty and unpredictable factors in ship operational performance prediction procedure. Kim et al. [6] had proposed a reliable methodology to estimate the ship speed loss of a container ship in specific sea conditions of wind and waves. The study showed the capability of the 2-D and 3-D potential methods and Computational Fluid Dynamics (CFD) to calculate the added resistance and ship motions in regular waves in various wave heading. It also demonstrated that the proposed methodology can estimate the impact on the ship operation speed and the required sea margin in irregular seas. Prpić-Oršić et al. [7] had presented the influence of various parameters, such as ship heading angle and weather conditions on ship fuel consumption and CO2 emission. The analysis has been done for various ship heading angles. The voluntary speed loss is taken into account. The analysis result permits to the owner to estimate the economic benefit of various voyage regimes, taking into consideration ship safety and mission. In [8], a simple method to estimate the wind loads on ships was developed, using in the first phase a couple of information, ship type and ship length overall, and additional ship breadth if possible, six or seven parameters representing above water structural features of ships. In the second phase, estimated the wind load coefficients using eight structural parameters, known and estimated in the first phase, and the procedure of precise method developed by the authors.

This paper is organized as follow: Sect. 1 is an introduction for this paper. In Sect. 2, the study approach materials and methods are presented. In Sect. 3, the operational measures impact on energy-saving and GHG emission, including simulations and presentation of the obtained result. In Sect. 4, the implementation of autonomous ship is proposed as an alternative for a sustainable shipping industry. In Sect. 5, the paper is concluded.

2 Materials and Methods

To scrutinize the energy consumption and energy management, several ships of type container ship were visited. Documents, such as voyage performance reports, bunker delivery note, fuel analysis reports, EEDI and SEEMP have been studied in order to assess the impact of the implemented measures on energy efficiency and GHG emission. Specialized software has been used for trim optimization simulation. The analysis approach is based on real case study and real ship operation collected data.

3 Improvement of Energy Efficiency and GHG Emission Reduction at Ship Operation Stage

The implementation of the technical and operational measure has a considerable impact on energy efficiency and reduction of GHG emission. These measures consist of speed optimization, weather routing optimization, ship trim optimization, hull and propeller condition based maintenance.

3.1 Trim Optimization

Most ships are designed to carry a designated amount of cargo at a certain speed for certain fuel consumption under a specified trim condition. The trim has a significant influence on the resistance of the ship through the water. However, an optimal distribution plan of cargo and consumable combined with adequate ballasting of water ballast tanks result in an optimal combination of trim and draft. This helps in fuel saving and GHG emission reduction. The Fig. 1 shows the three cases of ship trimming, i.e. ship is trimming by bow, when the forward draft is greater than the aft draft, ship is on even keel when the forward draft and the aft draft are equal, and ship is trimming by the stern when the aft draft is greater than the forward draft. In fact, for any given draft there is one optimum trim that gives minimum ship hull resistances. To fulfil this, the pertinent information is provided to the Master and cargo planner to allows them to choose the optimal combination of draft and trim for the cargo deadweight, consumables they must carry and the tanks to be ballasted to achieve the fuel saving target. The trim and draft optimization permits to reduce the fuel consumption by 1 to 2%. Recently several trim and draft software tools are commercialized. These applications help the Master to adjust the quantity of ballast and consumables and their distribution to gain some energy efficiency improvement. In our case study, the trim optimization simulation is conducted on a container ship, using the software "seacos MACS3 by interschalt v.NET 1.1".

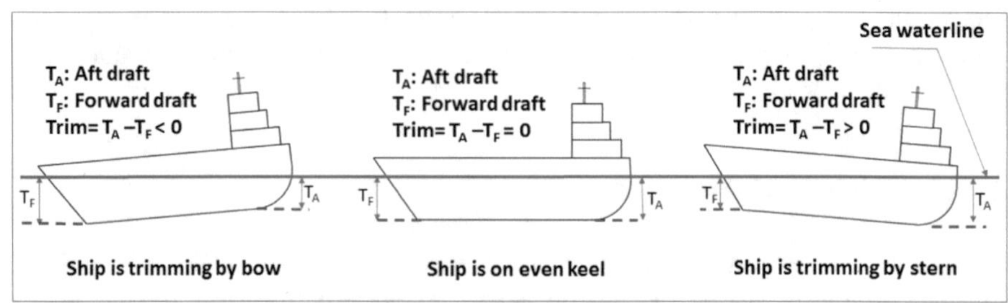

Fig. 1. Trim condition cases.

The container ship particulars are as follow:

Ship type: Container
Gross Tonnage: 41358 t
Length overall: 262.06 m
Breath: 32.25 m
Draught: 12.5 m
Main engine power: 36160 KW
RPM: 102 rpm
Speed: 24.1 kn
Main engine Fuel consumption at 90% of maximum continuous rating (MCR): 145 t/day.

In this case study, we assimilate two trim condition without changing the speed. The ship at even keel condition (trim is null) is taken as an initial condition. The current condition is with ship on the stern, presenting a trim of 0.78 m, and the third condition has been achieved by increasing the ship trim on stern to 1.25 m, by ballasting operation (Table 1).

Table 1. Trim simulation cases.

Particulars	Reference conditions	Current condition	Condition achievable by ballasting
Speed	19.0	19.0	19.0
Draught	10.67	10.67	10.45
Trim	0.04	0.78	1.25

The Table 2, Figs. 2 and 3, show the impact of trim modification on main engine produced power and consequently on energy-saving.

The change of ship trim from reference trim to current trim has been resulted in a decrease of main engine produced power of 1%. This decrease in power has resulted in reduction of the fuel consumption by 0.6 t/day. Based on fuel price of 600$/t, the fuel cost has been reduced by 346$/day. The decrease in fuel consumption has resulted in reduction of CO_2 production by 1.8%. In the second simulated case, the trim has been

Table 2. Trim simulation result.

Saving type	Current saving vs reference	By ballasting saving vs reference
% of main engine power	1.0	1.6
Fuel (t/day)	0.6	1.0
CO_2 (t/day)	1.8	3.1
Fuel cost ($/day)	346.0	585

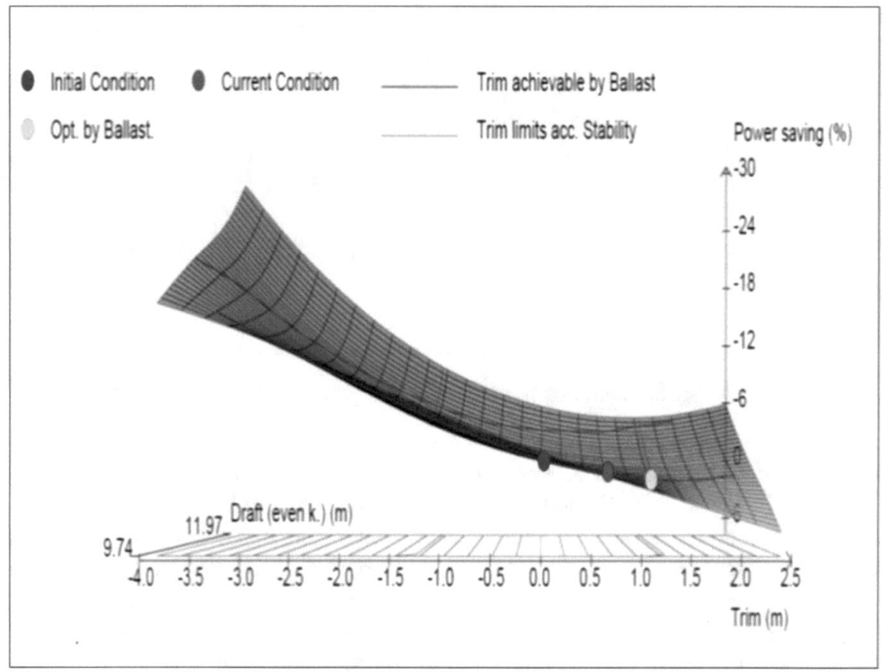

Fig. 2. Trim modification impact on produced power-saving

increased to 1.25 m. In this case, the main engine produced power has been reduced by 1.6%, the fuel consumption by 1.0 t/day, resulting in fuel cost reduction by 585$/day and decrease of CO2 production by 3.1 t/day. This demonstrates the effect of trim optimization on energy efficiency and environment pollution prevention.

3.2 Voyage Optimization

The voyage optimization consists of the selection of an optimal navigation route based on the prediction of the ship performance in various sea states, current and wind condition (Fig. 4). This voyage optimization results in the increase of the energy efficiency and reduction of GHG's emission. In the past decades, the captain has opted for safest and fastest route, whereas recently a safest and energy-efficient route prevails. Weather routing optimization aims at arriving at destination with the minimum fuel

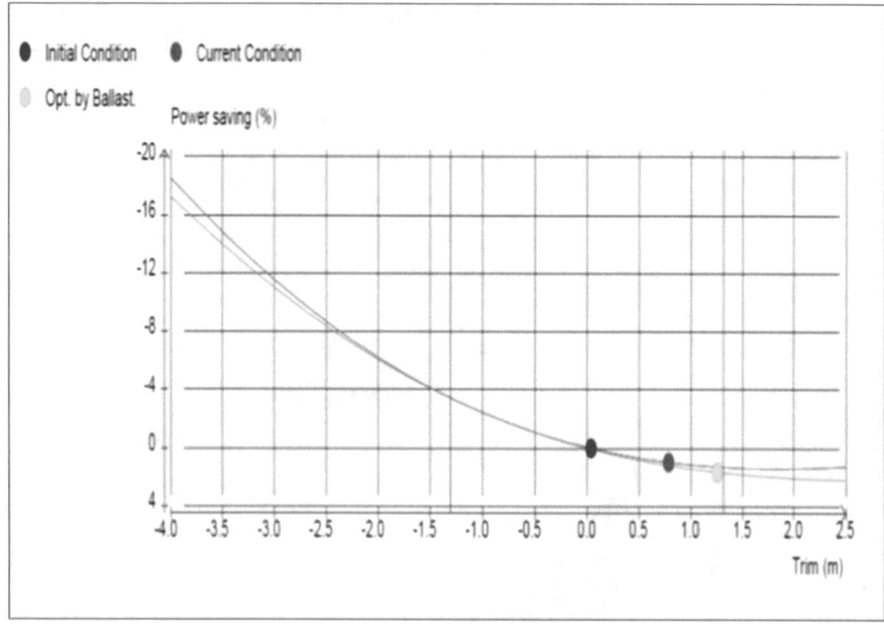

Fig. 3. Trim optimization produced power-saving curve

consumption and sailing time. The shortest distance between two ports of call not always the fastest as it is affected by the sea and weather condition, i.e. wave, current and wind depending on weather severity and voyage length. The weather forecasting and voyage simulation evaluation have been supported by computers for at least 50 years. The implementation of the weather routing and optimization of route may result in an energy-saving of 0.1% to 4% and in a reduction of GHG's emission. To demonstrate the voyage optimization energy efficiency and environmental impact, we take as a case study, three voyages performance reports of a container ship with the following ship particulars,

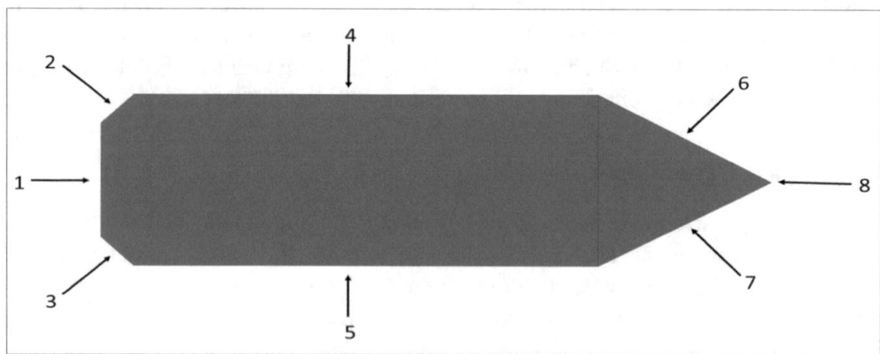

Fig. 4. Wind directions

Ship type: Container ship
Gross tonnage: 25535 t
Length overall: 199.98
Breath: 29.80 m
Draught: 10.1 m
Main engine power: 19810 KW
RPM: 108 rpm
Speed: 21.14 kn
Main engine Fuel consumption at 90% maximum continuous rating (MCR) is 81.5 t/day.

The ship performance in term of the wind direction and force impact on running distance, ship's speed, main engine load, and fuel consumption, has been analyzed for three voyages, with the same port of departure, i.e. port of Tanger Med and the same port of destination, i.e. port of Valencia. The recorded voyages performance is presented in Table 3.

Table 3. Voyage performance abstract.

Voy. no.	Date	Run. dist. (NM)	Ship speed (Kn)	M/E load (%)	Average wind		Sea day (day)	FO cons. (t/day)	CO_2 prod. (t/day)
					Dir	Force			
VOY01	22/07/2017	403	13.5	42	7	4	1.24	42.5	131.75
VOY02	24/11/2017	409	12.7	25	5	4	1.34	35.63	110.45
VOY03	29/12/2017	407	13.3	26	2	5	1.27	35.00	108.5

Note: Voy.: Voyage, Run. Dist.: Running Distance, M/E: Main Engine, Dir.: Direction, FO Cons.: Fuel Oil Consumption, Prod.: Production

During the voyage VOY01, the ship has taken the wind from the fore starboard bow (wind direction 7). Compared to the voyages VOY02 and VOY02, the wind resistance was bigger, the main engine load has increased resulting in an increase of the fuel consumption and CO2 production with less running distance. Whereas, during the voyage VOY03, the ship has taken the wind from the portside aft, the main engine load has decreased, at the same speed, resulting in a decrease of the fuel consumption and reduction of CO2 production. During the voyage VOY02, the ship has taken the wind from starboard side, the running distance and sea days have slightly increased, whereas the fuel consumption has been increased compared to the fuel's consumption during VOY03 and decreased compared to the fuel consumption during voyage VOY01. This study shows that the choice of an optimal route has a great impact on fuel-saving and reduction of GHG's emission reduction.

3.3 Ship Speed Optimization

The ship high speed is a key element in marine transportation, providing economic benefits, such as reducing cargo delivery time, increasing trade volume per unit time, and lowering inventory costs. Nowadays, the shipping companies feel under pressure due to the constant widening gap between charter rate and bunker price and the IMO environmental pollution prevention regulations. Because fuel consumption has a big impact on the ship operation cost, this has brought a new management perspective motivating shipowners to implement projects with strong focus on energy efficiency, i.e. low steaming. The low steaming consists of the reduction of ship speed to cut down the fuel consumption and to reduce the greenhouse gases emission. The determination of the optimal speed takes into consideration all parameters affecting the voyage plan, i.e. sailing speed, fuel economy and market demands. For example, a container ship travelling at a usual speed of around 21 to 24 Knots may travel at a low steaming speed of around 15 to 18 Knots resulting in engine power reduction and fuel oil consumption, which lead to a significant decrease in carbon emission. The major slow steaming benefits are higher fuel savings, reduction in carbon emission (CO2, NOx, and Sox), reliability improvement, and efficiency increase. However, the relationship between speed and fuel consumption is non-linear. It is a third function with the power output required for the propulsion. If the ship speed is doubled, the required power is increased by a factor of 8 [9]. Conversely, if the speed is decreased to 90% of the design speed, the required power is about 75% of the design power. Thus, a reduction of speed by 10% results in a fuel saving of 20% [9]. As per the case study result (Table 4), for a distance of 24000 nautical miles, a slowing steaming from a usual speed of 22 Kn to 17.3 Kn, for a fuel oil price of 600 $, a 2452 TEU container vessel can save approximately 1,131.18 tons of fuel oil resulting in financial savings of 678,708.00 $, leading to the reduction of GHG emission.

Table 4. Low steaming impact on FO consumption.

Engine load (%)	Engine speed (RPM)	Engine produced power (KW)	FO consumption		Ship speed (Kn)
			Hourly (MT)	Daily (MT)	
25	68	5443	0.994	23.851	13.3
50	85.7	10885	1.928	46.270	17.3
75	98.1	16328	2.848	68.340	20.0
90	104.3	19593	3.431	82.349	21.3
100	108	21770	3.489	83.731	22.0

3.4 Ship Condition Based Maintenance

The ship condition maintenance of hull, machinery and propeller impacts significantly the ship performance and results in an increase of fuel consumption and GHG's emission production. The ship condition based maintenance and monitoring of the hull surface roughness, propeller roughness and machinery performance contribute considerably in energy-saving and environmental pollution prevention.

3.4.1 Hull Roughness

The hull surface roughness has a great impact on this performance, resulting in a decrease of ship speed and increase of fuel consumption (Fig. 5). The hull roughness may be physical, i.e. steel plating waviness, welding quality, mechanical damage and corrosion or biological, i.e. fouling (Fig. 6). Both physical and biological hull roughness have effect on a ship's frictional resistance and consequently on fuel consumption. The hull surface should be kept smooth and free of structural damage. This is fulfilled by regular cleaning of hull at the scheduled drydocking or by in water cleaning by divers, application of high quality antifouling coating paint and repair of the corroded, buckled and indented steel plating. It is reported that an appropriate hull cleaning and maintenance combined with high-quality coating can yield an average reduction up to 3 to 4% in fuel consumption. Recoating a rough hull can yield 10 to 12% decrease in fuel costs [10].

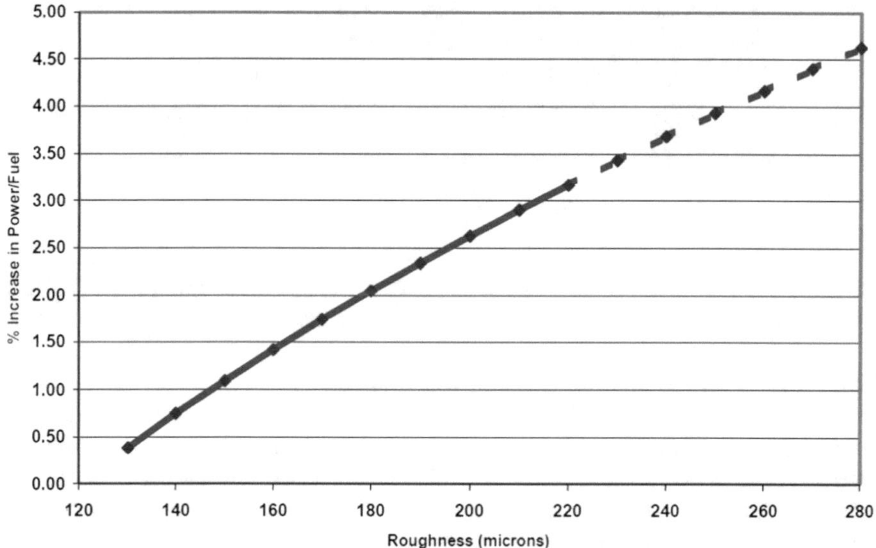

Fig. 5. Hull physical Typical Increase in power/fuel required to maintain vessel speed of a fast-fine ship vs increasing hull roughness (International Paint Ltd. Credit.)

3.4.2 Propeller Roughness

As the ship hull, the ship propeller blade surface smoothness can be affected by cavitation, fouling and damage. Regular cleaning and reconditioning of propeller surface, either at dry dock or underwater cleaning during ship operation (Fig. 7) improve the ship energy efficiency and reduce the GHG emission. Propeller surface roughness causes a greater energy loss per unit area than the condition of the hull surface. Propeller performance monitoring and condition-based maintenance permit to assess effectively the propeller surface condition and to decide either to proceed with an underwater polishing or to wait till next drydocking to clean and polish the propeller.

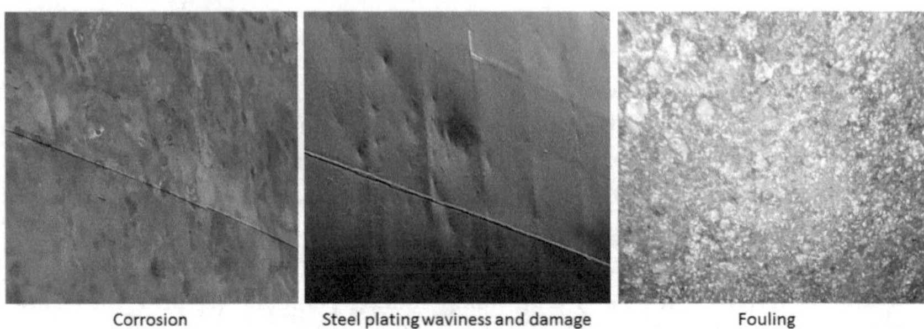

Fig. 6. Hull physical and biological roughness.

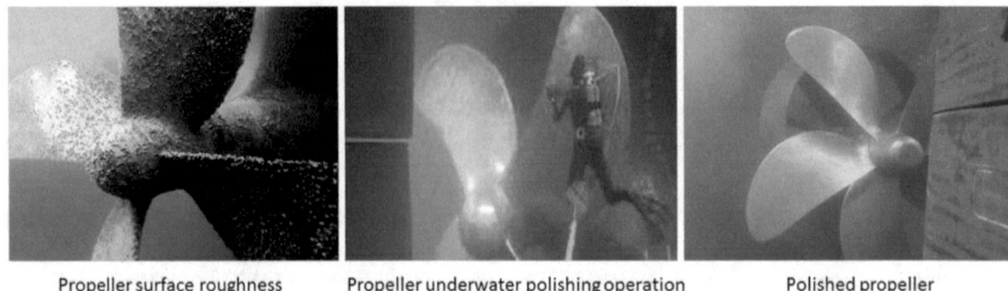

Fig. 7. Propeller roughness.

Cleaning and polishing of the propeller result in a reduction in propulsion fuel consumption up to 6%, leading to the reduction GHG emission [10].

3.4.3 Machinery Condition-Based Maintenance

Ship machinery maintenance is the key priority for energy-saving and GHG emission reduction. Due to normal wear and tear of machinery components, miss-adjustment, overload, long period running without routine maintenance and leakages, the demand in energy increases, leading to an increase of the fuel consumption and GHG emission production. An optimal tuning of main engine may result in reduction of fuel consumption by up to 1% even in extreme cases [11] and one bar increase in maximum cylinder pressure leads to about 0.1–0.2 g/Kwh reduction in fuel consumption [12]. The effectiveness of the shipping company maintenance policy attenuates these impacts and permits to operate the ship efficiently. However, in shipping industry, safety is first, the maintenance management may alternate between condition-based maintenance and preventive maintenance, depending in the criticality of the equipment and the level of redundancy.

3.4.4 Ship Maintenance Case Study

In order to assess the impact of the maintenance management on ship performance, a shipping company fleet performance has been analyzed. This fleet comprises five sister ships. The Table 5 depicts for each ship the performance in term of fuel consumption and speed. Theses performance are benchmarked against a performance reference curve (Fig. 8). The ship 'A' has consumed 26 t/day with an increase of 3,3 t/day compared to the reference fuel amount for the same speed, ship 'B' has consumed 28 t/day instead of 23.50 t/day, the ship C has consumed 29.50 instead of 27.00 t/day and the ship 'D' has consumed 28.50 t/day which is almost the same as the reference performance. Whereas, the ship 'E' has consumed 33.00 t/day which less than the reference performance. Based on the analysis of the maintenance record of each ship of the studied fleet, it found that ships 'A', 'B', and 'C' are due for drydocking to clean the hull and propeller and to overhaul the main engine units. While the ship 'D' was drydocked 6 months ago, and ship 'E' has just left the drydock, where all necessary maintenance has been carried out. This fleet study demonstrates the impact of the ship maintenance on energy-saving and reduction of GHG emission.

Table 5. Fleet operation performance abstract.

Ship	Ship A	Ship B	Ship C	Ship D	Ship E
Speed (Kn)	13.6	13.83	14.83	14.97	15.83
Reference fuel consumption (t/day)	22.70	23.50	27.00	28.00	33.50
Current fuel consumption (t/day)	26.00	28.00	29.50	28.50	33.00

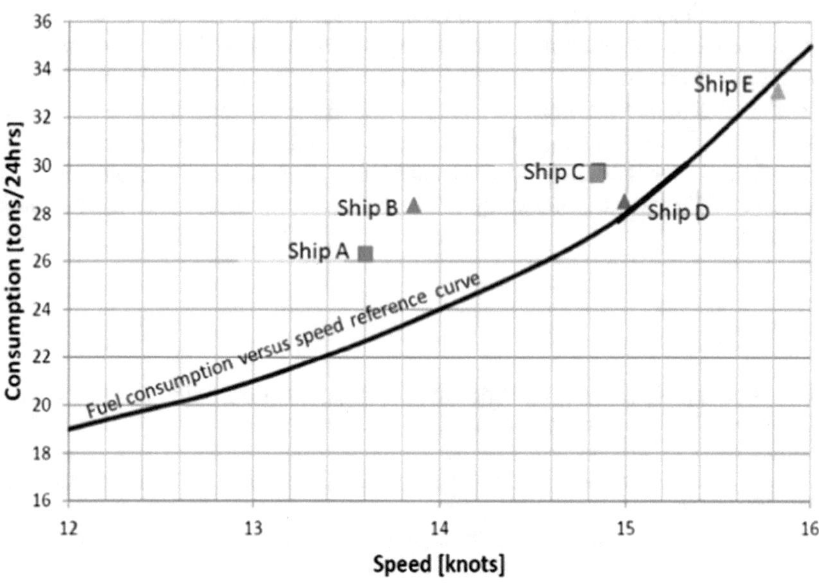

Fig. 8. Fleet operation performance.

4 Implementation of Autonomous Ship as Alternative for Energy-Saving and GHG Emission Reduction

The concept of autonomous ship consists of the elimination of crew onboard and to operate autonomously or to be remotely controlled from the shore control center based on shore. It is assumed that the AS will use clean energy for its propulsion such as liquefied natural gas (LNG), distillated FO, GO, and electrical propulsion. In addition, the AS design will be "non-ballast design". The implementation of autonomous ship will have a great impact on energy saving and reduction GHG's emission. As the autonomous ship, will operate without crew onboard, the accommodation and crew life being facilities are no longer needed. This will reduce considerably the energy consumption and production of garbage and sewage. The benchmark of autonomous ship and conventional ship in term of energy saving shows a saving of 74.5% of the energy consumed by adoption of new design and elimination of non-needed equipment [13].

5 Conclusion

In a world, where a drastic change in climate happens, the shipping companies must reconsider the way they are operating their ships, in compliance with the environment regulations and tough market competitiveness. Shipowners are called to implement innovative operating measures to ensure the maritime industry sustainability in term of environment, energy-saving and cost-effectiveness. In this paper, the impact of the implementation of operating measures has been demonstrated through the study of the ships operation performance and simulation of ship operation conditions. The result shows that ship operation measures may improve significantly the energy efficiency and reduce the GHG emission. This encourages the shipowners to adhere to this policy and to change their approach to ship operation and shipping competitiveness. The perspective of this work is the study of the energy optimization on board of autonomous ship.

References

1. United Nations Conference on trade And Development (UNCTAD); Review of maritime transport 2017; UNCTAD/RMT/2017, UNITED NATIONS PUBLICATION (2017) ISSN 0566-7682
2. DNVGL; Maritime forecast to 2050, Energy transition outlook; DNVGL, November 2017. http://dnv.com/eto
3. The International Council on Clean Transportation (ICCT). The end of the era of heavy fuel oil in maritime shipping. http://www.theicct.org/blogs/staff/end-era-heavy-fuel-oil-maritime-shipping
4. International Maritime Organization (IMO). Second IMO GHG Study 2009, Published in 2009 by the International Maritime Organization, 4 Albert Embankment, London SE1 7 SR (2009)

5. Lu, R., Turan, O., Boulougouris, E.: Voyage optimization: prediction of ship specific fuel consumption for energy efficient shipping. In: Low Carbon Shipping Conference, London (2013)
6. Kim, M., Hizir, O., Turan, O., Day, S., Incecik, A.: Estimation of added resistance and ship speed loss in a seaway. Ocean Eng. **141**, 465–476 (2017)
7. Prpić-Oršić, J., Vettor, R., Soares, C.G., Faltinsen, O.M.: Influence of ship routes on fuel consumption and CO_2 emission. In: Soares, C.G., Santos, T.A. (eds.) Maritime Technology and Engineering. Taylor & Francis Group, London (2015). ISBN 978-1-138-02727-5
8. Ueno, M., Kitamura, F., Sogihnara, N., Fujiwara, T.: A simple method to estimate wind loads on ships. In: The 2012 World Congress on Advances in Civil, Environment, and Materials Research (ACEM'12), Seoul, Korea, August 2012
9. Fathom, Ship performance management, edition (2014)
10. ABS. Ship energy efficiency measures, status and guidance; ABS ship energy efficiency advisory, TX 05/13 5000 13015
11. Second IMO GHG Study. International Maritime Organization (IMO), London (2009)
12. Armstrong, N.V.: Review - ship optimisation for low carbon shipping. Ocean Eng. **73**, 195–207 (2013)
13. Ait Allal, A., Mansouri, K., Youssfi, M., Qbadou, M.: Toward energy saving and environmental protection by implementation of autonomous ship. In: 19th IEEE Mediterranean Electronical Conference IEEE MELECON 2018, 2nd–4th May 2018, Marrakech Morocco (2018)

An Automatic System for Water Meter Index Reading

Naim Ayman[(✉)], Abdessadek Aaroud, and Said Saadani

Laboratory LAROSERI, Department of Computer Science, Faculty of Science, Chouaib Doukkali University, El Jadida, Morocco
aymannaim10@gmail.com, Said.saadani@gmail.com, aaroud.a@ucd.ac.ma

Abstract. Water meter is used as a tool to calculate water consumption. This tool works by utilizing water flow and shows the calculation result with mechanical digit counter. Practically, in everyday use, an operator will manually check the digit counter periodically. The operator makes logs of the number shows by water meter to know the water consumption. This manual operation is time consuming and prone to human error. Therefore, we propose in this article an Android mobile application that calculates the customer's water consumption in real time. By having a Smartphone supporting applications that run on Android, the customer can access his water bill at any time by creating his own subscriber account. Once the subscriber account has been created, the subscriber can take an image of his water meter which will subsequently be sent to the server for processing, the level of consumption as well as any alerts will be transmitted to the citizen on his Smartphone according to the image sent. The customer will, of course, be connected to the Internet to use this application.

Keywords: Character recognition · Viola Jones method · Eigenfaces method

1 Introduction

1.1 A Subsection Sample

Water meter is commonly found in a household or industry to calculate their water consumption. It is used by the water company to know how much water consumption of every household or industry. It shows the water consumption using mechanical digit. Figure 1 shows an example of water meter. The water company checks the number shown in the water meter panel periodically, usually every month. They subtract the number in every period with the last period to know the water consumption in each period. The subtracted number is multiplied by the rates to know how much each household or industry will be charged for their water consumption. Practically, it is checked by an operator each period. The operator visits each household or industry who use the water service. The operator takes a picture of the water meter just for the documentation. They also write down the number shown in water meter. The operator reports the findings back to the water company and the water company will calculate how much each customer will be charged. All the process here is time consuming and

M. Ezziyyani (Ed.): AI2SD 2018, AISC 912, pp. 320–327, 2019.
https://doi.org/10.1007/978-3-030-12065-8_29

prone to human error. Therefore if the image of water meter panel can be processed using computer to read the digit automatically it can save much work time.

Fig. 1. Example of water meter.

This kind of mechanical digit is still widely used in the world, especially in developing countries. It is used not just in water meter but also for gas meter and electricity meter. We found some works which takes the topics in reading the mechanical digit from digital image. The article is organized as follows. Section 1 talks about an introduction to this research. Section 2 contains related work, Sects. 3 and 4 contains the method proposed and Implemented system. The results and discussions are presented in Sect. 5. Section 6 concludes the results.

2 Related Work

There is a series of publication by a group research that focusing on reading a meter counter from gas meter. The first was by Vanetti et al. [1] who take the reading problem from gas meter. They used a neural models for detect the location of digits as they proposed in [2]. A Support Vector Machine is used to recognize each digit character. The overall accuracy is good but the complexity of the calculation is high. The example of digit counter that are shown in the test images have different background color and contrast with the panel background. It really helps in detecting the sequence digit area. Gallo et al. [3] proposed an angle invariant gas meter reading. It is focused on reading the gas meter digit sequence form images that is taken not perpendicular to the panel. They used region-based algorithm and multi-layer perceptron are utilized to localize the digit area. It produced good result, even in degraded images. But, they use the red area of the digit sequence to detect the sequence location. It cannot be used in general meter digit because it utilized the specific color in the digit sequence. There are some researches about text detection and recognition in literature [4, 5] but almost all of them are focusing in detecting text in natural scene and video [6–9]. There is also a focus in detecting license plate from vehicles which have closer characteristics with meter counter in term of shape and specific text location [10–13]. However, characters from digit counter have special characteristic which is different from license plate or text in natural scene. It is always has a boundary in each digit/character because of the effect of counter mechanics system. The mechanics system also makes the digit does not always appear in full or complete form. It makes the reading is difficult. From the previous research by Vanetti et al. [1] and Gallo et al. [3], there are some problems that need to be solved to improve the accuracy, such as the use of specific color to detect the sequence. Therefore, we propose in this article an Android mobile application that calculates the customer's water consumption in real time.

3 Proposed Methodology

3.1 Hardware

A device that can capture images from water meters before transmitting them to a server. By having a smartphone supporting applications that run on Android, citizens can access their water bill at any time by creating their own subscriber account. Once the subscriber account has been created, the subscriber can take an image of his water meter which will be sent to the server for processing.

3.2 Software

The image taken by the camera contains the meter dials as well as other surrounding parts of the meter. These other parts will hinder character recognition, therefore, only the numbers on the meter dials need to be extracted from the image. The first step and the detection of the area of the digits by the method of Viola Jones, after Eigen face for the extraction of the characteristic parameters and CFS for the selection of the characteristics and the Euclidean distance (KNN) for the classification of the digits. At this point the information is ready to be transmitted to the server.

3.3 Character Recognition

Recognising characters from an analogue water meter is unlike normal text, as numbers can be visible in various stages as the dials roll between numbers. Character recognition software has to be found that can easily be trained for a custom character set and thus be trained to recognize partially visible numbers. The character recognition software will then be trained with sufficient samples of numbers in various stages and the recognition results evaluated in terms of accuracy.

4 Implemented System

4.1 Implemented System Architecture

See Fig. 2.

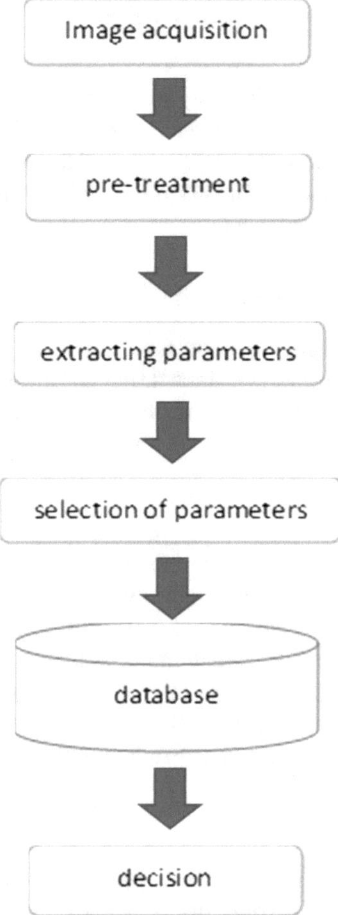

Fig. 2. Implemented system architecture.

4.2 Acquisition of the Image

Capturing is the first step in the process. We must succeed in capturing information relevant without noise. The image in this step will be in a raw state which creates a risk of noise, that can degrade system performance.

4.3 Pretreatment and Detection of the Digit Area

The second step, is the detection of the area of the digits, which consists in detecting the zone numbers in a digital image. This is a specific case of object detection, where one seeks to detect the presence and precise location of this area in the image. After this step, the detected area must be recorded without noise, this is the role of the preprocessing step, by elimination of pests caused by the quality of optical or electronic devices, when acquisition of the input image. The goal is to keep only the essential information, and so prepare the image for the next step.

4.4 Extraction and Selection of Characteristics

This step represents the heart of the recognition system, we extract from the image the information that will be saved in memory for later use in the decision. The choice of this useful information must be discriminating and not redundant by the application of certain criteria for the selection of characteristics. The analysis is called extraction of characteristics. The effectiveness of this step has a direct influence on the performance of the system recognition of numbers.

4.5 The Comparison of Characteristics

In this step, the system is to find the model that best fits for each digit from those stored in the database, the comparison algorithms different. Several approaches can be found in the literature, the simplest of which is the calculation of distance (similarity search).

5 Results and Discussion

I will present in this chapter the methods of extraction, selection and classification that I used to classify numbers in a counter image. As well as I will draw up the various tests carried out and results obtained in this work. Identifying the numbers in an image, a water meter capture would give much more reliable results. In this part I will demonstrate how to identify numbers in images. Before recognizing the objects in an image, the problem must be evaluated. Since my ultimate goal is to recognize the numbers on the image of a water meter, I followed the following steps.

5.1 Binarization

Binarization is an operation that produces two classes of pixels, usually they are represented by black pixels and white pixels as shown in the following Fig. 3.

Fig. 3. Phase of binarization.

5.2 Location of the Numbers Area

This can be done using detection following the principle of Viola and Jones method since there is enough contrast in the image (Fig. 4).

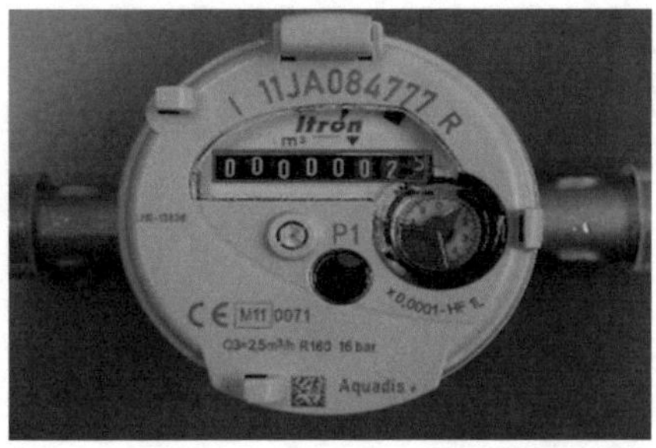

Fig. 4. Detection of the area of the numbers in the image.

5.3 Extraction of Numbers

The result image from the previous steps allows me to find cutouts and look for contours with a rectangular shape (Fig. 5).

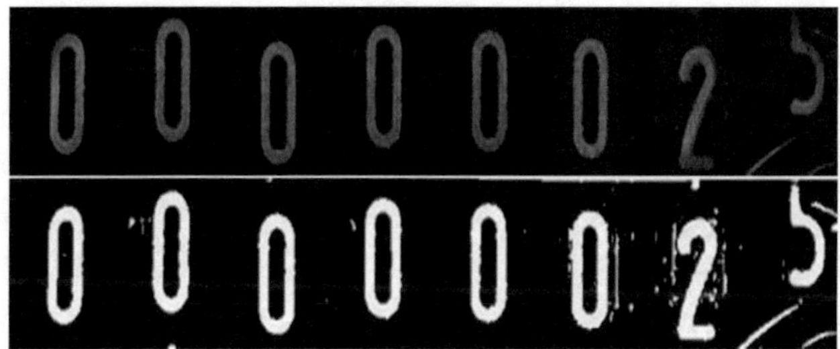

Fig. 5. Extraction and binarization of the numbers area.

5.4 Extraction of Digit Regions

Once we have isolated the numbers I can focus on extracting the numbers. Because seems to be contrast between the figure regions and the bottom of the display we are sure the thresholding and morphological operations can accomplish this (Fig. 6).

Fig. 6. Digit extraction.

5.5 Digit Identification

The identification of actual digit with the KNN classification method will involve the calculation of Markoviski distances and more precisely the Euclidean distance (Fig. 7).

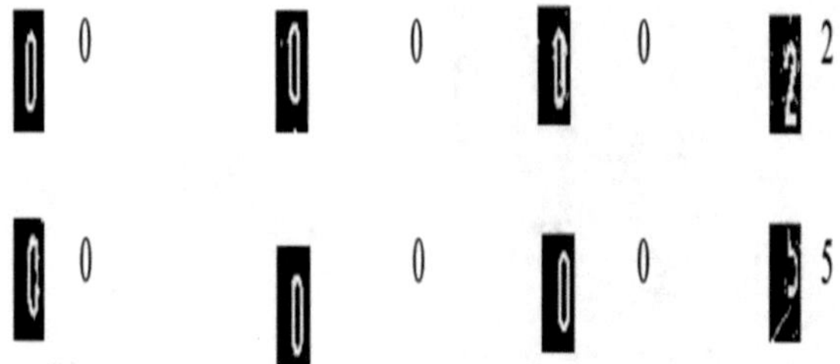

Fig. 7. Digit recognition.

5.6 Final Result

We did several tests to test the performance of our application, finally we found an accuracy of 96%, we were able to accomplish this five-phase process to recognize digits in an image, we will try to develop our system to increase accuracy for future work (Fig. 8).

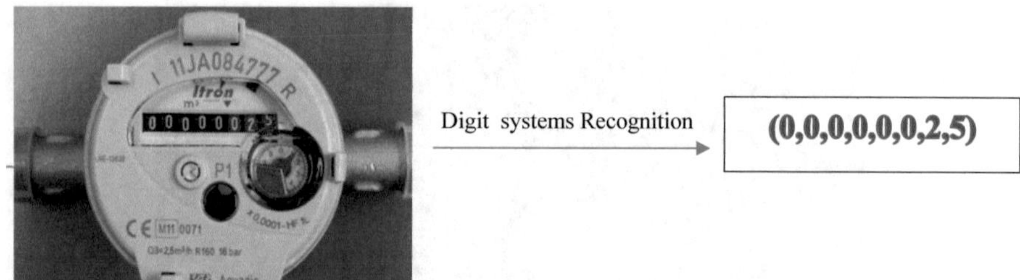

Fig. 8. Final result.

6 Conclusion

We have presented in this article an android application that allows to calculate the customer's water consumption in real time by taking a catch of the meter, this is done using the Viola Jones method to detect area numbers, eigenfaces for extraction of characteristic parameters and CFS for selection characteristics and Euclidean distance (KNN) for the classification of digits. The result shows a promising result that can be improved.

References

1. Vanetti, M., Gallo, I., Nodari, A.: GAS meter reading from real world images using a multinet system. Pattern Recogn. Lett. **34**, 519 (2013)
2. Nodari, A., Gallo, I.: A multi-neural network approach to image detection and segmentation of gas meter counter. In: Proceedings of the IAPR Conference on Machine Vision Applications, Nara, Japan (2011)

3. Gallo, I., Zamberletti, A., Noce, L.: Robust angle invariant GAS meter reading. In: Proceedings of the International Conference on Digital Image Computing: Techniques and Applications (DICTA), p. 1 (2015)
4. Ye, Q., Doermann, D.: Text detection and recognition in imagery: a survey. IEEE Trans. Pattern Anal. Mach. Intell. **37**, 1480 (2015)
5. Zhang, H., Zhao, K., Song, Y.Z., Guo, J.: Text extraction from natural scene image: a survey. Neurocomputing **122**, 310 (2013)
6. Gonzalez, A., Bergasa, L.M.: A text reading algorithm for natural images. Image Vis. Comput. **31**, 255 (2013)
7. Sun, L., Huo, Q., Jia, W., Chen, K.: A robust approach for text detection from natural scene images. Pattern Recogn. **48**, 2906 (2015)
8. Liu, J., Su, H., Yi, Y., Hu, W.: Robust text detection via multi-degree of sharpening and blurring. Signal Process. **124**, 259 (2015)
9. Minetto, R., Thome, N., Cord, M., Leite, N.J., Stolfi, J.: SnooperText: a text detection system for automatic indexing of urban scenes. Comput. Vis. Image Underst. **122**, 92 (2014)
10. Azam, S., Islam, M.M.: Automatic license plate detection in hazardous condition. J. Vis. Commun. Image Represent. **36**, 172 (2016)
11. Wang, Y., Ban, X., Chen, J., Hu, B., Yang, X.: License plate recognition based on SIFT feature. Optik Int. J. Light Electron Opt. **126**, 2895 (2016)
12. Neto, E.C., Gomes, S.L., Filho, P.P.R., de Albuquerque, V.H.C.: Brazilian vehicle identification using a new embedded plate recognition system. Measurement **70**, 36 (2015)
13. Tian, J., Wang, R., Wang, G., Liu, J., Xia, Y.: A two-stage character segmentation method for chinese license plate. Comput. Electr. Eng. **46**, 539 (2015)
14. Jawas, N.: Image based automatic water meter reader. J. Phys. Conf. Ser. **953**, 012027 (2018)

Control of a Proportional Resonant Current Controller Based Photovoltaic Power System

Soukaina Essaghir(✉), Mohamed Benchagra, and Noureddine El Barbri

ISERT Laboratory, Univ Hassan 1, ENSA, Khouribga, Morocco
essaghirsoukaina@gmail.com, m.benchagra@gmail.com, elbarbri.noureddine@yahoo.fr

Abstract. This paper presents a power factor control of PV system connected to the grid. A Proportional-Resonant (PR) controller is used for replacing the conventional Proportional-Integral (PI) controller in this system. By comparison with the conventional PI control method, the PR control can introduce an infinite gain at the fundamental frequency and hence can achieve zero steady-state error. In order to examine the effectiveness of the suggested control, a simulation using the Matlab/Simulink software has been done and it's concluded from the simulation results that the presented control by using the PR controller can be able to maintain maximum active power and to keep always a unity power factor despite variation load.

Keywords: PV system · Proportional Resonant PR · Proportional Integral PI · VSI · Unity power factor · Grid

1 Introduction

Nowadays, more and more interest of photovoltaic (PV) energy has been focused on interconnection between the PV power systems and the grid.

These power systems assume a quality of energy, flexibility and cost effectiveness. Also, the control voltage, the correction of the power factor and harmonic filtering should be provided [1, 2]. The control strategies applied to distributed systems become of high interest. And how to improve the performance of grid inverter, increase the switching frequency and power density to meet power quality requirements, has become a research hotspot in recent years.

The connection of the photovoltaic system to the grid generates a phase shift between the voltage and the current injected into grid, and moreover, during a variation load, fluctuations appear in the voltage and in the output power and the studied system become unstable. For this reason, we need a power factor control which allows the achievement of unity power factor and the adjustment of the output power as required. Two controllers which are used in current controlled PV inverters are the conventional PI controller and the PR controller and the purpose of this work is focused on a comparison between PI and PR current controllers of a Grid-Connected Photovoltaic System Under Load Variation.

M. Ezziyyani (Ed.): AI2SD 2018, AISC 912, pp. 328–336, 2019.
https://doi.org/10.1007/978-3-030-12065-8_30

The rest of this paper is organized as follows. Section 2 introduces a description on the studied system. Section 2.5, explain the philosophy of the proposed PR controller. Then, the obtained results using Matlab/Simulink are presented in Sect. 3. Finally, conclusion is giving in Sect. 4.

2 System Description

The block diagram of PV system connected to grid is shown in Fig. 1. It consists of PV module, DC-DC boost converter, DC-AC voltage source inverter, PWM generator, Current controller and grid.

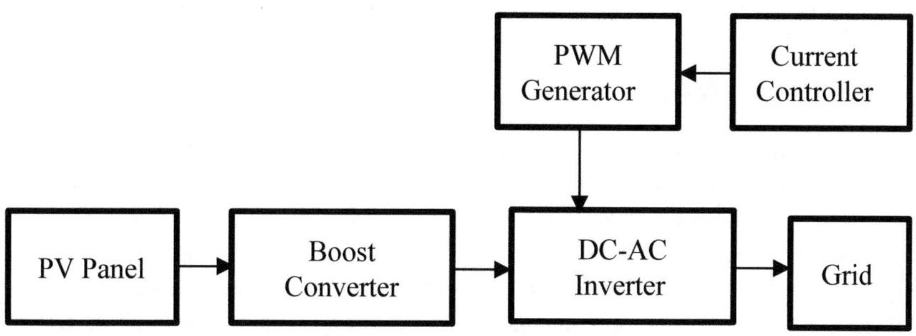

Fig. 1. Block diagram of PV system connected to grid

2.1 PV Panel

The PV array occurs of many cells connected in series and parallel to provide needed output voltage and current [3, 4]; in this paper the PV system consists of 66 strings of 5 series-connected modules connected in parallel. The parameters used in the mathematical modeling of the PV system are as follows:

Ipv: Photoelectric current (A)
Io: Diode saturation current (A)
Vt: Junction thermal voltage $\mathrm{Vt} = \frac{\mathrm{k} \times \mathrm{T}}{\mathrm{q}}$
q: Electron charge $[1.602 \times 10^{-19}\mathrm{C}]$
k: Boltzmann constant $[1.38 \times 10^{-23}\ \mathrm{J/K}]$
T: Cell temperature (K)
V: Terminal voltage (V)
Rs: Cell series resistance (Ω)
Rp: Cell shunt resistance (Ω)
A: Diode ideality factor

The PV circuit model is based on the following equation:

$$\mathrm{I} = \mathrm{Ipv} - \mathrm{Io}\left[\exp\left(\frac{\mathrm{V} + \mathrm{Rs} \times \mathrm{I}}{\mathrm{Vt} \times \mathrm{A}}\right) - 1\right] - \frac{\mathrm{V} + \mathrm{Rs} \times \mathrm{I}}{\mathrm{Rp}} \tag{1}$$

The principal specifications for this model are given in Table 1 (Appendix).

2.2 DC-DC Converter

The proposed method for controlling the power factor consists of two conversion systems; a boost converter and a voltage inverter are used to ensure a good connection to grid and a better quality of the generated waves [5–9], in this study we resort to using a boost converter.

2.3 Three Phase Inverter (DC-AC Converter)

Inverters are used in a large number of power applications. A Voltage Source Inverter (VSI) is one which takes a fixed dc voltage and converts it into independently controlled ac output. The main purpose of these topologies is to provide a three-phase voltage source, where the amplitude, phase, and frequency of the voltages should always be controllable. In our case, the three-level Voltage Source Inverter (VSI) regulates DC bus voltage and keeps unity power factor under variation load. In fact, the control system uses two control loops, an external control loop which regulates DC link voltage and an internal loop which regulates active and reactive currents components (I_d and I_q). The output voltage of the converter and the inverter is controlled by the PWM modulator [10, 11].

2.4 Current Controller

The DC/AC inverter is considered as the core of the whole system because of an important role in grid-connected operation, for this reason the current controller can have a significant effect on the quality of the current supplied to the grid by the PV inverter, therefore, it is important that the controller provides a high quality sinusoidal output by limiting the harmonics generated by these inverters in order to avoid the adverse effects on the grid power quality [12]. For this reason, we compare in this study two types of controllers: the conventional PI controller and the PR controller.

2.5 Proposed Controller

The PR current controller $G_{PR(s)}$ is represented by the following equation:

$$G_{PR} = K_P + K_I \frac{s}{s^2 + \omega_0^2} \tag{2}$$

Where, K_P is the proportional gain term, K_I is the integral gain term and ω_0 is the resonant frequency [13, 14].

The ideal resonant term on its own in the PR controller provides an infinite gain at the ac frequency ω_0 and no phase shift and gain at the other frequencies [15, 16]. The K_P term determines the dynamics of the system; bandwidth, phase and gain margins.

Unfortunately, the ideal PR controller acts like a network with an infinite quality factor, which is hard to implement the PR controller in reality. Firstly, the infinite gain introduced by PR controller leads to an infinite quality factor which cannot be achieved in either analog or digital system.

Secondly, the gain of PR controller is much reduced at other frequencies and it is no adequate to eliminate harmonic influence caused by grid voltage. Therefore, an approximating ideal (non-ideal) PR controller, is given by (3), using a high-gain low-pass filter is used to solve the problems mentioned above [17, 18].

Equation (2) represents an ideal PR controller which can give stability problems because of the infinite gain. To avoid these problems, the PR controller can be made non-ideal by introducing damping as shown in (3) below:

$$G_{PR} = K_P + K_I \frac{2\omega .s}{s^2 + 2\omega_c .s + \omega_0^2} \tag{3}$$

Where, ω_c is the bandwidth around the ac frequency of ω_0. With "(3)" the gain of the PR controller at the ac frequency ω_0 is now finite but it is still large enough to provide only a very small steady state error. This equation also makes the controller more easily realizable in digital systems due to their finite precision [19, 20].

3 Simulation Results

The obtained results using Matlab/Simulink are discussed in this section, Fig. 2 shows the full detailed model which contains a PV array delivering to the grid a maximum power of 100 kW at 1000 W/m^2 sun irradiance.

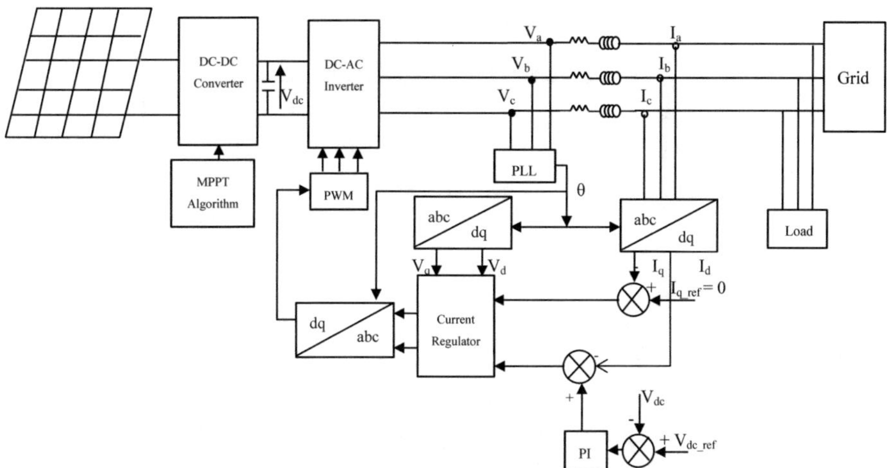

Fig. 2. Full detailed model using Matlab/Simulink

The detailed model contains:

- PV array delivering a maximum power of 100 kW at 1000 W/m^2 sun irradiance.
- Boost converter increases voltage from PV voltage (273.5 VDC at maximum power) to 500 VDC
- Switching duty cycle of the boost converter is optimized by the MPPT controller that uses the P&O technique.
- 1980 Hz three-level three-phase VSI. The VSI converts the 500 VDC link voltages to 260 VAC and keeps unity power factor.
- Current regulator which provide a unity power factor using the PI and PR regulators.
- 100kVA, 260 V/25 kV three-phase coupling transformer.
- Grid specifications (25 kV distribution feeders and 120 kV equivalent transmission systems).

Figures 3a and 3b show the frequency analysis of the grid current and its THD value by using the PI and PR controller, respectively.

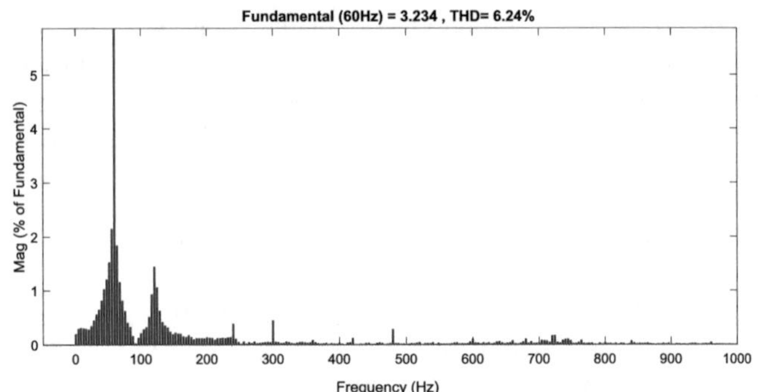

Fig. 3a. FFT analysis and THD value of grid current by using PI controller

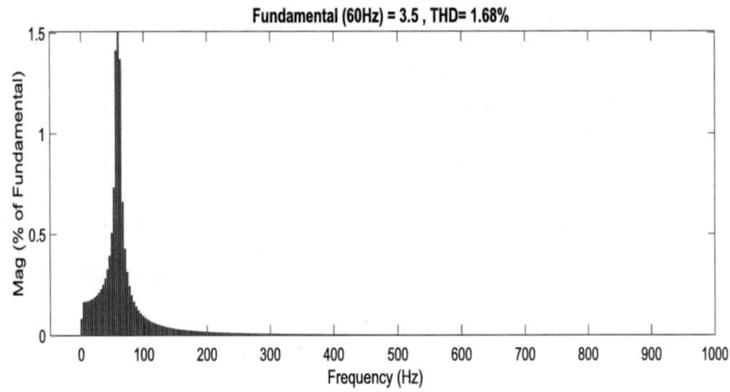

Fig. 3b. FFT analysis and THD value of grid current by using PR controller

The simulation results show that the Total Harmonic Distortion (THD) of the PR controller is much less than that of the PI controller, consequently the PR controller can track the sinusoidal reference and mitigate the harmonics better than PI controller.

Using a PI regulator, Fig. 4a shows that, after the addition of the inductive load at time t = 0.3 s, the current injected into grid becomes in phase delay compared to the voltage but after a few milliseconds, the two quantities V and I become in phase, unlike when using a PR regulator, as shown in Fig. 4b, the current and the voltage is always in phase despite the addition of the inductive load.

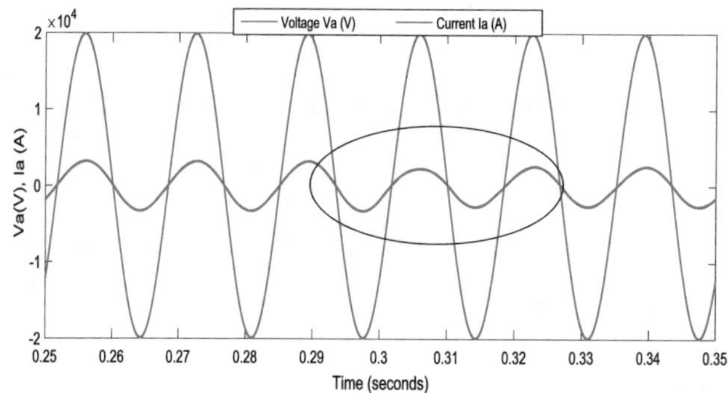

Fig. 4a. Voltage and current injected to grid with variation load by using PI controller

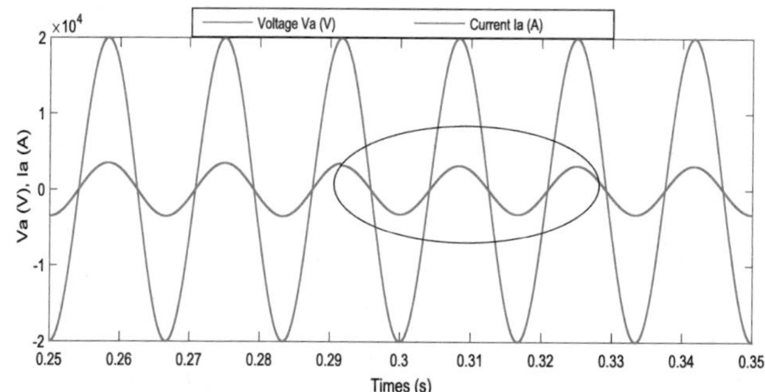

Fig. 4b. Voltage and current injected to grid with variation load by using PR controller

Similarly, Fig. 5a shows that when using a PI corrector, and at the moment when the inductive load is added (t = 0.3 s), the reactive power Q becomes positive and after a few milliseconds it retains its initial value, therefore, by using the corrector PR, the reactive power always keeps its value which is equal to zero which explains the single injection of the active power.

Moreover, it is the same case when we add a capacitive load (as its shown in Fig. 5b), by using a PR controller, the reactive power always keeps its value which is equal to zero which explains the single injection of the active power. For this reason, these figures show the importance of the choice of this type of corrector in order to

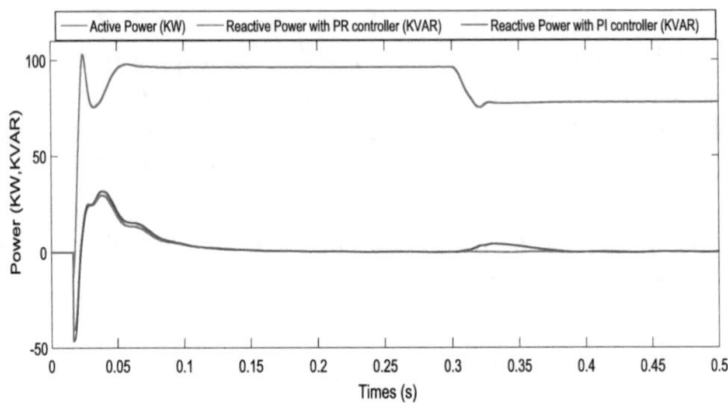

Fig. 5a. Active and reactive power with variation load by using PI ad PR controller (Inductive Load)

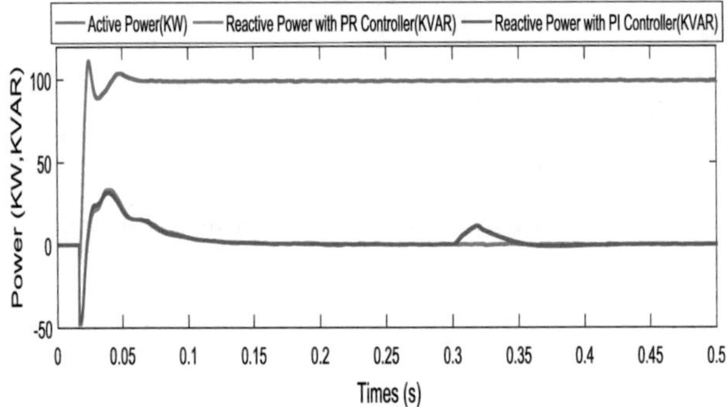

Fig. 5b. Active and reactive power with variation load by using PI ad PR controller (Capacitive Load)

ensure a synchronization between the voltage and the current injected into grid and also to maintain a single injection of the active power and to keep always a unity power factor despite variation load.

Therefore, the simulation gave us the expected results and the PR corrector allows us to have a unit power factor despite the addition of the reactive load and especially without influencing the pace of the reactive power and without introducing any instantaneous phase difference between the voltage and the current injected into grid.

4 Conclusion

In this paper, providing a power factor control of a PV system connected to grid consist on a comparison between conventional PI and PR current controllers in grid-connected PV systems. After the modeling of the three-phase grid connected PV systems and the power factor control schemes, the obtained results are satisfactory and show that the proposed model with the PR controller correctly describes the dynamic performance of the PV system studied, consequently, this corrector can track the sinusoidal reference and mitigate the harmonics better than PI controller, moreover, it allows us to have a unit power factor despite the addition of the reactive load and especially without influencing the pace of the reactive power and without introducing any instantaneous phase difference between the voltage and the current injected into grid.

Appendix

Table 1. Photovoltaic parameters

Temperature	T	25	°C
Open circuit voltage	V_{oc}	64.2	V
Short circuit current	I_{sc}	5.96	A
Voltage, maximum power	V_{max}	54.7	V
Current, maximum power	I_{max}	5.58	A
Maximum power	P_{max}	305	W

References

1. Hongpeng, L., Shigong, J., Wei, W., Dianguo, X.: The maximum power point tracking based on the double index model of PV cells. In: 6th International Power Electronics and Motion Control Conference IPEMC (2009)
2. Villalva, M.G., Gazoli, J.R., Filho, E.R.: Comprehensive approach to modeling and simulation of photovoltaic arrays. In: IEEE Transactions on Power Electronics (2009)
3. Kadri, R., Gaubert, J.P., Champenois, G.: An Improved maximum power point tracking for photovoltaic grid-connected inverter based on voltage-oriented control. IEEE Trans. Ind. Electron. **58**(1), 66–75 (2011)
4. Kim, I.-S., Kim, M.-B., Youn, M.-J.: New maximum power point tracker using sliding-mode observer for estimation of solar array current in the grid-connected photovoltaic system. IEEE Trans. Ind. Electron. **53**(4), 1027–1035 (2006)
5. Su, J., Chien, C., Chen, J., Wang, C.: SIMULINK behavior models for dc-dc switching converter circuits using PWM control ICs. Int. J. Eng. Educ. **22**(2), 315–322 (2006)
6. Soto, D., Green, T.C.: A comparison of high-power converter topologies for the implementation of FACTS controllers. IEEE Trans. Ind. Electron. **49**(5), 1072–1080 (2002)
7. Rodríguez, J., Lai, J.S., Peng, F.Z.: Multilevel inverters: a survey of topologies, control and applications. IEEE Trans. Ind. Electron. **49**(4), 724–738 (2002)
8. Wu, T.F., Nien, H.S., Shen, C.L., Chen, T.M.: A single-phase inverter system for PV power injection and active power filtering with nonlinear inductor consideration. In: IEEE Transactions on Industry Applications (2005)
9. Ouatman, H.: Modeling and Control of a Grid-Connected PV Energy Conversion System, vol. 10, pp. 484–492 (2015)
10. Khanna, V., Das, B.K., Bisht, D.: Matlab/simelectronics models based study of solar cells. Int. J. Renew. Energy Res. **3**(1), 30–34 (2013)

11. Rezgui, W., Mouss, L.H., Mouss, M.D.: Modeling of a photovoltaic field in malfunctioning. In: International Conference on Control, Decision and Information Technologies (CoDIT) (2013)
12. Perera, B.K., Ciufo, P., Perera, S.: Point of common coupling (PCC) voltage control of a grid-connected solar photovoltaic (PV) system. In: IECON Proceedings, Industrial Electronics Conference, no. IECON, pp. 7475–7480 (2013)
13. Zammit, D., Staines, C.S., Apap, M.: PR Current Control with Harmonic Compensation in Grid Connected PV Inverters, vol. 8, no. 11, pp. 1591–1597 (2014)
14. Mari devi, S., Punitha, K.: Resonant current controller based THD reduction in AC Micro. Eng. Electron. Eng. Electron. **22**(2), 425–429 (2016)
15. Teodorescu, R., Liserre, M., Rodriguez, P.: Grid Converters for Photovoltaic and Wind Power Systems. Wiley, Chichester (2011)
16. Zhang, N., Tang, H., Yao, C.: A systematic method for designing a PR controller and active damping of the LCL filter for single-phase grid-connected PV inverters. Energies **7**(6), 3934–3954 (2014)
17. Xiaoqiang, G., Qinglin, Z., Weiyang, W.: A single-phase grid-connected inverter system with zero steady-state error. In: Power Electronics and Motion Control Conference 2006. IPEMC 2006 (2006)
18. Teodorescu, R., Blaabjerg, F., Liserre, M., Loh, P.C.: Proportional-resonant controllers and filters for grid-connected voltage-source converters. IEE Proceedings Electric Power Appl. **150**(2), 139–145 (2003)
19. Zmood, D.N., Holmes, D.G.: Stationary frame current regulation of PWM inverters with zero steady-state error. In: IEEE Transactions on Power Electronics (2003)
20. Essaghir, S., Benchagra, M., El Barbri, N.: Power factor control of a photovoltaic system connected to grid under load variation. In: International Conference on Electrical and Information Technologies (ICEIT) (2017)

Dimensioning of the Coldroom of a Solar Adsorption Cooling System Using Moroccan Climate Data

Hanane Abakouy[1(✉)], Hanae El Kalkha[2], and Adel Bouajaj[2]

[1] Laboratory of Innovative Technologies, National School of Applied Sciences, Tangier, Tangier, Morocco
hananeabakouy@gmail.com

[2] Department of Industrial and Electrical Engineering, National School of Applied Sciences, Tangier, Tangier, Morocco
elkalkha_hanae@yahoo.fr, dbouajaj@yahoo.fr

Abstract. The present study proposes a contribution to the optimization of solar adsorption cooling system, which is a process that produces cold using solar thermal heat energy, this process is based on the phenomenon of adsorption that occur when a balance is established between an adsorbate/adsorbante. This kind of system can be used to save the quality of perishable goods such as meat and vegetables. In this paper the dimensions of the coldroom which is the most important part of this system is studied.

The aim of this study is to determine the influence of the insulation thickness of the coldroom, in order to increase system performance.

The model developed in this paper helps to determine the optimal insulation thickness using the ambient temperature, the thermal conductivity and the cooling load, through different kind of insulators of different conductivity values, and using Moroccan climate data. Taking into account these parameters the insulation thickness of 12 cm with wood wool which is an environment friendly material will makes it possible to reduce the heat transferred to the cold room.

Keywords: Thermal energy · Cold room · Insulation thickness

1 Introduction

Due to its ability to use low temperature heat source and environment friendly materials, the solar adsorption cooling machine gained significant attention, and a considerable numbers of researchers are focusing on the study of this kind of system. The advantage and development of adsorption cycle have been widely studied by Meunier in 1998 [1]. Later researchers have made development to adsorption technology, some have paid attention to the improvement of the COP values whiles others are focusing on the system cooling load and the improvement of mass and heat transfer. A lot of researchers have published their work: Advanced cascaded cycle (1986) [2], thermal wave cycles by Shelton (1990), with experimentation (2015) [3] have been introduced for the enhancement of COP values. While mass recovery cycle by Wang (2001) [4] and Akahira (2005) [5] is for the improvement of system cooling load. Moreover,

M. Ezziyyani (Ed.): AI2SD 2018, AISC 912, pp. 337–347, 2019.
https://doi.org/10.1007/978-3-030-12065-8_31

several works in this field have been successful [6, 7] and refrigerators have been realized [8, 9].

As a part of adsorption cooling system, the coldroom is the element where the cold is produced. This part is used to store perishable goods such as meat and vegetables to slow down their deterioration and preserve them as fresh as possible for as long as possible. Heat accelerates their deterioration so the products are cooled down by removing the heat. In this paper we had paid attention to this part of this machine and the dimensions of the coldroom is studied.

2 System Description

2.1 Description of Solar Adsorption Cooling System

The solar adsorption refrigeration machine as Fig. 1 shows is constituted by three main elements:

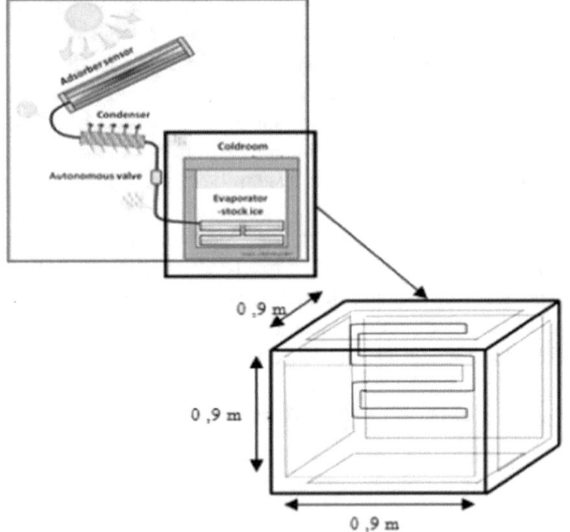

Fig. 1. Example of a solar adsorption cooling system shows the cold room

- The collector-adsorber: a reactor (adsorber) enclosed in a solar collector, containing the working pair, in this part, where the phenomenon of adsorption and desorption are produced;
- The condenser: This element serves to condense the adsorbate vapors desorbed in the collector-adsorber, so in this part where the refrigerant is liquefied;
- The coldroom that contain the evaporator, in which the refrigerant evaporates, producing cold; This element is the useful part of the refrigerator composed of the isolated chamber and the evaporator comprising the adsorbate in liquid and solid form.

2.2 Cold Room Description

In this paper we propose to design the coldroom which provide the proper cold to preserve the quality of perishable goods such as meat and vegetables. This coldroom shown in Fig. 1 contains the evaporator that allows its cooling. To minimize the heat transfer between the atmosphere and the coldroom, we propose to isolate its walls with different kind of insulators: glass wool, wood wool and polyurethane. In the following, we will calculate the optimum coldroom insulation thickness.

3 Thermodynamic Modelization

For the thermodynamic modelization, we have developed an algorithm (Fig. 2) to calculate the variation of the optimal isolation thickness of the coldroom according to the cooling load, in order to minimize the thermal losses.

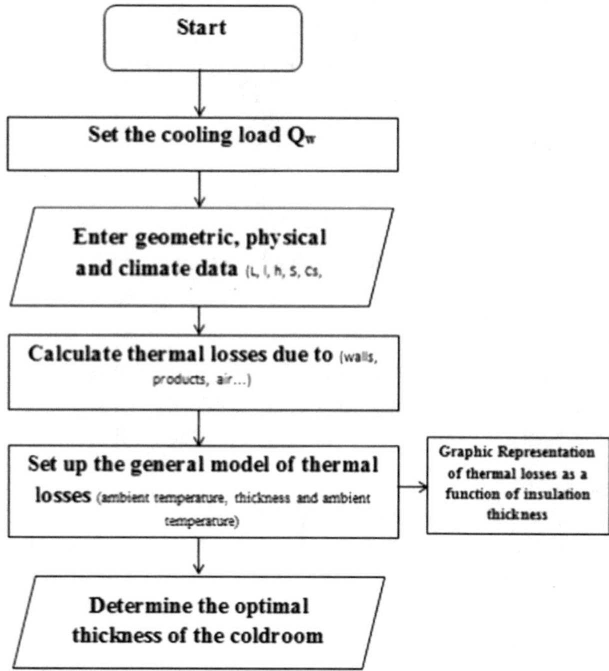

Fig. 2. Calculation algorithm

3.1 Cold Room Heat Sources

Transmission Loads

According to the first thermodynamic law, heat always flows from hot to cold, and the inside of the cold room (Tc) is plainly a lot colder than its surrounding (Ta), (Fig. 3), so heat is always trying to enter the space because of that difference of temperature (Tc is inferior of Ta). If the cold room is exposed to direct sunlight then the heat transfer will be higher, for that reason we propose to place the cold room inside a building and the solar collector on the roof.

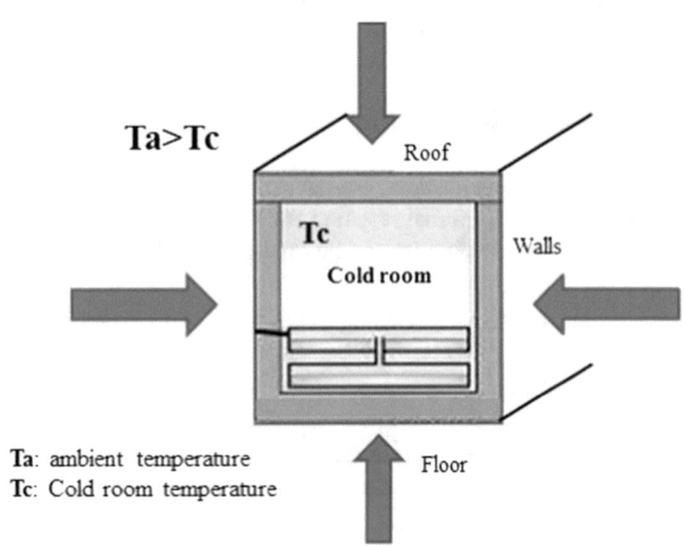

Fig. 3. Transmission load to the cold room

Product Loads
This accounts for the heat that enters the cold room when new products are introduced. It's also the energy required to cool down the products. This part takes the highest percentage of the heat that we have to remove from the cold room.

Infiltration Load
One other thing that we need to consider is the infiltration load which adds a percentage to the cooling load. This occurs when the door opens so there is a transfer of heat into the cold room through the air.

3.2 Dimensioning of the Coldroom

The insulation thickness of the coldroom is calculated step by step following the algorithm shown in Fig. 2, first, we will set the cooling load, then we establish a thermal balance which lead us to set the general model of thermal losses, to determine the relationship between the cooling load, insulation thickness, insulator conductivity and the ambient temperature.

To be able to model our refrigerator, these assumptions are taken into account:

- All walls are at the same temperature
- The ground temperature is 10 °C.
- The temperature difference between the wall and the atmosphere is 5 °C
- The area of the coldroom is calculated by the outside dimensions of the coldroom
- The cold room is placed inside a building to minimize heat transfer.

With these assumptions and the equations of heat transfer [10], we were able to obtain the following models:

- Model of heat loss through the walls of the coldroom

$$Q_{walls} = kS\Delta T \tag{1}$$

With:

k: coefficient of heat transfer (W/m^2.K)
S: coldroom area (m^2)
ΔT: Difference between the air temperature inside the room and the ambient external air temperature (K)

- Model of heat loss by air

$$Q_{air} = V \times \Delta h \times \varphi \times n \tag{2}$$

With:

V: coldroom volume (m^3)
n: cold room opening number
Δh: the enthalpy difference between the ambience in the coldroom and the atmosphere (Wh/kg)
φ: air density (kg/m^3)

- Model of the quantity of heat to be extracted from products

$$\mathrm{Q_{produit}} = \mathrm{P \times C_s \times \Delta T + P \times 1.4} \tag{3}$$

With:

P weight of the products (kg)
Cs: specific heat of the products (Wh/kgK)
ΔT: Difference between the temperature at the arrival of the products and the temperature of storage (K)

4 Results

The convection coefficient, the cold room dimensions and other design parameters that were used in the current study are summarized in Table 1.

Table 1. Parameters value used in the simulation

Parameter	Value	Unit
Natural convection coefficient	8	W/m^2
Convection coefficient in touch with the air	30	W/m^2
Coldroom width	0.9	m
Coldroom length	0.9	m
Coldroom height	0.9	m
Coldroom volume	0.73	m^3

4.1 Climate Data

The results of numerical simulation are obtained using the hourly climate data (ambient temperature) corresponding to a clear typical day of July in Tetouan (Morocco, $35°35'N, 5°23'W$), from the climatological database [11] (Fig. 3).

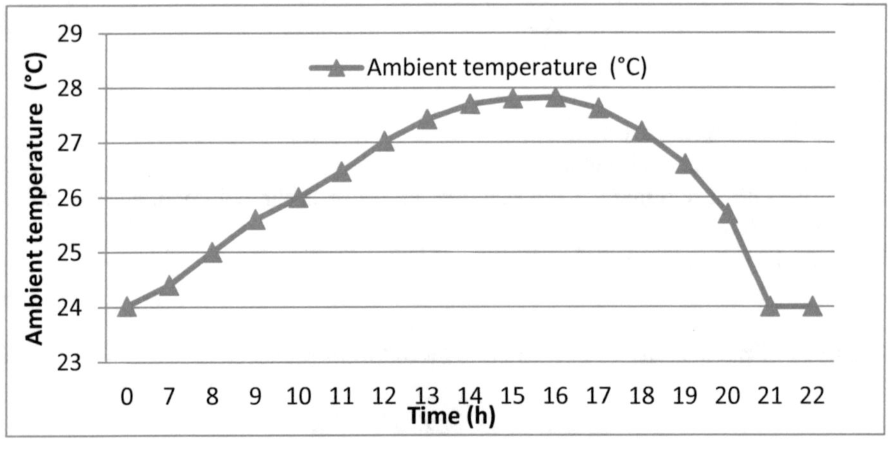

Fig. 4. Climate data used in the simulation [11]

To minimize the heat transferred to the coldroom, the machine must be placed inside a builiding and the collector on the roof. In the following, the temperature used in the calculation of the influence of thermal conductivity is the highest temperature reached, which is 28 °C (Fig. 4).

4.2 Optimal Insulation Thickness

The produced quantity of cold to be installed in the refrigerator was calculated by the following general model:

$$Q_t = \frac{4,05 + 18ep + 20ep^2) \times (T_{amb} + 2) + 1,11 \times 10^2 + 7,01 \times 10^2 \times \frac{ep}{\lambda}}{0,158 + \frac{ep}{\lambda}} \quad (4)$$

We can see that a lot of parameters can influence the quantity of cold produced the main parameters that have been highlighted in the general model are:

- The ambient temperature T_{amb}
- The insulation thickness ep
- The Thermal conductivity of the coldroom walls
- The coldroom dimensions

The influence of each parameter on the quantity of the produced cold leads us to choose the optimal value of insulation thickness. So the results show that the optimum insulation thickness for a cooling load of the evaporator of 760 W must be in the range of 7 to 12 cm.

In the following, we will study the influence of the insulation thickness on the cold room thermal resistance and the thermal losses, using different insulators of different conductivity (Table 2).

Table 2. Optimal insulation thickness using different types of materials

Insulator	Conductivity (W/mK)	Insulation thickness (cm)
Wood wool	0,038	12
Glass wool	0,036	11
Polystyrene	0,028	7,5

As we can see in the following Figs. 5, 6, and 7, the added insulation thickness increases the resistance of the cold room walls, so the heat transferred to the cold room decrease with the insulation thickness.

Influence of the Thermal Conductivity

Polystyrene. Its conductivity is 0,025 W/mK [12], Fig. 5 represents the variations of the thermal losses with the added insulation thickness, so the heat transferred to the cold room without the insulation thickness of 6 cm is 166 W, and the coldroom walls resistance is 0,504 K/W.

Wood Wool. Its conductivity is 0,038 W/mK [12], Fig. 6 represents the variations of the thermal losses with the added insulation thickness, so the heat transferred to the cold room without the insulation thickness of 12 cm is 242 W, and the coldroom walls resistance is 0,607 K/W.

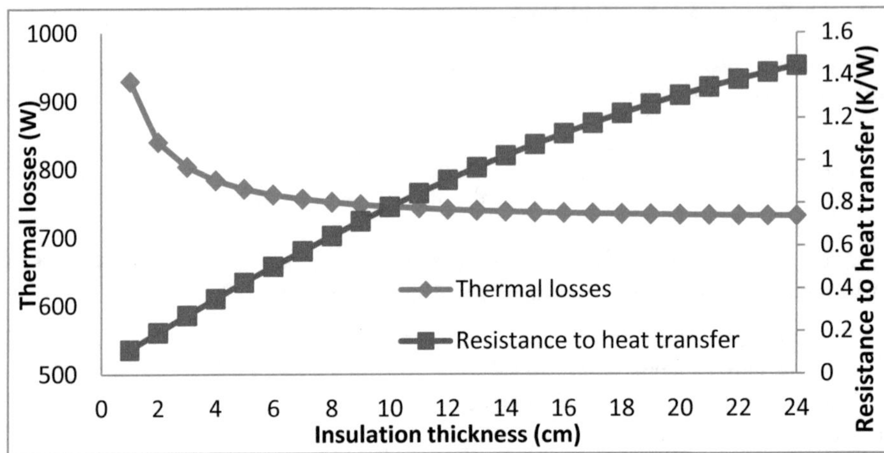

Fig. 5. Influence of the insulation thickness on the coldroom walls resistance to heat transfer and thermal losses with polyurethane

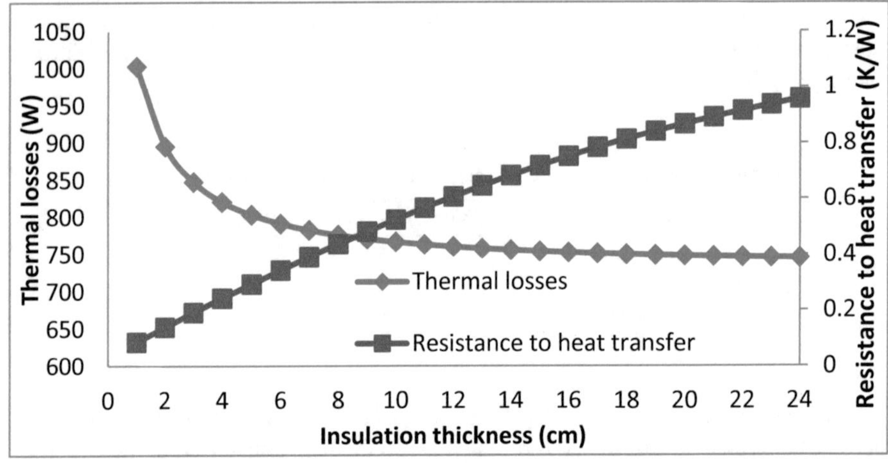

Fig. 6. Influence of the insulation thickness on the coldroom walls resistance to heat transfer and thermal losses with Wood wool

Glass Wool. Its conductivity is 0,036 W/mK [12], Fig. 7 represents the variations of the thermal losses with the added insulation thickness, so the heat transferred to the cold room without the insulation thickness of 11 cm is 232 W, and the coldroom walls resistance is 0,596 K/W.

We can see from these results that the best insulators that provide the effective cooling load with the minimum insulation thickness is the polyurethane with 6 cm of insulation thickness, but since we are working on environment friendly materials, the wood wool is the best ecological insulator among all the materials used in this study.

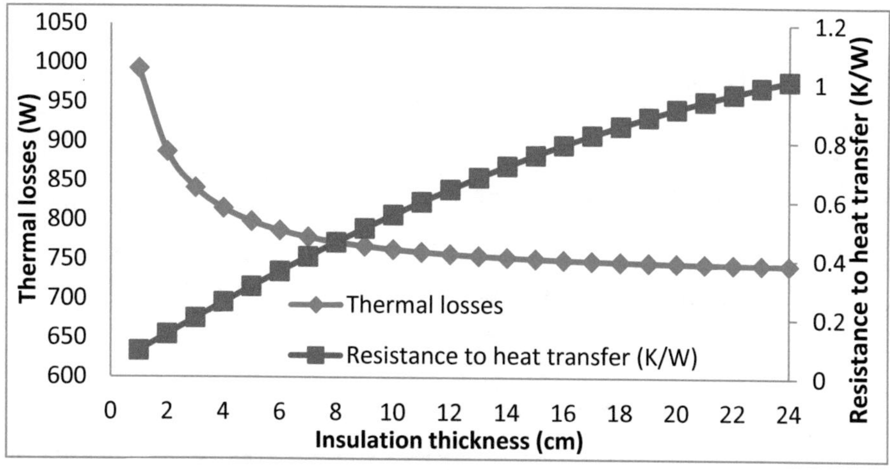

Fig. 7. Influence of the insulation thickness on the coldroom walls resistance to heat transfer and thermal losses with glass wool

So the optimum insulation thickness using the wood wool is 12 cm. This insulation helps to reduce heat transfer and improve system performance.

So the dimensions of the coldroom with the added insulation are summarized in Table 3.

Table 3. Cold room dimensions

Parameter	Value	Unite
External width	1,02	m
External length	1,02	m
External height	1,02	m
Coldroom total volume	1,06	m^3
Coldroom total area	6,24	m^2

Influence of the Ambient Temperature

The Fig. 8 shows the influence of the ambient temperature on the insulation thickness, when the ambient temperature is higher we have to increase the insulation thickness to reduce the heat transferred to the cold room, with the wood wool we have to reach 18 cm in high temperature 34 °C.

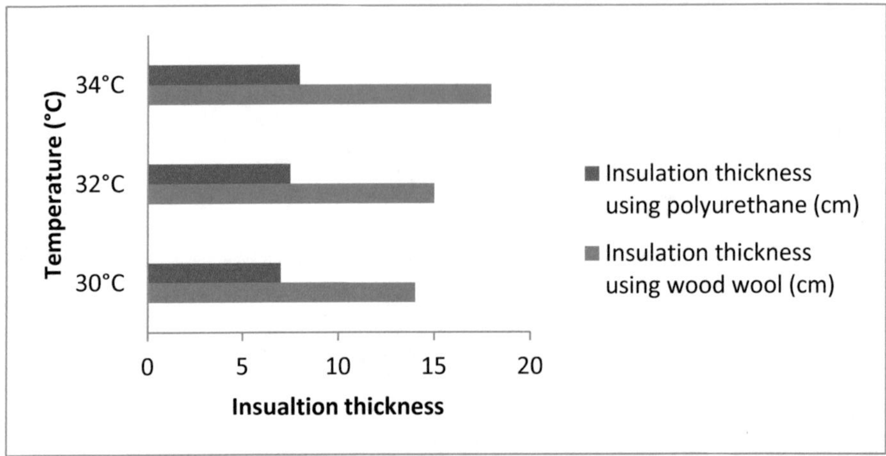

Fig. 8. Influence of the ambient temperature on the insulation thickness of the cold room

5 Conclusion

The solar adsorption cooling system is an innovative process that produces cold using solar thermal heat, as a part of this system the coldroom is the mean element where the cold is produced.

In this study, we had been able to determine the optimal insulation thickness of the coldroom in this kind of system, to reduce heat transferred to the coldroom and improve system efficiency, the calculation is based on the cooling load, we have compared different types of materials, so the optimal insulation thickness of the cold room built with the best ecological material tested in this study is 12 cm of the wood wool, to complete this work it's interesting to add an economic study and we propose as a future work to dimension the other components of our system: the solar collector, the condenser and the evaporator.

References

1. Meunier, F.: Solid sorption heat powered cycles for cooling and heat pumping applications. Appl. Therm. Eng. **18**(9–10), 715–729 (1998)
2. Douss, N., Meunier, F.: Experimental study of cascading adsorption cycles. Chem. Eng. Sci. **4**(2), 225–235 (1989)
3. Grzebielec, A., Rusowicz, A., Laskowski, R.: Experimental study on thermal wave type adsorption refrigeration system working on a pair of activated carbon and methanol. Chem. Process. Eng. **36**(4), 395–404 (2015)
4. Wang, R.Z.: Performance improvement of adsorption cooling by heat and mass recovery operation. Int. J. Refrig. **24**(7), 602–611 (2001)
5. Akahira, A., Alam, K.A., Hamamoto, Y., Akisawa, A., Kashiwagi, T.: Mass recovery four-bed adsorption refrigeration cycle with energy cascading. Appl. Therm. Eng. **25**(11–12), 1764–1778 (2005)

6. Tchernev, D.I.: Solar energy application of natural zeolites. NASA STI/Recon Technical report A, vol. 79, pp. 479–485 (1978)
7. Benelmir, R., Merabtine, A., Descieux, D.: Energy tri-generation: combined gas cogeneration/solar cooling. Int. J. Therm. Environ. Eng. **4**(2), 157–163 (2012)
8. Merabtine, A., Mokraoui, S., Benelmir, R., Laraqi, N.: A bond graph model validation of an experimental single zone building. J. Fluid Dyn. Mater. Process. (FDMP) **8**(2), 215–240 (2012)
9. Boubakri, A.: A new conception of an adsorptive solar-powered ice maker. Renew. Energy **28**(5), 831–842 (2003)
10. Sakadura, J.F.: Initiation aux Transferts Thermiques, (Introudction to Thermal Transfers), Edition Techniques et Documentation (2000)
11. Meteotest: Meteonorm version 6.1 – handbook. www.meteonorm.com. Last Accessed 05 Apr 2018
12. Oliva, J.P.: L'isolation écologique: Conception, matériaux, mise en œuvre, (Ecological Insulation: Conception, materials, Implementation). Edition Terre Vivante (2001)

Turbulent Forced Convective Flows in a Horizontal Channel Provided with Heating Isothermal Baffles

Kamal Amghar(✉), Louhibi Mohamed Ali, Salhi Najim, and Salhi Merzouki

Laboratory of Mechanics and Energy, Faculty of Sciences, Mohammed First University, 6000 Oujda, Morocco
amgharkamaler@gmail.com

Abstract. This work presents numerical results of turbulent forced convection in a horizontal channel provided with heating two baffles mounted on its lower wall. The upper and lower surfaces are maintained at the high temperature and the fluid inlet temperature is lower than the temperature of the walls. Calculations are made by using a finite volume method and an efficient numerical procedure is introduced for studying the effect of inclination angles on heat transfer and flow field for air and high Reynolds number. Results are reported in terms of streamlines, isotherms and local Nusselt numbers. Overall, we can conclude that the results of the study show that the inclination angles of the heated plate alter significantly the temperature distribution, the flow field and the heat transfer in channel.

It was found that the total heat transfer in a horizontal channel mounted the obstacles is increased under effect increase of inclination angle of baffle.

Keywords: Heat transfer · Heated baffles · Forced convection · Turbulent flow · Numerical study

1 Introduction

Heat exchangers are used in a wide range of engineering applications, such as, power generation, auto and aerospace industry, electronics. The main purpose of a heat exchanger is to efficiently transfer the heat from one fluid to the other, which in most of the cases are separated by a solid wall. In many applications such as, power generation, the efficiency of the heat exchangers plays an important role in controlling the overall performance of the system. Computational fluid dynamics (CFD) is now an established industrial design tool, offering obvious advantages [1]. In this study, a full CFD model of shell and tube heat exchanger is considered. By modeling the geometry as accurately as possible, the flow structure and the temperature distribution inside the channel are obtained. Several works have been published on this subject and various phenomena (forced convection phenomena with laminar and turbulent flows) relative to the latter were studied [2–5]. Numerous experimental and numerical investigations have been

---

The original version of this chapter was revised: The correction in author's name "Kamal Amghar" has been incorporated. The correction to this chapter is available at https://doi.org/10.1007/978-3-030-12065-8_39

M. Ezziyyani (Ed.): AI2SD 2018, AISC 912, pp. 348–356, 2019.
https://doi.org/10.1007/978-3-030-12065-8_32

carried out on the problem of heat transfer enhancement from surface mounted heated baffles. Demartini *et al.* [6] investigated air flow through a rectangular channel with two plate baffles. A comprehensive analysis of the velocity profiles and pressure gradients was carried out in this work. While Demartini's approach used a rectangular channel with plate baffles, the question remains whether baffles of different shapes will achieve the same results. Other authors studied in detail the effect of the size of baffles and orientations on the heat transfer enhancement in the tubes heat exchangers. These authors [7–11] have recently studied numerically the fluid flow and heat transfer induced by forced convection in a horizontal channel equipped with two isothermal baffles attached to its isothermal walls.

The objective in the present article consists in studying to quantify numerically the heat transfer by forced convection for different inclination angle of two baffles for analyzed the effects on heat transfer and flow field.

The paper is organized as follows: in the next section, the presentation of the problem and mathematical formulation is supplied; Sect. 3 is devoted to an overview of the numerical procedure. In Sect. 4, our calculations are presented results and discussed. It is followed by a conclusion drawn from obtained results.

2 Mathematical Formulation

2.1 Definition of Geometry

The system considered in this study is sketched in Fig. 1. It consists of a horizontal infinite channel of height *H,* with the upper and lower walls surface maintained at a constant hot temperature T_w. Heating two baffle inclined (of height h) mounted on lower wall surface and also the baffle maintained same wall temperature. A fluid flow enters the tube with a uniform axial velocity u_{int}, temperature T_{int}. The governing equations are written in the coordinate system (*x*, *y*).

2.2 Assumptions and Governing Equations

For analyzing and resolving linear equations of the fluid flow (air) in a horizontal channel, we formulate the following simplifying assumptions have been used in the analysis:

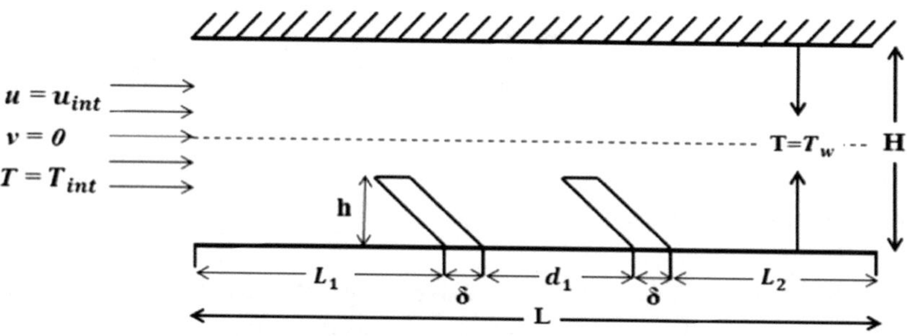

Fig. 1. Physical model

- The flow is considered to be incompressible;
- The problem is two dimensional and steady;
- The heat transfers by the radiation are negligible;
- The temperatures applied to the walls of the duct are considered constants;
- The thermo physical characteristics of the fluid are assumed constant.

The dimensionless equations for the fluid flow and heat transfer are listed as follow:

- Continuity equation:

$$\frac{\partial}{\partial x_i}(\rho u_i) = 0 \tag{1}$$

- Momentum equation

$$\frac{\partial}{\partial x_j}(\rho u_j u_i) = -\frac{\partial P}{\partial x_i} + \frac{\partial}{\partial x_j}\left[(\mu + \mu_t)(\frac{\partial u_i}{\partial x_j} + \frac{\partial u_j}{\partial x_i})\right] \tag{2}$$

- Energy equation

$$\frac{\partial}{\partial x_j}(\rho u_j T) = \frac{\partial}{\partial x_j}\left[\left(\frac{\mu}{P_r} + \frac{\mu_t}{\sigma_T}\right)\frac{\partial T}{\partial x_i}\right] \tag{3}$$

- Shear-stress transport k–ω turbulent model

$$\frac{\partial}{\partial x_i}(\rho u_i k) = \frac{\partial}{\partial x_j}[(\mu + \frac{\mu_t}{\sigma_k})]\frac{\partial k}{\partial x_j} + G_k - Y_k + S_k \tag{4}$$

$$\frac{\partial}{\partial x_i}(\rho u_i w) = \frac{\partial}{\partial x_j}[(\mu + \frac{\mu_t}{\sigma_w})]\frac{\partial w}{\partial x_j} + G_w - Y_w + D_w + S_w \tag{5}$$

Where: Γ_k, Γ_w are the effective diffusivity of k and ω, G_k is the turbulent kinetic energy generation due to mean velocity gradient; G_w is the kinetic energy generation due to buoyancy; Y_k, Y_w are the dissipation of k and ω S_k, S_w are the source term for k and ω.

Local Nusselt number on the channel walls are computed with the following equation:

$$N_\mu(x) = \frac{h(x) \times D}{k} = \frac{q_w \times D}{k(T_w - T_b(x))} \tag{6}$$

3 Numerical Method

Numerical methods must be employed to solve this problem due to the nonlinear nature of the governing partial differential equations. In this order approximations are made to convert partial derivatives to algebraic expressions using the finite volume method [1], while using square mesh. The pressure-velocity fields have been linked by the simplec algorithm [12]. The fully implicit method is adopted. In the flow direction, the derivative terms are discretized with first order backwards differencing scheme while the transversal derivative terms by Quick scheme [10-1]. Hence, we arrange and resolution the system of the discretization algebraic equations coupled with the boundary conditions.

Besides that, a special care was made to ensure the accuracy of the numerical computation, by generating a non-uniform grid in both directions. Accordingly, the grid is refined at the interface of the upper and lower wall, also near the two baffles.

The resulting systems of the discretization equations were solved by an iterative procedure based on a preconditioned conjugate gradient method. A numerical solution is assumed to converge when maximum residual for different physical quantities become lower than ε (see Fig. 2).

This criterion is defined by:

$$\frac{\sum_{i,j}\left|\varphi_{i,j}^{m+1}-\varphi_{i,j}^{m}\right|}{\sum_{i,j}\left|\varphi_{i,j}^{m+1}\right|}\leq\varepsilon \quad (7)$$

Where m is an integer that counts the number of iterations and $\varepsilon = 10^{-8}$ for temperature and 10^{-5} for velocity components.

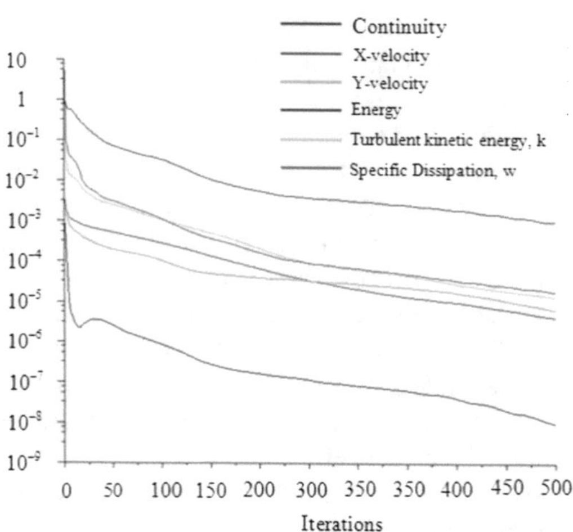

Fig. 2. Convergence test

4 Results and Discussion

In this paper, the process of heat transfer by turbulent forced convection along a channel is analyzed. In all cases, the channel length and hydraulic diameter are respectively L = 0.554 m and H = 0.146 m. The flow of air is entered with inlet pressure P_{inlet} = 1 atm, inlet temperature T_{inlet} = 25 °C and velocity of air particles at the inlet U_{inlet} = 7.8 m/s.

4.1 Flow and Temperature Fields

Figure 3 shows the computational results in the channel through velocity plots at Reynolds number, $Re = 8.73 \times 10^4$ and for three inclinations angle α = 45°, 60° and 90°. Note, that the Reynolds number is kept greater, for this work to restrict the study to the two-dimensional (2D) flow. For α = 45°, the flow is steady in Fig. 3a and b show an almost symmetric distribution (about the channel centerline) of the velocity profiles. Two recirculation zones between the baffles are clearly visible and also the formation of the vortex observed in the vicinity of the upstream first baffle about 0.218 are smaller than those observed of the downstream second baffle.

As α increase the maximum vortex number was increased between the baffles and the core flow penetrates more between the blocks and then the vortices downstream the second baffle became very important to those of the first case, while the vortex was located close to the solid baffles.

Note that for very small values of Re (not presented here), a creeping flow is obtained for each baffle, and the flow around an obstacle is unaffected by the presence of the neighbouring obstacles. Indeed, when α increases, the vortices upstream of the first baffle are removed, and the recirculation zones are still observed between the baffle because the fluid flow is blocked.

Figure 4 shows the effect of the inclination angle α on the computed isotherms for $Re = 8.73 \times 10^4$. In general, the symmetry of the isotherms structure is destroyed. Indeed, in the turbulent forced convection, one notes that for inclination angles α = 45° and 60°, the major part of the air entering inside the channel passes on the side of the baffle, above which a cell forced convection is installed, while, One notes a bad heat exchange especially on the level near the baffle. On the other hand, the increase of the inclination angle to α = 90° shows that the air passes entirely in channel and especially blocked between the baffle while causes increasing the heat transfer, and becomes more intense and increases in volume. This is confirmed by the isotherms structure near the hot walls and the baffles of the channel.

Finally, this solution leads to different thermal behaviors for each inclination angles and the heat transfer is expected to be more important in the case of the vertical baffles (α = 90°) and the main heat transfer is slightly affected by the variation of the inclination angle.

The variation of velocity for the three cases appears clearly on contours and their scales which present positive and negative values. For studying this dependence well, we plotted the velocity distribution for these cases. Therefore one to say little that the use of two baffles purely vertical (plane) results in a significant delay, then it offers time necessary to us in order to ensure the heat exchange desired by contribution the other steepness.

Therefore, the use of the baffles is significant for the improvement of heat exchange but in return, the presence of the baffle in the flow causes an additional pressure drop. So this is a duality to controlling between the thermal performance and the hydraulic performance.

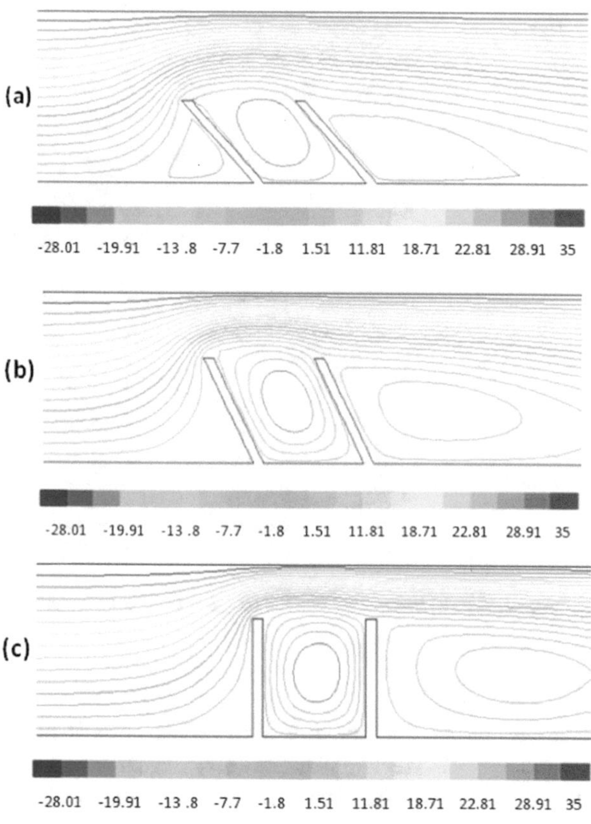

Fig. 3. Streamlines: (a) $\alpha = 45°$, (b) $\alpha = 60°$, (c) $\alpha = 90°$

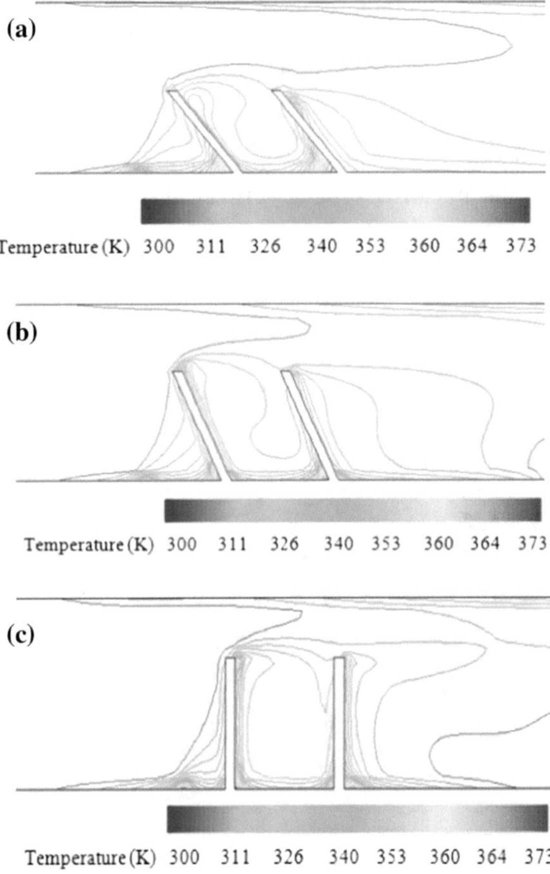

Fig. 4. Isotherms: (a) $\alpha = 45°$, (b) $\alpha = 60°$, (c) $\alpha = 90°$

4.2 Local Nusselt Number

An important outcome of the computation is the local Nusselt number distribution along the channel mounted two rectangular baffles separated by distance d_1. The effect of the inclination angle α on the local Nusselt number distribution for $Re = 8.73 \times 10^4$ is presented in Fig. 5.

Figure 5 show difference between the curves of local Nusselt number for the following cases 45°, 60° and 90 while we notes that in downstream the second baffle we will have increase of heat transfer especially in case rectangular baffle. For the angle $\alpha = 90°$, we note that the peak of the Nusselt number is located approximately at X = 0.45 m, which occurred at a location corresponding to the maximum vorticity, this case is more important compared to other cases.

Moreover, the plots shows that for different values of inclination angle the Nusselt number curves fall in a wide band indicating that there is significant effect of the baffle inclination on the heat transfer augmentation.

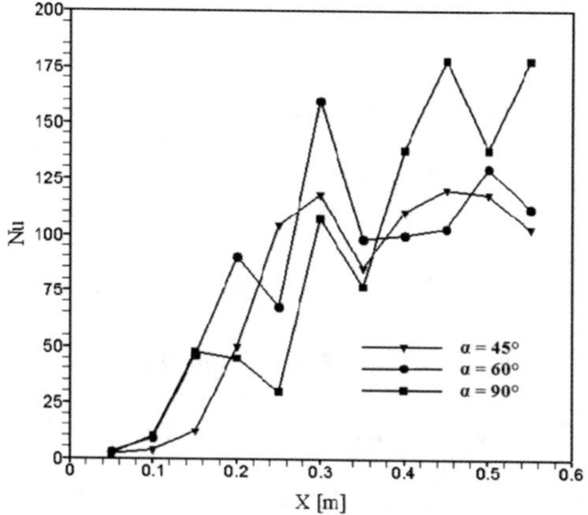

Fig. 5. Local Nusselt number along the channel for $Re = 8.73 \times 10^4$

5 Conclusion

A numerical study of forced convection induced in a horizontal channel provided with heating isothermal two rectangular baffles on its lower wall is performed. A particular interest has been given to the effect of the inclination angles of the baffles on the flow structure and heat transfer inside the channel. Results are reported in terms of isotherms, streamlines and local Nusselt-numbers. The following main conclusions can be drawn.

- Whether in the forced convection, the inclination angle φ affects strongly the structures of the isotherms and the streamlines.
- The variation of the inclination angle has a noticeable influence on the local Nusselt number along the channel, but it is seen that when inclination angles increases, the local Nusselt number becomes important especially the vertical baffle ($\alpha = 90°$).
- The maximum enhancement of heat transfer is about 50% for $Re = 8.73 \times 10^4$ in the presence of the vertical baffle.
- The maximum enhancement of heat transfer is about 48% in the presence of the vertical baffle.

Generally, for these solutions, the quantities of heat evacuated from each case of inclination angles of the baffle present important differences. Although not considered in this study, further points should be investigated: (1) Increasing the height of the baffle and it effect on the pressure drop and heat transfer characteristics. (2) Effect of the aspect ratio on heat transfer and structure of the flow.

Hence it can be concluded that our code is very efficiency and we could study others configuration in a heat exchanger for example: a channel with the square blocks and we try to develop the code for numerical modeling of geometries in 3d for studying all forms of industrial exchangers.

References

1. Versteeg, H.K., Malalasekera, W.: An Introduction to Computational Fluid Dynamics the Finite Volume Method. Addison Wesley Longman Limited, London (1995)
2. Wilfried, R., Deiying, W.L.: Effect of baffle/shell leakage flow on heat transfer in shell-and-tube heat exchanger. Exp. Thermal Fluid Sci. **8**(1), 10–20 (1994)
3. Rajendra, K.: Experimental study of heat transfer enhancement in an asymmetrically heated rectangular duct with perforated baffles. Int. Commun. Heat Mass Transf. **32**(1–2), 275–284 (2005)
4. Gupta, B.B., et al.: A helical baffle for cross-flow micro filtration. J. Membr. Sci. **102**, 31–42 (1995)
5. Lei, Y.-G., et al.: Design and optimization of heat exchangers with helical baffles. Chem. Eng. Sci. **63**(17), 4386–4395 (2008)
6. Demartini, L.C., et al.: Numeric and experimental analysis of the turbulent flow through a channel with baffle plates. J. Braz. Soc. Mech. Sci. Eng. **26**(2), 153–159 (2004)
7. Nasiruddin, M.H., Kamran, S.: Heat transfer augmentation in a heat exchanger tube using a baffle. Int. J. Heat Fluid Flow **28**(2), 318–328 (2007)
8. Saim, R., Bouchenafa, R., Benzenine, H., et al.: A computational work on turbulent flow and heat transfer in a channel fitted with inclined baffles. Heat Mass Transfer **49**(6), 761–774 (2013)
9. Saim, R., Benzenine, H., et al.: Turbulent flow and heat transfer enhancement of forced convection over heated baffles in a channel: effect of pitch of baffles. Int. J. Numer. Methods Heat Fluid Flow **23**(4), 613–633 (2013)
10. Amghar, K., Louhibi, M.A., Salhi, N., Salhi, M.: Numerical simulation of forced convection turbulent in a channel with transverse baffles. J. Mater. Environ. Sci. **8**(4), 1417–1427 (2017)
11. Amghar, K., Louhibi, M.A., Salhi, N., Salhi, M.: Numerical modeling of the effect baffle inclination angle on flow and heat transfer along a horizontal channel. World J. Model. Simul. **14**(3) (2018, in press)
12. Patankar, S.V.: Numerical Heat Transfer and Fluid Flow, p. 131. McGraw–Hill, New York (1980)

Modeling the Flow Behavior of Phosphate-Water Slurry Through the Pipeline and Simulating the Impact of Pipeline Operating Parameters on the Flow

Hamza Belbsir(✉) and Khalil El-Hami

Faculty of Khouribga, Laboratory of Nanosciences and Modeling, Hassan First University, Khouribga, Morocco
h.belbsir@uhp.ac.ma

Abstract. In this study, we were interested in the flow of phosphate-water slurry into the pipeline linking the Khouribga mine pole (Morocco) and the Eljorf-Lasfar industrial platform in El Jadida (Morocco). We carried out a mathematical modeling on the linear head losses, the hydraulic gradient, the friction factor and the geographic profile followed by the pipeline. The results of the modeling were grouped in a program in MATLAB. The modeling allowed us to know the values of the pressures in all the points of the pipeline according to the operating parameters. The agreement of the experimental results with the results of our modeling pushed us to make simulations through our theoretical model. The simulations focuses on the impact of flow rate, density, viscosity of the slurry and choke on the behavior of head losses. This simulation allowed us to present perfectly the impact of the physical properties of the slurry and the operating parameters of the pipeline on the behavior of head losses. This study includes the case of continuous pumping in slurry and the batch case.

Keywords: Modeling · Simulation · Pipeline · Head losses · Geographic profile · Pressure · Density · Viscosity · Choke

1 Introduction

Between the Khouribga mining basin and the industrial platform of JorfLasfar, the OCP Group built, over 187 km, the Slurry Pipeline to convey phosphate slurry for valorization. This transport infrastructure is revolutionizing both upstream and downstream of phosphate mining in Morocco. The objective of the transport of the phosphate slurry through the pipeline is to reduce the energy bill and to increase the production capacity. This mode of slurry transport essentially depends on two constraints: The geographic profile of the route (Khouribga- Jorflasfar), and the physical properties of the phosphate slurry.

The "Slurry pipeline" project consists of a main line of **187 km** in length and **90 cm** in diameter, linking the head station (pumping station) to the terminal station, and 4 secondary pipelines feeding the head station In slurry from phosphate laundries [1–3, 8].

M. Ezziyyani (Ed.): AI2SD 2018, AISC 912, pp. 357–370, 2019.
https://doi.org/10.1007/978-3-030-12065-8_33

The fluid studied is phosphate slurry, which is a mixture of very fine grains of phosphate and water, the mixture consists of **60%** phosphate and **40%** water. In this study, we consider that the phosphate slurry behaves like a pseudo plastic fluid characterized by its density $\boldsymbol{\rho}$ and its viscosity $\boldsymbol{\mu}$ [9], the flow regime of the slurry is always turbulent to avoid the sedimentation of solid aggregates down the pipe.

The main objective of the study in this paper is to establish a theoretical model that simulates the operation of the pipeline's operating parameters. This study uses linear interpolations to model the geographical profile followed by the pipeline, In order to have a function $\mathbf{Z(x)}$ which manages the behavior of the geographical profile, and we also rely on the formulas of 'Darcy' and 'Swamee and jain' [4–7], to model the linear head losses in order to have a function $\mathbf{H(x)}$ which manages the head losses and the behavior of the hydraulic grade line during the flow.

We also adopt an experimental approach to compare the real results with the results generated by the theoretical model. We begin with the experimental approach to describe the components of the pipeline installation and to mention experimental values of pressure and hydraulic gradient in the pipeline.

2 Experimental Approach

2.1 Description of the System

The Hydraulic System
A typical pipeline system consists of a tank, a suction line with a valve, a pump, an exhaust pipe and a tank [8].

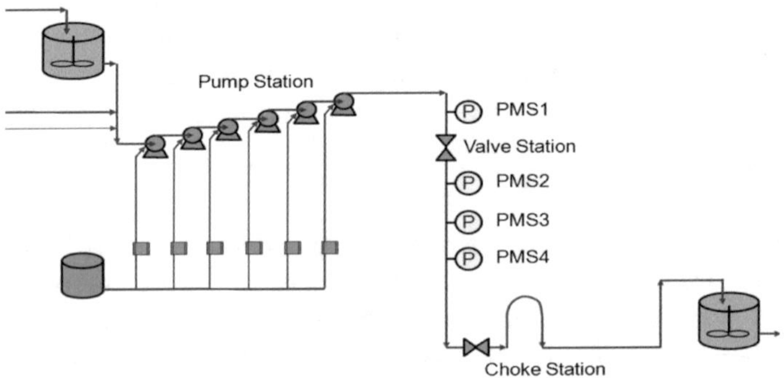

Fig. 1. The components of the hydraulic pipeline system.

The head station consists of 4 slurry storage tanks and a pumping station, this pumping station is equipped with 6 centrifugal pumps mounted in series, to increase the pressure and keep the same flow. The terminal station is composed of 8 slurry storage tanks and a (choke station), the role of the choke station is to brake the pipeline from its end to adjust the pressure and flow rate, because of this choke the hydraulic head increases in the pipe. Along the path are installed pressure and monitoring stations (PMS) marked in (Fig. 1). The PMS1 is located 46 km from the head station, the PMS2 is located 101 km from the head station. The PMS3 is located 129 km from the head station, the PMS4 is located 161 km from the head station. For the station of the valves this station is located at 68 km of the head station, it also contains sensors for measurement of pressure [8] (Fig. 2).

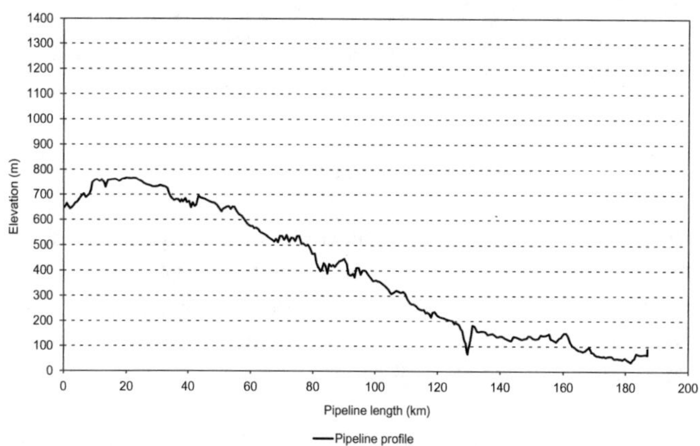

Fig. 2. The geographical tracing Khouribga-Jorf Lasfar

The Geographical Tracing (Pipeline Profile)
The geographical tracing is the path followed by the pipeline, taking into account at sea level height. The geographical tracing represents the profile of the pipeline, linking the head station to the terminal station.

The height of the head station (Khouribga) in relation to sea level is 645 m.

The height of the terminal station (jorflasfar) in relation to the sea level is 66 m.

The geographical tracing presents the profile of the pipeline over the 187 km (altitude as a function of the distance).

The head station is at the point (x = 0), and the terminal station is at the point (x = 187 km).

We note that this is a slope between the head station and the terminal station, therefore the energy required for the pumping is optimized by the gravity.

2.2 Measurement of Operating Parameters and Physical Properties of the Slurry

The head station contains an internal laboratory which aims to measure the plastic viscosity of the slurry; and analyzing the particle size of the particles contained in the mixture (water-phosphate), the diameter of the phosphate particles is always lower to **250 µm**, for measuring the density of the slurry, the pipeline is equipped with Sensor with gamma ray, this sensor is permanent, it delivers us the value of the density of the slurry at every moment. Measurement of flow in the pipeline is carried out through an ultrasonic flow meter and pressure measurement through ABB2600T sensors. These pressure and flow sensors are contained in all of the (PMS) mentioned previously (Fig. 3).

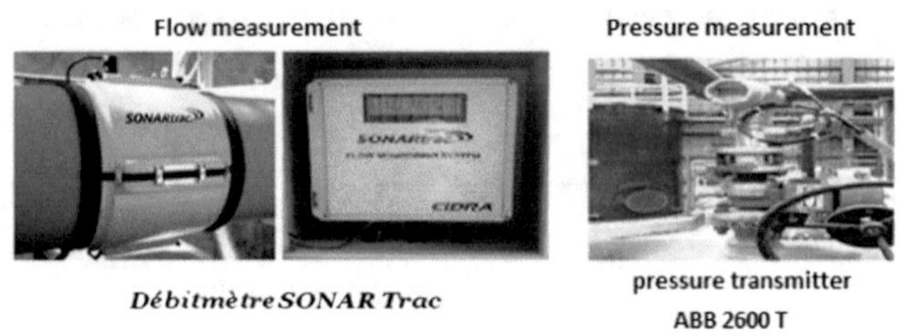

Fig. 3. Measuring instruments

All these instruments are linked to a control and data acquisition system (SCADA), which displays in the control room the general behavior of the transport process via pipeline. This allows operators to track, and control the operating parameters. The values measured by all the metrology instruments that we have already mentioned are displayed in an interface in the control room.

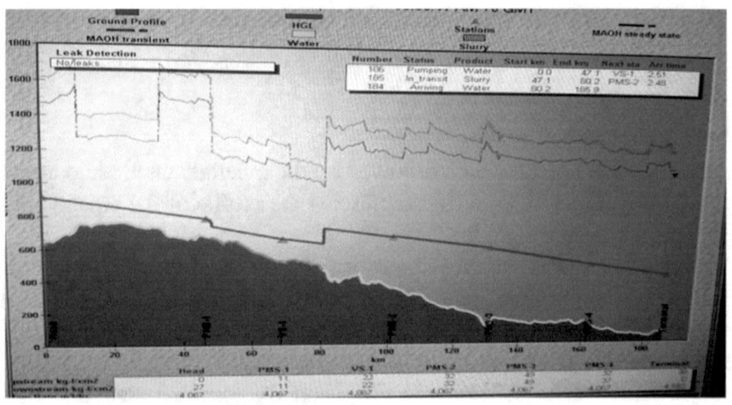

Fig. 4. Real supervision system (SCADA)

The information contained in this screenshot (Fig. 4) are the pressures in all (PMS) in bar, flow rate in m3/h, hydraulic head of output pumping station, and the drawing of the hydraulic grade line (behavior of the pressure drops), it is the line in dark blue. We notice clearly in (Fig. 4) that part of the pipeline is colored orange and the other two parts are colored in sky blue, the part colored in orange is the part filled with slurry for the other part is filled with water, this section of slurry is pushed by the water to the terminal station (piston effect), this is the case for (Fig. 4). The slurry section is called a batch.

We also notice that the behavior of the hydraulic grade line for the part filled with slurry is not the same for the part filled with water, which is normal because the head losses are due to the physical properties of the fluid (Table 1).

The pressure values detected in the case of (Fig. 4) are shown in the following table:

Table 1. The pressure values detected by the various pressures measuring stations

	Head station	PMS 1	Vanes station	PMS 2	PMS 3	PMS 4	Terminal station
Pressure (in bar)	27,89	11,07	22,11	32,51	49,92	37,6	38,85

This test was carried out for a flow rate of 4000 m3/h and a slurry density ρ = 1600 kg/m3 and a plastic viscosity of slurry μ = 0.0102 Pa.s.

The hydraulic gradient of the flow for the slurry filled part is: J = 0.0038 m/m of the pipeline.

Throughout this study, we are interested in the fact that the hydraulic gradient of the flow of the part fills in slurry.

(The plastic viscosity of the phosphate slurry varies in a design range between 0.006 Pa.s and 0.0102 Pa.s.) [9].

3 Theoretical Approach

3.1 Modeling of the Geographical Profile Followed by the Pipeline

The problem that exists is that the profile of the pipeline given by the constructor is not continuous; it does not contain all the points of the path between the head station and the terminal station. It contains only the coordinates of 303 points belonging to the route. This profile is discontinuous.

Table 2. The coordinates of the 303 points belonging to the pipeline profile

Point	Distance (m)	Elevation (m)
1	0,0	645,8
2	428,5	651,5
3	1094,7	656,7
4	1782,3	649,0
5	2354,1	643,6
↓	↓	↓
302	187252,3	66,4
303	187252,3	93,2

In order to have a continuous pipeline profile we will precede to an interpolation method between the points given in Table 2, this method consists in establishing linear functions of the form,

$$Z\,(x) = ax + b \tag{1}$$

between each point and which follows. The **a** and **b** are determined from the coordinates of the points in the Table 2. That is, between point 1 of coordinates (0; 645) and point 2 of coordinates (428; 651.5):

$$\begin{aligned} Z(x) &= \left(\frac{651,5 - 645}{428 - 0}\right)x + 645 \\ &= (0,\, 01330)\ x + 645 \ \text{ for } x \in [0; 428] \end{aligned}$$

Between 2 → 3: $Z(x) = (0,\,00780)\ x + 648,\,1577$ for $x \in [428;\ 1094.7]$

Between 3 → 4: $Z(x) = (-0,\,01119)\ x + 668,\,9418$ for $x \in [1094.7;\ 1782.3]$

Between 4 → 5: $Z(x) = (-0,\,00944)\ x + 665,\,8220$ for $x \in [1782.3;\ 2354.1]$

↓ ↓ ↓

Between 302 → 303: $Z(x) = (-0.00099)\ x + 251{,}7980$ for $x \in [186747.4\ ;\ 187252.3]$

Now the geographical profile is continuous over the interval [0; 187252.3], and manage by the function **Z**(**x**). The function **Z**(**x**) is defined by part on the interval [0; 187252.3]. (187252, 3) is the length of the pipeline in meters.

3.2 Modeling the Behavior of Head Losses

Case of Continuous Pumping in Slurry

The hydraulic grade line is the line describing the linear head losses along the pipe; we will determine the function **H**(**x**) which manages this line. For continuous pumping of the slurry the hydraulic grade line keeps the same inclination along the entire pipeline. In general, for turbulent flows, the friction factor **f** is determined from the MOODY diagram and the COLBROOK-WHITE formula, but for the case where: ($\mathbf{10^{-6} < k/D < 10^{-2}}$ **and** $\mathbf{5 \times 10^3 < Re < 1 \times 10^8}$), which is our case; we used the equation "Swamee and Jain". The coefficient of pressure drop **J** or (hydraulic gradient) from Darcy's formula [4–7]:

$$J = \left(\frac{\Delta H}{\Delta x}\right) = \frac{2fV^2}{gD} \quad (2)$$

With

$$f = \frac{0.33125}{\left[\ln\left(\frac{k}{3.7D} + \frac{5.74}{Re^{0.9}}\right)\right]^2} \quad (3)$$

f: the friction factor; **V**: the flow velocity of the fluid. **D**: The internal diameter of the pipeline (D = 0.85 m); **g**: acceleration of gravity; **k**: The roughness of the pipe. The friction factor **f** is related to the Reynolds number **Re** and the roughness **K** by the ratio (k/D) as indicated by the Eqs. (2) and (3). The inner layer of the pipeline is made of a plastic material (H.D.P.E) having a roughness $k = 2 \times 10^{-5}$. For an elementary displacement on the pipeline:

$$\mathbf{lim}_{\Delta x \to 0} \frac{\Delta \mathbf{H}}{\Delta \mathbf{x}} = \frac{\mathbf{dH}}{\mathbf{dx}} = \frac{\mathbf{2fV^2}}{\mathbf{gD}} \quad (4)$$

So: $\mathbf{dH} = \frac{\mathbf{2fV^2}}{\mathbf{gD}} \times \mathbf{dx}$; Integrating this formula: $\int \mathbf{dH} = \int \frac{\mathbf{2fV^2}}{\mathbf{gD}}.\mathbf{dx}$ We will arrive at: $\mathbf{H(x)} = \left(\frac{\mathbf{2fV^2}}{\mathbf{gD}}\right)\mathbf{x} + \mathbf{cte}$

Determining the integration constant: at x = 0 $\mathbf{cte} = \mathbf{H(x = 0)}$. $\mathbf{H(x = 0)}$ is the head supplied by the pumping station (**hi**) + the head due to the choke (**C**). (**x** = **0** is the position of the pumping station).

So the analytical function that manages the hydraulic grade line for the case of continuous pumping in slurry:

$$\mathbf{H(x)} = \left(\frac{\mathbf{2fV^2}}{\mathbf{gD}}\right)\mathbf{x} + \mathbf{H(x = 0)} \quad \forall x \in [0; 187252] \quad (5)$$

In this case the pumping is continuous in slurry (the slurry, fills the entire pipeline).

Batch Case

For the batch case the pumping is not continuous in slurry, the batch does not cover the entire pipeline, it occupies part of the pipeline and the other parts are occupied by water to keep continuity in the pipeline, and pushed The batch to the terminal station (piston effect). Water assures the separation between the different volumes of slurry in the case of several batches.

In this case we have two types of functions that manage the hydraulic grade line, function in the part filled by the slurry and function for the water filled parts. The position of the batch tail in the pipeline is (i) and the position of its head is (j).

$$\mathbf{j > i \text{ and } (i; j) \in [0; 187252]^2}$$

The behavior of the hydraulic grade line along the pipeline in this case is managed in this way:

$$\text{For } x \in [0;\ i]: \qquad \mathbf{H}(\mathbf{x}) = \left(\frac{\mathbf{2f''V^2}}{\mathbf{gD}}\right)\mathbf{x} + \mathbf{H}(\mathbf{x} = \mathbf{0}) \tag{6}$$

(f'' is the friction factor for water flow)

$$\text{For } x \in [i;\ j]: \qquad \mathbf{H}(\mathbf{x}) = \left(\frac{\mathbf{2fV^2}}{\mathbf{gD}}\right)\mathbf{x} + \mathbf{H}(\mathbf{x} = \mathbf{0}) \tag{7}$$

(f is the friction factor for slurry flow)

$$\text{For } x \in [j;\ 187252]: \qquad \mathbf{H}(\mathbf{x}) = \left(\frac{\mathbf{2f''V^2}}{\mathbf{gD}}\right)\mathbf{x} + \mathbf{H}(\mathbf{x} = \mathbf{0}) \tag{8}$$

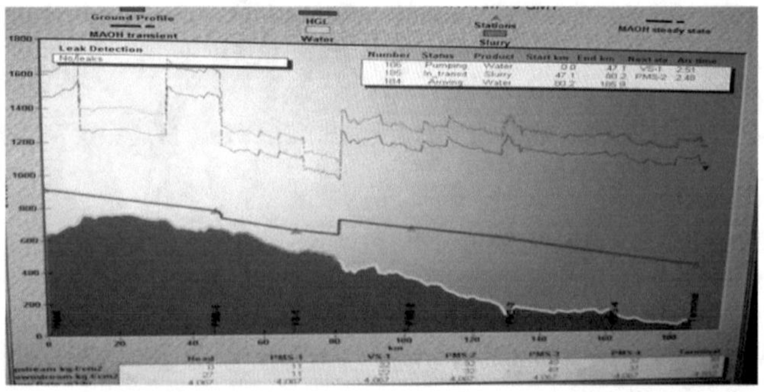

Fig. 5. Batch operation

The friction factor **f** depends on the density $\boldsymbol{\rho}$ and on the viscosity $\boldsymbol{\mu}$ of the fluid, so the inclination of the hydraulic grade line for the water flow does not have the same behavior as for the flow of the slurry.

Construction of the Program on MATLAB

We have obtained functions to model the behavior of the geographical route (pipeline profile), and to model the head losses (hydraulic grade line) along the pipeline, either for continuous pumping or for batch. And this in order to have a program in MATLAB that provides the value of the pressure in any point of the pipeline, for given conditions of operation. Condition of operation: (flow rate, viscosity, density, pumping station outlet head, …). The structure of the program is based on the function $\mathbf{Z}(\mathbf{x})$ which characterizes the behavior of the pipeline profile and on the function $\mathbf{H}(\mathbf{x})$ which characterizes the hydraulic head losses along the pipeline (hydraulic grade line). Are the functions obtained in the preceding paragraphs, and we will also rely on the method of calculating the pressure at a point (M) belonging to the pipeline.

P(M) = [**H**(M) − **Z**(M)] × **ρg** (Bernoulli's theorem), it is a pressure relative. In this study, we have interest that this pressure. ρ: density of the fluid; g: acceleration of gravity; P (M): the pressure at point (M), H (M): The head in point (M), Z (M): height of (M) with respect to sea level.

4 Results and Discussion

4.1 Comparison of Program Results with Real Results

The plot displayed by the program is similar to the plot displayed on the real control system in the control room, the difference between the two plots is that the first plot (Fig. 6) is based on theoretical calculations and the second plot (Fig. 5) is based on experimental measurements. The operating conditions in this case are: the position of the batch tail (in km): 47, the position of the batch head (in km): 80, the flow rate (in m3/h): 4000, the slurry density (in Kg/m3): 1600, plastic viscosity of slurry (in Pa.s): 0.0102, the hydraulic head at pumping station output (in meters): 240, the hydraulic head caused by choke (in meters): 35.

We carried out a series of experiments to compare the actual results with the theoretical results given by the program. These tests consist in comparing the pressure values given by the (PMS) with our model in MATLAB.

The following table shows the result of the test closer to reality, the actual values of the pressures marked in this table are the values already quoted in the paragraph (experimental approach) (Table 3).

Table 3. A comparison of real results and results delivered by our model

	Head station	PMS 1	Vanes station	PMS 2	PMS 3	PMS 4	Terminal station
Real results (in bar)	27,89	11,07	22,11	32,51	49,92	37,6	38,85
Theoretical results (in bar)	27,18	12,47	22,63	33,11	51,23	37,3	39,41

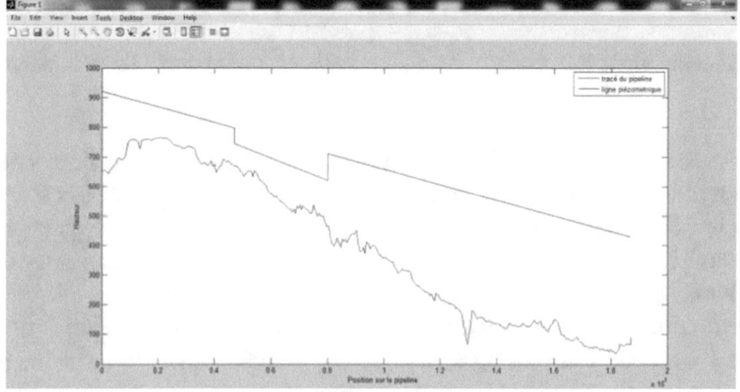

Fig. 6. The plot of the geographical profile and hydraulic grade line delivered by the program

In this test the position of the tail of the batch was in the kilometer 47 and its head in the kilometer 80. This test was made for a running flow of 4000 m3/h and a slurry density ρ = 1600 kg/m3 and a plastic viscosity of slurry μ = 0.0102 Pa.s, the hydraulic gradient for the flow of the slurry in this simulation is: J = 0.0038 meters/meter of the pipeline, it is the same gradient obtained experimentally. We notice the approximate agreement of the real results with the theoretical results, which validates and values our theoretical model.

4.2 Simulation Through Our Theoretical Model

We have proved the validity of our model, and at what level the results of the model are in agreement with the experimental results. It is what encouraged us to do some simulations has shortcoming our program, to know the influence of certain operating parameters on the behavior of the process.

This simulation consists of having all the parameters fixed, and of varying one of them in order to know its impact on the pressures and the head losses. These parameters are: (flow rate, slurry density, slurry viscosity, and the hydraulic head that associates the choke to pipeline).

Influence of Flow Rate

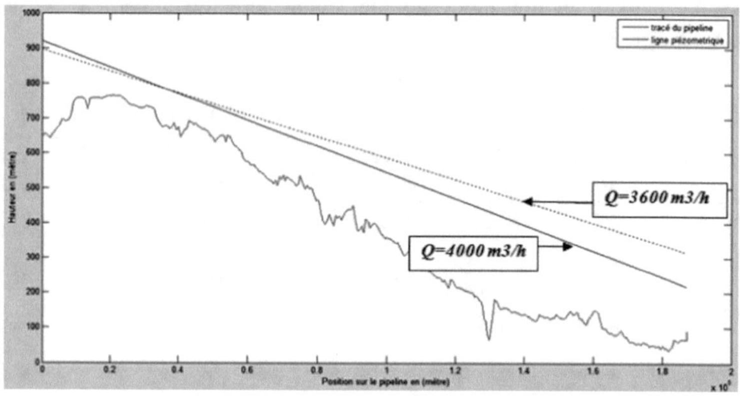

Fig. 7. Hydraulic grade line for flow rates 3600 m3/h and 4000 m3/h, for continuous pumping in slurry

Concerning continuous pumping of slurry, in the case where the flow rate is 4000 m3/h, the hydraulic grade line is more inclined at the bottom than the case where the flow rate is 3600 m3/h, the head losses for (Q = 4000 m3/h) are greater than for (Q = 3600 m3/h). It can be concluded that the higher the flow rate increases the greater the head losses (Fig. 7).

The hydraulic gradients shown in our program are J = 0.0038 m/m for (Q = 4000 m3/h) and J = 0.0031 m/m for (Q = 3600 m3/h). The flow rate has the same influences for the case of batch. The red line is the geographical tracing of the pipeline, and the blue line is the hydraulic grade line (Fig. 8).

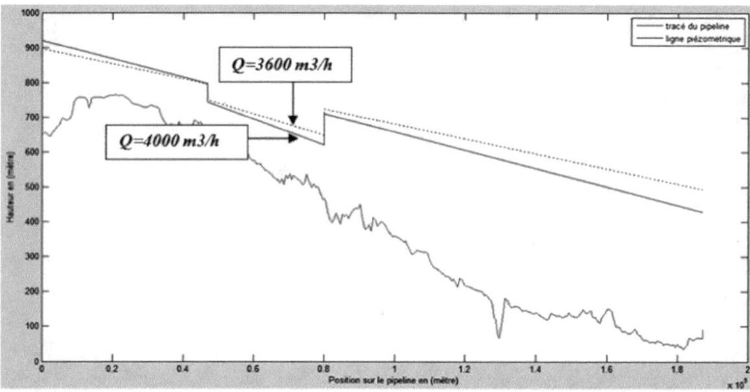

Fig. 8. Hydraulic grade line for the flow rates 3600 m3/h and 4000 m3/h, for the batch case

Influence of Slurry Density

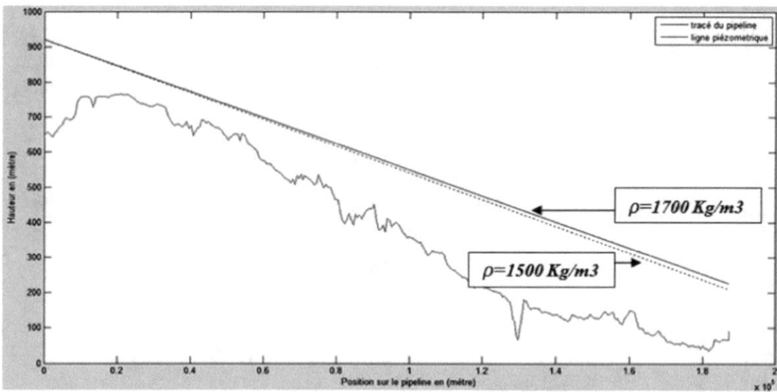

Fig. 9. Hydraulic grade line for densities 1700 kg/m3 and 1500 kg/m3, for continuous pumping

Concerning continuous pumping, in the case where the slurry density is ρ = 1500 kg/m3, the hydraulic grade line is more inclined at the bottom than the case where the slurry density is ρ = 1700 kg/m3, so the head losses of (ρ = 1500 kg/m3) are greater than for (ρ = 1700 kg/m3), normally we must have the opposite tendency, but here we have to consider that the viscosity is constant between the two cases. In general, we notice from the curves that the slurry density has no significant influence on the behavior of the pressure drops. The hydraulic gradients shown in our program are J = 0.0037 m/m for (ρ = 1500 kg/m3) and J = 0.0038 m/m for (ρ = 1700 kg/m3). The slurry density has the same influences on the batch case (Figs. 9 and 10).

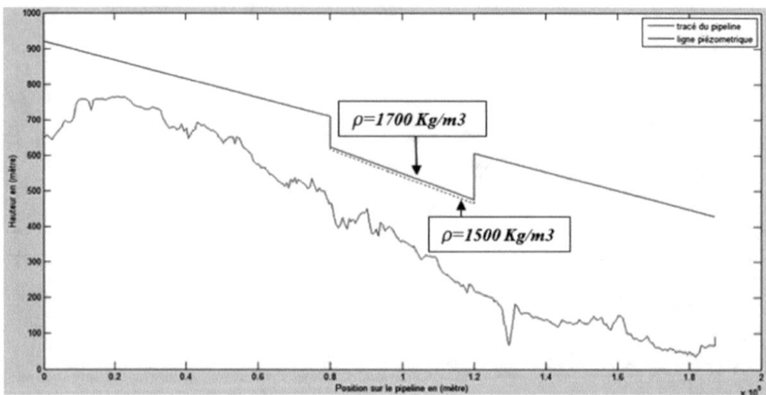

Fig. 10. Hydraulic grade line for densities 1700 kg/m3 and 1500 kg/m3, for the batch case

Influence of the Plastic Viscosity of the Slurry

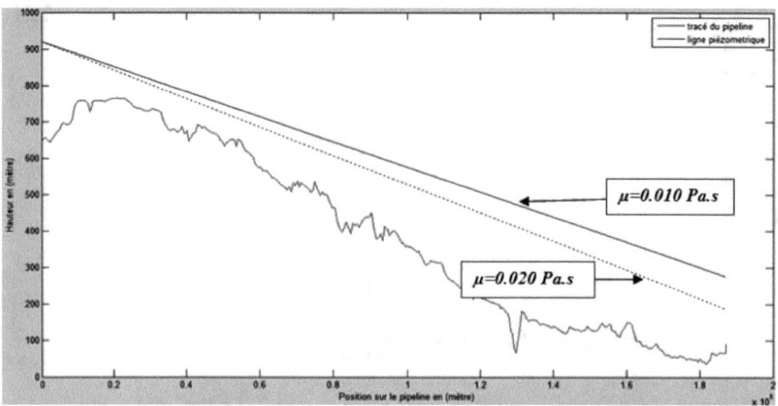

Fig. 11. Hydraulic grade line for viscosities 0.010 Pa.s and 0.020 Pa.s for, continuous pumping in slurry.

Concerning continuous pumping, in the case where the slurry viscosity is $\mu = 0.020$ Pa.s, the hydraulic grade line is more inclined at the bottom than the case where the slurry viscosity is $\mu = 0.010$ Pa.s, so the head losses for ($\mu = 0.020$ Pa.s) are greater than for ($\mu = 0.010$ Pa.s). Then it can be concluded that the higher the slurry viscosity increases, the greater head losses. The hydraulic gradients shown in our program are $J = 0.0034$ m/m for ($\mu = 0.010$ Pa.s) and $J = 0.0039$ m/m for ($\mu = 0.020$ Pa.s). The slurry density is assumed constant between the two cases (Figs. 11 and 12).

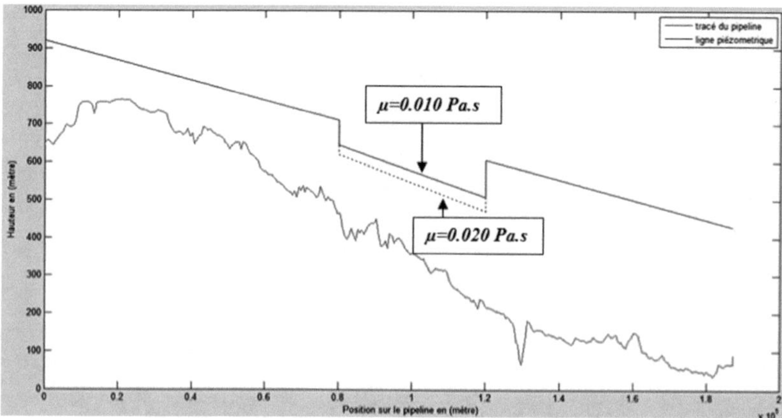

Fig. 12. Hydraulic grade line for viscosities 0.010 Pa.s and 0.020 Pa.s, for batch case

Influence of Hydraulic Head Caused by Choke

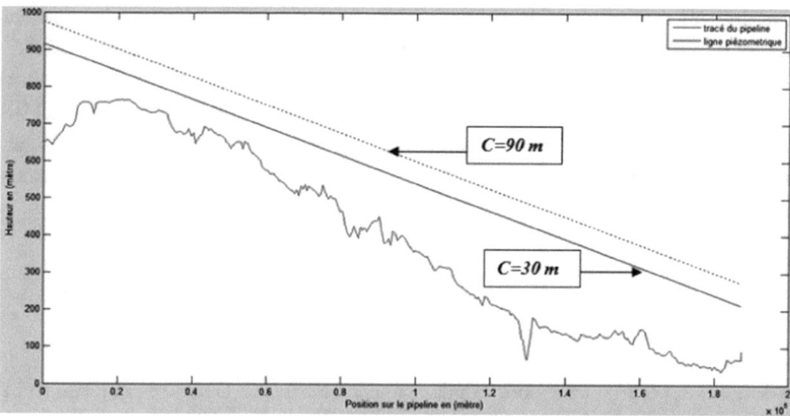

Fig. 13. Hydraulic grade line for choke at 30 m and 90 m, for continuous pumping in slurry

Concerning continuous pumping in slurry, in the case where the hydraulic head due to the choke is C = 30 m, the hydraulic grade line retains the same inclination as the case where C = 90 m, so the choke has not Influence on the head losses. The higher the choke level, the more the hydraulic grade line is translated at the top without changing the inclination. It can be concluded that the choke increases the hydraulic head in the pipeline without influencing the pressure drops. In the case of batch the choke has the same influences (Figs. 13 and 14).

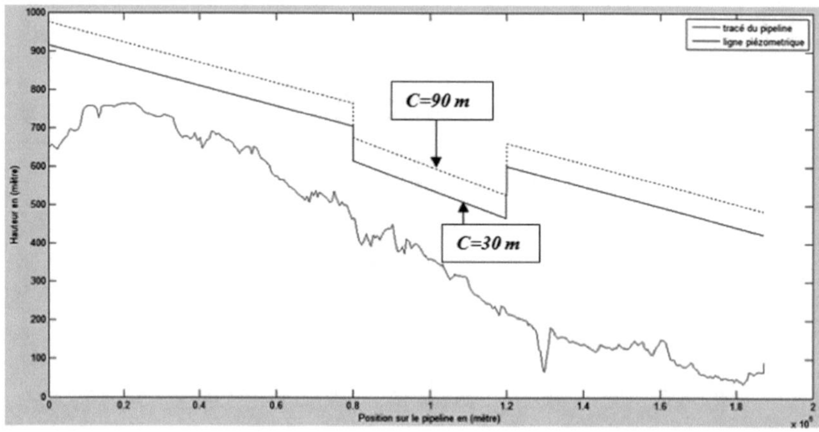

Fig. 14. Hydraulic grade line for choke at 30 m and 90 m, for the batch case

5 Conclusions

A mathematical modeling of a number of parameters influencing the process of conducting phosphate through the pipeline was carried out. Thus, linear head losses, hydraulic gradient, friction factor, pipe roughness and fluid mechanics of the phosphate slurry were optimized and the result was grouped into a program in MATLAB.

The simulation allowed us to understand perfectly the work of the pipeline OCP and to know the influence of certain parameters on the operation of the process.

References

1. Jacobs, B.E.A.: Design of Slurry Transport Systems, pp. 285–286 (1991)
2. Abulnaga, B.: Slurry Systems Handbook. McGraw-Hill, New York (2002)
3. Miedema, S.A.: Slurry Transport: Fundamentals, a Historical Overview and The Delft Head Loss & Limit Deposit Velocity (2016). Edited by Robert C. Ramsdell
4. Gray, J.J.: Johann Heinrich Lambert, mathematician and scientist. Hist. Math. **5**(1), 1728–1777 (1978)
5. Moody, L.F.: Friction factors for pipe Flow. Trans. ASME **66**(8), 671–684 (1944)
6. Guyon, É., Hulin, J.-P., Petit, L.: Ce que disent les fluides, Belin (2005)
7. Saatdjian, E.: Les bases de la mécanique des fluides et des transferts de chaleur et de masse pour l'ingénieur. Éditions Sapientia (2009)
8. Rusconia, J., Lakhouajab, A., Kopuzc, M.: The design and engineering of the 187 km Khouribga to Jorf Lasfar Phosphate Slurry Pipeline. In: SYMPHOS 2015 - 3rd International Symposium on Innovation and Technology in the Phosphate Industry, pp. 142–150. Procedia Engineering, Elsevier (2016)
9. Belbsir, H., El-Hami, K., Soufi, A.: Study of the rheological behavior of the phosphate-water slurry and search for a suitable model to describe its rheological behavior. Int. J. Mech. Mechatron. Eng. IJMME-IJENS **18**(04), 73–81 (2018)

Optimized Energy Management of Electric Vehicles Connected to Microgrid

Youssef Hamdaoui(✉) and Abdelilah Maach

Department of Computer Science, Mohammadia School of Engineers (EMI), Mohammed V University, Rabat, Morocco
youssefhamdaoui@research.emi.ac.ma, maach@emi.ac.ma

Abstract. As an important part of renewable energy utilization and smart distribution, the increasing penetration of electric vehicles for the rapid growth in load demand and especially electric vehicle can be a smart solution for the islanded nanogrid or in an emergency outage case. The electric vehicle is not a new concept and has been conceptually and practically available for the last century and will change the financial as well as environmental attractiveness of on-site generation (e.g. PV, or fuel cells). In islanded mode, based on the power balance between renewable electric sources and loads, the energy management and dispatch of EVs and PV are optimized to minimize the operational cost and maximize the benefit of islanded microgrid. This paper present an efficient power management based on mobile power stored in electrical vehicle and power produced by solar panel in a residential distribution and focuses on the analysis of the optimal interaction of electric vehicles with householders network, which may include photovoltaic (PV) for an efficiency energy management system. The effectiveness of the proposed strategies for the optimized operation of EVs is validated by case studies and performances analysis.

Keywords: Nanogrid · Distribution network · Distributed energy resources · Electric vehicle · Energy management

1 Introduction and Related Work

Traditionally, the existing power grids are generally used to carry power from a few central generators to a large number of customers [1] and islanding is one of smart grid advantages based on integration of distributed generations and evolution of communication infrastructure. Islanding or Outages are caused in major cases by weather and affect an important population in the word. Consequently, renewable resources can be a solution to supply demands in islanded region and can have the role of main utility duration outage period. The best way to have a smart distribution system is to break the grid into clusters or micro grids. The approach given in this paper also provides ideas about the optimal operation of a future smart grid with the booming of EVs.

A Smart Grid (SG) is a modernized electric system that uses sensors, monitoring, communications, distribution system automation, advanced data analytics, reliability, efficiency and safety of the electricity system, It increase consumer choice by allowing them to better control their electricity use in response to prices or other parameters.

M. Ezziyyani (Ed.): AI2SD 2018, AISC 912, pp. 371–387, 2019.
https://doi.org/10.1007/978-3-030-12065-8_34

A SG includes diverse and distributed energy resources ad accommodates electric vehicle charging. In short, it brings all elements of the power system (production, transmission, distribution and consumption) closer together to improve overall system operation for the benefit o consumers and the environment. Integrating all components (SG, Electric Vehicles (EV), Infrastructure communication technologies, distributed generation …) to meet the needs of consumers and regions, is the concept of Smart energy network.

The MG is an electrical power network with DGs, controls and storage designed to work over a relatively small area, Common agreement is that it is has a capacity lower than 10 MW that can be operated in a islanded Mode.

Distributed generation (DG) can improve energy utilization efficiency and energy supply reliability. When a fault is detected in a distribution network, DG is required to provide a continuous and stable power and must have the potential to supply adequate power during grid outages. In this condition, it is important for the micro-grid to continue to provide a constant power to the islanded zone load [2]. The distribution network can form some local island systems to effectively restore some emergency loads. DG can significantly increase the electricity resiliency.

An efficient matching between non distributed energy resources (NDER) and distributed energy resources (DER) into the electrical distribution power grid allow reduction in power cost, transmission cost, distribution cost and will protect the environment. The successful application of DER in the electrical distribution system present major opportunities to change the existing grid concept, response future energy demand and to reduce events like the 2003 blackout [3].

The peak may be many times as large as the off-peak load. This poses challenges to power suppliers because they have to meet the peak demand, resulting low utilization during off-peak times. One of the goals of utilities is to reduce peak load so as to reduce the operating costs and defer the new investment to the power grid and integrate the new distributed energy resources like photovoltaic (PVs) and EVs.

There are numerous advantages of using EVs as an alternative method of transportation. However, an increase in EV usage in the existing residential distribution grid poses problems such as overloading the existing infrastructure. Today the cost of electricity to drive the EV is becoming competitive with the cost of fossil fuel required to drive the same distance [4].

The price of an EV is still higher than its traditional counterpart running on fossil fuel but the total cost is lower in the long term due to less maintenance being needed for EVs and overall price spent on fuel per mile. EVs also reduce the consumption of natural fossil fuels and make the environment clear by reducing the Green House. The coordination of EV charging not only reduces the difference between the load peak and valley with a higher load rate, but also decreases system loss and improves electric power quality (PQ). It is also envisioned to help cut down the system operating cost and charging costs of EV users [5–9].

Figure 1 shows the state of charge (SOC) for an EV connected to an office building. It is clear that the commercial building benefits from energy (The zone in yellow color) that has a carbon foot print that is related to a period when the EV is not connected to the commercial building. This becomes even more complicated if the EVs are connected to different buildings during a certain period of time.

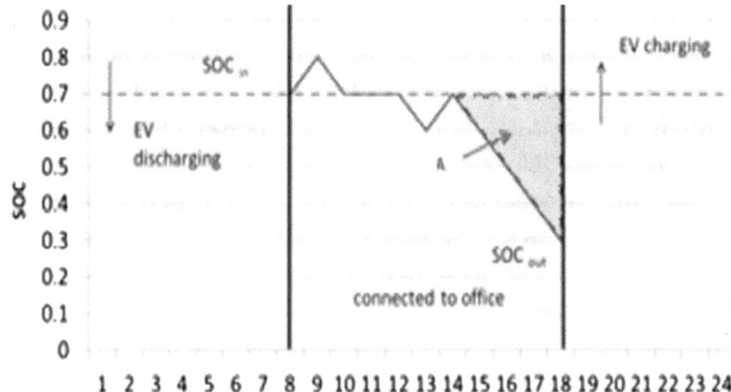

Fig. 1. Electric vehicle charging and discharging profile

The definition of an outage or Intentional Islanding by research papers [10, 11] is a condition in which a portion of the power grid, which contains both load and DG is isolated from the remainder of the utility system resulting from extreme weather or other emergency situations or when the demand in the peak hours exceeds the supply [12, 13]. The duration of interruptions might be from minutes to hours depending on the severity of the fault that occurred. The major energy outage is mostly related to weather, maintenance problems, substation failures and transmission lines failures are low causes of the outages.

An intensive research on the transmission system islanding scheme have be done to resolve islanding problem. The ordered binary decision diagram and the algorithms combine slow coherency theory and the multilevel recursive bisection algorithm and the spectral clustering method [13, 14]. The difference in infrastructure and in network between transmission and distribution system require some novel smart islanding algorithm to increase efficiencies using waste heat and reduction of line losses and enhance customer reliability and manage smartly the DER-based distribution. Distribution system provides major opportunities for smart grid because of its important role of distributing electrical power from the main grid to users, it has the potential to supply electricity based on renewable resources during grid outages [15].

The flexibility of microgrid and its deep penetration levels of renewable energy sources provide more opportunities for the interaction between EVs and microgrid. As the controllable load and dispersed energy-storage units, EVs and Batteries Storage BS can mitigate load variability and improve microgrid economics. The reliability of the power supply can also be improved by taking full advantage of vehicle-to-grid (V2G) technology. Their participation in the ancillary service of microgrid allows more acceptance of renewable energy sources as well [16–19].

This paper proposes an new optimized model of EV in a MG and resolve islanding issue based on using mobile resources. GridLAB-D is an open-source power system modeling and simulation tool developed by pacific Northwest National Laboratory

(PNNL) with the funding of the Department of Energy (DOE) [20]. GridLAB-D is a discrete event-based power systems simulator which employs an agent-based simulation approach to model and simulate the distribution power grid. used to illustrate the effects of the different EV penetration rates in the power grid in order to validate an optimized energy management system.

The rest of the paper is organized as follows: in Sect. 2, describes the distribution grid model. Section 3, we present the potential of EV integration with some profile of EV charging. A novel smart grid architecture, case of study, simulation and results are presented in Sect. 4 and finally Sect. 5 concludes the paper.

2 Novel Distribution Grid Model

2.1 Novel Smart Microgrid Structure

MGs are integrated energy systems consisting of inter connected loads and distributed energy resources as shown in Fig. 2 which as a system can operate in parallel with the main grid or in an outage mode. Dynamic islanding is a key feature of a smart micro grid. Numerous benefits accrue from this ability to island for events like faults and voltage sags. Smart islanding can greatly enhances the value proposition for the utility and the customer [21].

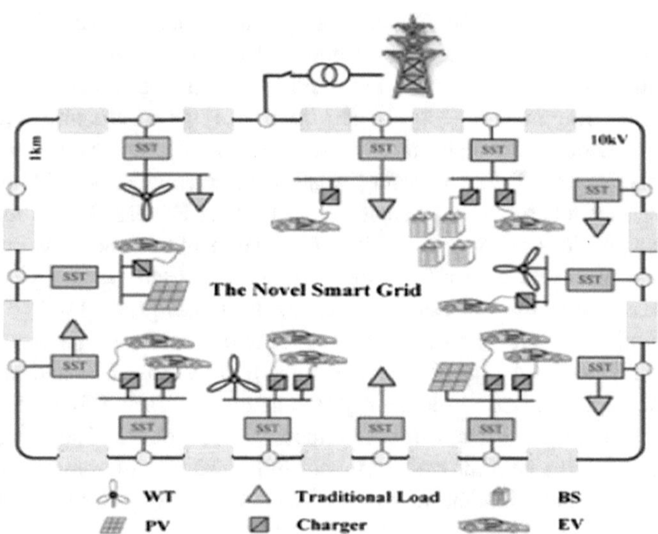

Fig. 2. Novel smart microgrid structure

2.2 Distributed Energy Renewable Benefits

The majority of studies confirm that using a single renewable energy generation is difficult to provide a continuous and stable power supply all the time, for example PV system is flexible; ranging from a few watts to hundreds of megawatts and its technologies depends on how they capture and distribute solar but the efficiency of electric-power generation is very low and depends on weather conditions. Wind turbines are affected by system fault at t = 52,000 s that means power output of wind generation is reduced to zero. Consequently, the hybrid energy system with a DER combination can be effective solution to unstable effects of electricity supply. A hybrid solar-wind-battery system can provide 100% of power supply for costumers, thus greatly decreasing the energy costs and increasing the reliability of power supply [19].

2.3 Distribution Grid Model

Figure 3 present the MG architecture model of our approach; it contains different distribution network level from the DER resource (EVs, Solar) to the end users (residential houses).

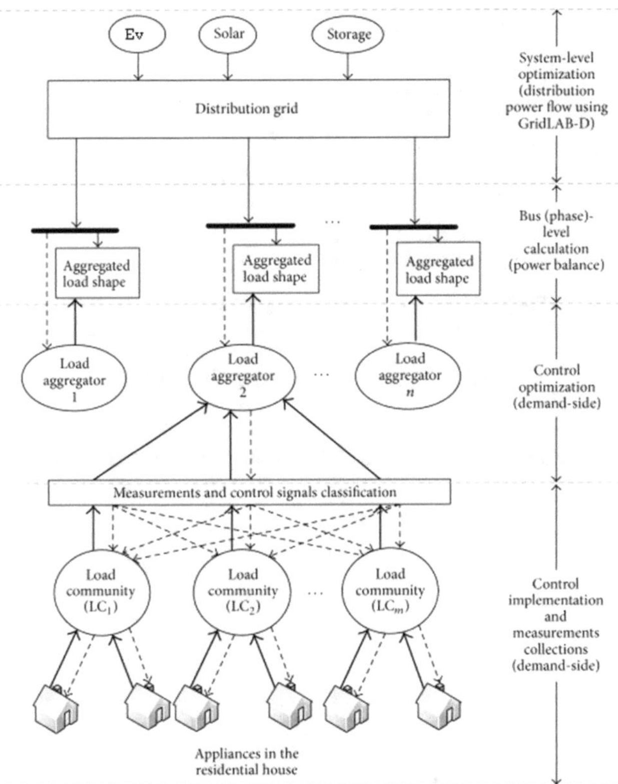

Fig. 3. MG architecture layer Model

2.4 Distribution Grid Model

Figure 4 present the declaration of an object house instance with some house parameters object definition in Gridlab-D simulator (GLM file) with the ZIPLoad as meter object, floor area, cooling_cop, air_temperature…..

```
}
object house {
      name b1m1_house;
      parent b1m1_house_meter;
      thermal_integrity_level LITTLE;
      hvac_power_factor 0.97;
      cooling_COP 3.90;
      floor_area 1040;
      cooling_setpoint 75;
      thermostat_deadband 2;
      air_temperature 72.5;
      object ZIPload {
            schedule_skew -685;
            base_power LIGHTS*1.8752;
            power_fraction 0.600000;
            impedance_fraction 0.400000;
            current_fraction 0.000000;
            power_pf -0.780;
            current_pf 0.420;
            impedance_pf -0.880;
            heat_fraction 0.91;
      };
      object ZIPload {
            schedule_skew -685;
            base_power CLOTHESWASHER*0.4354;
            power_fraction 1.000000;
            impedance_fraction 0.000000;
            current_fraction 0.000000;
            power_pf 0.970;
            current_pf 0.970;
            impedance_pf 0.970;
            heat_fraction 0.70;
      };
      object ZIPload {
```

Normal text file

Fig. 4. GridLAB-D house object code

Figure 5 present the GridLAB Model of house connected to node in a microgrid.

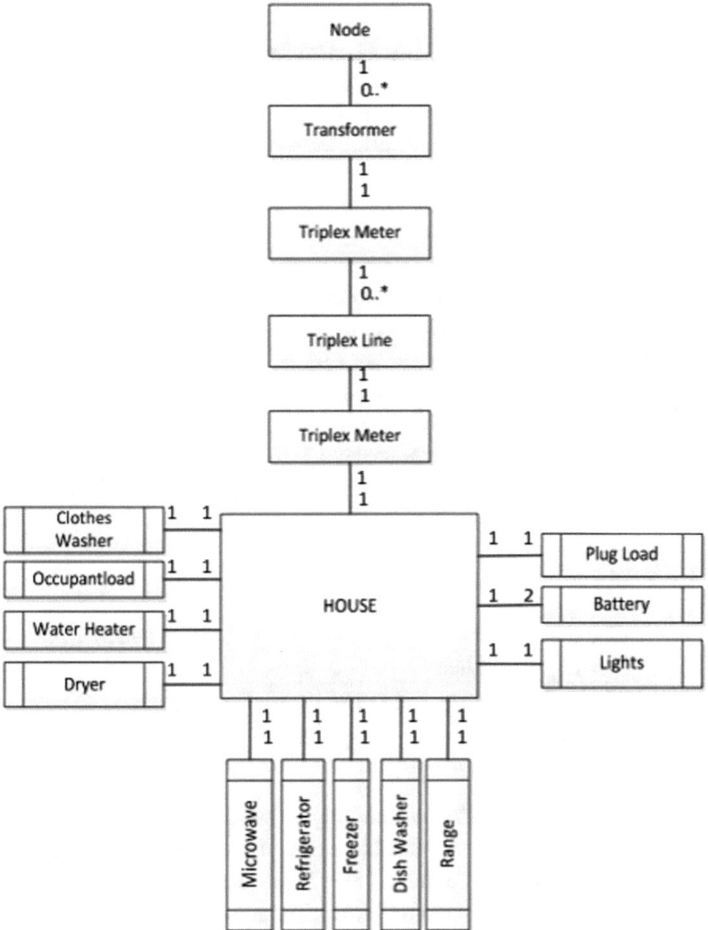

Fig. 5. Structure of residential microgrid model using GridLAB-D

3 Potential of EV and Operational Strategy in V2G Mode

EVs parked in the microgrid can be fed by chargers that collect information on the battery from the battery-management system equipped in EVs [22]. Meanwhile, the chargers display messages, such as the estimated charging prices in the coming hours and microgrid conditions. Then, the optimized charging strategy will be produced to commence the smart charging after users input their foreseen parking time [23].

EV batteries can transfer electricity to the residential building and vice versa provided they are connected to it (Eq. 1). The building energy management system (EMS) can use this additional battery capacity to lower its energy bill and whenever possible, economically attractive energy from a renewable energy source at the building can be used to offset EV charging at home [24].

$$S_u + S_{DER} + S_{St} + S_{EV} + V = D_B + D_{St} + D_{EV} \quad (1)$$

S_u electricity supplied by utility
S_{DER} electricity supplied by distributed energy resources
S_{St} electricity supplied by local Storage
S_{EV} electricity supplied by EVs
D_B electricity demand from the building
D_{EV} electricity demand from EVs
D_{ST} electricity demand from local storage

Table 1 present an EV battery characteristic example and the EV average capacity on the market.

Table 1. EV battery specification

Charging efficiency	0.95
Discharging efficiency	0.95
Battery hourly decay (related to stored electricity)	0.001
Capacity	16 khw

Figure 6 shows the curve of forecasted power output of DERs on a typical day with the workday and weekend load curves. The error of the installed capacity for the day-ahead forecast of wind power is estimated 13%. The mean absolute percentage error in the forecast of solar power is estimated to be 8.96%. The load forecast error factor is assumed at 3% [25]. Thus, it is readily derived that the minimum system reserve is 141.76 kW.

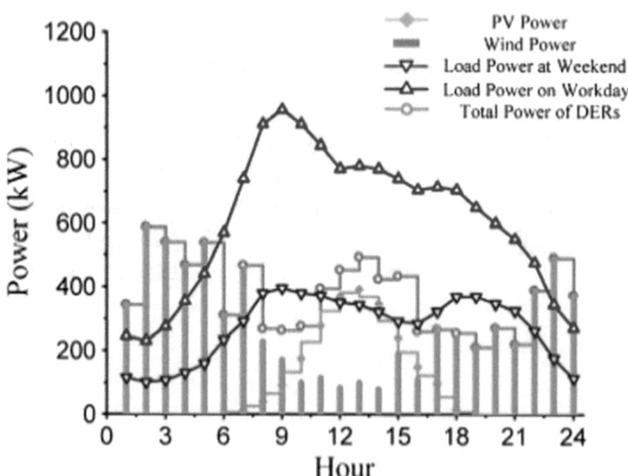

Fig. 6. DER forcasted power output on a day

In Fig. 7, the average of vehicles parked in MG is estimated that 20% of EVs have charging demand. Owing to the randomness in the battery's SOC that conforms to the normal distribution, the Monte Carlo simulation test is adopted to determine the initial SOC of each EV [26]. Ideally, the nominal capacity of the lithium-ion battery is 64 kWh and the rated power is 16 kW.

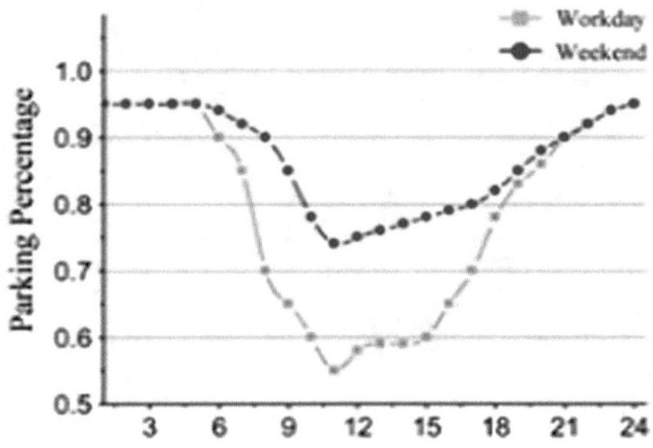

Fig. 7. Average of vehicles in MG.

Figure 8 Illustrate the EV uncoordinated charging, considering of number of EVs parking percentage, the charging load can coincide with system peak load. A significant amount of EVs in peak hours on workdays boosts network total load up to 1.20 MW that exceeds the rated capacity of the microgrid. The reasons why there is nearly no charging in midnight are twice amount.

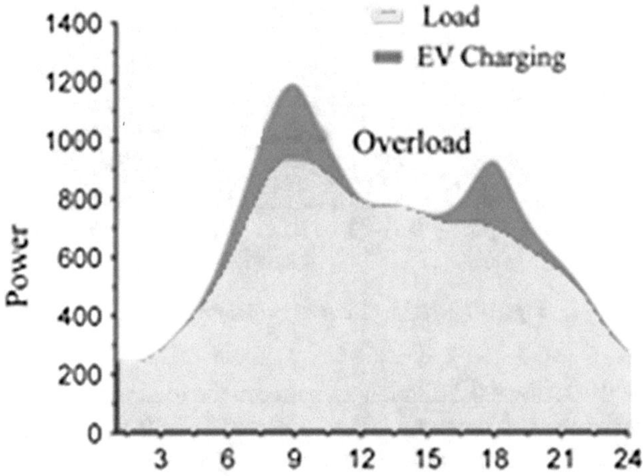

Fig. 8. Uncorrdonate charging of EVs

The second EV charging strategy is coordinated charging and is show in Fig. 9. The hourly price is more pushed up when the load power increases. EVs are encouraged to get charged at low cost and favorable prices in valley hours.

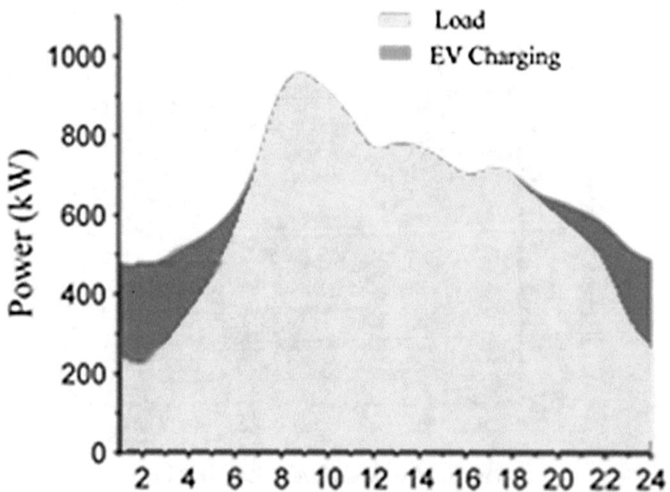

Fig. 9. Wordays EVs coordination charging

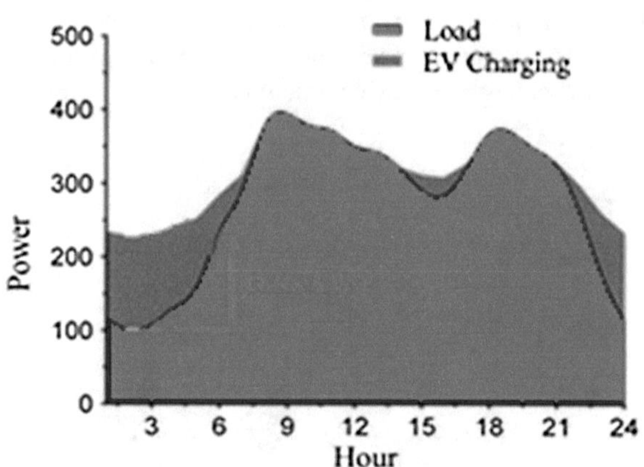

Fig. 10. Weekend Evs coordinating charging

Figures 9 and 10 shows the differences between the load peak workdays and on the weekend and are narrowed down by 242.52 kW and 165.89 kW, respectively. Thus, it can be due that the deep penetration of EVs served with charging coordination does not burden the MG in peak hours but rather absorbs surplus DER power at night.

4 Case of Study, Simulation and Results

The study case is an IEEE13 Bus as shown in Fig. 11 connected to main utility and contains a PV Solar Panel as DER and the EV mobile resources integration is supposed by PLUG connector vehicle to grid (V2G).

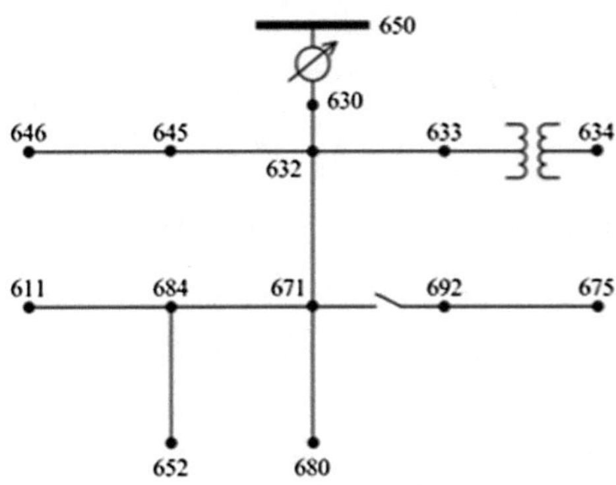

Fig. 11. IEEE 13-bus improved test system.

Table 2. Solar panel characteristic

Solar penetration	25%
Solar size (KVa)	6
Climate region	Seattle, WA
Season	Summer
Weather	Sunny

Table 2 present the solar panel characteristic and the capacity generated in summer season and sunny day with real SEATTLE data weather and with solar penetration 0.25 and solar size is 6KVA:

Results of energy consumed and produced calculated by the simulator (Total energy consumed at the substation: 15,423.697 kWh and the total energy from solar panels: 214.6077 kWh). Figures 12, 13 and 14 presents the results graph of our Gridlab simulation of our model IEEE13 to show the substitution load, solar production and the last figure present the voltage of the node 646 balanced between the main power and the solar production power for one day simulation. Others simulation is explained to have a smart islanding approach based on balancing between DER productions and based on MG context parameters are cited in [27–31].

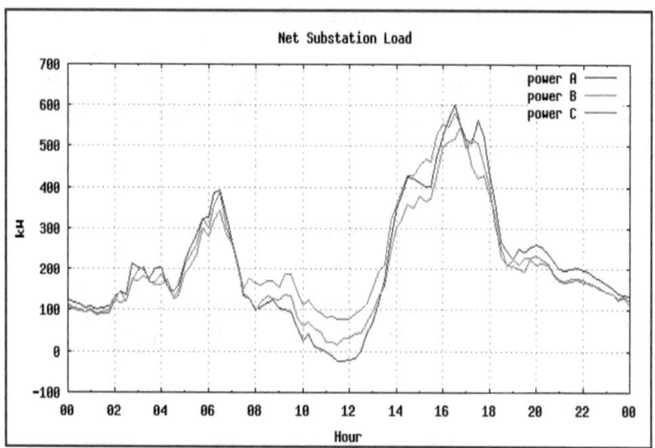

Fig. 12. Net substation load

The structural model of our residential distribution grid consists as presented in the Fig. 11 modeling with tool GridLAB-D by using IEEE 13 node. The primary voltage of feeder is 33 kV and the secondary voltage is 2.4 kV. The power rate for the substation transformer is 5MVA.

We have simulated 1000 houses in the distribution network among the nodes of the IEEE 13 node feeder in which every 3 to 7 houses connect to a step-down node with two kind of houses.

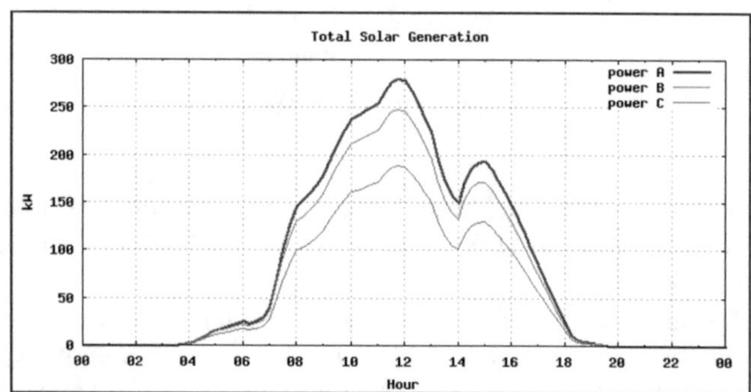

Fig. 13. Total solar generation

We have assumed that all EVs come back house following the Gaussian probability distribution model in which the mean is 5:30 PM and the standard deviation is 1 h. Table 3 presents the battery capacity is either 25 kWh or 40 kWh with a charging Amp 30A and charging volt 240 V and Fig. 15 shows the arrival and departure of the EV.

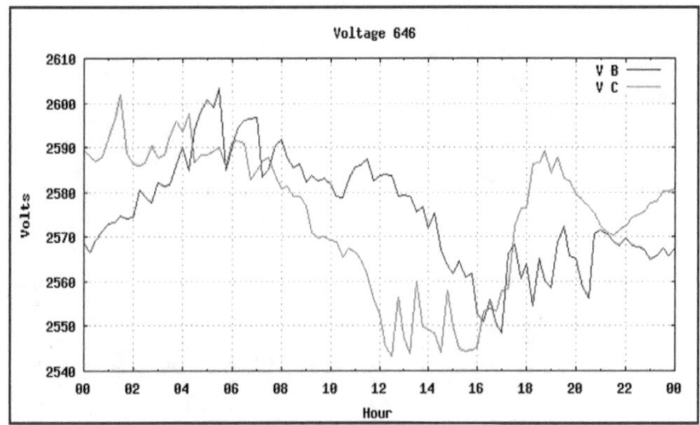

Fig. 14. Voltage graph for node 646

Table 3. EV Model caractéristique

House	Batterie size	Kms	SOC
Type1	25 kwh	120	20%
Type2	40 kwh	225	25%

We considered four different penetration rates of EVs in the residential domain (0%, 10%, 30%, and 50%). we have arbitrary distributed EVs among houses; for example, 100 EVs were distributed among 1000 houses for a 10% penetration rate. Figure 16 shows the average power output. For 30% and 50% penetration, the average transformer output is 1 during peak load period.

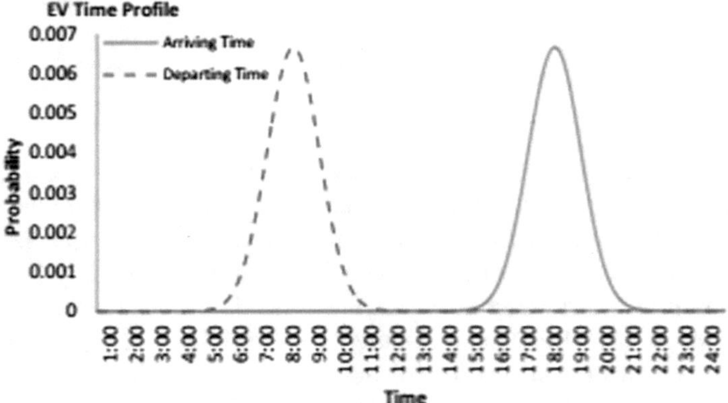

Fig. 15. Arrival and departure of EVs (Gaussian Probability)

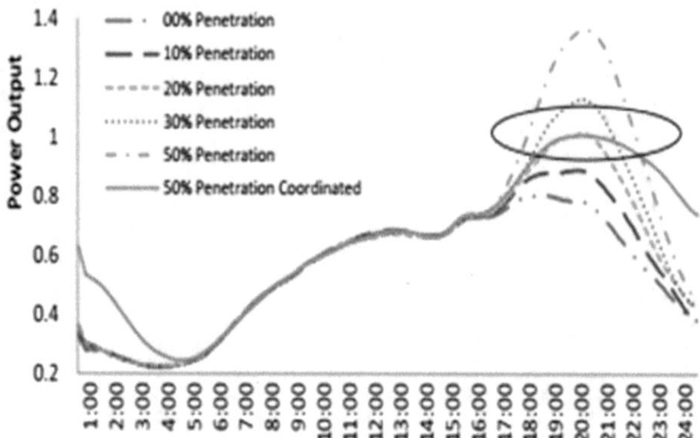

Fig. 16. Transformer-level power output

The output power does not exceed the maximum of transformer by coordinating the EV charging. The results of our simulation are summarized in Table 4. According to this table for 50% penetration rate, 60% of transformers are overloaded by EV charging and the maximum duration is about 5 h. The overloading time is more than 200 h for 50% penetration. The EV charging effect can be mitigated by coordinating the EV charging. By increasing the penetration rate, the number of overloaded residential transformers increases during evening, its normal because all EVs arrive.

Table 4. Transformers overload

EV penetration	Overloaded transformers	Maximum duration (sec)	Overloading time (hour)
0	0	0	0
10	3	8220	29340
30	26	14400	284640
50	60	19620	780360

Figure 17 presents the second simulation results with using battery storage (BS) to store the surplus production, the power exchange between microgrid and the main utility in each time interval on workdays and on weekends. It can be seen that the optimized strategy reduces the dependence of microgrid on the utility in peak hours demands and provides power support for the utility in converse, which relieves the system burden for peak-load control. During load time of EVs charge and BSs charge to absorb the surplus production. It is demonstrated that the charging cost of EVs is considerably cut down. This simulation is done to validate the benefit of using EVs as mobile resource with a combination of Batteries storage (BS).

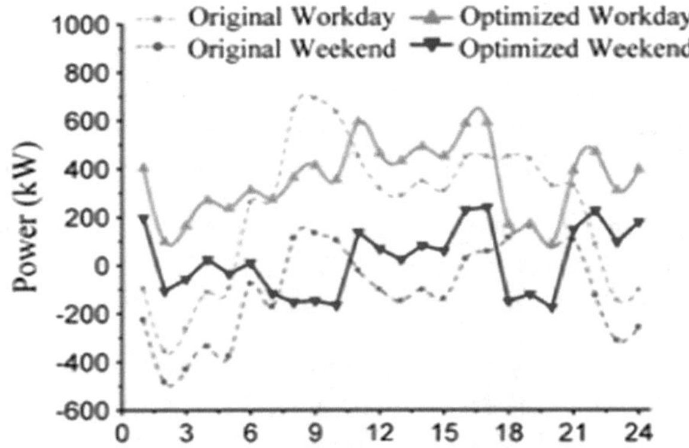

Fig. 17. Power interchange between the utility and microgrid.

5 Conclusion

This paper presents the approach of EV penetration to optimize the current distribution network and have an efficient management system in a smart Microgrid and especially when we have an outage or an emergency islanding case. The paper proposes a novel distribution model based on distributed energy resources. We demonstrate in this paper a high rate of overloading on step-down transformers during peak load time in residential distribution network by integrating the EV strategy of charging. Our approach can be a solution for nanogrid like houses in a far region that the transmission lines cost very expensive and can be replaced by a hybrid combination based on mobile resources like electric Vehicles that present a revolution in our current smart grid.

References

1. Farhangi, H.: The path of the smart grid. IEEE Power Energ. Mag. **8**(1), 18–28 (2010)
2. Haque, M.E., Negnevitsky, M., Muttaqi, K.M.: A novel control strategy for a variable-speed wind turbine with a permanent-magnet synchronous generator. IEEE Trans. Ind. Appl. **46**(1), 331–339 (2010)
3. Horowitz, S., Phadke, A., Renz, B.: The future of power transmission. IEEE Power Energy **8**(2), 34–40 (2010)
4. Kock, B.: Why EVs are so important to the future of the smart grid. Smart Grid News (2013)
5. Kiviluoma, J., Meibom, P.: Methodology for modelling plug-in electric vehicles in the power system and cost estimates for a system with either smart or dumb electric vehicles. Energy **36**(3), 1758–1767 (2011)
6. Rotering, N., Ilic, M.: Optimal charge control of plug-in hybrid electric vehicles in deregulated electricity markets. IEEE Trans. Power Syst. **26**(3), 1021–1029 (2011)
7. Wu, D., Aliprantis, D.C., Ying, L.: Load scheduling and dispatch for aggregators of plug-in electric vehicles. IEEE Trans. Smart Grid **3**(1), 368–376 (2012)
8. Cao, Y., Tang, S., Li, C., Zhang, P., Tan, Y., Zhang, Z., Li, J.: An optimized EV charging model considering TOU price and SOC curve. IEEE Trans. Smart Grid **3**(1), 388–393 (2012)

9. Suand, W., Chow, M.: Performance evaluation of an EDA-based largescale plug-in hybrid electric vehicle charging algorithm. IEEE Trans. Smart Grid **3**(1), 308–315 (2012)
10. Hamdaoui, Y., Maach, A.: A smart approach for intentional islanding based on dynamic selection algorithm in microgrid with distributed generation. In: International Conference on Big Data Cloud and Applications (BDCA) (2017). https://doi.org/10.1145/3090354.3090410
11. Hamdaoui, Y., Maach, A.: Energy efficiency approach for smart building in islanding mode based on distributed energy resources. In: Ezziyyani, M., Bahaj, M., Khoukhi, F. (eds.) AIT2S 2017. LNNS, vol. 25, pp. 36–49. Springer, Cham (2018). https://doi.org/10.1007/978-3-319-69137-4_4
12. Hamdaoui, Y., Maach, A., El Hadri, A.: Autonomous power distribution system through smart dynamic selection model using islanded micro grid context parameters and based on renewable resources. Int. J. Mech. Eng. Technol. (IJMET) **9**(11), 1755–1780 (2018)
13. Ding, L., Gonzalez-Longatt, F.M.: Two-step spectral` clustering controlled islanding algorithm. IEEE Trans. Power Syst. **28**, 75–84 (2013)
14. Zhuan, Y., Liu, T., Jiang, D.: A new searching method for intentional islanding of distribution network. In: Proceedings of the Asia-Pacific Power and Energy Engineering Conference (APPEEC), Shanghai, China, 1–4 March 2012
15. Hamdaoui, Y., Maach, A.: Smart islanding in smart grids. In: 2016 IEEE Smart Energy Grid Engineering (SEGE), pp. 175–180 (2016). https://doi.org/10.1109/SEGE.2016.7589521
16. Sortomme, E., El-Sharkawi, M.A.: Optimal scheduling of vehicle-to-grid energy and ancillary services. IEEE Trans. Smart Grid **3**(1), 351–359 (2012)
17. Bessa, R.J., Matos, M.A., Soares, F.J., Lopes, J.A.P.: Optimized bidding of a EV aggregation agent in the electricity market. IEEE Trans. Smart Grid **3**(1), 443–452 (2012)
18. Su, W., Rahimi-Eichi, H., Zeng, W., Chow, M.: A survey on the electrification of transportation in a smart grid environment. IEEE Trans. Ind. Inf. **8**(1), 1–10 (2012)
19. Ma, Y., Houghton, T., Cruden, A., Infield, D.: Modeling the benefits of vehicle-to-grid technology to a power system. IEEE Trans. Power Syst. **27**(2), 1012–1020 (2012)
20. Teoh, W.Y., Tan, C.W.: An overview of islanding detection methods in photovoltaic systems. In: 58 International Conference of World Axademy of Science, Engineering and Technology, WASET 2011, Bali (INDONESIA), 26–28 October 2011, pp. 674–682 (2011)
21. Hamdaoui, Y., Maach, A.: An intelligent islanding selection algorithm for optimizing the distribution network based on emergency classification. In: 2017 International Conference on Wireless Technologies, Embedded and Intelligent Systems (WITS), pp. 1–7 (2017). https://doi.org/10.1109/WITS.2017.7934627
22. Ranjbar, A.H., Banaei, A., Khoobroo, A., Fahimi, B.: Online estimation of state of charge in Li-Ion batteries using impulse response concept. IEEE Trans. Smart Grid **3**(1), 360–367 (2012)
23. Sanchez-Martin, P., Sanchez, G., Morales-Espana, G.: Direct load control decision model for aggregated EV charging points. IEEE Trans. Power Syst. **27**(3), 1577–1584 (2012)
24. Stadler, M., Marnay, C., Sharma, R., Mendes, G., Kloess, M., Cardoso, G., Mégel, O., Siddiqui, A.: Modeling electric vehicle benefits connected to smart grids
25. Chen, S.X., Gooi, H.B., Wang, M.Q.: Sizing of energy storage for microgrids. IEEE Trans. Smart Grid **3**(1), 142–151 (2012)
26. Sandels, C., Franke, U., Ingvar, N., Nordstrom, L., Hamren, R.: Vehicle to grid-Monte Carlo simulations for optimal aggregator strategies. Presented at the Power System Technology Conference, Hangzhou, China (2010)

27. Hamdaoui, Y., Maach, A.: Dynamic balancing of powers in islanded microgrid using distributed energy resources and prosumers for efficient energy management. In: 2017 IEEE Smart Energy Grid Engineering (SEGE), IEEE (2017). https://doi.org/10.1109/SEGE.2017.8052792
28. Hamdaoui, Y., Maach, A.: A novel smart distribution system for an islanded region. In: Ezziyyani, M., Bahaj, M., Khoukhi, F. (eds.) AIT2S 2017. LNNS, vol. 25, pp. 269–279. Springer, Cham (2018). https://doi.org/10.1007/978-3-319-69137-4_24
29. Hamdaoui, Y., Maach, A.: A cyber-physical power distribution management system for smart buildings. In: Ben Ahmed, M., Boudhir, A.A. (eds.) SCAMS 2017. LNNS, vol. 37, pp. 538–550. Springer, Cham (2018). https://doi.org/10.1007/978-3-319-74500-8_50
30. Hamdaoui, Y., Maach, A.: Prosumers integration and the hybrid communication in smart grid context. In: Networked Systems. LNCS, vol. 9466. Springer, Cham (2015)
31. Hamdaoui, Y., Maach, A.: Ontology-based context agent for building energy management systems. In: Ezziyyani, M. (ed.) AI2SD 2018, vol. 912, pp. 131–140. Springer, Cham (2020)

Hybrid Techniques to Conserve Energy in WSN

Zouhair A. Sadouq[1(✉)], Mostafa Ezziyyani[2], and Mohamed Essaaidi[1]

[1] Information Systems and Telecommunications Laboratory, FST-Tetuan, Abdelmalek Essaadi University, Tetuan, Morocco
sadouq@hotmail.com, essaaidi@ieee.org

[2] Computing, Systems and Telecommunications Laboratory, FSTT-Tangier, Abdelmalek Essaadi University, Tetuan, Morocco
ezziyyani@gmail.com

Abstract. Cloud Wireless Sensor Networks consist of a large number of miniature devices called sensor nodes scattered over a geographical area called sensor field. These nodes are attempted to collect information or data which is forwarded through gateways called base stations. The communication scenario through sensor nodes leads to some amount of energy wasting. Therefore we need to design suitable techniques and protocols in order to optimize the energy consumption and increase the network lifetime. In this paper we propose a novel energy-aware framework for a long-lived sensor network. Our framework is based on clustering architecture and achieves a good performance in terms of lifetime by minimizing energy consumption for in-network communications and balancing the energy load among all the nodes. In fact, it's an energy optimization approach based on cross-layer for wireless sensor networks, joining optimal design of the physical, medium access control, and routing layer.

Keywords: WSN · Clustering · Self-organization · Power control

1 Introduction

Nowadays, the development of mobile technology applications such as web browsing, online banking, online gaming and social media, has stimulated the wide spread usage of wireless network. Therefore, wireless networks have become almost a necessity and a vital component of contemporary daily life.

In this context, Wireless Sensor Network applications are expected to experience an enormous rise in the next few years, as well as the number and variety of sensors deployed in each WSN. They have a wide-range of applications. From homes to factories, from military surveillance to disaster prediction, WSN have attracted a lot of attention from researchers in the military, industry and academic fields.

WSN raise a growing interest among industries and civil organizations where monitoring and recognition of physical phenomena are a priority. Their possible applications are extremely versatile and are expected to be intensely applied in different domains such health, agriculture, habitat monitoring, routing traffic, security and military. WSN represent a significant technology that attracts more and more considerable

M. Ezziyyani (Ed.): AI2SD 2018, AISC 912, pp. 388–406, 2019.
https://doi.org/10.1007/978-3-030-12065-8_35

research attention in recent years. It has emerged as a result of recent advances in low-power digital and analog circuitry, low-power RF design and sensor technology.

WSN consist of a large number of nodes which communicate over wireless channels and perform distributed sensing and collaborative data processing. Recent research in domain considers WSN as a collection of smart mobile nodes, which form a dynamic and autonomous system. These nodes communicate wirelessly in a self-organized, self-configured and self-administered manner [1].

By correlating their output, they can provide functionality that an individual node cannot.

In a WSN, nodes are dispersed over an operational area where the phenomena of interest may appear. Sensors are deployed to monitor target region, to produce a measurable response to changes in specific physical conditions and to bridge the gap between the physical and the virtual world. The primary aim of WSN is the collection of data from an external environment, regardless of the underlying reasons for deploying a sensor network.

Harnessing the capability of such inexpensive and power-efficient nodes leads to the formation of a sensing network using different architectures called network topologies. Several topologies are possible: star, cluster-tree and mesh. Nodes act as sensors that collect information and transmit it or can perform more computational or routing tasks [2].

The increase of WSN requirements in terms of services and application constraints complicates the efficient modeling of sensor networks and the methodological development of dependable application software [3]. WSN systems are dynamic by nature: sensor node configuration needs to adapt to environment changes, which occurs many times during system execution [4].

In order to achieve effective coordination between nodes, it is important to address the problems of sensor network organization and the subsequent reorganization and maintenance [5].

In WSN, the communication scenario through sensor nodes leads to some amount of energy wasting. Therefore we need to design suitable techniques and protocols in order to optimize the energy consumption and increase the network lifetime [6]. Routing protocols are the main building block in route establishment and traffic delivery, which must be accomplished anywhere and anytime, between a pair of source and destination [7]. Constantly changing network topology, limited bandwidth and energy issues make the task of routing in WSN a challenging one.

We address the energy-consumption efficiency as a major design challenge in succeeding the vision of self-organized WSN. Energy is a very critical resource and must be used very sparingly. Therefore, energy efficiency is one of the determining factors for survivability and lifetime of WSN. WSN survivability is one of the critical issues leading the optimization of the energy efficiency to be a major research topic and leaving the other performance metrics like security, QoS and real-time performance as secondary objectives [7]. We propose a new architecture that aims reduces node energy consumption and contributes to extending the lifetime of the entire network.

Tanja framework is based on the GSM model [8]. It can be considered as a special kind of clustering architecture that extends the network life by efficiently using every node's energy and distributes management tasks to support the scalability of the management system in densely deployed sensor networks. However, it is more systematic, more robust and more scalable. In our solution we propose dynamic construction of clustering. The network is partitioned into clusters or cells. A cluster is composed with nodes, where every node can play one of three roles: source or sensing role as a slave, router, or a master as a cluster head and a gateway to the external world. Distribution of management tasks in sensor nodes is an energy efficient approach and utilizes node resources effectively in a large scale WSN [9] Therefore, Sensor nodes will take more management responsibilities and decision making in order to achieve a self-managed network.

In order to minimize this energy consumption, several techniques were introduced. Idle listening, high transmission power, and retransmission are the major source of energy waste which result from collision and suboptimal utilization of the available resource. One of techniques that have been introduced in S-MAC protocol is duty cycling mechanism. The goal of this technique is to reduce the energy consumption of idle listening. The majority of approaches of controlling transmission power, while keeping network connectivity, aim to decrease both, the unnecessary transmission energy consumption, and the interference among nodes. The power aware routing protocols approach, save significant energy. This approach works by selecting the adequate route which is based on the available energy of nodes or energy demand of transmission paths. The WSN requires the reduction of the energy consumed or the wasted energy such as idle, reception, transmission, etc., in all states. Indeed, all approaches cited above will be applied in the WSN.

The reminder of this paper is organized as follow: Sensor node architecture is described in Sect. 2. Section 3 focuses on the Networking challenges. Section 4 presents design details of the proposed framework. It focuses on the computation of optimal transmission power, routing, and duty-cycle schedule that optimize the WSNs energy-efficiency and by the way, reduces node energy consumption and contributes to extending the lifetime of the entire network. In Sect. 5, simulations and performance evaluation of the approach are presented. The approach is evaluated using the NS2 simulator. Simulation results show that it is an energy-efficient approach and able to achieve significant performance improvement as well. Section 6 discusses our architecture by highlighting a few significant features. Section 7 provides a brief review of related work in the literature. Section 8 is about future works suggested. Finally, Sect. 9 concludes the paper.

2 Sensor Node Architecture

Basically, each sensor node is composed by a Processor, Sensor(s)/Actuator(s), Analogue-to-Digital Converters ADC, Radio Frequency Transceiver, Data Storage, Controllers that tie the pieces together and a Power Supply. The sensor node has three other complementary sub-units which operate according to the application: (1) location finding system that allows to accomplish the routing techniques and find the sensing tasks, (2) mobilizer which can move the sensor node and (3) power generator that is assimilated to a small box of matches which allows to supply the component for as long time as possible (as shown in Fig. 1).

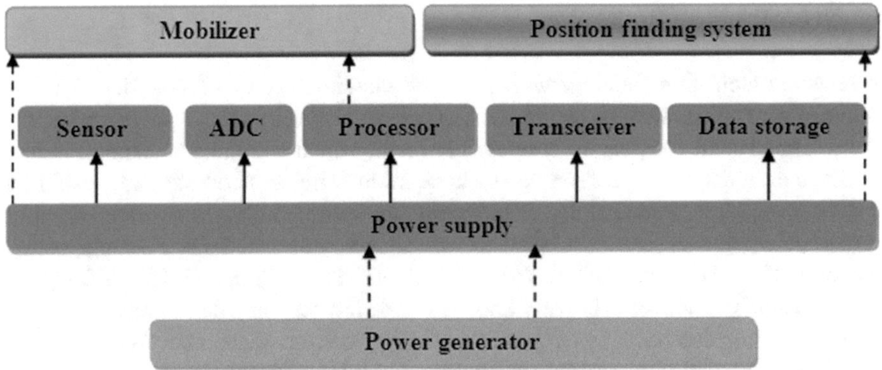

Fig. 1. Cloud Sensor node components.

- Processor: is the core of a sensor node. It retrieves data coming from the sensor, processes it and decides when and where to send it.
- Analog-to-Digital Converter: The continual analog signal produced by the sensors is digitized by an analog-to-digital converter and sent to controllers for further processing.
- Data storage: there is a need for RAM (Random Access Memory) to store intermediate reading, packets from other nodes and sensors, but the main drawback of RAM is that it loses its contents if power supply is interrupted. Flash memory and EEPROM (Electrical Erased Programmable Read-Only-Memory) can serve as intermediate storage when RAM is insufficient or there is no power supply.
- Transceiver: connects the node to network via RF, and is linked to an omnidirectional antenna that allows node to effectively communicate in all directions. The main task of a transceiver is to convert a bit stream (or sequence of bytes or frames) coming from the micro-controller to radio waves end. There are some low cost transceivers commercially available that incorporates all the circuitry required for transmitting-receiving, modulation, demodulation, amplifier, filter, mixer and so on.
- Sensor: the design of sensors in WSN takes into consideration several criteria including the volume of complete sensor, power consumption, and packing requirements [10]. Many applications call for multi-mode sensing, so each device may have several sensors on board. The specific sensors are highly dependent on the application, for example they may include temperature, humidity, pressure and chemical sensors or even low-resolution imager.
- Power supply: sensor nodes can be powered from energy storage or energy scavenging. In the first method, a variety of tiny batteries are being developed including thin film vanadium oxide and molybdenum oxide [11] that are fabricated using micro-machined cavities containing an electrolyte, in addition to chemical energy storage. There is another technique energy storing which uses radioactive isotropes. In the second case, scavenging energy from the environment will allow the sensor network to operate indefinitely. The most important energy source is solar radiation and vibration.

3 Networking Challenges

In telecommunication field, networks can be classified in two categories. The first category is wired networks, which rely on physical links such as wires and optical fibers. The second category is wireless networks, which use radio transmission techniques to establish links between nodes. In addition, wireless networks can be split into two classes, as presented in Fig. 2, Infrastructure based wireless networks, which use fixed access points as gateways between wired and wireless area. For example, cellular networks (2G, 3G, and LTE), WiFi (IEEE 802.11), WiMax (IEEE 802.16) and infrastructure less networks broadly known as Ad Hoc networks do not rely on any pre-established infrastructure consequently Ad Hoc networks are self-organized, self-configured and self-administered. Furthermore, Ad Hoc Networks are single-hop like Bluetooth or multi-hop like Wireless Sensor Networks (WSN), Wireless Mesh Network (WMN) and Mobile Ad Hoc Network (MANET).

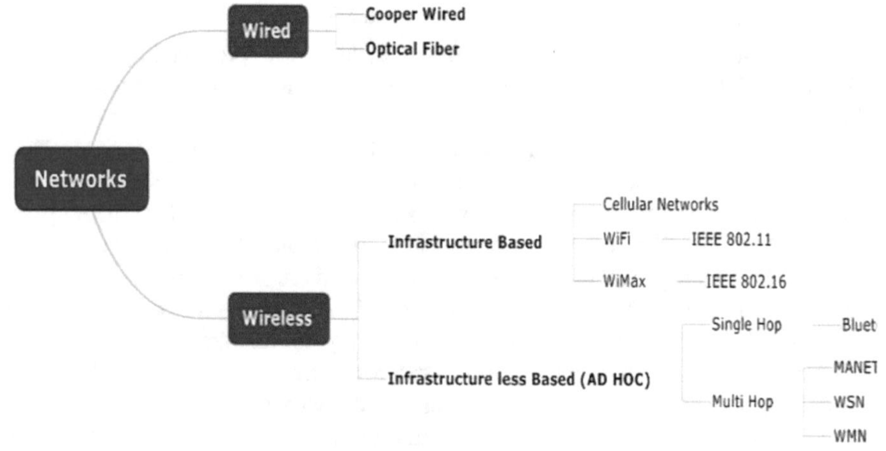

Fig. 2. Network categories.

WSN networking is a multi-level issue because of its autonomous operations; hence network layer should adapt its routing operations to several network constraints, such as nodes mobility, nodes Energy, scarce bandwidth and network size to establish efficient paths for data communication. In this context, many routing protocols has been designed in order to deal with different constraints and guarantee the quality of service required by WSN applications [1, 2].

WSN are a set of smart mobile nodes, which form a dynamic and autonomous system and where each node within the network has the ability to change its location and configure itself on the fly. These nodes are able to establish routes, anywhere and anytime, between a pair of source and destination using routing protocols. In addition, nodes mobility, bandwidth-constrained, energy-constrained, and limited security of shared medium of WSN make designing process of routing protocols most important and difficult instead of design process in fixed network. Due to the dynamic nature of mobile nodes, WSN experience frequent link failures, which cause frequent network topology change, this has led to the design of various routing protocols. The purpose of each protocol is to solve problems for a specific WSN topology condition; therefore, the designer should have a prior knowledge about the condition or the context of the network targeted with routing protocol design. Therefore, different routing protocols perform differently with different networks' conditions such as level of mobility, size of the network in terms of connected nodes number or type of packets being routed through the network [1–3].

The main building bloc in WSN is routing. Therefore, designing routing protocols have attracted the interest of researchers. Since several routing protocols have been proposed in order to meet required functionalities related to a specific application field. As a result, there is no routing protocol that could fit to all WSN contexts [6, 7]. Routing protocols can be classified using several approaches, depending of the purpose or the goal for which the protocol is designed. See illustration in Fig. 3.

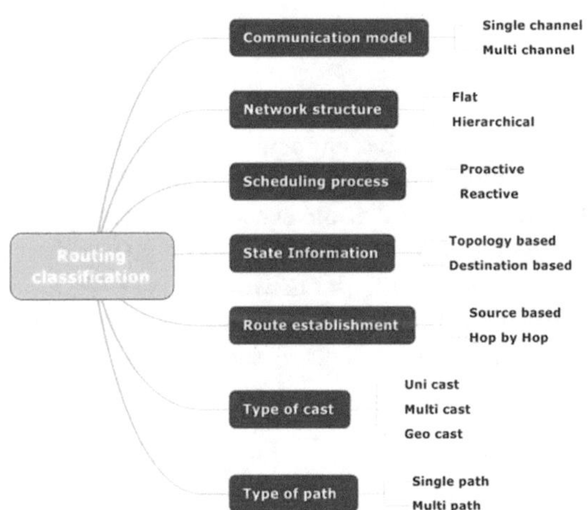

Fig. 3. Routing classification

There are different criteria for classifying routing protocols in WSN:

- Communication Model
- Network structure
- Scheduling model
- State Information
- Route establishment
- Type of Cast
- Type of path

The online version of the volume will be available in LNCS Online. Members of institutes subscribing to the Lecture Notes in Computer Science series have access to all the pdfs of all the online publications. Non-subscribers can only read as far as the abstracts. If they try to go beyond this point, they are automatically asked, whether they would like to order the pdf, and are given instructions as to how to do so.

Please note that, if your email address is given in your paper, it will also be included in the meta data of the online version.

4 Tanja Framework

Within the class of wireless ad hoc networks that Tanja framework addresses, transmitter nodes can be installed at specific locations or be placed randomly. They can be thrown out of an airplane and, on landing, they are capable of self-organizing into a network. After the initial deployment the network has to be easily scalable, since new sensor nodes may be added, removed or replaced during the network lifetime, which affects node location, density, and overall topology. Moreover, once deployed, sensors are prone to failures due to the manufacturing defects, environmental conditions or battery.

4.1 Topology

The topology of a sensor network has an important impact on several network aspects, including power consumption, battery life and routing mechanisms. In this kind of network, the fundamental idea is that the sensor network is partitioned into a set of clusters or cells (as shown in Fig. 4). Because of the large number of sensor nodes that are expected to populate a wireless sensor network, direct node-level addressing cannot be considered as a feasible approach, and thus clustering is recognized as an effective means of interacting with dense networks of sensors.

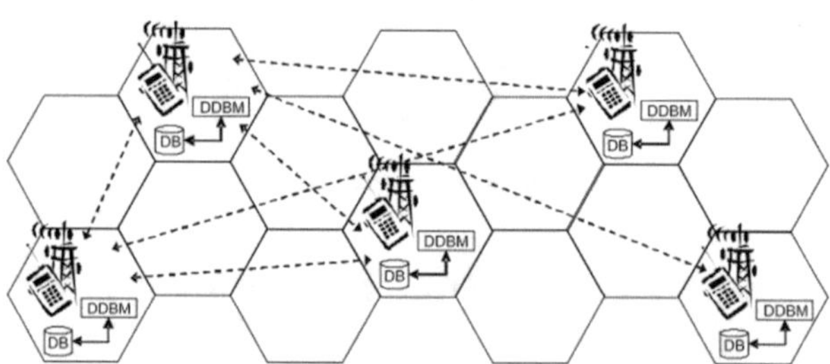

Fig. 4. Mobile/Wireless Ad Hoc Network

Another motivating factor for the establishment of clustering is that after the deployment of sensor nodes, it is likely that multiple neighboring nodes will observe the same phenomenon and obtain the same codes, thus being reported as identical hosts to the base station. To allow individual nodes to directly report their observations to their base station represents a potential waste of energy caused by the transmissions of duplicated data. In fact, the master node may eliminate duplicate readings and improve the quality of the data by confirming shared observations substantiated by multiple sources. As in the GSM model, the cluster head or the master acts like a Virtual Base Station or BTS; its role consists of managing the nodes that belong to its cluster by refining and compressing the sensed data from the cluster, which reduces the network traffic caused by redundant transmissions, reduces the energy output of its cluster members, and contributes to extending the lifetime of the entire network. Clusters may exchange data with each other to further refine the sensed data across cluster boundaries. This behavior permits high-density sensor deployments to be considered, and treated, as progressively density hierarchies of clusters.

Each cluster is composed with nodes, where every node can play one of three roles: source or sensing role when it acts as a slave; router, or master, acting as a cluster head; and a gateway for interacting with the external world (see Fig. 5).

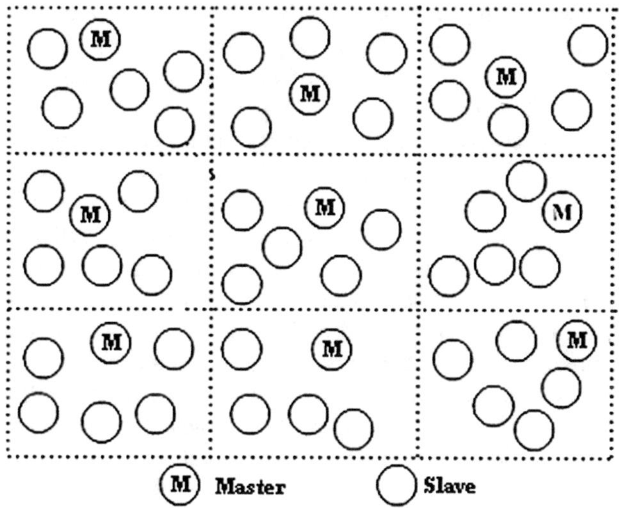

Fig. 5. WSN cluster architecture

Once established, the topology usually does not change when there is no node mobility. However, as nodes perform their assigned tasks, they deplete and eventually exhaust their energy store, causing them to die. The sensor network must be able to detect and recognize failures as well as dynamically adapt itself to changing conditions and maintain the correctness of operations. The topology may be refreshed by the periodic addition of new nodes to the clusters.

Because of cost and energy constraints, only the master nodes are generally able to transmit data from the sensor network to the "outside world" by means of a longer-range connection. It makes use of a remote Base Station as a gateway to link the external world -such as the Internet or Satellite- to the wireless sensors network (as shown in Fig. 6).

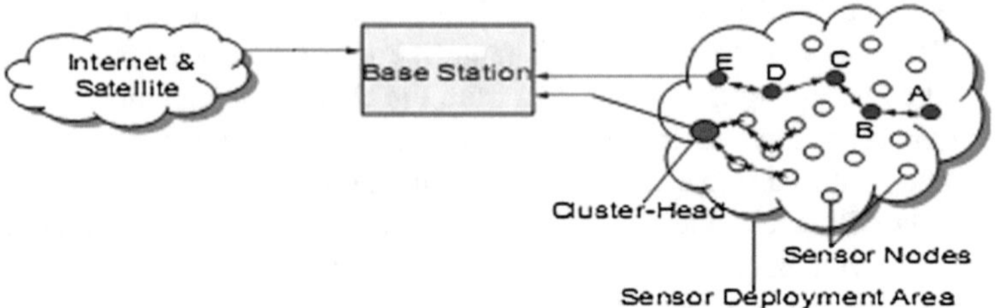

Fig. 6. General WSN architecture

4.2 Low-Power Approach and Transmission Power Control

Clouds Sensor nodes are typically battery-driven; however, they are too small and too numerous for battery replacement or charging. Moreover, micro sensor networks are often deployed in remote or dangerous environments. Hence, the increase of sensor node lifetime becomes a major design and implementation challenge. The necessary lifetime has a high impact on the required degree of energy efficiency and robustness of the nodes, thereby requiring the minimization of energy expenditure. We concern the energy-consumption efficiency as a major design challenge in succeeding the vision of self-organized WSN and hence we address it in the following sections.

Power control is necessary to overcome what is known as the near-far problem that is as some slave nodes are closer to cluster head it must reduce its power to avoid causing interference to other slave nodes. The transmission power within the WSN needs to be set the right level dynamically with spatial and temporal change. Temporal and Spatial factors affect the transmission power between Cluster Head and slave nodes. Temporal factors include surrounding environmental changes in general, such as weather conditions, while spatial factors include the surrounding environment, such as terrain and the distance between the transmitter and the receiver. To control the transmission power dynamically, a Received Signal Strength Indicator RSSI was introduced specifying the transmission power level during runtime.

Every slave node is called to find the minimum transmission power level to communicate with its Cluster Head successfully. The Cluster Head measures the received signal strength in the uplink. The Cluster Head compares the received signal strength to target signal strength. If this one is below the target the cluster head will request the slave node to increase its power and decrease it if above the target [12].

4.3 Duty-Cycle Scheduling

In order to improve energy consumption, duty cycle mechanisms have been introduced. Sensor MAC (S-MAC) uses new procedures to decrease energy consumption and support self-configuration. S-MAC is based on contention. Each sensor node follows a periodic synchronized listen/sleep schedule. Nodes in the S-MAC exchange their sleeping schedule and before going to sleep nodes broadcast their schedule to their neighbors as a SYNC packet. For S-MAC, energy consumption in idle listening is to be reduced by allowing neighboring nodes of transceiver and receiver to sleep periodically during transmission, by doing so this scheme put nodes into low duty cycle. Figure 7 reflects SMAC listen sleep schedule. Periodically sleeping is good in low traffic cases. If a node can sleep for longer time it consumes less energy [13].

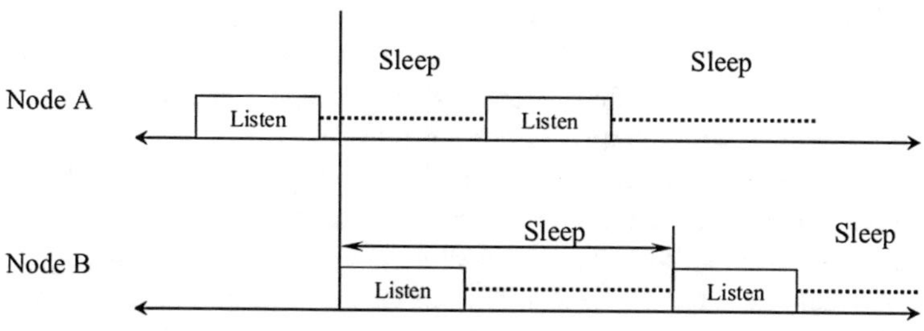

Fig. 7. S-MAC listen/sleep schedule

4.4 Signaling Systems

The signaling system is the nervous system of the network. A great deal of information needs to be passed back and forth between nodes while data transferring as well as in the servicing of specialized features. Here, data refers mainly to the sensed data. Sometimes it also refers to the network infrastructure information concerned by the applications. Four main types of signals handle this passing of information: supervisory signals, alerting signals, connection signals and discovery signals.

- Discovery signals are used after the deployment of the sensor network in order to setup up and self-organize the whole network. They can be later used for discovering new nodes. Discovery signals convey information consisting of ID, location and energy level of nodes.
- Supervisory signals handle energy and location status of nodes composing the network. A master is always monitoring its slaves or its cluster members which send back to it their updates, in return, including their node ID and energy level.
- Connection signals are used in the handover mode. When an acoustic signal strength detected by the master exceeds a predetermined threshold, the active master then broadcasts an information solicitation message, asking sensors in its vicinity to join the cluster and provide their sensing information.

Alerting signals are used in the downlink way from the master nodes to nodes belonging to its cell. Those signals perform which nodes will perform sensing and which will go to sleep.

4.5 Network Setup and Self-organization

The enormous number of nodes in sensor networks requires sophisticated solutions for the automatic organization of the network. A manual boot procedure by an administrator is nearly impossible. Therefore, the middleware residing on each node has to autonomously set-up an operating network infrastructure by interacting with its neighboring nodes. For self-organizing networks, the knowledge of the current context (context awareness) is important. During setup phase, infrastructure context (perception of network bandwidth and reliability) and domain context (relation between the network participants) are primarily important. The current system context is necessary for a sensor node to operate correctly. This context can change permanently because of the mobility of nodes, therefore updates mechanisms have to be considered [14, 15].

Upon deployment, the WSN self organizes. Given the large number of nodes and their potential placement in difficult locations, it is essential that the network is able to self-organize; manual configuration is not feasible [16, 17].

Moreover, nodes may fail (either from lack of energy of from physical destruction), and new nodes may join the network. Therefore, the network must be able to reconfigure itself periodically. We propose dynamic construction of clustering. Instead of assuming the same role for all the sensors, the sensor network is composed of sensors assuming the role of a cluster head or master upon triggering by certain signal events and sensors whose function is to monitor their environment and provide sensor information to masters upon request [18, 19].

To facilitate scalable operations in sensor networks, sensor nodes should be aggregated to form clusters based on their power levels and proximity. We consider sensor networks where each sensor node is aware of its own location. The network can use location services such as to estimate the locations of the individual nodes, and no GPS receiver is required at each node. Every node sends a discovery message that consist of its node ID, its location and its energy level. The node with the highest life time or energy will be appointed as a cluster master. Sometimes, a secondary parameter is considered in the operation of master selection that is the node's proximity to its neighbors. A master is elected from the sensor nodes belonging to the same cluster, and is responsible for acquiring data from the sensor nodes in its cluster. Masters keep changing in each cluster in order to extend network lifetime. The master broadcasts a supervisory signal to its slaves or its cluster members which in return send their updates including their node ID and energy level. The master maintains a virtual database table, called NODES, whose columns contain information such as sensor location, sensor node ID, and remaining battery power. Upon receiving updates from slaves, the master decides if it keeps its status as a master or must change to the slave role and join the new cluster which belongs to the master that contains the highest energy level. The accomplishment of those tasks leads to the formation of a true WSN [20].

The master node then decides which nodes will continue performing sensing and which ones will go to sleep. It should enable nodes with more battery, processing, or memory resources to participate more in the network coordination, data aggregation and processing, and data dissemination. A widely employed energy saving technique is to place nodes in sleep mode, corresponding to low power consumption and reduced operational capabilities. Initially, a small number of active nodes participate in routing, being the rest in passive mode. Nodes in passive mode regularly change to test mode and return to active mode when there are not many neighbor nodes and the packet loss is high.

WSN benefits from the fact that clustering attempts to extend network battery lifetime by rotating the role of masters and from protocols that enable sensor nodes to take turns in turning off their transceivers.

4.6 Data Routing and Delivery

The data collected by sensor nodes are pre-processed to obtain partial results and can be stored temporarily before their transmission to the master nodes. These data are then collected by the master node for performing an additional processing to get the final result. Sensors may transfer data in a single-hop from the source node to master node or may instead use multiple hops over several nodes. It is widely accepted that multi-hop data routing provides a greater level of efficiency, and contributes to the longevity of the sensor network. This paradigm allows sensor nodes far away from master nodes to transmit data to neighboring sensor nodes, which in turn forward the data towards the intended master node. The forwarding process may cause that multiple sensor nodes on the path between the source node and the collection point get involved. Thus, this paradigm uses a centralized, multi-hop communication model. Regardless of the length of the path, the data eventually reaches the collection point. Coordination among nodes in routing the data to the end point is part of this paradigm [22].

5 Related Works

Self-organization is one of the most significant research topic in the wireless networks. It philosophy involves abstracting the communicating entities into an easily controllable network infrastructure. Clustered or connected dominating set CDS, grid, tree, or mesh based organization are key terms in self-organization.

A self-organized wireless node can be grouped or clustered into an easily manageable network infrastructure [5]. There are several methods to form clustering. Cluster is formed in two stages by nodes. In the first stage, a header is selected among the nodes by election algorithm. In the second stage, the cluster or group is formed due to the interaction between the nodes and the headers [23].

The scheme proposed in [24] is based on cluster formation. The idea of the scheme proposed is partition of the network into different groups or clusters and then a cluster head is appointed for each group. The cluster head perform major tasks and should contains more resources than other cluster members. We should keep in mind that the data can only be transmitted to other clusters through cluster heads. Limitation of accessibility to the nodes under its supervision exists when there is a failure of a cluster head.

The proposed framework, permits sensor nodes to autonomously determine their management role based on the node real time capability like energy. Clustering has been used to address various issues i.e. routing, energy efficiency, management and huge-scale control. In addition, other approaches, concerning cross-layer management, have been introduced for WSN. They can ameliorate the performance of clustering in the Wireless Sensor Networks. Indeed, those approaches can reduce significantly the consumption of energy in WSN communications. They can be nearly divided into three different groups to in terms of interaction or modularity among physical PHY, medium access control MAC, routing, and transport layers.

MAC+PHY: The energy consumption for physical and MAC layer was discussed in [25]. The cross layer solution among the application layer, MAC layer, and the application layer for Wireless Sensor Networks is introduced by [26].

MAC+Routing: In the literature, the receiver-based routing is exploited for MAC and routing cross-layer modularity [27, 28].

PHY+MAC+Routing: The MAC, PHY layer, and joint routing optimization are introduced in [29], which adopts a variable-length TDMA scheme and MQAM modulation. The optimization of transmission power, transmission rate, and link schedule for TDMA-based WSN was discussed in [30, 31].

6 Simulation and Results

6.1 Data Routing and Delivery

To evaluate our approach, we use a simulation tool dedicated to wireless networks and considered a crucial asset search, called NS2 simulator. Then s-2 network simulator is the most widely used simulator in networking research. Architecturally a discrete event simulator, it allows for simulation of most modern TCP, routing and multicast network protocols over both wired and wireless network links. The ns2-allinone-2.29 version is used in the simulations.

On the implementation side, it is possible to run ns-2 with a real-time scheduler to allow for interactions between real and simulated network components. ns-2 and related tools are available at http://www.isi.edu/nsnam/ns/.

We conduct computer simulations to assess the performance of the proposed methodologies. We consider three plans to measure the performance of our approach:

- Transmission Power Control and Duty-Cycle scheduling TF.
- Non-Transmission Power Control and Duty-Cycle scheduling NTPC. In this plan, a static transmission power is used in the approach, which is set to the average value between maximal and minimal values.
- Non-Duty-Cycle scheduling and Transmission Power Control NDC. In this plan, nodes keep idle listening without transmitting data, using CSMA mechanism to perform a non-duty-cycle.

6.2 Results and Discussion

Energy Consumption. Figure 8 shows that TF is the most energy-efficient solution. In this simulation, we consider that energy consumption is the most significant metric performance to evaluate. The total energy consumption is measured, when all slave nodes send effectually their data packets to Cluster Head.

Network Lifetime Analysis. The amount of living nodes over time is illustrated in Fig. 9. One may observe that the number of damaged nodes increases during the execution time. One may observe also that the number of living nodes with TF is more than with NDC and NTPC. This can be explained by the fact that passive slaves nodes sleep during transmission activity. Those slave nodes don't have to use their energy allowing the extension of their lifetime, the whole cluster lifetime and WSN lifetime by the way. Thus, low message transmission will help conserve node energy, and consequently it prolongs the network lifetime. However, in NDC, slave nodes are set in receiving mode prepared to receive data sent by Cluster Head at any time whenever they don't transmit data packets, while in NTPC, a static transmission power is used, leading to a non optimal use of energy during transmission activity. During the run-time, adopting NTPC and NDC approaches will consequently consume valuable node energy and shorten the network lifetime.

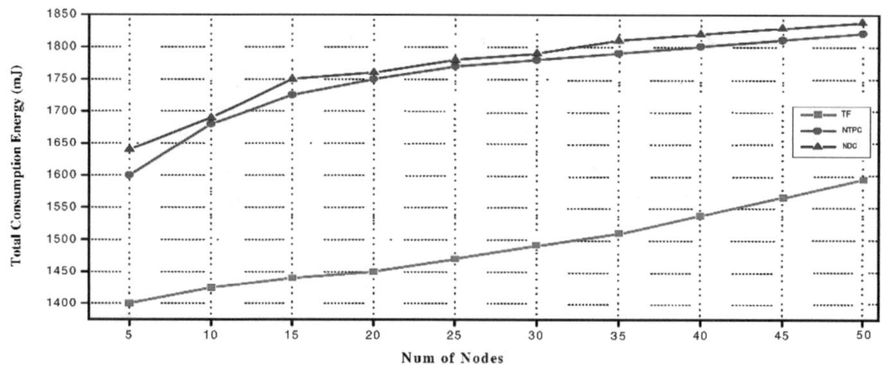

Fig. 8. Total consumption energy according to the number of source nodes

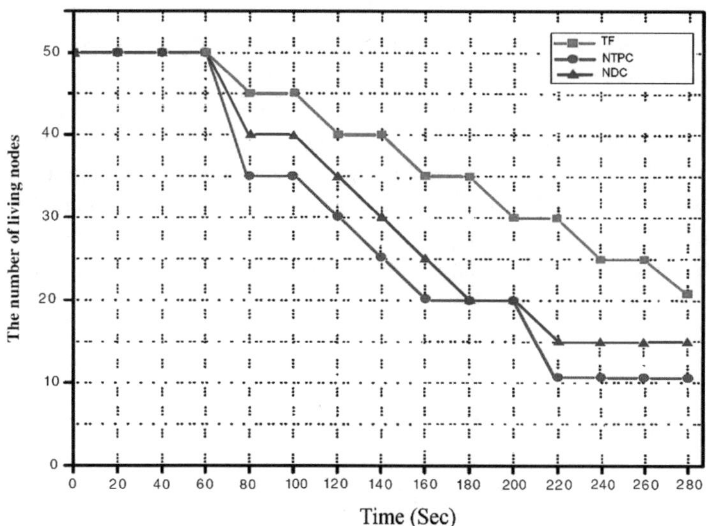

Fig. 9. The number of living nodes over time

End to End Delay. The end-to-end delay is one of the most essential and crucial issues for wireless sensor networks. Many applications of sensor networks require an end-to-end delay guarantee for time sensitive data, such as in security environments and real-time phenomena. Figure 10 shows that the end-to-end delay is closely banded with number of nodes forming clusters. The challenge is to minimize the end-to-end packet delivery delay. The End-To-End Delay can by calculated by:

$$\text{EED} = \frac{\textit{Time spent to deliver packets}}{\sum \textit{received packets}}$$

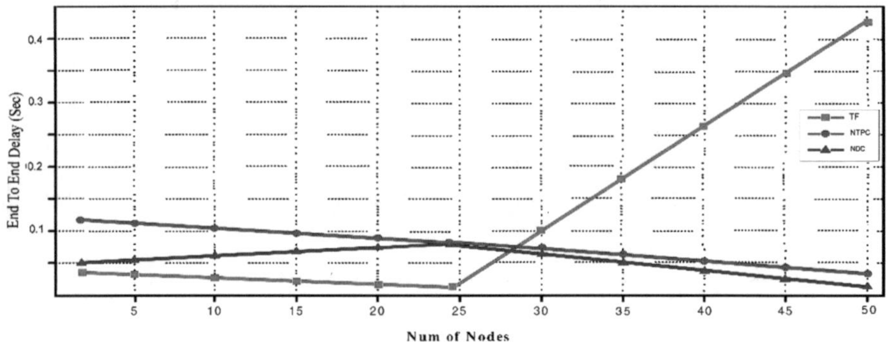

Fig. 10. End to end delay

Packet Delivery Ratio. Packet Delivery Ratio is one of the main parameters in WSN. Packet Delivery Ratio is deteriorated by Packet Loss that happens during the lifetime of the sensor nodes. Packet Loss and energy consumption represent some of basic challenges in the design of WSN. Packet Loss in WSN is due to several reasons. Buffer overflow occurs when the queue in the receiving node is full. The queue size is limited by hardware characteristics. Link failure happens when the link between the sender and receiver is failed or broken due to channel fading, shadowing, interference, power failure, and sensor nodes mobility.

Since the packet delivery ratio is reflected by packet loss, this ratio can be analyzed through transmitted packet loss as:

$$\text{Transmitted Packet} = \text{Received Packets} + \text{loss Packets}$$

Thus, the packet delivery ratio can be defined as:

$$\text{PDR}(\%) = 100 \times \frac{\sum \textit{Received Packets}}{\textit{Generated Packets}}$$

TF sends data to Cluster Heads in an efficient manner based on Transmission Power Control and Duty-Cycle scheduling. It increases the efficiency of data transmission by reducing the number of sensors attempting to transmit data in clusters and minimizing total data packet loses consequently. Figure 11 shows that the proposed method is more accurate being simple and energy efficient at the same time. However, when the network size is increased, the Packet Delivery Ratio is decreased. We recognize that our approach is more suitable for a small size clusters.

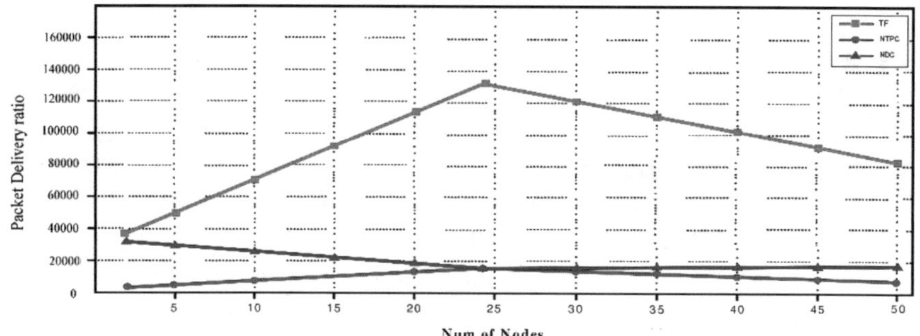

Fig. 11. Packet delivery ratio

6.3 Formulas

Displayed equations or formulas are centered and set on a separate line (with an extra line or halfline space above and below). Displayed expressions should be numbered for reference. The numbers should be consecutive within each section or within the contribution, with numbers enclosed in parentheses and set on the right margin.

$$x + y = z \tag{1}$$

Equations should be punctuated in the same way as ordinary text but with a small space before the end punctuation mark.

7 Discussion

Cluster-based hierarchical framework have proven to be an effective method to provide better data aggregation and scalability for the sensor network while conserving limited energy. It facilitates the efficient use of limited energy in WSN and therefore extends the lifetime of the network. It increases the efficiency of data transmission by reducing the number of sensors attempting to transmit data in clusters and minimizing total data packet loses consequently. Within clusters, and in normal conditions, Slave nodes are susceptible to transmit data over short distances compared to Cluster Heads that are obligated to deliver data to the Base Station over long distances. In this case, more energy is drained from Cluster Heads compared to other sensor nodes in the cluster.

The possibility of the re-election of Cluster Heads in accordance with the residual energy still an interesting technique to balance the power consumption of each cluster.

Adjusting transmission powers and signaling rates on the existing links to reduce energy consumption. Sleep-wake scheduling is another effective mechanism that helps to prolong the lifetime in WSN. In fact, energy consumption in idle listening is to be reduced by allowing slave nodes to sleep periodically during transmission when there are no events or in low traffic cases. If a node can sleep for longer time it consumes less energy.

8 Future Works

Self-organization is one of the most significant research topic in the wireless networks. It philosophy involves abstracting the communicating entities into an easily controllable network infrastructure. Clustered or connected dominating set CDS, grid, tree, or mesh based organization are key terms in self-organization.

9 Conclusion

WSN has presented various research challenges and one major challenge is to design an efficient self-organized architecture of wireless sensor network. In this chapter we have presented a new component framework for developing WSN applications. We described a new self-organized architecture based on GSM model for wireless sensor networks. We divide the whole network into multiple cells, where each cell comprises of a group of nodes. In the reference operational setting, based on a self-organized architecture of sensor clusters, each one is governed by a master node that is elected by cluster nodes autonomously. The approach is able to use the physical layer's transmission power as metric to optimize energy consumption, and use S-MAC protocol in duty-cycle.

References

1. Akyildiz, F., Su, W., Sankarasubramaniam, Y., Cayirci, E.: Wireless sensor networks: a survey. Comput. Netw. **38**(4), 393–422 (2002)
2. Aakvaag, N., Mathiesen, M., Thonet, G.: Timing and power issues in wireless sensor networks- an industrial test case. In: Proceedings of the International Conference on Parallel Processing Workshops ICPP 2005, Oslo, Norway, 14–17 June, pp. 419–426 (2005)
3. Sadouq, Z., Essaaidi, M.: Design challenges of wireless sensor networks based on real-time middleware. In: Information and Communication Technologies International Symposium, ICTIS 2007, 3–4 April–5 Fez (2007)
4. Tuñón, M., López, W.M., Sadouq, Z.: Characteristic analysis of wireless sensor networks with real-time middleware. In: Proceedings of the II Congreso Español de Informática CEDI 2007, 12–14 September, Zaragoza, pp. 21–28 (2007)

5. Kochhal, M., Schwiebert, L., Gupta, L.: Role-based hierarchical self-organization for wireless ad hoc sensor networks. In: Proceeding of the 2nd ACM International Conference on Wireless Sensor Networks and Applications, pp. 98–107 (2003)
6. Yebari, M., Addali, T., Sadouq, Z., Essaaidi, M.: Energy conservation challenges in communication protocols for wireless sensors networks: the state of the art. In: Proceedings of the International Journal on Information and Communication Technologies, vol. 1, No. 1–2, pp. 29–35, January-June 2008
7. Demirkol, I., Ersoy, C., Alagöz, F.: 2006 MAC protocols for wireless sensor networks: a survey. IEEE Commun. Mag. **44**(4), 115–121 (2006)
8. Sadouq, Z., Tuñón, M., Essaaidi, M.: Tanja: a framework for modelling Wireless Sensor Networks. In: Workshop on Sensor Networks and Applications, WSeNA 2008, Gramado, 1–5 September (2008)
9. Yu, M., Mokhtar, H., Merabti, M.: A survey of network management architecture in wireless sensor network. In: Proceedings of the Sixth Annual Post-Graduate Symposium on the Convergence of Telecommunications, Networking and Broadcasting (2006)
10. Ilyas, M., Mahgoub, I.: Handbook of Sensor Networks: Compact Wireless and Wired Sensing Systems. CRC Press (2005). LLC 2005, ISBN 0-8493-1968-4
11. Baltes, H., et al.: Micromachined thermally based CMOS microsensors. Proc. IEEE **86**, 1660–1678 (1998)
12. Lin, S., Zhang, J., Zhou, G., Gu, L., He, T., Stankovic, J.A.: ATPC: adaptive transmission power control for Wireless Sensor Networks. In: Proceeding of SenSys 2006 (2006)
13. Ye, W., Heidemann, J., Estrin, D.: An energy-efficient MAC protocol for wireless sensor networks. In: 21st Conference of the IEEE Computer and Communications Societies (INFOCOM), vol. 3, pp. 1567–1576, June 2002
14. Bharathidasas, A., Anand, V.: Sensor networks: an overview. Technical report, Dept. of Computer Science, University of California at Davis (2002)
15. Estrin, D., Govindan, R., Heidemann, J., Kumar, S.: Next century challenges: scalable coordination in sensor networks. In: Proceedings of ACM Mobicom, Seattle, Washington, USA, pp. 263—270, August 1999
16. Akyildiz, I.F., Su, W., Sankarasubramaniam, Y., Cayirci, E.: A survey on sensor networks. IEEE Commun. Mag. **40**, 102–114 (2002)
17. Estrin, D., Girod, L., Pottie, G., Srivastava, M.: Instrumenting the world with wireless sensor networks. In: Proceedings of the International Conference on Acoustics, Speech and Signal Processing ICASSP 2001 (2001)
18. Biagioni, E.S., Sasaki, G.: Wireless sensor placement for reliable and efficient data collection. In: Proceedings of 36th International Conference on Systems Sciences, Hawaii, January 2003
19. Krishnan, R., Starobinski, D.: Efficient clustering algorithms for self-organizing wireless sensor networks. Ad Hoc Networks (2005)
20. Savvides, A., Han, C., Strivastava, M.: Dynamic finegrained localization in ad-hoc networks of sensors. In: Proceeding of ACM MOBICOM 2001, pp. 166–179. ACM Press (2001)
21. Doherty, L., El Ghaoui, L., Pister, K.S.J.: Convex position estimation in wireless sensor network. In: Proceeding of IEEE INFOCOM 2001, Anchorage, AK, April 2001
22. Heinzelman, W.R., Chandrakasan, A.P., Balakrishnan, H.: Energy-efficient communication protocol for wireless microsensor networks. In: Proceedings of the Thirty Third International Conference on System Sciences HICSS 2000, Maui, January 2000
23. Chen, J.L., Lu, H.F., Lee, C.A.: Autonomic self-organization architecture for wireless sensor communications. Int. J. Network Manag. **17**(3), 197–208 (2007)

24. Tai, A.T., Tso, K.S., Sanders, W.H.: Cluster-based failure detection service for large-scale ad hoc wireless network applications in dependable systems and networks. In: DSN 2004 (2004)
25. Haapola, J., Shelby, Z., Pomalaza-Raez, C., Mahonen, P.: Cross-layer energy analysis of multi-hop wireless sensor network. In: Proceeding of European Conference on Wireless Sensor Networks EWSN 2005, Istanbul, 31 January–2 February, pp. 33–44 (2005)
26. Vuran, M.C., AKyildiz, I.F.: Spatial correlation-based collaboration medium access control in wireless sensor network. IEEE/ACM Trans. Networking **14**(2), 316–329 (2006)
27. Skraba, P., Aghajan, H., Bahai, A.: Cross-layer optimization for high density sensor networks: distributed passive routing decision. In: Nikolaidis, I., Barbeau, M., Kranakis, E. (eds.) ADHOC-NOW 2004, vol. 3158, pp. 266–279 (2004)
28. Zorzi, M., Rao, R.: Geographic random forwarding GeRaF for ad hoc and sensor networks: multihop performance. IEEE Trans. Mob. Comput. **2**(4), 337–348 (2003)
29. Cui, S., Madan, R., Goldsmith, A., Lall, S.: Joint routing, MAC, and link layer optimization in sensor networks with energy constraints. In: Proceeding of IEEE ICC 2005 (2005)
30. Madan, R., Cui, S., Lall, S., Goldsmith, A.: Cross-layer design for lifetime maximization in interference-limited wireless sensor networks. In: Proceeding of IEEE INFOCOM 2005 (2005)
31. Le, D., Kumar, R., Nguyen, G., Chatterjee, J.M.: Cloud Computing and Virtualization. Wiley, Chichester (2018)

Optimal Power Control Strategy of a PMSG Using T-S Fuzzy Modeling

A. Benkada[1(✉)], H. Chaikhy[2], M. Monkade[1], and M. Kaddari[2]

[1] Department of Physical, University Chouaib Doukkali, El Jadida, Morocco
Abdennasser.bk@gmail.com
[2] Department of Electrical and Energetic, University Chouaib Doukkali, El Jadida, Morocco

Abstract. This article offers two different method control strategies to have the maximum power from wind turbine (WT) based on the Permanent Magnet Synchronous Generator (PMSG). The first control strategy is composed of standard proportional-integral (PI) regulators. The PI controllers are tuned for a specific operation mode. However, since the system is nonlinear, for different operating conditions, the values of the PI parameters may not be optimal. The second approach presents a new fuzzy tracking control method using Takagi-Sugeno (T-S) fuzzy of the WT, to achieve improved speed performance under different operating points. Finally, simulation results are provided to demonstrate the validity and the effectiveness of the proposed method.

Keywords: Maximum power · Wind turbine · PMSM · PI controller · Takagi-Sugeno fuzzy

1 Introduction

The main solution to decrease greenhouse gases is to use renewable green energy, this has become the principal solution to environmental pollution. For this reason, power generation systems based on renewable energy are contributing to total energy production. The wind power is among the different kinds of renewable energy that has shown rapid progress around the world. Wind power has the proper energy aspect, is omnipresent, and is freely available [1, 2].

Renewable energy sources has been the main solution for the world over the last decade. Variable Speed Wind Energy Generation (VS-WEGS) is among various varieties of renewable energy, for that this (VS-WEGS) has been noted as the most important growing technology and becomes the most competitive form of renewable green energy [1, 2]. In the area of VS-WEGS, Permanent Magnet Synchronous Generators (PMSG) possess some attractive qualities such as high power density, implementation facility, absence of the gearbox and efficient energy production, compared with other generator technologies [3, 4]. In large wind power system, PMSG is known as the most appropriate for application in large wind power system. In order to extract the optimal accessible power given by the WECS, several studies have presented multiple control schemes [3]. The simplest techniques for searching the maximum power point are based on PI controller but this method is still classical and lack of

M. Ezziyyani (Ed.): AI2SD 2018, AISC 912, pp. 407–416, 2019.
https://doi.org/10.1007/978-3-030-12065-8_36

performance, the perturb & observe method is the simplest algorithm for searching the maximum power point but the duty of the power converter perturbs rapidly and causes energy loss near the maximum power point [4, 5]. The neural network method estimates the wind speed from the measured turbine power and generator speed and thus decides the maximum power generator speed command or torque reference for the operational point tracker [6]. In addition, fuzzy logic methods have been widely adopted and they have presented better effectiveness [7]. Nevertheless, the main issues of most of these methods are lack of stability and strict theoretical analysis so that the maximum power point varies over a wide range.

In literature, the T-S fuzzy model-based control has been fully mentioned for example [8, 9]. Thus, the control based on (T-S) model has been more popular as one of the most successful techniques for systems and control applications due to its reliability and effectiveness. The principal advantage is that the controller is systematically realized by using parallel distributed compensation (PDC) and a linear matrix inequality (LMI) technique [10]. The fuzzy model of Takagi-Sugeno allows modelling the nonlinear system by representing local dynamics by linear models. As a result, the global system model is achieved by a combination of the different linear models. Then, a linear feedback control has to be constructed for every local linear model. Accordingly, the consequent overall nonlinear controller is once more a fuzzy combination of each distinct linear controller [11].

This paper is organised as following: the second section will present a dynamic modelling of the WECS. In the section that follows, we will present control strategy. Afterword, we will present fuzzy modelling of the system. We explain the control design and we will give stability conditions finally. In Sect. 4 a numerical simulations is presented which illustrate the effectiveness of the proposed control topology. A conclusion will be presented.

2 Modelling of the Wind Turbine Generator

2.1 Wind Turbine Model

The Fig. 1 shows a schematic overview of the WECS. The system supplies a resistive load and consists of a wind turbine rotor, PMSG, rectifier, and a boost converter. Wind turbine converts the wind energy into mechanical energy, which then runs a generator to create electrical energy.

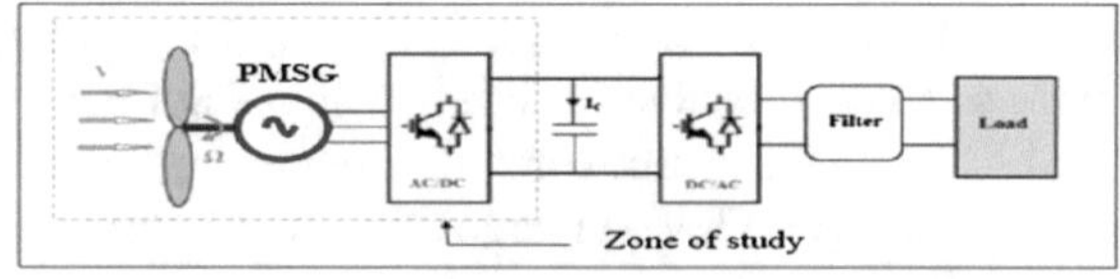

Fig. 1. Structural diagram of WECS.

The amount of power captured by the wind turbine is given as:

$$P_m = \frac{1}{2}\rho\pi R^2 V^3 C_P(\lambda) \tag{1}$$

Where, ρ is the air density (kg/m^3), R is the blade radius (m), v is the wind velocity (in m/s), and $C_p(\lambda)$ is the power coefficient usually given as a function of the tip-speed ratio λ. The tip-speed ratio λ is defined as:

$$\lambda = \frac{R\omega_m}{V} \tag{2}$$

The wind turbine mechanical torque output Tm given as:

$$T_m = \frac{1}{2}\rho\pi R^5 \frac{C_P(\lambda)}{\lambda^3}\omega_m^2 \tag{3}$$

Many different versions of fitted equations for C_p have been used in previous studies. This paper defined C_p based on the following [12, 13]:

$$C_P = -0.212\lambda^3 + 0.0856\lambda^2 + 0.2539\lambda \tag{4}$$

The maximum of C_p that is $C_{P\max} = 0.15$, is reached for $\lambda_{opt} = 0.78$. Hence, there is one particular $\lambda = \lambda_{opt}$, optimal power coefficient, and an optimal turbine rotating velocity ω_{m_opt}, under which C_p takes a maximum value $C_{p\_\max}$. λ_{opt} is the optimal value of the tip speed ratio, which is dependent on the characteristics of the turbine system. Also, the maximum power can be captured from the wind. Accordingly, the turbine system works in maximum power point tracking (MPPT) [15], as depicted in Fig. 2.

The Fig. 3 presents as a function of. According to the figure, there is only one optimal point, denoted by, where is maximum. Continuous operation of the wind turbine at this point guarantees that it will obtain the maximum available power from the wind at any speed, as shown in Fig. 4.

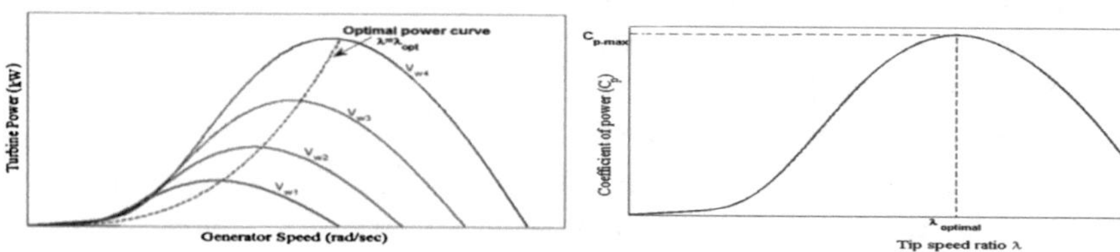

Fig. 2. Characteristics of turbine power as a function of the rotor speed

Fig. 3. The characteristic of the power coefficient as a function of tip speed ratio

PMSG Model

In order to simplify the dynamic model of the electrical part of PMSG, applying a change of landmarks which rotates at the same speed as the rotor (Two bobbins dq rotate with the rotor and generating the same effect as the three fixed coils) [15]. Then the electrical model of the PMSG is defined in the rotating reference fame d-q as follows:

$$V_{sd} = R_s I_{sd} + L_d \frac{dI_{sd}}{dt} - \omega_e L_q I_{sq} \tag{5}$$

$$V_{sq} = R_s I_{sq} + L_q \frac{dI_{sq}}{dt} + \omega_e (L_d I_{sd} + \psi_m) \tag{6}$$

Where V_{sq}, V_{sd} are the direct and quadrature components of the PMSG voltages, R_s, L_d and L_q, respectively, are the resistance, the direct and the quadrature inductance of the PMSG winding, $\psi_m(wb)$ represents the magnet flux, ω_e (rad/s) is the electrical rotational speed of PMSG, I_{sd}, I_{sq} (A) are the direct and quadrature components of the PMSG currents, respectively.

The dynamic equation of the wind turbine is described by:

$$\frac{d\Omega_t}{dt} = \frac{1}{J_T} T_e + \frac{D_T}{J_T} \Omega_t - \frac{1}{J_T} T_t \tag{7}$$

The electromagnetic torque of a p-pole machine is obtained as [16]:

$$T_{em} = \frac{3}{2} n_p (\psi_m I_{sq} + (L_d - L_q) I_{sd} I_{sq}) \tag{8}$$

Where T_{em} (Nm) is the electromagnetic torque, n_p is the number of pole pairs, D_T is the damping coefficient and J_T is the moment of inertia.

3 Control Strategy

3.1 The Control Strategy of the MPPT

The Fig. 4 present the control scheme treated in the current work in the aim to maintain generated power at its maximum value. MPPT is a process that searches the operating point which allows extracting the maximum available produced power at any wind speed level. The control algorithm requires to measure reference signals to be tracked for the rotor speed, the direct and quadrature stator current. $\dot{x}(t) = Ax_i(t) + B_i u(t) + E_i T_m(t)$ Then, the MPPT control is realized by acting on the rectifier side by generating control signals (U_{dr} and U_{qr}) to track the peak power point. The rotor speed is measured according to the following equation:

$$\Omega_r = \frac{\lambda_{opt} V_v}{R} \tag{9}$$

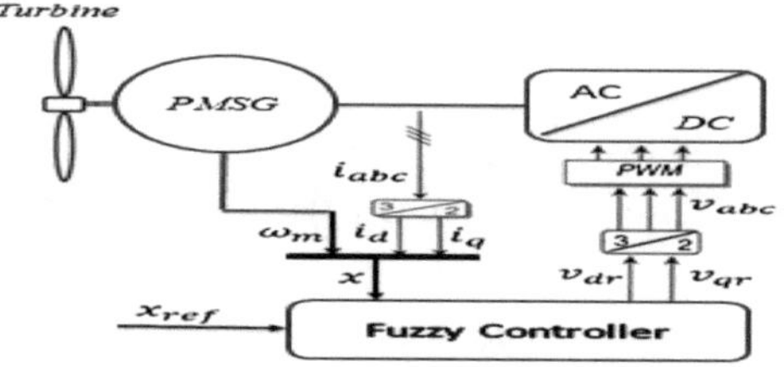

Fig. 4. Fuzzy tracking control diagram

In this study, the d-axis reference current is chosen as: $i_{sdr} = 0$ we can calculate the reference of the current in quadrature given, once the rotor speed reference is determined by:

$$i_{qref} = \frac{C_{em_{ref}}}{k_t} = Cte.\frac{\Omega_t^2}{k_t} \tag{10}$$

Where $k_t = \varphi_f \times p$ and $Cte = \frac{\rho \pi R_t^5 C_{p\max}}{2\lambda_{opt}^3}$

3.2 T-S Fuzzy Model of Wind Energy Conversion System

Considering $x = (\Omega_m \quad i_{sd} \quad i_{sq})^T$ are the state variables, $u(t)$ are the control inputs, A_i, B_i et E_i are the state matrices of the sub-system, hence the state presentation of the system can be written such as:

$$\overset{\bullet}{x}(t) = A(x(t))x(t) + Bu(t) + ET_m(t) \tag{11}$$

Where:

$$x = (\Omega_m \quad i_{sd} \quad i_{sq})^T \tag{12}$$

$$A(x(t)) = \begin{pmatrix} -\frac{D_T}{J} & \frac{p\psi}{J} & 0 \\ -\frac{p\psi}{L_q} & -\frac{R_s}{Lq} & -p\omega_m \\ 0 & p\omega_m & -\frac{R_s}{L_d} \end{pmatrix}, B = \begin{pmatrix} 0 & 0 \\ \frac{1}{L_q} & 0 \\ 0 & \frac{1}{L_d} \end{pmatrix}, u = (V_q \quad V_d)^T,$$

$$E = \left(-\tfrac{1}{J} \quad 0 \quad 0\right)^T$$

Where V_d and V_q represent the average control signals of the inverter in d-q frame. Moreover, the nonlinear model (12) can be presented by a T-S fuzzy model with r = 2^1 fuzzy If-Then rules as follows:

Model rules i: $z(t)$ is F_1^1 and then

$$\overset{\bullet}{x}(t) = Ax_i(t) + B_iu(t) + E_iT_m(t) \tag{13}$$

$$y(t) = Cx(t) \tag{14}$$

Where $z(t) = \omega_m(t)$ is the premise variable and F_i^1 is the membership functions, Let define $M_j = \max(z_j(t))$ and $m_j = \min(z_j(t))$, so $z_j(t) = [m, M]$.

$$F_j = \frac{z_j(t) - m_j}{M_j - m_j} \tag{15}$$

$$\bar{F}_j = 1 - F_j \tag{16}$$

Hence, the local subsystem matrices are given by:

$$A_1 = \begin{pmatrix} -\frac{D_T}{J} & \frac{p\psi}{J} & 0 \\ -\frac{p\psi}{L_q} & -\frac{R_s}{Lq} & -pM \\ 0 & pM & -\frac{R_s}{L_d} \end{pmatrix}, B_1 = B_2 = \begin{pmatrix} 0 & 0 \\ \frac{1}{L_q} & 0 \\ 0 & \frac{1}{L_d} \end{pmatrix},$$

$$A_2 = \begin{pmatrix} -\frac{D_T}{J} & \frac{p\psi}{J} & 0 \\ -\frac{p\psi}{L_q} & -\frac{R_s}{Lq} & -pm \\ 0 & pm & -\frac{R_s}{L_d} \end{pmatrix}, E_1 = E_2 = \begin{pmatrix} -\frac{1}{J} & 0 & 0 \end{pmatrix}$$

After defuzzification, the fuzzy system of a PMSM can be expressed as:

$$\dot{x}(t) = \sum_{i=1}^{r} h_i(z(t))(A_i x(t) + B_i u(t) + E_i T_m(t)) \tag{17}$$

Where

$$h_i(z(t)) = \frac{F_j^1(z(t))}{\sum_{j=1}^{r} F_j^1(z(t))} \tag{18}$$

$\forall t > 0,\ h_i(z(t)) \geq 0$ and $\sum_{i=1}^{r} h_i(z(t)) = 1$.

3.3 Error State Model

In the aim to ensure a smooth tracking of the references parameters, a new state is introduced which is corresponding to an integral action on the error: $e_I = \int e$ as presented in Fig. 5 [3].

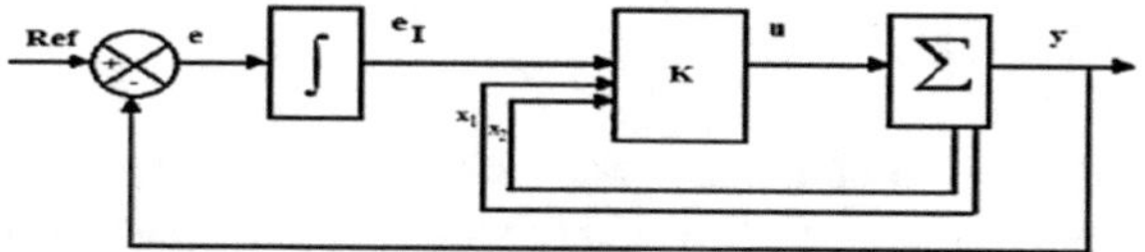

Fig. 5. Diagram of the closed loop system

We present a state reference vector to ensure a smooth tracking of the references:

$$x_{opt} = \begin{pmatrix} \Omega_{opt} & i_{sd} & i_{sq} \end{pmatrix}^T \tag{19}$$

Defining $e(t) = x(t) - x_{opt}(t)$ as the state tracking errors, hence the new PDC fuzzy controller is designed such as [14]:

$$\mu(t) = -\sum_{i=1}^{2} h_i(z(t))K_i x_e(t) \tag{20}$$

Where $x(t) = \omega_m(t)$, $x_{opt}(t)$ is the desired value of ω_m, and $e(t)$ be defined as the tracking error and its time derivative is given by:

$$\dot{x}_e(t) = \dot{x}(t) - \dot{x}_{op}(t) \tag{21}$$

For this reason, augmented state model representation can be resented as:

$$\dot{e}(t) = \sum_{i=1}^{r} h_i(z(t))(A_i + B_i\mu(t) + A_i x_{opt}(t)) \tag{22}$$

Where $\mu(t) = \begin{pmatrix} \mu_d & \mu_q \end{pmatrix}^T$, τ_q, τ_d are new controller to be designed via LMIs approach, K_i denotes the control gain corresponding to each linear sub-model and $e(t)$ is the augmented state vector. Therefore, by replacing τ by its expression, the closed loop model is described by:

$$\dot{e}(t) = \sum_{i=1}^{r}\sum_{j=1}^{r} h_i(z(t))h_j(z(t))(A_i - B_iK_j)e(t) \tag{23}$$

With: $G_{ij} = (A_i - B_iK_j)$
The x_e can be written as:

$$\dot{x}_e(t) = \sum_{i=1}^{r}\sum_{j=1}^{r} h_i(z(t))h_j(z(t))G_{ij}x_e(t) \tag{24}$$

4 Results and Discussion

In this section, a numerical simulation was carried out by using MATLAB/Simulink, the system described in Fig. 4. First, we have simulated the system without controllers. The wind speed is given as a sum of several harmonics (Fig. 6):

$$V_{vent} = V_0 + \sum_{i=1}^{n} V_i . \sin(\omega_i . t) \tag{25}$$

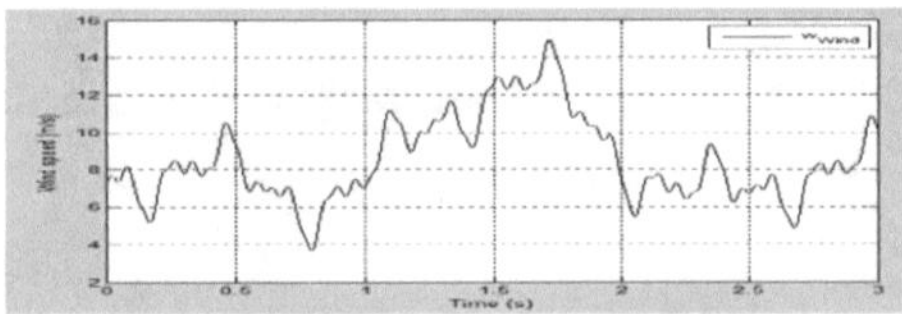

Fig. 6. Wind speed variation

The proposed fuzzy controller results in the MPPT responses shown in Figs. 7 and 8, where the trajectory of the speed of the PMSM is tracking its reference signal and the Power coefficient is quickly achieved. It means the maximum wind power can be extracted by the proposed controller during the change of operation points. Next, to carry out comparisons with the traditional control method, we apply proportional-integral (PI) control to track the Rotor speed and the Power coefficient Cp (Figs. 9, 10, 11 and 12).

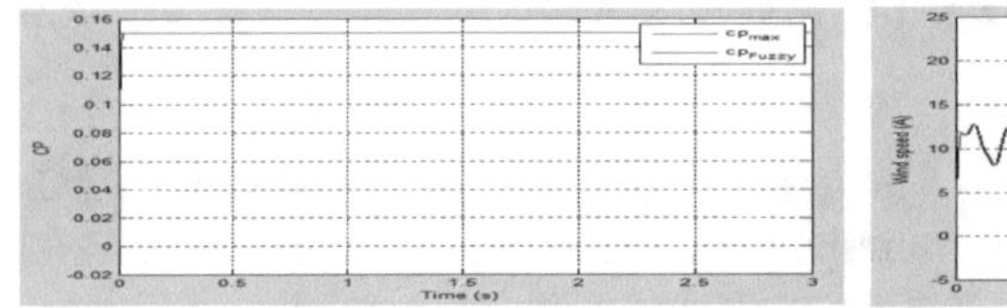

Fig. 7. Power coefficient Cp

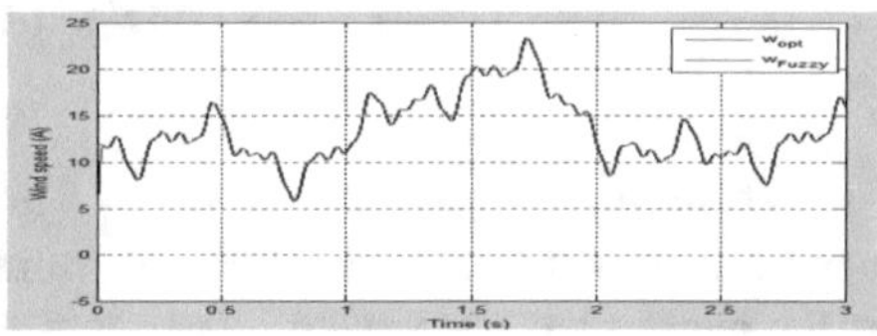

Fig. 8. Rotor speed tracking

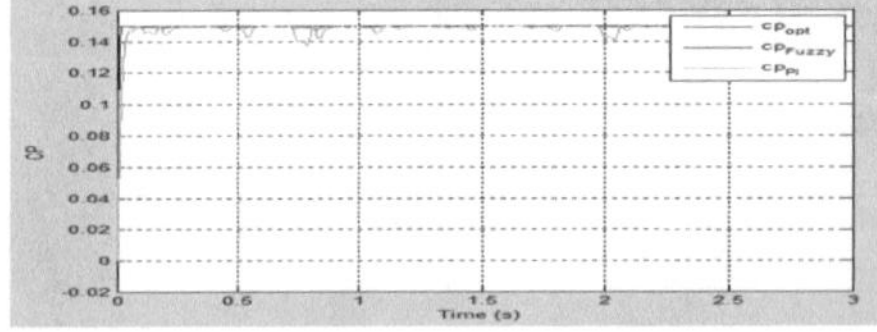

Fig. 9. Response of power coefficient Cp

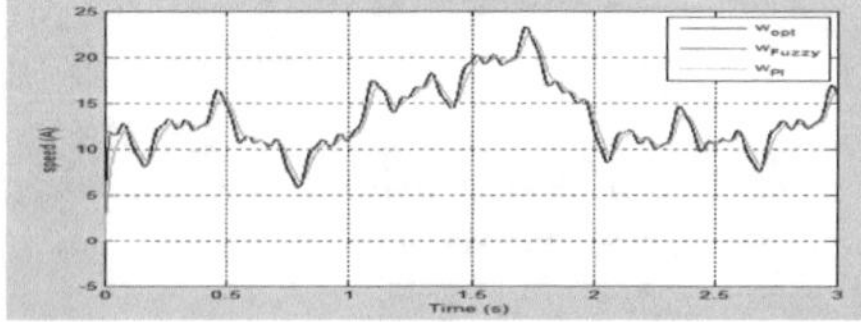

Fig. 10. Response of rotor speed tracking

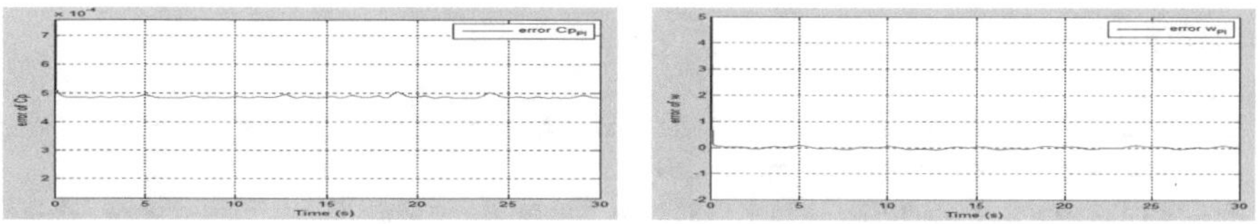

Fig. 11. The power coefficient error CP_opt – Cp of: - the fuzzy MPPT control, and - PI based MPPT control

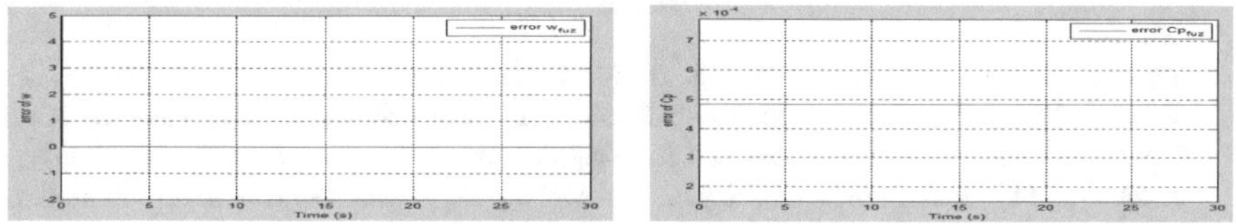

Fig. 12. The rotor speed error w_opt – w of: -the fuzzy MPPT control, and - PI based MPPT control

This Figs. 13 and 14 present the responses of active power P and the reactive power Q, where the trajectory of the power of the PMSG is tracking its reference signal.

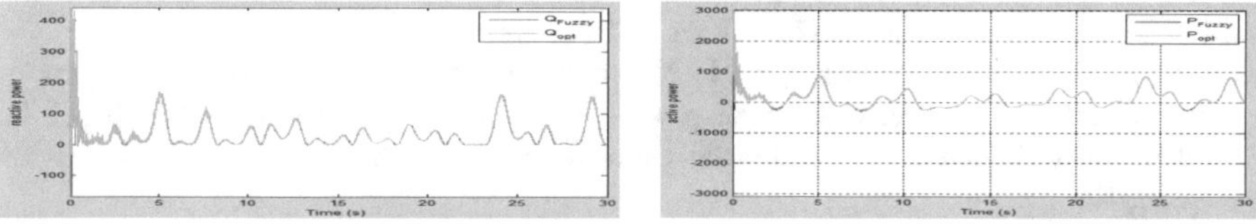

Fig. 13. Response of reactive power Q

Fig. 14. Response of active power P

5 Conclusion

The main objective to ensure maximum peak power tracking (MPPT) has achieved, the proposed method control using Takagi-Sugeno (T-S) fuzzy of the WT is presented, Simulation results show that the proposed wind system has been compared with the PI controller demonstrate that the system with the synthesized T-S fuzzy controller more stable response without perturbation behavior and gives best performance and were found the faster than the PI controller.

References

1. Benkada, A., Chaikhy, H., Monkade, M.: MPPT control for wind energy conversion system based on a T-S Fuzzy, vol. 9, no. 2 (2018)
2. Luo, F., Meng, K., Dong, Z.Y., Zheng, Y., Chen, Y., Wong, K.P.: IEEE Trans. Sustain. Energy **6**, 253–262 (2015)
3. Errami, Y., Ouassaid, M., Maaroufi, M.: Int. J. Electr. Power Energy Syst. **68**, 180–194 (2015)
4. Harrabi, N., Souissi, M., Aitouche, A., Chaabane, M.: Intelligent control of wind conversion system based on PMSG using T-S fuzzy scheme. Int. J. Renew. Energy Res. IJRER **5**(4), 952–960 (2015)
5. Peftitsis, D., Adamidis, G., Fyntanakis, A.: Modulation of three phase rectifier in connection with PMSG for maximum energy extraction. In: Proceedings of the 13th European Conference on Power Electronics and Applications, pp. 1–10, Barcelona, Spain (2009)
6. Allouche, M., Souissi, M., Chaabane, M., Mehdi, D.: Takagi-Sugeno fuzzy control of induction motor. Int. J. Electr. Comput. Eng. **8**, 1263–1269 (2010). World Academy of Science, Engineering and Technology
7. Bahraminejad, B., Iranpour, M.R., Esfandiari, E.: Pitch control of wind turbines using IT2FL controller versus T1FL controller. Int. J. Renew. Energy Res. (IJRER) **4**(4), 1077 (2014)
8. Anju, M., Rajasekaran, R.: Power system stability enhancement and improvement of LVRT capability of a DFIG based wind power system by using SMES and SFCL. Int. J. Electr. Comput. Eng. (IJECE) **3**(5), 618–628 (2013)
9. Narimani, M., Lam, H.K.: Relaxed LMI-based stability conditions for Takagi-Sugeno fuzzy control systems using regional-membership function shape-dependent analysis approach. IEEE Trans. on Fuzzy Syst. **17**(5), 1221–1228 (2009)
10. Errami, Y., Hilal, M., Benchagra, M., Maaroufi, M., Ouassaid, M.: Nonlinear control of MPPT and grid connected for wind power generation systems based on the PMSG. In: International Conference on Multimedia Computing and Systems (ICMCS), pp. 1055–1060, May 2012
11. Lee, D.H., Park, J.B., Joo, Y.H.: A new fuzzy Lyapunov function for relaxed stability condition of continuous-time Takagi-Sugeno fuzzy systems. IEEE Trans. on Fuzzy Syst. **19** (4), 785–791 (2011)
12. Lamnadi, M., et al.: Modeling and control of a doubly-fed induction generator for wind turbine-generator systems. IJPEDS Int. J. Power Electron. Drive Syst. **7**(3), 982–995 (2016)
13. Mena Lopez, H.E.: Maximum power tracking control scheme for wind generator systems. Master Thesis, A&M University, Texas (2007)
14. Narayana, M., Putrus, G.A., Jovanovic, M., Leung, P.S., McDonald, S.: Generic maximum power point tracking controller for small-scale wind turbines. Renew. Energy **44**, 72–79 (2012)
15. Ounnas, D., Ramdani, M., Chenikher, S., Bouktir, T.: A fuzzy tracking control design strategy for wind energy conversion system. In: Proceedings of 4th International Conference on Renewable Energy Research and applications, Palermo, Italy, 22–25 November 2015, pp. 777–782 (2015)

Influence of Glass Properties in the Performance of a Solar Cooling Ac-Nh3 Adsorption Machine

Hanae El Kalkha[1(✉)] and Abelaziz Mimet[2]

[1] Laboratory of Innovative Technologies, National School of Applied Sciences, Tangier, Morocco
elkalkha_hanae@yahoo.fr
[2] Faculty of Sciences, University Abdelmalek Essaidi, BP 2121, 93000 Tetouan, Morocco
abdelaziz.mimet@uae.ma

Abstract. This work presents the results of the development of a dynamic model aiming to contribute to the design and performance evaluation of an ammonia-activated carbon adsorption solar cooling systems using a sensor with two different types of glass, one normal and other selective. Model takes into account the transient behavior of input variables as solar radiation and ambient temperature and it calculates, according certain initial parameters and a given solar flux, the internal system temperatures, the adsorbed mass and the pressure of the reactor. This allows us to calculate with good accuracy the cycled mass of refrigerant, the quantity of cold produced in the machine and the performance coefficient of solar refrigerating machine. The solar collector used for converting solar energy to heat is a solar flat plate collector. The model is applied to the performance evaluation and the calculation of amount of cold produced of this kind of devices in different locations at Morocco.

Keywords: Solar energy · Adsorption coolin · Amount of cold produced · Performance coefficient

1 Introduction

The production of cold seems to be, after the heating of the water, the application of the most promising solar energy at the moment. During the last twenty years, great efforts have been devoted to solar refrigeration using photovoltaic panels or thermodynamic cycles, the competition between these two channels being still relevant.

Solar refrigeration using adsorption cycles has been widely studied [1–9]. This approach has allowed the marketing of refrigeration units in some countries. However, the dimensioning of these units requires tests under several real weather conditions and comparisons of several technical parameters that influence their yields.

In fact, the easiest way is to model the refrigerating machine using models for simulating operation based on the actual solar data.

M. Ezziyyani (Ed.): AI2SD 2018, AISC 912, pp. 417–426, 2019.
https://doi.org/10.1007/978-3-030-12065-8_37

It is proposed in this paper to simulate the operation of an adsorption refrigerator ammonia on activated carbon. Precisely, to calculate the amount of cold that can produce an adsorption solar refrigeration unit using a flat glass solar collector, by comparing two different types of glazing in different regions of Morocco.

Different adsorbent/adsorbate pairs have been studied for their use in solar cooling machines [10]. The activated carbon-ammonia pair has been chosen by several authors [11, 12], since activated carbon (AC) is a microporous adsorbent, it has a large specific surface area. Ammonia (NH3/R717) in turn has a good adsorbate according to its several advantages mentioned in different works [13, 15].

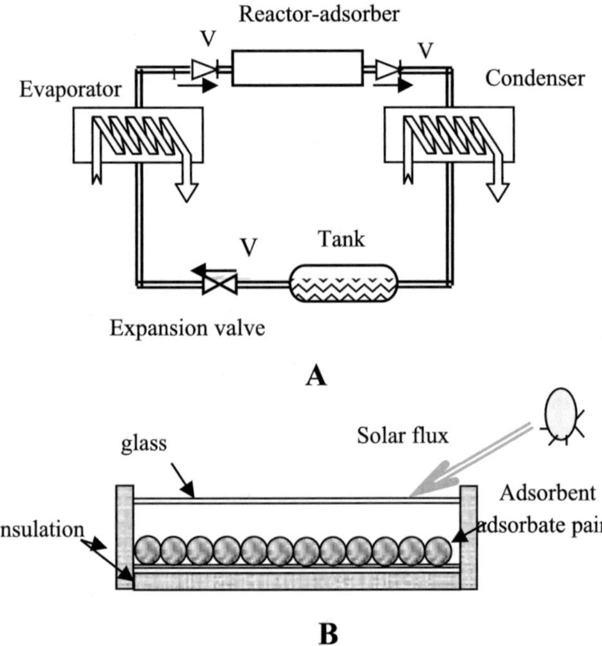

Fig. 1. A: solar adsorption refrigeration machine. B: cut of the solar collector

2 Description of the System

As shown in Fig. 1, the refrigerator studied mainly includes; a reactor (adsorber) enclosed in a solar collector, containing the adsorbent-adsorbate mixture, where the adsorption and desorption reactions can be carried out, connected to a condenser and an evaporator, and finally to the non-return and expansion valves.

The adsorption and desorption of the refrigerant are generated by cooling and alternately heating the reactor located inside the flat solar collector. The machine only works with solar energy, according to the principle of adsorption-desorption.

The planar solar collector, allows to convert the electromagnetic radiation into heat, to transfer it to the porous medium during the day, and to dissipate during the night. The thermal energy received by the adsorbent/adsorbate mixture and also that released by the adsorption phenomenon.

When the adsorbent (at the temperature T) is in exclusive contact with the adsorbate vapor (at the pressure P), an amount of adsorbate mass is trapped in the micropores in an almost liquid state. This adsorbed mass is a function of T and P according to a divariant equilibrium m = f (T, P). Moreover, at constant pressure, m decreases with the increase T, and at constant adsorbed mass, P increases with T. Thus, the refrigeration cycle consists of a heating/desorption/condensation period, followed by a period of cooling/adsorption/evaporation.

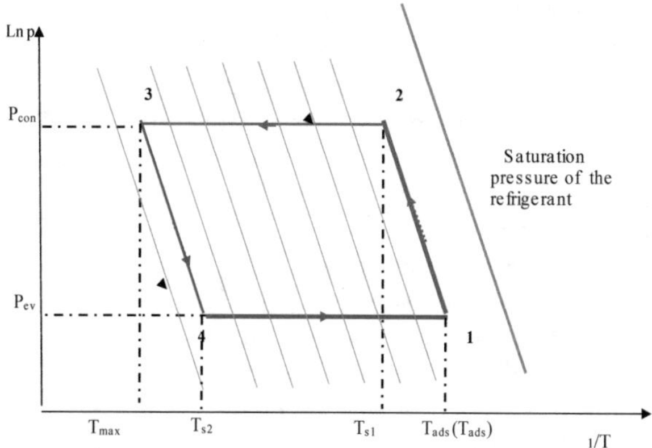

Fig. 2. Thermodynamic cycle of adsorption refrigerating machine

Heating/desorption/condensation: the generator is heated from T_{ads} to T_{max}. When the pressure in the generator reaches the condensation pressure P_c, condensation begins. The heat input during this step is Q_s while the heat output in the condenser is Q_C.

Cooling/adsorption/evaporation: the generator is cooled from T_{max} to T_{ads}. When the pressure in the evaporator reaches the evaporation pressure P_{ev}, the evaporation begins. The heat input of this phase to the evaporator is Q_F (Fig. 2).

3 Mathematical Modeling

We give below the different equations of the model, to be resolved, taking into account the initial conditions equations and boundary equations giving the wall temperature Tp and the glass temperature Tv.

3.1 Equation of Heat and Mass Transfer

$$\begin{aligned} & \left[(1-\varepsilon)\rho_s C_s + (\varepsilon-\alpha)\rho_g C_g + \alpha\rho_a C_a\right]\frac{\partial T}{\partial t} - \frac{\partial\left((\varepsilon-\alpha)\rho_g\right)}{\partial t}\frac{P}{\rho_g} \\ & - \rho_{app}\frac{P}{\rho_a}\frac{\partial m_a}{\partial t} - \Delta H_{ads}\rho_{app}\frac{\partial m_a}{\partial t} = \lambda_e \Delta T \end{aligned} \quad (1)$$

With:

T	absolute temperature (°C)
P	pressure in the reactor (bar)
m_a	adsorbed mass in kg of ammonia per kg of activated carbon (kg/kg-CA)
C_s	specific heat of solid phase (J/kg K)
C_g	specific heat of gaz phase (J/kg K)
C_a	specific heat of adsorbed phase (J/kg K)
ΔH_{ads}	adsorption heat of ammonia on activated carbon (J/kg)
ε	porosity of activated carbon
ρ_{app}	apparent density of activated carbon (kg/m^3)
ρ_s	density of the solid phase (kg/m^3)
ρ_g	density of the gaz phase (kg/m^3)
ρ_a	density of the adsorbed phase (kg/m^3)
α	volume fraction of the adsorbed phase
(1 − ε)	volume fraction of the solid phase
(ε − α)	volume fraction of the gaseous phase
λe	equivalent thermal conductivity

3.2 The Dynamic Equations of Heat Transfer the Glazing and the Wall

$$C_p\frac{dT_p}{dt} = q_p - h_{pv}\left(T_p - T_v\right) - h_{pa}\left(T_p - T_a\right) + h_{pm}\left(T_p - T_m\right) \quad (2)$$

$$C_v\frac{dT_v}{dt} = q_v - h_{va}\left(T_v - T_a\right) - h_{vs}\left(T_v - T_s\right) + h_{pv}\left(T_p - T_v\right) \quad (3)$$

With:

T_v	temperature of the glass (°C)
T_p	temperature of the wall (°C)
T_m	temperature of mixed adsorbent-adsorbate (°C)
T_a	ambient temperature (°C)
T_s	sky temperature (°C)
C_v	thermal capacity of the glass (J/K. m^2)
C_p	thermal capacity of the wall (J/K. m^2)
h_{va}	coefficient of heat exchange between the glass and the atmosphere (W/K. m^2)

h_{vs} coefficient of heat exchange between the glass and the sky (W/K. m^2)
h_{pv} coefficient of heat exchange between the wall and the galss (W/K. m^2)
h_{pa} coefficient of heat exchange between the wall and theatmosphere (W/K. m^2)
h_{pm} coefficient of heat exchange between the wall and the mixture (W/K. m^2)
q_v net flow energetic of the glass (W/m^2)
q_p net flow energetic of the wall (W/m^2)

The Eq. (1) of the model constitute a nonlinear and coupled system of equations. The numerical algorithm that has been used for the resolution is based on the finite difference method using the implicit scheme. Then we obtained a system of algebraic equations which is solved using the Gauss Seidel method. A calculation program in FORTRAN 90, allowed us to achieve results of the spatial and temporal distribution of temperature, pressure and adsorbed mass.

3.3 Quantity of Cold Produced

It is the useful effect produced by the evaporator, which is equal to the latent heat of vaporization of refrigerant unless the sensible heat required to cool the refrigerant condensing temperature to the temperature of evaporation, its expression is:

$$Q_f = \Delta m \left[L(T_{ev}) - \int_{T_{ev}}^{T_{cond}} C_l dT \right] \tag{4}$$

With:

Q_f quantity of cold (J/m^2)
Δm cycled mass of ammonia (kg/kg-CA)
$L(T_{ev})$ latent heat of ammonia at evaporation temperature (J/kg)
T_{ev} evaporation temperature (°C)
T_{cond} condension temperature (°C)
C_l liquid specific heat (J/kg K)

The expression of the mass cycled is given by:

$$\Delta m = m_a(T_{ads}, P_s(T_{ev})) - m_a\left(T_g, P_s(T_{cond})\right) \tag{5}$$

With:

Δm cycled mass of ammonia (kg/kg-CA)
m_a adsorbed mass in kg of ammonia per kg of activated carbon (kg/kg-CA)
T_{ev} evaporation temperature (°C)
T_{cond} condension temperature (°C)
T_g temperature at start of regeneration(°C)
T_{ads} adsorption temperature (°C)
P_s pressure of saturation (bar)

Where $m_a(T, P)$ is the mass of ammonia adsorbed at temperature T and pressure P, calculated using the BET model.

3.4 Solar Coefficient of Performance

The performance evaluation of solar machine is determined from the amount of heat Qf and Qs where Qs is the amount of global irradiation received by the collector surface of the sensor, its expression is given by (Fig. 3):

$$Q_s = S_c \int_{Sunrise}^{Sunset} G(t)dt \tag{6}$$

With:

Q_s quantity of heat (J/m^2)
S_c collecting area sensor (m^2)
G globale irradiation (W/m^2)
t time (s)

The solar coefficient of performance is given by

$$COP_{sol} = \frac{Q_f}{Q_s} \tag{7}$$

With:

COP_{sol} solar coefficient of performance
Q_s quantity of heat (J/m^2)
Q_f quantity of cold (J/m^2)

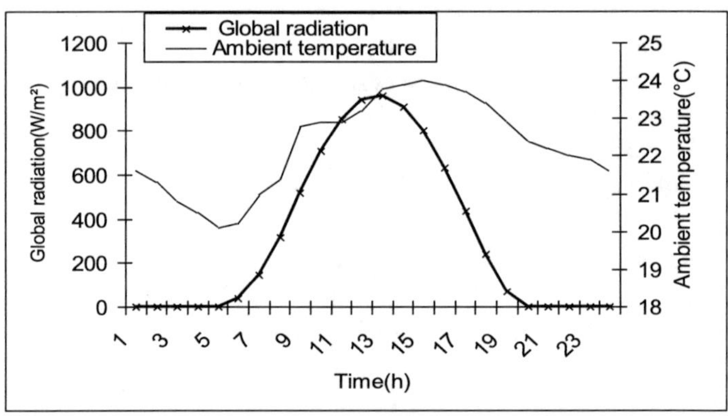

Fig. 3. Example for a Global radiation and ambient temperature (used clear typical day)

4 Improved Glazing of the Adsorber by Selective Layer

Since we used a sensor plan glass in our model, therefore we sought to improve our coating layer by selective glazing. This type of glass has the advantage a good absorption of solar radiation by emitting little infrared (Table 1).

Table 1. Parameters of glass used in simulation

Support of the glass	Solar absorption factor (a_v)	Infrared emissivity factor (ε_v)
Normal	0,06	0,9
Steel and black nickel on nickel	0,95	0,07

The different values of the solar absorption factor and the infrared emissivity factor are introduced in solar contributions and different heat exchange coefficients which are written as follows:

$$q_v = a_v.G; \tag{8}$$

is the glass absorbed radiations

$$q_p = a_p.G; \tag{9}$$

is the wall absorbed radiations

With:

q_v net flow energetic of the glass (W/m^2)
q_p net flow energetic of the wall (W/m^2)
a_p net absorption coefficient of the wall
a_v net absorption coefficient of the glass.

The radiative heat exchange coefficient of glass – sky h_{vs} is given, according to the temperatures of the glass and the sky, by the following expression [16]:

$$h_{vs} = \varepsilon_v.\sigma\left(T_s^2 + T_v^2\right).(T_s + T_v) \tag{10}$$

With:

ε_v glass Infrared emissivity factor
σ Boltzmann constant = 5.68 10^{-8} ($W/m^2.\ K^{-4}$).

The radiative heat exchange glass-adsorber, and convective, one writes [16]:

$$h^r_{pv} = \sigma.\frac{\left(T_p^2 - T_v^2\right).(T_p - T_v)}{\frac{1}{\varepsilon_p} + \frac{1}{\varepsilon_v} - 1} \tag{11}$$

Where:

ε_v glass Infrared emissivity factor.
ε_p wall Infrared emissivity factor.

5 Results and Discussion

We have used the hourly solar data and climate (ambient temperature and global radiation on inclined surface) corresponding to a clear a typical day, from the climatological database [17]. The inclination sensor is equal to the approximate altitude to receive maximum solar energy [18]. The software METEONORM uses the Perez model for calculating the radiation on an inclined surface from the values of global horizontal radiation [17].

The condensation temperature is that corresponding to the ambient temperature recorded the same day, the evaporation rate is set at 0 °C.

Once developed the model, we have applied it in a systematic way to locations of Morocco having, according reference database, the requested meteo inputs allowing system performance simulation. Two cases have been considered: the use of selective and normal glazing on solar collector allocating adsorber elements.

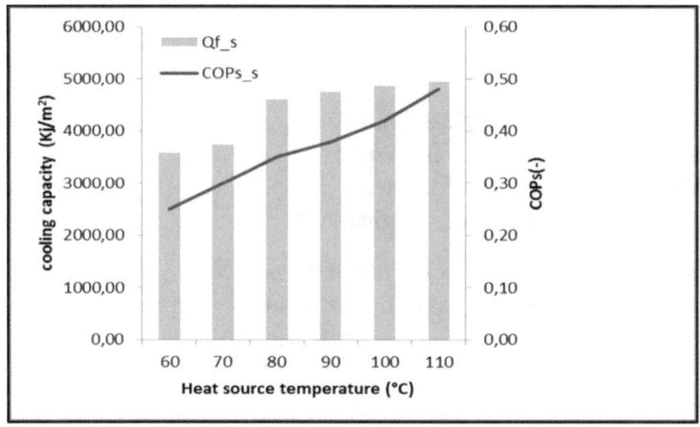

Fig. 4. Variation of daily cooling capacity and COPs with heat source (used normal glazing on solar collector allocating adsorber)

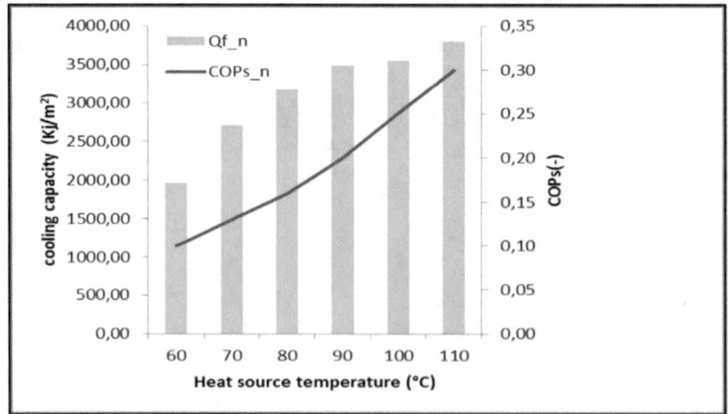

Fig. 5. Variation of daily cooling capacity and COPs with heat source (used selective glazing on solar collector allocating adsorber)

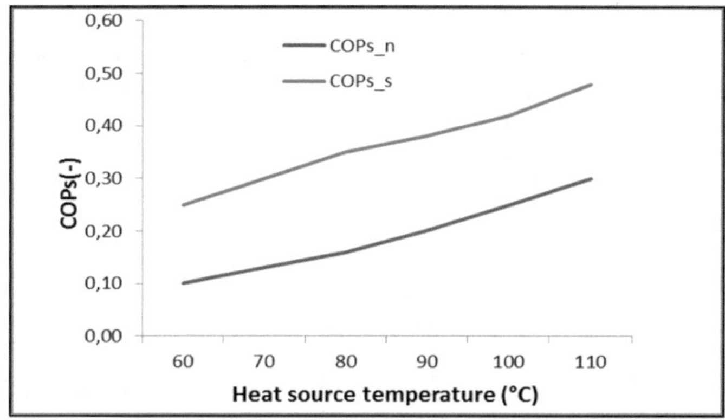

Fig. 6. Variation of COPs with heat source using two cases selective and normal glazing

We present in Fig. 4 the variation of daily cooling capacity and COPs with heat source calculated in the evaporator of the adsorption machine working with activated carbon-ammonia pair equipped with a flat plate collector with normal glazing. The Fig. 5 shows the same results with a flat plate collector with selective glazing. And the Fig. 6 Displays the variation of COPs with heat source using two cases selective and normal glazing.

The results are given for a north region of Morocco according to the solar radiations maps for a typical clear days of the months July.

We note an improvement on the variation of the quantity of cold produced and the solar coefficient of performance by selective glazing used in a flat plate collector given by the better absorption of solar radiation and infrared emissivity decrease compared to a normal glazing.

The quantity of cold produced and the solar coefficient of performance depend on the values of global radiation received in one day and on initial temperature of the morning.

6 Conclusions

This work presents the results of a study of the transient behavior of the adsorber of the solar adsorption refrigeration machine using the activated carbon ammonia pair. The model used is based on the coupled equations of heat transfer and mass in porous medium consisting of the activated carbon-ammonia mixture. The analysis of the results shows a significant improvement in the amount of cold produced and the coefficient of solar performance. We have therefore shown the influence of physical parameters by two different types of glazing on the solar performance of the solar adsorption refrigeration machine.

We applied the model on selective glass to determine the solar performance map using the solar radiation data for the different Moroccan regions.

Once the cooling effect was calculated based on the available data, the ArcGIS™ software was used to evaluate the geographic distribution of cold-run model application results, based on the results obtained. we can then have a distribution of the coefficient of performance on a Moroccan map.

References

1. Dieng, A.O., Wang, R.Z.: Literature review on solar adsorption technologies for ice-making and airconditioning purposes and recent developments in solar technology. Renew. Sustain. Energy Rev. **5**, 313–342 (2001)
2. Boubakri, A., Guilleminot, J.J., Meunier, F.: Adsorptive solar powered ice maker: experiments and model. Sol. Energy **69**(3), 249–263 (2000)
3. Al Mers, A., Mimet, A., Boussouis, M.: Numerical study of heat and mass transfer in a cubical porous medium heated by solar energy "Boubnov-Galerkin method". In: Crolet, J.M. (ed.) Computational Methods for Flow and Transport in Porous Media (2000)
4. El Fadar, A., Mimet, A., Azzabakh, A., Pérez-Garcia, M., Castaing, J.: Study of a new solar adsorption refrigerator powered by a parabolical trough collector. Appl. Therm. Eng. **29**(5–6), 1267–1270 (2008)
5. El Fadar, A., Mimet, A., Pérez-Garcia, M.: Modeling and performancesludy of a continuous adsorption refrigeration system driven by parabolical trough solar collector. Sol. Energy **83** (6), 850–861 (2008)
6. Louajari, M., Mimet, A., Ouammi, A.: Study of the effect of finned tube adsorber on the performance of solar driven adsorption cooling machine using activated carbon–ammonia pair. Appl. Energy **88**(3), 690–698 (2010)
7. Louajari, M., Mimet, A., Ouammi, A.: Sustainable development of a solar adsorption cooling machine. Manag. Environ. **21**, 589–601 (2010)
8. Al Mers, A., Azzabakh, A., Mimet, A., El Kalkha, H.: Optimal design study of cylindrical finned reactor for solar adsorption cooling machine working with activated carbon–ammonia pair. Appl. Therm. Eng. **26**, 1866–1875 (2006)
9. El Kalkha, H., Ezzarfi, A., Mimet, A., Perez-Garcia, M., Ganaoui, M.: Evaluation of a cold map for a Moroccan climate by a solar adsorption refrigeration machine working with activated carbon–ammonia pair. J. Fluid Dyn. Mater. Process. (2011)
10. Wang, L.W., Wang, R.Z., Oliveira, R.G.: A review on adsorption working pairs for refrigeration. Renew. Sustain. Energy Rev. **13**, 518–534 (2009)
11. Mimet, A.: Etude théorique et expérimentale d'une machine frigorifique à adsorption d'ammoniac sur charbon actif, Thèse de Doctorat, FPMs, Mons, Belgique (1991)
12. Critoph, R.E., Gong, F.: A rapid cycling ice-maker for use in developing countries. In: Proceeding of the 2nd World Renewable Energy Congress, UK (1992)
13. Institut Internationnal du foid, Tables et diagramme pour l'industrie du froid. Propriétés thermodynamiques du R717, Paris (1981)
14. Meunier, F.: Le froid solaire par adsorption, Cahier. AFEDES, no. 5 (1978)
15. Modélisation de l'adsorption par les charbons microporeux: Approches théorique et expérimentale, 'université de Namur, thèse de Doctorat (2002)
16. Ong, K.S.: Thermal performance of solar air heaters: mathematical model and solution procedure. Sol. Energy **55**(2), 93–109 (1995)
17. Meteotest: Meteonorm version 6.1 – handbook (2008). www.meteonorm.com
18. Aggour, M.: Mesures et correlation du rayonnement solaire et des caractéristiques des panneaux photovoltaïques, thése de doctorat, Faculté des sciences, Rabat (1987)

Secured Remote Control of Greenhouse Based on Wireless Sensor Network and Multi Agent Systems

Kamal Moummadi, Rachida Abidar(✉), Hicham Medromi(✉), and Ahmed Ziani(✉)

Hassan II University, ENSEM, BP 8118, Oasis, Casablanca, Morocco
{kamal.moummadi, rachida.abidar, hicham.medromi, ahmed.ziani}@ensem.ac.ma

Abstract. QueryAgent-oriented formalisms are now increasingly used in artificial intelligence. Their success is partly due to their easy adaptation to the needs of distributed real-time applications. This paper explains the design and implementation of a novel platform called Secured Remote Control of Greenhouse (SRCG) for the remote control of the inside and outside climatic and also soil parameters that influence the production in greenhouses such as temperature, humidity, CO2 and soil moisture…. A Wireless Sensor Network (WSN) provides pertinent information that is used to supervise ventilation, heating and pump…. The use of SRCG avoids the needed to perform the monitoring actions on site. The platform described in this paper is simple to be installed and used by farmers who do not have knowledge in computer skills. Thus, all farmers can control their greenhouses from a distance device in an easy and an ubiquitous manner. They can control actuators to adjust these parameters (fan, heater, drip irrigation…). The architecture of the platform is based on Multi agent systems (MAS) and a Distributed Constraint Satisfaction Problem (DCSP). MAS gather, integrate, and deliver the collected climate's parameter information from distributed sensors, and synchronize this information with a remote supervisor computer. Proposed SRCG has advantage that can handle situations in the far away area from the farms through PDA (Personal Digital Assistant) and mobile device, which shortens time, expense and supports agricultural decision-making. The prototype is built in Java employing general interfaces of both MAS and constraint programming (CP) platforms, using JADE and CHOCO libraries.

Keywords: Greenhouse · Decision-making · Wireless sensor network · Multi agent systems · Control · Monitoring · JADE · CHOCO

1 Introduction

In the last decades, all over the world we try to have a rational use of water. Hence the idea of our platform, called SRCG, this hardware/software platform implements a scalable architecture based on MAS in compliance with CP paradigms and a WSN, deployed in a greenhouse. The greenhouse separates the crop from the environment,

M. Ezziyyani (Ed.): AI2SD 2018, AISC 912, pp. 427–439, 2019.
https://doi.org/10.1007/978-3-030-12065-8_38

thus providing some way of shelter from the direct impact of the external weather conditions [1]. This allows agricultural production that otherwise would not be produced at that location. So, it leads to, less use of protective chemicals, higher crop yield, prolonged production period, and better quality. This paper presents the design and development of a platform based on different nodes that integrates remote control functions rooted in a WSN. SRCG allows the acquisition of different climatic parameters through sensors and the regulation of these parameters through actuators, in an agricultural greenhouse.

The proposed architecture is implemented in open source and ensure, the reliability as the core software used is reliable (Net beans, Apache…), the viability as it is standards-based computing (J2EE, PHP…); the interoperability because of the compatibility with different devices and finally the portability to operate on all operating systems; optimization because it uses a model called Controller-Agents for Constraints Solving (CACS). This model is based on special kind of agents called Controllers and it is intended to be used for solving DCSP which is an emerged field from the integration between two paradigms of different nature: Constraint Satisfaction Problem (CSP), where all constraints are treated in central manner as a black-box, and MAS that is characterized by the autonomy and the distribution of its entities. The main role of controller agents is to enclose and verify some constraints assigned to it. This model allows grouping constraints to have a subset that will be treated together as a local problem inside controller agents. Based on CACS, a prototype of DCSP solver is built. This paper presents the implementation of that prototype. The prototype is made in Java programming language; it uses general interfaces of both MAS and CP platforms, employing JADE and CHOCO libraries.

In the next sections we reference some related works of similar platforms. After that, we describe our approach for mixing both paradigms MAS and CP. Then we give an outline onto WSN used to route the flow of data in a network in a secured manner, and the communication platforms used to synchronize data between a central server and mobile devices with ANDROID OS, in order to remote control greenhouses or parcels through mobile devices. This article ends by a realization, conclusions and perspectives.

2 Related Works

Many researchers related that, the green house technology is being an important part of agriculture engineering. The integration of WSN and artificial intelligence in the field of agriculture are the recent concepts. These mixture and integration leads to precision agriculture. [2] stipulated that, the green house technology can be used to increase the quality agricultural yield by monitoring soil and environment properly. They besides noticed that, in early stage of WSN, farmers were reluctant to deploy it, because of high cost. Latterly, technological development has shortened the cost. [3] reviews the need of WSN in agriculture, the wireless sensor network technology and their applications in different aspects of agriculture and report existing system frameworks in agriculture domain. Such frameworks are developed in black box manner or have a model layer implemented in a central way, which doesn't ensure the reliability, availability, and efficiency of these solutions.

3 Multi-agent Approach

The word agent has many definitions in computer engineering. [4] defines an agent as a software system that is:

- Positioned in an environment,
- Capable of autonomous actions in order to meet its objectives.
- Capable of communicating with other agents.

So, we can summarize this definition as follows, an agent is an entity that can act and react in his environment and interact with other agents.

A multi agent system is composed of a set of several agents that exist at the same time, share common resources and communicate with each other. For short a MAS can be observed as a network of agents lightly coupled. These agents work together to solve problems that are beyond their individual skills and capacities [5].

The simulation of the common resources management poses the problem of correlation between groups of agents and dynamic resources. In MAS paradigm, we look at the simulated system from a distributed and cooperative point of view [6].

4 Constraint Logic Programming

4.1 Introduction

Constraints are logical relations between variables, each variable taking value from a specific domain, and a domain is composed of several values. Hence, a constraint restricts the possible values that a variable can have. Constraints are:

- Declarative: they specify a relationship between variables (entities) without resolving a specific computational procedure.
- Rarely independent: commonly constraints share variables.
- Additive: we are interested in the conjunction of constraints and not in the order in which they are imposed.

Accordingly constraints are a natural way for commonality to explicit problems in many fields.

Constraint programming has been applied to several domains, such as: operations research (optimization problems, scheduling), business applications (options trading analysis), electrical engineering (diagnosis), warehouse location, crew rotation, planning, healthcare and many more others [7].

4.2 Constraint Satisfaction Problem

The concept of constraint satisfaction problem is specifieded as a set of n distinct variables X = {x1; x2; …; xn}, a set of m domains each one associated with a variable D = {d1; d2; …; dm} and a set of relationships, namely constraints, that restrict the values that can be assigned to them simultaneously. A solution is an assignment of values to the problem variables that satisfy the conditions imposed by the constraints [8].

There are two reasons, which encourage adopting such representation. The first one is that it is convenient to the natural problem statement, since variables act as entities and the constraints do not have to be translated, just stated over the entities they concern. The second reason is that constraint satisfaction problem algorithms are in many cases more efficient in solving such problems than other approaches.

Numerous techniques have been proposed for finding solutions. Most of them focus on removing inconsistent values from the problem variables domains in order to decrease the treatment time and the search space of the problem. Variables generally are composed of graph nodes and constraints as arcs between the nodes forming the commonly named constraint network or graph. The proposed algorithms perform constraint propagation by repeatedly eliminating values from more values can be eliminated. These are classified into three fundamental categories, that is, node consistency, arc consistency and path consistency according to the degree of consistency checks that they perform.

A constraint programming package can be implemented as either a library that is used with a conventional programming language (C, C++, and Java), we can mention for example, the ILOG Solver, or the open source CHOCO platform.

4.3 Negotiation or Filtering Algorithm

We choose an agent with a leadership role, in which concentrated all the information about the variables, their domains and constraints. The agent “leader” will direct all activities of other agents in the process of solving the problem [9].

In our architecture this leader agent is called controller agent.

The purpose of this algorithm is to reduce the cardinal sets D1, D2,… Dm, by excluding inconsistent values.

Each agent Ai is responsible for a variable xi and knows its domain Di and constraints on its own variable. All agents know the addresses of neighboring agents (those variables that occur in the same constraints as that of the agent).

For each agent Ai, filtering algorithm includes the following:

- Ai communicates with its neighbors the set Di.
- After receipt of a domain Dj executes filtering values of its own domain Di according to the pseudo code:

```
remove = true
  for each di in Di do
  while((remove = true) and (Dj <> Ø) do
     // x is a value of Dj
     Dj = Dj - {x}
     if di is consistent with x
        remove = false
     end if
  end while
 if remove then
     Di = Di - {di }
  end if
end for
```

In the end you can get in one of the following situations:

- Each domain contains only one value: so we found a solution.
- There exists at least one empty field: the problem is very constrained.
- There exists at least one field with more established value, and then proceed to filter [10].

5 Constraint Programming Approach and Multi-agent Systems

Problem solving is one of the essential applications of multi-agent systems. As an alternative to centralized problem solving, either because problems are themselves distributed, or because the distribution of problem solving between different agents reveals itself to be more efficient way to organize the problem solving, it can be flexible and allow failures in the system, or because, in some cases, it is the only way to solve the problem.

In many industrial applications a huge amount of time and effort is dedicated to developing complex and sophisticated software systems. The combination of constraint programming and multi-agent systems technology offers a robust platform for implementing industrial applications. Multi-agent system approach allows building complex, powerful and flexible systems that can easily and naturally adapt to any enterprise specification by encoding the needs of each unit in a different agent. Diversely, constraint technology provides an excellent platform to represent the properties and interconnections of the production units and provides an efficient evaluation of the constraints imposed.

Distributed constraint satisfaction problem is an appropriate abstraction for multi-agent cooperative problem solving [11]. It is specified by multiple reasoning agents making local independent decisions about a global common constraint satisfaction problem. In a distributed constraint satisfaction problem, the variables are distributed among agents. Yokoo [12] has proposed solving distributed constraint satisfaction problem by using an asynchronous backtracking algorithm. This can be done by allowing agents to run concurrently and asynchronously. Each agent give values for its own variables and communicates these values with relevant agents.

We propose in this paper a novel approach for distributed constraint satisfaction problem resolution. Our approach consists of mixing multi-agent system and constraint satisfaction problem. Our approach is based on a specific kind of agents named controller agent, this agent is responsible of verifying the constraints satisfaction using the negotiation algorithm described in the previous section.

6 Wireless Sensor Network

6.1 Wireless Sensor Network Technology

Wireless sensor network could advance many scientific pursuits and provide an effective tool for application research as new information technology. Agriculture has just become a typical and significant application field among the wide range of wireless sensor network applications. More and more information technologies have been applied for data acquisition in greenhouse production such as different kind of sensors, field bus and wireless communication. In the last decade, the apparition of wireless sensor network has opened new perspectives in the matter of data acquisition and big data. It could provide dynamic, real-time data of a landscape about monitored variables that would enable scientists to measure properties that have not previously been viewed continuously.

While the wireless sensor network technology is still in discussion and development, some research groups and companies have found interest in it and began to work on the deployment and application in agriculture [13].

6.2 Global System Architecture

The system could be described as a two-part framework shown in Fig. 1. The first part is the wireless sensor network for data acquisition. Monitoring network could sense the environmental parameters including temperature, humidity and soil moisture…. After acquisition, data is routed to a special sink node such as arduino card or raspBerry PI for more security of the secured remote control of greenhouse. This last gather data and send them to the other side, remote information system controller based on the available network interface. The communication between the systems is ensured by the appliance described in the next section.

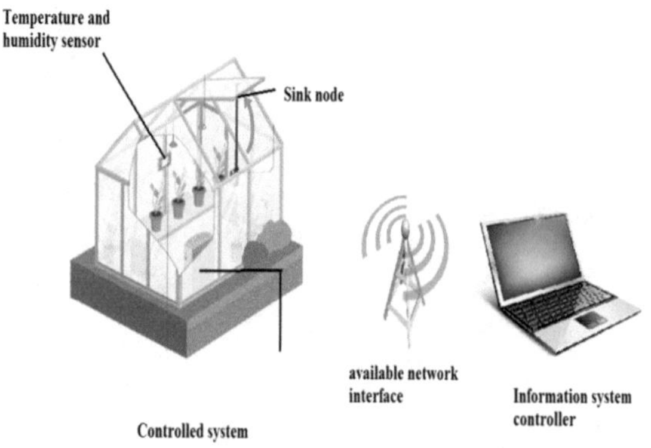

Fig. 1. Global system architecture

7 Data Synchronization and Communication Between a Central Server and Mobile Devices

Abidar et al. [14] has proposed a new appliance called EAS IDS, to synchronize and communicate a central server and mobile devices in a secured manner [15]. This platform is used to implement the security, synchronization and communication module in SRCG. The exchanged files between a central system and a distant device have XML or JSON format and pass through EAS IDS appliance for more security [16].

Figure 2 illustrates the open source mobile platform that offers many features as:

- Automatic and real-time synchronization between a central system and mobile devices.
- Use of the mobile technologies UMTS, GPRS, GSM, 3G and 4G [17].
- Encoding exchanged data by https protocol.
- Manage access security by authentication.

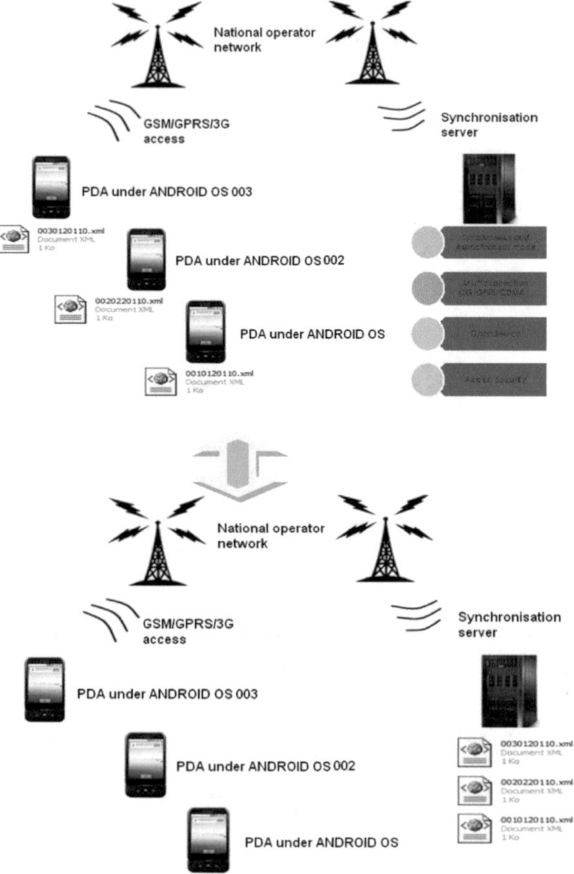

Fig. 2. Synchronization and communication mechanism

8 Proposed Architecture

Secured Remote Control of Greenhouse is organized into three levels as it is illustrated perfectly in Fig. 3:

Layer 0: intra-greenhouses, Layer 1: supervision, and Layer 2: treatment.

Intra-greenhouse layer is composed of:

- Acquisition Modules as sensors, which measures real-time parameters including temperature, humidity, lighting and content of CO2 in the atmosphere…
- Actuators can include activating the ventilation, open the roof of the greenhouse to reduce or increase these parameters.

Supervision layer is formed by:

- Server with a management interface to monitor and supervise parameters.
- Modular system expandable and secure access via Internet.
- Database to organize and store data.

A knowledge base that defines the set of rules by plant type, season, different modes of water consuming: agricultural (crop et livestock), domestic, industrial, different types of water resources (water-points), some of them are owned by farm owner and others are shared between several farmers, several means of water transport, several irrigation techniques.

Treatment layer is provided by:

- Notifications and alerts the system administrator on detection of critical cases at the system level.
- Communication between the modules and the central server will be through a secure wireless link.

Secured Remote Control of Greenhouse is implemented to optimize and resolve two main constraints, energetic and security problems [21–23], in order to reduce the consumption of energy inside the appliance nodes and also prevent it from malwares and intrusions.

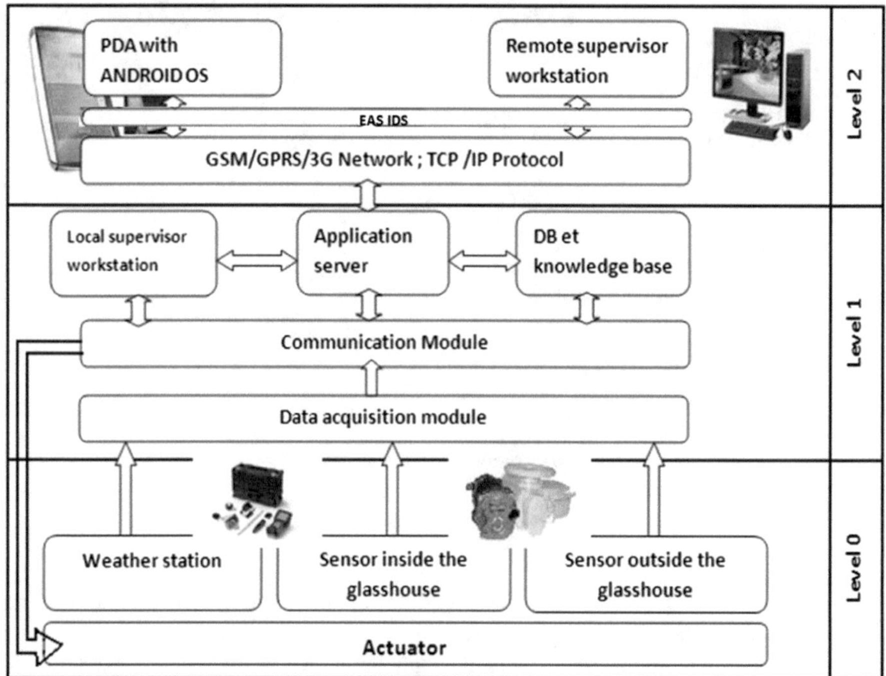

Fig. 3. Proposed architecture

Secured Remote Control of Greenhouse is designed to use clean renewable energy to power the system that ensures energy efficiency.

9 Realization and Results

JADE stands for Java Agent DEvelopment Framework [18], it is a software Framework compliant with the FIPA [19] specifications and is fully built in Java language. It simplifies the implementation of multi-agent systems through a set of graphical tools that supports the debugging and deployment phases. As a constraint satisfaction problem platform, we have extended Choco jar. Choco [20] is a library for constraint satisfaction problem, constraint programming and explanation-based constraint solving. It is built on an event-based propagation mechanism with backtrackable structures. It is open-source software implemented in java also. It allows the use of multiple solvers for different problem separately.

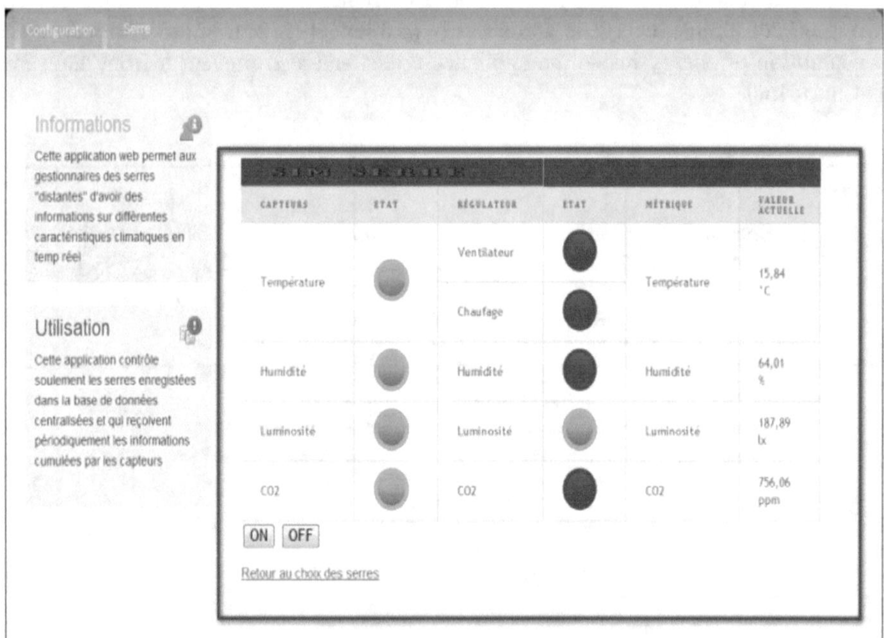

Fig. 4. Greenhouse supervisor dashboard

Figure 4 illustrates a secured remote control of greenhouse web component that monitors environmental parameter of several instances from a remote management center, and regulate these parameters through actuators immediately based on the satisfaction of constraints like min max ones, to have the best possible crop.

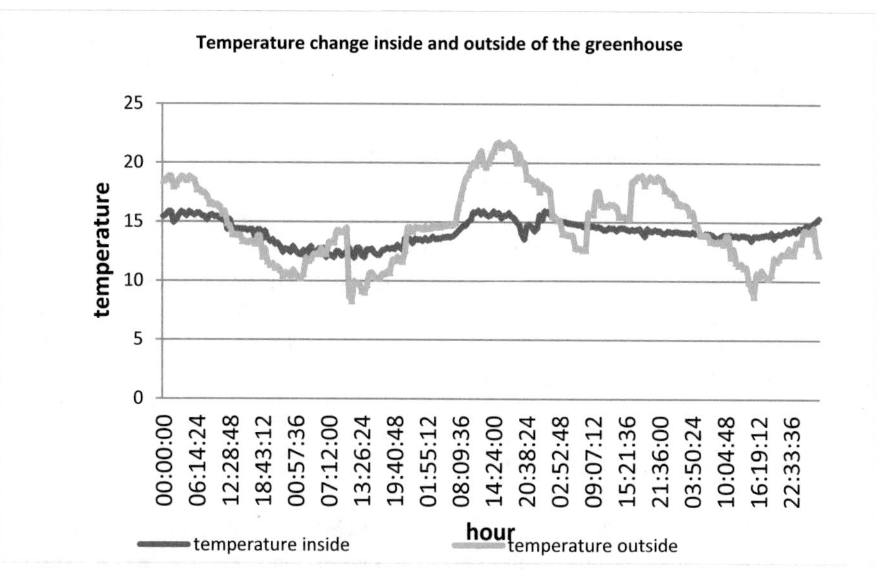

Fig. 5. Variation of the temperature inside and outside the greenhouse

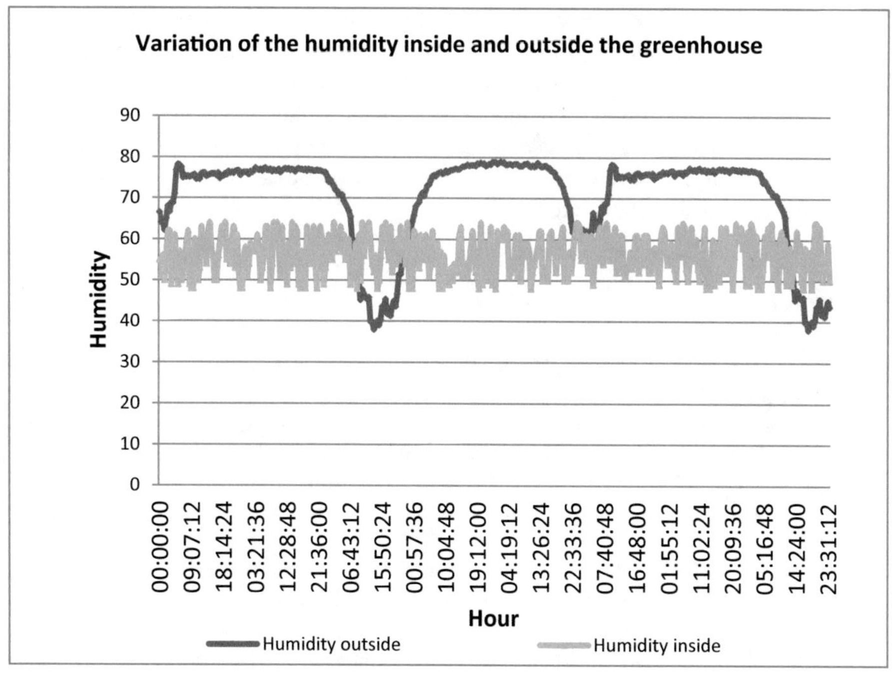

Fig. 6. Variation of the humidity inside and outside the greenhouse

Figure 5 presents the evolution of the external temperature and the controlled internal temperature. The value of the outside temperature is in the interval [8,5 °C, 22 °C]. Whereas, the temperature inside the greenhouse is maintained in the desired range [12 °C, 16 °C]. Moreover, we note that the evolution of internal temperature is influenced by the evolution of external temperature. Indeed, these two parameters display their minimum and maximum values at the same time. The experimental result shows that the optimum of the internal temperature for the photosynthesis can be maintained constant for a relatively long time.

We can make the same remarks with regard to the evolution of external humidity as it is shown in Fig. 6, which varies within the range of [38%, 79%]. It can be seen that the relative humidity inside the greenhouse is influenced by the inside temperature and varies in the interval of [48,62%, 64,78%].

10 Conclusion and Perspectives

The proposed architecture uses wireless sensor network for mobiquitous monitoring of environmental parameters in greenhouses. It is generic and portable and is based on multi-agent system to benefit from its main characteristics, which are: intelligence and autonomy. It is based likewise on constraint satisfaction problem platform for the resolution of problems; this approach allows the separation between the treatment of constraints and the other functionalities of the agents in the system. Also, it allows dividing a general constraint problem into multiple sub-problems easier to be treated. This architecture can be used to realize several applications.

Our current work focuses on the implementation and integration of a decision support system to help farmers make better decisions, and on the design and the use of big data in order to improve crop quality. We aim also to extend this work in other use cases, such as e-health application.

References

1. Yick, J., Mukherjee, B., Ghosal, D.: Wireless sensor network survey. Comput. Netw. **52**(12), 2292–2330 (2008)
2. Blackmore, S.: Precision farming: an introduction. Outlook Agric. **23**(4), 275–280 (1994)
3. Aqeel-ur-Rehman, Abbasi, A., Islam, N., Shaikh, Z.: A review of wireless sensors and networks' applications in agriculture. Elsevier, April 2011
4. Wooldridge, M.: Agent-based software engineering. IEEE Proc. Softw. Eng. **144**(1), 26–37 (1997)
5. Durfee, E.H., Montgomery, T.A., MICE: a flexible testbed for intelligent coordination experiments. In: Proceedings of 9th International AAAI Workshop on Distributed Artificial Intelligence, pp. 25–40 (1991)
6. Moummadi, K., Abidar, R., Medromi, H., Moutaouakkil, F.: Network alert management based on multi agent systems for surveillance and supervising software and hardware components. IRECOS **9**(6) (2014)
7. Wallace, M.: Practical applications of constraint programming. Constraints J. **1**, 139–168 (1996)

8. Moummadi, K., Abidar, R., Medromi, H.: A real time platform to supervise and control climate's parameters and manage drip fertigation in greenhoses based on multi-agents system. In: 4th International Conference – SIIE 2011, Marrakech 17–19 February (2011)
9. Yokoo, M., Ishida, T., Durfee, H., Kuwabara, K.: Distributed constraint satisfaction for formalizing distributed problem solving. In: 12th IEEE International Conference on Distributed Computing Systems, pp. 614–621, June 1992
10. Moummadi, K., Abidar, R., Medromi, H., Mobile device and multi agent systems: an implemented platform of real time data communication and synchronization. In: International Conference on Multimedia Computing and Systems (ICMCS 2011) (2011)
11. Havens, S.: NoGood caching for MultiAgent backtrack search. American Association for Artificial Intelligence, June 1997. www.aaai.org
12. Yokoo, M., Ishida, T., Durfee, H., Kuwabara, K.: The distributed constraint satisfaction problem: formalization and algorithms. IEEE Trans. Knowl. Data Eng. **10**(5), 633–685 (1998)
13. Wang, N., Zhang, N., Wang, M.: Wireless sensors in agriculture and food industry-resent development and future perspective. Comput. Electron. Agric. **15**, 1–14 (2006)
14. Moummadi, K., Abidar, R., Medromi, H., SBAA: Conception et réalisation d'une plateforme de communication et de synchronisation temps réel à base des systèmes multi agents entre les terminaux mobiles sous ANDROID et un serveur central, 2èmes Journées Doctorales en Technologies de l'Information et de la Communication, JDTIC'10, Fès-Morocco, 15–17 Juillet 2010
15. Abidar, R., Moummadi, K., Medromi, H.: Multi-Agent System for work orders management based on android operating system. Int. J. Eng. Res. Technol. (IJERT) **4**(1), January 2015 www.ijert.org. ISSN: 2278-0181 IJERTV4IS010620
16. Abidar, R., Moummadi, K., Medromi, H.: Intelligent and pervasive supervising platform for information system security based on multi-agent systems. IRECOS **10**(1) (2015)
17. Poslad, S., et al.: CRUMPET: creation of user-friendly mobile services personalised for tourism. In: Second International Conference on 3G Mobile Communication Technologies, London, UK, March 2001
18. JADE Homepage. http://jade.tilab.com
19. Foundation for Intelligent Physical Agents (FIPA), The FIPA 2000 Specifications. http://www.fipa.org
20. Choco Homepage. http://choco.sourceforge.net/
21. Shankar, T., Shanmugavel, S., Karthikeyan, A.: Modified harmony search algorithm for energy optimization in WSN. Int. J. Commun. Antenna Propag. **3**(4), 214–220 (2013)
22. Shankar, T., Shanmugavel, S., Karthikeyan, A.: Hybrid approach for energy optimization in wireless sensor networks using PSO. Int. J. Commun. Antenna Propag. (IRECAP) **3**(4), 221–226 (2013)
23. Telagarapu, P., Govinda Rao, L., Srinivasa Rao, D., Devi Pradeep, P.: Analysis of mobile user identification inside the buildings. Int. J. Commun. Antenna Propag. (IRECAP) **1**(2), 196–203 (2011)

Correction to: Turbulent Forced Convective Flows in a Horizontal Channel Provided with Heating Isothermal Baffles

Kamal Amghar, Louhibi Mohamed Ali, Salhi Najim,
and Salhi Merzouki

Correction to:
Chapter "Turbulent Forced Convective Flows in a Horizontal Channel Provided with Heating Isothermal Baffles" in:
M. Ezziyyani (Ed.): *Advanced Intelligent Systems for Sustainable Development (AI2SD'2018)*, AISC 912,
https://doi.org/10.1007/978-3-030-12065-8_32

In the original version of this chapter, the following belated correction has been incorporated: The author's name was corrected from "Amghar Kamal" to "Kamal Amghar". The corrected book has been updated with the change.

The updated version of this chapter can be found at
https://doi.org/10.1007/978-3-030-12065-8_32

M. Ezziyyani (Ed.): AI2SD 2018, AISC 912, p. C1, 2019.
https://doi.org/10.1007/978-3-030-12065-8_39

Author Index

M. Ezziyyani (Ed.): AI2SD 2018, AISC 912, pp. 441–442, 2019.
https://doi.org/10.1007/978-3-030-12065-8

Printed in the United States
By Bookmasters